MOMENTUM DISTRIBUTIONS

MOMENTUM DISTRIBUTIONS

Edited by

Richard N. Silver
Los Alamos National Laboratory
Los Alamos, New Mexico

and

Paul E. Sokol
The Pennsylvania State University
University Park, Pennsylvania

Plenum Press • New York and London

Library of Congress Cataloging in Publication Data

Momentum distributions / edited by Richard N. Silver and Paul E. Sokol.
    p.   cm.
    Proceedings of the workshop on Momentum Distributions, held October 24–26, 1988, at Argonne National Laboratory.
    Includes bibliographical references.
    ISBN 0-306-43364-8
    1. Momentum distributions—Congresses. 2. Neutrons—Scattering—Congresses. 3. Quantum liquids—Congresses. I. Silver, Richard N. II. Sokol, Paul E. III. Workshop on Momentum Distributions (1988: Argonne National Laboratory)
QC173.4.M67M66 1989   89-22841
530—dc20   CIP

Proceedings of a workshop on Momentum Distributions,
held October 24–26, 1988, at Argonne National Laboratory,
Argonne, Illinois

© 1989 Plenum Press, New York
A Division of Plenum Publishing Corporation
233 Spring Street, New York, N.Y. 10013

Printed in the United States of America

PREFACE

This volume presents the proceedings of the *Workshop on Momentum Distributions* held on October 24 to 26, 1988 at Argonne National Laboratory. This workshop was motivated by the enormous progress within the past few years in both experimental and theoretical studies of momentum distributions, by the growing recognition of the importance of momentum distributions to the characterization of quantum many–body systems, and especially by the realization that momentum distribution studies have much in common across the entire range of modern physics.

Accordingly, the workshop was unique in that it brought together researchers in nuclear physics, electronic systems, quantum fluids and solids, and particle physics to address the common elements of momentum distribution studies. The topics discussed in the workshop spanned more than ten orders of magnitude range in characteristic energy scales. The workshop included an extraordinary variety of interactions from Coulombic to hard core repulsive, from non–relativistic to extreme relativistic. Nevertheless, the discussions were united around common themes which include the methods to perform *ab initio* calculations of momentum distributions, the impulse approximation used to extract momentum distributions from scattering experiments at high energy and momentum transfers, the corrections to the impulse approximation such as due to final state interactions and the excitation of internal degrees of freedom, and the interpretation of momentum distribution data in terms of what they reveal about the underlying wave functions and interaction potentials in strongly correlated many–body systems. Judged by the spirited exchanges among scientists from diverse areas of physics which took place at the meeting, and by the efforts which the contributors to this volume have made to make their papers intelligible for non–specialists, we believe the conference was very successful in meeting its objectives. These were to survey the state–of–the–art in momentum distribution studies and to provide cross–disciplinary fertilization.

This volume begins with an introduction to momentum distributions in the overview paper by P. E. Sokol, R. N. Silver, and J. W. Clark. The intent is to establish a common language, to identify unifying themes and to compare momentum distribution studies in various subfields of physics. We hope the overview will help the reader to appreciate the more specialized articles which follow in these proceedings. The workshop consisted of a set of invited talks, which are organized in this volume in the same order as they were presented at the meeting, while contributed papers were presented in a poster session. The workshop summary by R. O. Simmons at the end of the proceedings is intended to provide readers with one expert's view of the significant new developments in momentum distribution studies as revealed by the workshop. Other events at the workshop, which regretably we are not able to include in these proceedings, included a lively discussion session chaired by J. W. Clark, an entertaining history of the "Quest for n(p)" presented by E. C. Svensson, and invited talks presented by S. E. Koonin and J. H. Hetherington.

We wish to thank the staff of Argonne National Laboratory who helped to organize and support this Workshop. Financial support was graciously provided by the Intense Pulsed Neutron Source through its Director, Bruce Brown, by the Division of Educational Programs, through Harold Myron, and by the University of Chicago, through Joel Snow. We thank Mariam Holden and the conference support staff at Argonne for their able assistance in logistical arrangements for the meeting. We thank Plenum Publishing Corporation, and especially Melanie Yelity, for their expert work in producing these proceedings. We gratefully acknowledge the advice of J. W. Clark regarding the selection of invited papers. We also thank the able chairpersons of the sessions: P. A. Whitlock, A. S. Rinat, W. Stirling, R. B. Wiringa, A. Griffin, S. K. Sinha, and J. M. Carpenter.

Finally, we would also acknowledge our individual research grants. RNS is supported by the OBES/DMS funding of the Los Alamos Neutron Scattering Center at Los Alamos National Laboratory. PES is supported by NSF grant DMR–8704288 and OBES/DMS support of the Intense Pulsed Neutron Source at Argonne National Laboratory under DOE grant W–31–109–ENG–38.

<table>
<tr><td>Los Alamos, New Mexico
University Park, Pennsylvania</td><td>Richard N. Silver
Paul E. Sokol</td></tr>
</table>

CONTENTS

---

* Denotes Speaker

WRAP-UP

MOMENTUM DISTRIBUTIONS: AN OVERVIEW

P. E. Sokol

Department of Physics
The Pennsylvania State University
University Park, PA 16802

R. N. Silver

MS B262 Theoretical Division
Los Alamos Neutron Scattering Center
Los Alamos National Laboratory
Los Alamos, New Mexico 87545

J. W. Clark

McDonnell Center for the Space Sciences
and Department of Physics
Washington University
St. Louis, MO 63130

INTRODUCTION

While all systems of classical particles have single-particle momentum distributions $n(p)$ of Maxwell-Boltzmann form, the momentum distribution plays a role central to our understanding of systems of quantum particles. An outstanding example is the low-temperature superfluid behavior of the Bose liquid, ^{4}He, where the superfluidity is associated with Bose condensation of a macroscopic fraction of the ^{4}He atoms into a zero-momentum state. The momentum distribution is complementary to other characterizations of many-body systems and can be more informative. The pair correlation function of liquid ^{4}He is very close to that of a hard-sphere classical fluid, whereas the momentum distribution reveals the quantum behavior in the form of a δ-function spike in $n(p)$ at $p = 0$ due to the Bose condensate. Momentum distributions are equally fundamental to the description of Fermi systems. The Fermi-liquid properties of ^{3}He and electrons in metals are associated with a discontinuity in the momentum distribution at the Fermi momentum, k_F, which defines a Fermi surface for a three-dimensional system. A detailed description of the, often complex, Fermi surfaces in metals is essential to understanding their transport, optical, and magnetic properties. At low temperatures, the transition to superfluid behavior of ^{3}He and the transition to superconducting behavior of electrons is associated with the disappearance of this Fermi surface. An outstanding problem in nuclear physics is how the quasi-exponential high-p tails observed in the $n(p)$ of nucleons in nuclei are related to the short-range real-space correlations of nucleons due to the strongly repulsive core of the nucleon-nucleon potential. The momentum distributions of quarks are found to be different inside nuclei and inside free nucleons, which suggests a possible role for quarks in the description of nuclear forces. These varied illustrations attest to the importance of the momentum distribution as a revealing probe of the wave functions of quantum many-body systems.

Momentum distributions are of interest for most of the subjects of research in modern physics, including: systems of atoms, in solid and liquid phases; interacting electron systems, such as conduction electrons in metals; systems of nucleons in atomic nuclei and nuclear matter; and systems of quarks in high-energy physics. We contend that the study of momentum distributions shares many common elements across this diverse range of energy and length scales, forces, and system types. The general features of $n(p)$ for systems of given statistics are in principle quite similar. In practice, theorists in the various areas of physics face similar problems when they attempt to predict the momentum distribution accurately, and experimentalists face similar problems when they make measurements of quantities related to the momentum distribution and attempt to extract $n(p)$ from their data. In all fields there has been a rapid advance in theory and experiment (in particular, the development of large-scale facilities for scattering experiments), which is presenting exciting new scientific opportunities.

There have been several excellent reviews of momentum-distribution research in particular subject areas of physics such as electronic systems[1] and nuclear systems.[2] However, it is the commonality of interests, difficulties, and prospects across all of physics, along with certain pivotal advances, which led to the organization of an interdisciplinary Workshop on Momentum Distributions held at Argonne National Laboratory on October 24-26, 1988. The purpose of this overview is to explain why scientists with such diverse backgrounds have been brought together at this meeting, to introduce and discuss the common elements of momentum-distribution studies, and to establish a common language. We hope to facilitate an appreciation of the more specialized articles which follow in these proceedings.

We begin by summarizing the general properties of momentum distributions. Differences and similarities of atomic, electronic, and nuclear many-body systems are examined, in terms of characteristic lengths and energies, relative importance of exchange, and the nature of the two-particle interactions. We continue with a brief commentary on the microscopic methods used to calculate $n(p)$ from first principles. Thereafter the discussion focuses on the ideas, techniques, and issues involved in the experimental determination of the momentum distribution: deep-inelastic scattering, the impulse approximation, Y-scaling, final-state effects, and scale breaking. Finally, some typical examples of theoretical and experimental momentum distributions will be presented and compared, for a variety of systems.

FUNDAMENTALS

The momentum distribution $n(p)$ of a quantum-mechanical system is the average number of particles with momentum p, determined by the expectation value

$$n(p) = <\Psi| \sum_{\sigma} a_{p\sigma}^{\dagger} a_{p\sigma} |\Psi> \quad . \tag{2.1}$$

In this expression, $|\Psi>$ is the unit-normalized N-particle state of the system and $a_{p\sigma}^{\dagger}$ and $a_{p\sigma}$ are creation and annihilation operators for a particle with momenta p and spin projection σ. Usually, one deals with the momentum distribution of particles having a *given* spin projection, defined by removing the spin sum in eq (2.1) . At finite temperatures, Eq. (2.1) is replaced by an ensemble average over all N-particle states.

In the quantum systems of interest to us, the de Broglie wavelength for single-particle motion can be of the order of the interparticle spacing and in some cases much larger. This implies large exchange effects. The type of quantum statistics obeyed by the particles, Bose or Fermi, then has an important bearing on the character of the momentum distribution. For Bose particles, the many-body wave function must be symmetric and there is no restriction on the occupancy of any given one-body momentum state. For Fermi particles, the overall wave function must be antisymmetric. Consequently, the occupancy of any chosen one-body momentum state cannot be greater than one.

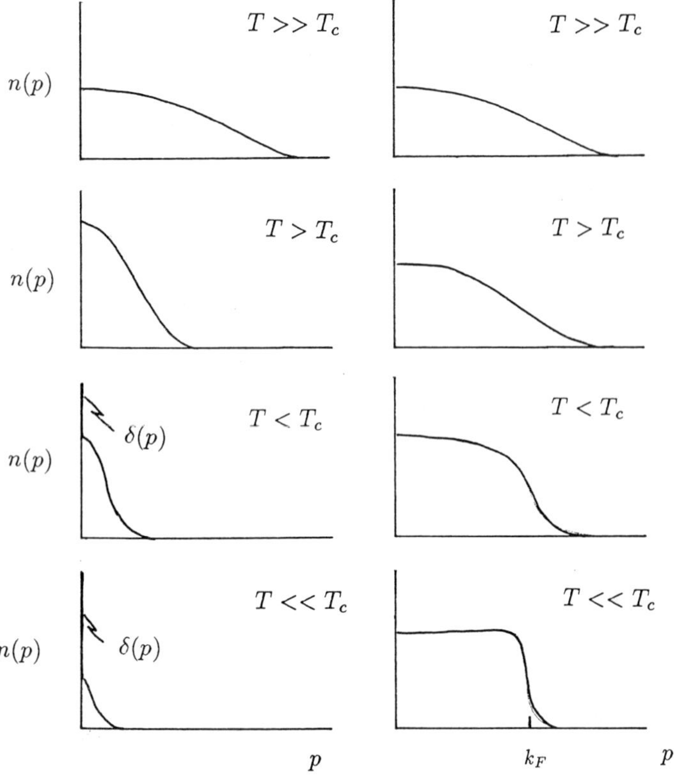

Fig. 1 Typical behavior of the momentum distribution as a function of temperature for the ideal Bose and Fermi gasses. The characteristic temperature T_c is defined in the text.

The behavior of non-interacting Bose and Fermi gases as the temperature is decreased from very high to very low values, or equivalently as the density is increased, is shown in Fig. 1. At high temperatures, the shape of the momentum distribution approaches a Gaussian, the classical Maxwell-Boltzmann form, and the occupancy of any particular momentum state is always much less than one. In this limit, the width of the momentum distribution is proportional to the thermal energy of the particles and the mean kinetic energy is $3k_B T/2$.

Upon lowering the temperature (or increasing the density), the occupancy of some of the one-body momentum states begins to approach one and the effects of statistics begin to emerge. The momentum distributions begin to deviate from the Maxwell-Boltzmann form. For non-interacting particles where the single-particle motion can be described by states of definite momentum, the symmetry requirement for Bose particles leads to the familiar Bose-Einstein momentum distribution, while the antisymmetry requirement for Fermi particles leads to the familiar Fermi-Dirac distribution.

In the Bose case, the momentum distribution becomes more peaked around $p = 0$, as the particles take advantage of the lack of any restriction on multiple occupation. The behavior in the Fermi case is quite different. The distribution remains flat and close to one in the small-p region as the temperature is lowered, as if there were a repulsion between particles trying to pile up at the same momentum. This behavior is a direct consequence of the exclusion principle which forbids multiple occupation of a particular momentum state.

The effects of quantum statistics become dominant below a characteristic temperature, T_c. For both Fermi and Bose systems, we may define such a characteristic temperature by the condition that the thermal de Broglie wavelength of the particles is equal to some appropriate measure of the mean interparticle spacing.

The characteristic temperature for a Bose gas may be taken as the Bose-Einstein condensation temperature $T_c = T_{BE} = (2\pi\hbar^2/1.897mk_B)\rho^{2/3}$, where ρ is the density of the gas, m is the particle mass and k_B is Boltzmann's constant. At temperatures higher than T_{BE}, the occupancy of any particular one-body momentum state remains finite. Below this temperature, a macroscopic (i.e. of order the total number of particles in the system, N) occupation of the $k = 0$ momentum state develops, which is called "Bose-Einstein condensation." This is reflected in the appearance of a term in $n(p)$ proportional to a Dirac delta function $\delta(p)$, with a coefficient which determines the (finite) fraction of particles residing in the Bose condensate. The condensate fraction increases as the temperature is lowered further until finally, at $T = 0$, all of the particles are in the condensate. The condensate particles occupy a single quantum state with a well defined momentum, namely $p = 0$. Bose condensation represents a novel macroscopic manifestation of quantum principles.

The characteristic temperature for a Fermi gas is determined by the Fermi energy. By definition this is the energy of the highest single-particle level occupied at $T = 0$, all of the lower levels each being filled with ν particles, where ν is the single-particle level degeneracy arising from spin (and possibly isospin) degrees of freedom. Thus $T_c = T_F = (\hbar^2/2mk_B)(6\pi^2\rho/\nu)^{2/3}$. As the temperature and hence the available thermal energy declines, the particles attempt to reduce their energies by occupying lower energy levels, but their readjustments are constrained by the exclusion principle. When the temperature has dropped substantially below T_c (say to $T_c/5$ or $T_c/10$), the lowest-lying single-particle levels will be completely filled with their retinue of fermions and $n(p)$ will approach one. However, levels near the Fermi surface defined by $k = k_F = (6\pi^2\rho/\nu)^{1/3}$ will be only partially occupied because of thermal excitation. The momentum distribution then exhibits the characteristic Fermi-Dirac shape, as shown in Fig. 1. With further decrease in temperature, the fall-off near the Fermi wave number k_F steepens until, at $T = 0$, all the particles have condensed into the "Fermi sea." The sharp discontinuity that appears in $n(p)$ divides the momentum states below k_F, which are fully occupied, from those above, which are empty.

At $T = 0$ in non-interacting systems, the fraction of Bose particles which are in the zero-momentum state reaches one, and there is a discontinuity of the Fermi $n(p)$ at k_F equal to one. In interacting systems at $T = 0$, the features of macroscopic condensation at $p = 0$ in Bose systems and a finite discontinuity at k_F in Fermi systems are predicted to persist under rather general assumptions. However, due to the interactions, the fraction of Bose particles which condense is less than one, and, for Fermi particles, the discontinuity at k_F takes some value less than one. These "depletion" effects will be discussed more fully below.

The single-particle properties of many-body systems, both interacting and non-interacting, may also be fruitfully discussed in terms of the one-body density matrix. For a unit-normalized pure quantum state, this quantity is defined by

$$\rho_1(r_1, r_1') = N \int dr_2...dr_N \Psi^*(r_1 r_2...r_n)\Psi(r_1' r_2...r_N) \ , \tag{2.2}$$

where spin and other internal degrees of freedom have been suppressed for simplicity. The two-point function $\rho_1(r_1, r_1')$ measures the change of the wave function as a single particle is moved from r_1 to r_1', all the other particles remaining fixed. In general, the one-body density matrix will depend on r_1 and r_1' individually. However, in a homogeneous, isotropic fluid it can only depend on the magnitude of the separation vector:

$$\rho_1(r_1, r_1') = \rho_1(|r_1 - r_1'|) \ . \tag{2.3}$$

The single-particle density matrix of the fluid contains all the features of interest for this overview; hence we shall restrict our attention to that case.

The momentum distribution and the one-body density matrix are related by Fourier transformation. For a Fermi system, we may write simply

$$n(p) = \nu^{-1} \int \rho_1(r)e^{ipr} dr \ , \tag{2.4}$$

where ν is the level degeneracy. For a Bose system, the momentum distribution is traditionally separated into two components, a delta function term representing the zero-momentum condensate and a smooth component corresponding to occupation of the other single-particle states. Thus

$$n(p) = (2\pi)^3 \rho n_0 \delta(p) + n'(p) \ , \tag{2.5}$$

where the condensate fraction is determined by

$$n_0 = \lim_{r \to \infty} \rho_1(r)/\rho \tag{2.6}$$

and the non-condensate portion has the Fourier representation

$$n'(p) = \int [\rho_1(r) - \rho_1(\infty)]e^{ipr} dr \ . \tag{2.7}$$

Typical behaviors of both $n(p)$ and $\rho(r)$ for interacting Bose and Fermi systems at zero temperature are indicated in Fig. 2. To be definite, the particles are assumed to experience strong repulsive two-body interactions at small separations.

Consider first the one-body density matrix and momentum distribution of the ground state of the interacting Bose system. The dynamical short-range correlations due to the core repulsion, which govern the small-r behavior of $\rho_1(r)$ without regard to statistics, determine $n(p)$ at large p. The effects of statistical correlations are most apparent in $\rho_1(r)$ at large r and in $n(p)$ at small p. The condensate, which gives rise to a finite value of $\rho_1(r)$ at infinity, again manifests itself in $n(p)$ as a delta-function spike at $p = 0$ (not visible in the plot). The detailed behavior of both quantities at intermediate r, or p, is also significantly affected by the statistics. An interesting singular feature of the uncondensed component at small p results from the coupling of long-wavelength density fluctuations to the condensate. This feature leads to a finite intercept of $pn(p)$ at $p = 0$, as shown[3] in Fig. 2.

Now consider the ground-state momentum distribution of the interacting Fermi system, as sketched in Fig. 2. The step function $\theta(p - k_F)$ which gives $n(p)$ for the noninteracting Fermi gas is modified in the presence of interactions, but the general shape is preserved. The interactions promote some of the particles from single-particle states inside the Fermi sea, i.e., with momenta less than k_F, to states outside, thus depleting the Fermi sea and creating a tail at higher momenta. If the system remains "normal," meaning that the interactions are not such as to create a superfluid ground state, $n(p)$ retains its most characteristic Fermi feature, namely a discontinuity at the the Fermi wave number (cf. Fig. 2). The size of the discontinuity, denoted herein by Z_{k_F}, decreases as the interactions become stronger and thus provides an inverse measure of the strength of the interparticle coupling.

Since the type of statistics has little effect on $\rho_1(r)$ at small r, the one-body density matrix for the Fermi system is similar to that for the Bose system in that region, as seen in Fig. 2. However, at large r the behavior of this quantity is markedly different in the two systems: whereas the Bose $\rho_1(r)$ approaches a constant value, reflecting the existence of a condensate and therefore off-diagonal long-range order[4,5] (ODLRO), the Fermi $\rho_1(r)$ damps out to zero, in accordance with the absence of ODLRO at the one-particle level. The oscillatory behavior of the Fermi $\rho_1(r)$ is required to produce the discontinuity of $n(p)$ at k_F. The zeros of $\rho_1(r)$ are determined by the location of the Fermi surface, and the overall amplitude of the oscillations is determined by the magnitude of Z_{k_F}.

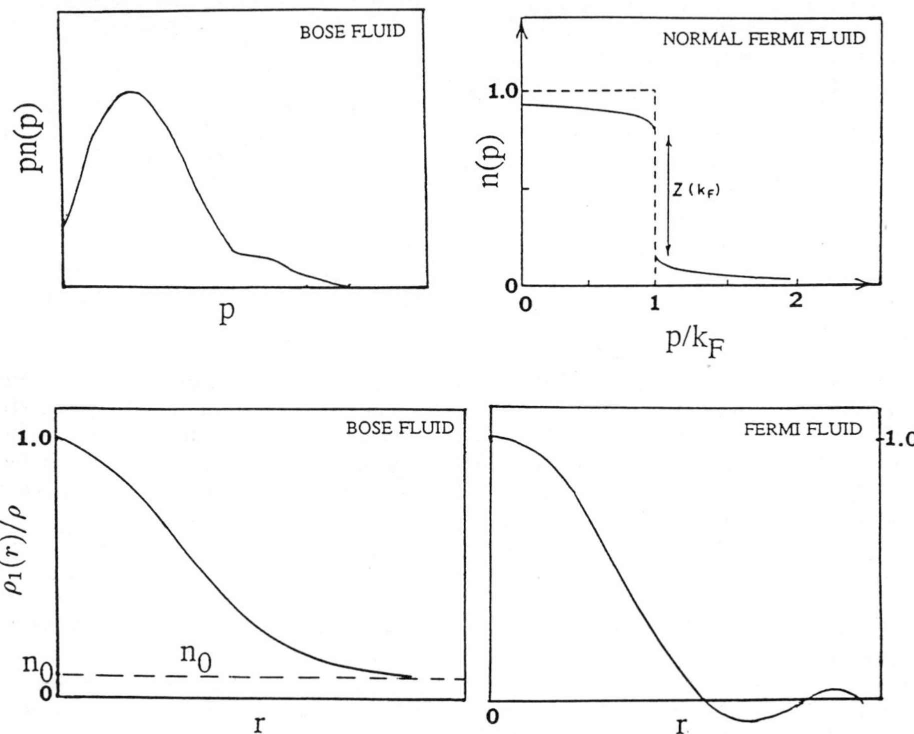

Fig. 2 Typical behavior of the momentum distribution and one-body density matrix in the ground state for interacting Bose and Fermi systems.

SCALES

Before proceeding to more concrete matters of calculation and measurement, it will be useful to formulate meaningful bases for comparison of the diverse many-body systems involved in our studies. In so doing we shall gain a better understanding of their similarities and differences and begin to establish a common language for the subsequent discussions.

Our overview will focus on three types of many-body systems. It will be convenient to refer to collections of atoms, like liquid or solid ^{4}He or ^{3}He or solid molecular hydrogen, simply as *atomic systems*. By *nuclear systems* we shall mean both finite nuclei and idealized, infinite nuclear matter. The third category, *electronic systems*, includes, narrowly, the system of electrons in a solid, and, more broadly, a wide variety of systems of great topical interest in condensed-matter physics (e.g. in high-temperature superconductivity). Other many-body systems could also be considered, such as collections of quarks in particle physics, and collections of electrons in atoms and molecules. Although lack of space prevents us from doing justice to these additional examples, the concepts we shall discuss are generally applicable to all momentum-distribution studies.

A length scale appropriate to microscopic description of a given many-body system may be taken as a typical interparticle spacing, while the binding energy per particle provides a reasonable energy scale. The three classes of many-body problems we have just delineated involve very different scales in energy and length, ranging over many orders of magnitude. To make a meaningful comparison of system properties and behavior, we need somehow to remove these large variations.

We begin by considering condensed systems of atoms: solids and liquids. A typical interparticle spacing for atomic systems is on the order of Å, setting the characteristic length scale. Typical binding energies for atomic systems are on the order of meV, setting the characteristic energy scale. Numerical values are shown in Table I for ^{3}He, a "representative" atomic system.

Focusing on Fermi examples, how important are exchange effects in determining the shape of the momentum distribution? A measure of the strength of exchange, or statistical correlations, is given by the Fermi energy E_F of the system. If E_F is large compared to the binding energy per particle, E_b, the Fermi statistics will have a profound influence on $n(p)$, whereas the condition $E_F << E_b$ implies that statistical effects are unimportant and $n(p)$ is well approximated by the classical result. For liquid ^{3}He we find $E_b/E_F \sim 0.5$, indicating that exchange plays a substantial but not overwhelming role in this system.

As we shall see, experimental determination of $n(p)$ involves inelastic scattering processes at energy and momentum transfers much larger than the characteristic energy and inverse length scales of the systems under study. In neutron scattering from atomic systems, momentum transfers Q up to ~ 30 Å^{-1} and energy transfers ω up to a few eV are currently attainable. Since these values are much larger than the Fermi energy E_F and Fermi momentum k_F for bulk atomic ^{3}He (see Table I), a measurement of the momentum distribution of this system would appear to be experimentally feasible.

We turn next to electronic systems. The unit of length conventionally adopted is the Bohr radius $a_o = 0.5292$ Å, also called the atomic unit (au). The radius r_s of the volume per particle, measured in au, lies in the range 2-6 for the conduction-electron subsystem in metals. Hence the characteristic length scale is comparable to that of systems of atoms like liquid helium. However, typical cohesive energies of metals, per atom, run to some tenths of Rydbergs (the conventional unit of energy, $R = 13.61$ eV), so the characteristic energy scale is three orders of magnitude larger than in the atomic case. Values for the Fermi energy and momentum associated with the conduction electrons in sodium are quoted in Table I. Comparing the binding and Fermi energies of this system we find $E_b/E_F \sim 0.3$. Thus, we expect statistics to have an effect on $n(p)$ comparable to that in liquid ^{3}He, though somewhat larger.

Experimentally, X-ray Compton scattering is used to study the dynamic structure of electronic systems. Typical momentum and energy transfers employed in current work are indicated in Table I. Just as in neutron scattering from a system of ^{3}He atoms, these can be much larger than the relevant Fermi momentum and energy. Thus we again infer that the momentum distribution is an experimentally accessible quantity.

TABLE I

SCALES

DEEP INELASTIC NEUTRON SCATTERING

1 Angstrom($\mathring{A}$) $= 10^{-8}$ cm
1 meV $= 11.6$ K

^{3}He $\qquad$ $E_F = 5.0$ K $\qquad$ $k_F = 0.789 \ \mathring{A}^{-1}$

$\qquad\qquad$ $E_b/E_F = 0.5$

$5\mathring{A}^{-1} \leq Q \leq 30\mathring{A}^{-1}$
20 meV $\leq E_i \leq 5000$ meV

X-RAY COMPTON SCATTERING

1 atomic unit (au) $= 5.29 \times 10^{-9}$ cm
$E = 13.61$ eV

Na $\qquad$ $E_F = 3.23$ eV $\qquad$ $k_F = .486$ p(au)

$\qquad\qquad$ $E_b/E_F = 0.3$

2.5 p(au) $\leq Q \leq 100$ p(au)
10 KeV $\leq E_i \leq 400$ KeV

QUASIELASTIC ELECTRON NUCLEUS SCATTERING

1 Fermi (fm) $= 10^{-13}$ cm $\qquad$ 1 GeV/c $= 5.06$ fm^{-1}

$\qquad\qquad$ $E_F = 38.4$ MeV $\qquad$ $k_F = 1.39$ fm^{-1}

$\qquad\qquad$ $E_b/E_F = 0.42$

$Q \leq 10$ fm^{-1} (2GeV/c) $\qquad$ *sometimes quote* $Q_4^2 = Q^2 - \omega^2$
500 MeV $\leq E_i \leq 4$ GeV

Nuclear matter has characteristic length and energy scales which are vastly different from those of the preceding examples. For nuclear systems the characteristic length is the Fermi, which is five orders of magnitude smaller than the Å scale of the atomic and electronic systems. The standard unit of energy is the MeV, which is nine orders of magnitude larger than for the atomic case. In infinite nuclear matter the binding energy per particle, E_b, is 16 MeV. The characteristic Fermi energies and momenta of nuclear systems, entered in Table I, also differ from those of the atomic and electronic cases by many orders of magnitude. Nevertheless, when we form the dimensionless measure E_b/E_F, we obtain a value 0.4, putting nuclear matter somewhere between atomic and electronic systems in the importance of quantum statistics.

Experimentally, electron scattering is used to probe high-momentum components of the nuclear wave function. The momentum transfers relevant to studies in the energy region of the quasielastic peak, prior to the onset of inelastic processes corresponding to the excitation of internal degrees of freedom of the nucleonic constituents, reach only to 10 fm^{-1}. This is just an order of magnitude larger than the characteristic k_F for nuclear systems. While the excess is not as large, in a relative sense, as in the other two cases, important aspects of the momentum distribution will be experimentally accessible here as well.

In summary: The similarities between atomic, electronic, and nuclear systems are striking. Their characteristic energy and length scales may differ by many orders of magnitude; yet they display a comparable balance of binding and exchange effects. They are also similar in the sense that it is feasible to perform scattering measurements for which the momentum and energy transfers are substantially larger than characteristic values of particle momentum and energy in the ground state. With some qualifications to be noted later, such experiments may be considered to measure the pertinent momentum distributions.

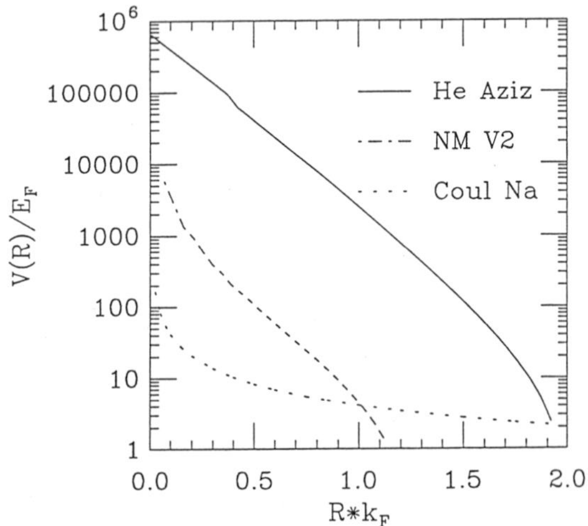

Fig. 3 Typical interaction potential for atomic, electronic, and nuclear, systems. The interaction strength has been scaled by the appropriate Fermi energy and the distance has been scaled by the inverse of the Fermi momentum.

To compare the three systems in another way, we may look at the basic interactions between the atomic, electronic, or nucleonic particles. As expected, these interactions generally differ by many orders of magnitude in their strengths and ranges. To make a sensible comparison, we need to scale the potentials with suitable energy and length measures for the different systems. For the energy measure, we may adopt the Fermi energy, and for the length measure, the inverse of the Fermi momentum is chosen. (Again, comparisons are to be made for a given type of statistics, hence the Fermi case.) Fig. 3 juxtaposes representative potentials for atomic (helium), nuclear, and electronic systems, scaled in this manner. Even when scaled, the potentials differ by orders of magnitude. The helium potential is the strongest at short distances, its core repulsion being very hard (but not infinite). The core of the nuclear potential is two orders of magnitude softer and the electronic (coulombic) potential is two orders of magnitude weaker still.

Both the helium and nuclear potentials have relatively weak attractive tails, which are not visible on the semi-log plot. The electronic potential, with a simple monotonic $1/r$ repulsive behavior, is quite different: while its core is very weak in comparison with the helium and nuclear examples, this potential falls off very slowly at larger r. As we look at larger distances the electronic potential becomes stronger than the nuclear potential and ultimately surpasses the atomic potential. These differences in the behavior of the basic two-body interactions have interesting consequences for the respective many-body systems. Whereas the strong-coupling limit is at high density in the helium and nuclear problems, it is at low density in the electronic case.

Thus, while there are important similarities between the three classes of systems, there are important differences as well.

CALCULATION METHODS

Many theoretical methods have been developed to study the momentum distribution of many-body systems. These may be divided into non-stochastic and stochastic approaches. By the former we mean the more traditional, analytically based procedures (like perturbation theory and hypernetted-chain methods) which typically involve manipulations with field-theoretic operators or wave functions and make heavy use of diagrams, before numerical work begins. By the latter, we mean computationally-intensive procedures (like variational, Green's function, and path-integral Monte Carlo methods) based on random-walk algorithms for evaluation of expectation values or thermal averages, or for solution of the Schrödinger equation. The various methods may be further classified according to their ability to handle systems with stronger interparticle couplings. At this point we shall make some general remarks on the strengths and limitations of the most prominent approaches. A more detailed review by Clark and Ristig,[6] including a rather complete set of references, appears later in this volume.

Non-Stochastic Methods

The most familiar examples of non-stochastic methods are ordinary perturbation theory in the bare interaction, starting from the noninteracting system, and variational methods based on independent-particle trial wave functions. Neither approach is useful for predicting the momentum distribution for the systems under study here, either because of the strength of the repulsive core (atomic and nuclear cases) or because of the long range of the interaction (electronic case). On the one hand, rearrangements or resummations of perturbation theory are necessary, and on the other, a viable variational treatment must incorporate dynamical correlations among the particles.

Electronic problems (particularly those involving a uniform electron gas) can often be successfully attacked with perturbation theory, *provided* the ring diagrams are summed to produce a screening of the long-range Coulomb force. This approach is usually framed in terms of Green's functions. The random-phase approximation (RPA), or some variant of it, is used to sum the ring or bubble diagrams in the perturbation expansion of the one-particle Green's function.

Another non-stochastic approach to weakly-interacting electronic systems is band-structure theory, which derives self-consistent one-electron wave functions using a variety of methods including the local-density approximation. This is an eminently practical and highly-developed method for treating the electronic structure of real solids where the lattice plays an essential role, and its successes within Compton-scattering and positron-annihilation studies of momentum distributions have been extensively reviewed elsewhere.[1,7] The limitations of band-structure theory become apparent in dealing with strongly-correlated electronic systems,[8] and our overview of calculational methods will focus on general approaches to strongly-correlated systems.

When applying perturbation-based theories, the main worry is convergence—does the expansion converge, and if so, how fast? For example, perturbative approaches which re-sum only ring diagrams deteriorate or fail as the short-range core of the interaction becomes stronger. They begin to deteriorate for strongly-coupled electronic systems, as represented, say, by the Hubbard model. They fail completely for nuclear and helium systems due to their strongly repulsive cores.

Perturbative techniques can, however, be extended to more strongly interacting systems. Historically, this was first done by Brueckner,[9] his ideas being systematized in terms of a Goldstone diagrammatic expansion.[10,11] When strong repulsive cores are present, it becomes imperative (as a minimum) to re-sum the particle-particle ladder diagrams. This leads to "hole-line expansions" for the quantities of interest, which have seen extensive use in nuclear-matter theory at not-too-high densities. However, the ladder and self-energy resummations which define Brueckner theory do not suffice for very strongly correlated systems like the helium liquids.

In a few cases, analysis to all orders within perturbation theory can be carried out to yield valuable *exact* results, even for very strong couplings. These results typically involve limiting conditions on one or more of the relevant variables, including density, distance, wave number, and temperature. One such result is the prediction of a $1/p$ singularity in the ground-state $n(p)$ of a Bose system[3] like liquid ^{4}He. Another is the prediction of a Fermi-surface discontinuity in the momentum distribution of all normal Fermi fluids.[12,13]

Self-consistent summation of rings and ladders, leading ultimately to parquet theory[14] offers hope for a comprehensive and quantitative microscopic theory of strongly-coupled systems within the perturbative framework. Unfortunately, this approach has proven exceedingly difficult to implement, especially for Fermi systems. In the interim, variational methods have come to the fore as the most practical means for evaluating the properties of the helium liquids and of nuclear matter at high density.

In the variational approach, a ground state-wave function is chosen on the basis of an intuitive understanding of the correlation structure of the many-body system. The energy expectation value is then minimized with respect to variational parameters or functions appearing in the trial state. The most common trial wave function is the Jastrow form, originally motivated by the requirement that short-range two-body correlations be included in the wave function. It is remarkable that this choice also provides for a correct description of the long-range correlations corresponding to virtual phonons. Although the Jastrow form leads to useful results for the systems under study, the predictions for properties like the ground-state energy and momentum distribution are generally only semi-quantitative.

More sophisticated wave functions, incorporating further aspects of the correlation structure (triplets, momentum-dependent backflow, spin-dependent correlations, noncentral correlations ...) are now in wide use. Quantitative results are obtained for the ground states of liquid ^{4}He and ^{3}He, and presumably also for the ground states of nuclear and neutron matter. Correlated variational methods have not reached the same state of refinement or popularity in application to the electron gas and other electronic systems. However, they have been rather successful in this context even at the simple Jastrow level (uniform $3D$ electron gas), or at the (simpler still) Gutzwiller level. Their most dramatic "electronic" success story is found in Laughlin's theory of the fractional quantum Hall effect.[15]

As described, the variational approach is limited to the ground state. However, variational ideas have been extended to excitations in the correlated random-phase approximation,[16,17] derived from the Dirac-Frenkel time-dependent variational principle. Moreover, the variational approach has been extended to the evaluation of equilibrium properties of quantum fluids at finite, but low, temperatures.[18]

In variational approaches to atomic, nuclear, and electronic systems, the counterpart of the convergence worry of perturbation theory is the problem of reliable calculation of expectation values and matrix elements with respect to correlated wave functions. This highly nontrivial problem is discussed by Clark and Ristig[6] and will not be considered here in any detail. Depending on the application and the sophistication of the wave functions, higher-order cluster diagrams contributing to correlated expressions may be either partially re-summed, approximated, or ignored. Current practice[19,20] involves the use of "scaling procedures" to simulate the effects of higher-order terms which are difficult or impossible to evaluate explicitly.

A deeper problem of more conceptual weight is the constraint which the choice of wave function imposes on the physical description. Lacking the right intuition, the proposed wave function will not have sufficient freedom, and interesting phenomena (for example, phase transitions) may be missed. There exist stability tests which make this drawback less serious, but these tests usually refer only to local stability. Recent work with shadow wave functions[21] introduces a welcome flexibility which reduces the reliance on intuition.

A related criticism of the variational approach is that improvements are not very systematic. This criticism may be answered by an extension of variational theory known as the method of correlated basis functions[22,23] (CBF). In CBF a basis of functions is generated by applying a correlation operator determined variationally, to a complete set of model wave functions suitable for a weakly interacting system (e.g. Slater determinants). This scheme combines the insights and techniques of the variational approach to strong interactions, with the formal advantages of perturbation theory and other approaches (e.g. RPA, BCS, etc.) designed for weak interactions. Of course, the method will also suffer (in lesser degrees) from the same kind of convergence worries which plague these underlying approaches. Although complicated in appearance, CBF is one of the more efficient, quantitative, and powerful of the non-stochastic approaches, yielding a variety of useful results for atomic, nuclear, and electronic systems.

Stochastic Methods

The available non-stochastic techniques all experience some degree of difficulty in handling very strong interactions. Such a difficulty is not intrinsic to the stochastic approaches, although some "hangover" may be experienced since the implementation of stochastic treatments usually relies on information provided by a prior non-stochastic study. Another (related) advantage is that stochastic methods are immune to the convergence problems which beset perturbative and variational procedures. On the negative side of the ledger, stochastic methods suffer from the well-known disadvantages of computer-intensive, "granular" simulations.

The variational Monte Carlo (VMC) method contains basically the same physics as the variational approach described under non-stochastic methods. However, a Metropolis Monte Carlo algorithm is used to evaluate the many-body integrals. This technique is superior to those employed in non-stochastic variational theory, such as hypernetted-chain re-summation, since higher-order cluster diagrams are automatically included. The statistical errors associated with the Metropolis algorithm can be effectively controlled by variance reduction techniques. Perhaps the major concern is that (for obvious practical reasons) the simulations are performed for a sample of a few dozen or a few hundred particles, instead of extended medium. Periodic boundary conditions are imposed in a finite cube, with side length adjusted to give the pre-assigned average particle density. Finite-box-size effects can be significant, especially for the long-range behavior of the one-body density matrix, though they are generally believed to be of little importance.

Needless to say, VMC suffers from the same intrinsic limitations as its non-stochastic counterpart. Unless the general nature of the interesting physics is known in advance, the variational approach may pass it by.

A more powerful alternative is the Green's function Monte Carlo (GFMC) method, which stochastically generates a solution for the ground-state wave function. The Schrödinger equation, in imaginary time, may be mapped into a diffusion equation in real time, which may in turn be solved by Monte Carlo techniques. In principle, this approach leads to the exact ground state. In practice, it works exceedingly well in application to Bose systems, such as liquid ^{4}He, where the ground-state wave function is positive. As usual, there are statistical errors (arising in the solution of the Schrödinger equation as well as in the evaluation of various quantities), but again these can be controlled by various techniques. There are also finite-box-size effects, since again one must work with a finite number of atoms. In particular, the size of the simulation box limits the accuracy of GFMC predictions of the large-r behavior of the exact $\rho_1(r)$ and consequently the singular behavior of $n(p)$ at small p is missed. Otherwise, state-of-the art predictions for the momentum distribution and the condensate fraction are obtained.[24]

While the treatment of Bose systems in GFMC is relatively easy, Fermi applications remain problematic. The antisymmetry requirement, a global property, is hard to build into the diffusion algorithm, which is by nature local. In straightforward application, the fact that the wave function is not of one sign leads to an exponential growth of the statistical error. Nevertheless, approximate realizations of GFMC–e.g. the fixed-node approximation and transient estimation–may be used to obtain quantitatively reliable results[24,25] for liquid ^{3}He. Even so, much work remains to be done to bring the fermion problem to an aesthetically satisfactory conclusion.

Apart from this specific Fermi difficulty, the GFMC treatment should eventually converge to the exact ground-state wave function of the fermion or boson system. However, in practice, with finite running time, this may not be the case. Importance sampling is used to speed the convergence of the calculation, but it may also prejudice the final results. An initial trial wave function, taken say from non-stochastic variational theory, is commonly used as an importance function to select the most likely configurations. If this wave function is not close to the true ground-state wave function, but represents instead some sort of metastable state, the method may not get the chance to find totally new or unexpected features of the many-body system.

Recently, in a beautiful implementation of Richard Feynman's view of quantum theory, Ceperley[26] has developed path-integral Monte Carlo (PIMC) methods to evaluate the equilibrium properties of quantum fluids at finite temperatures. The calculation begins with an accurate representation of the density matrix at high temperatures. A lower-temperature density matrix is then constructed from a path integral over products of high-temperature density matrices. Metropolis-type algorithms allow accurate results to be obtained for the properties of liquid ^{4}He over a broad range of temperatures and pressures.

IMPULSE APPROXIMATION

In previous sections we have compared and contrasted the theoretical issues which are important to momentum distribution studies in different systems. As we scan from atomic systems through electronic and nuclear systems all the way to particle physics, we have covered ten orders of magnitude range in energies, momenta, and interaction strengths. Nevertheless, we have found many common conceptual elements and calculational approaches. In the present section, we show that experimental studies of momentum distributions also share common themes across this extraordinary dynamic range. Momentum distributions are usually measured by scattering experiments in which the energy and momentum transferred are very high compared to the energies and momenta characteristic of ground-state properties and collective behavior. In this limit, the scattering law may be related to the momentum distribution by invoking the *impulse approximation*, which assumes that a single particle of the system is struck by the scattering probe, and that this particle recoils freely from the collision.

Such experiments began with the discovery of the Compton effect[27] in the scattering of X-rays from electrons in metals.[28] Today, neutron scattering at energies of hundreds of meV is used to measure $n(p)$ in atomic systems, photon scattering at energies of tens of KeV is used to measure $n(p)$ in electronic systems, and electron scattering at GeV energies is used to measure $n(p)$ in nuclei and inside nucleons. In all cases, the observation of appropriate scaling[29] (e.g. "Y-scaling") behavior of the cross section is presumed to indicate that conditions for the validity of the impulse approximation have been approached. Violations of this scaling are also of interest. They can arise from excitation of internal degrees of freedom ("scale breaking"), or they can be due to breakdowns of the impulse approximation such as final-state interactions of the struck particle, etc. In all cases, the measured quantity is a "Compton profile," with amounts to a projection of the momentum distribution onto one dimension. The various experimental settings entail common data analysis and methodological questions. In this section we shall elaborate on these common elements of the measurement of momentum distributions.

Deep-Inelastic Neutron Scattering

The scattering of neutrons by an atomic system is described by the double differential scattering cross section, which is the scattering per unit solid angle and per unit energy. This is calculated in first Born approximation, the interaction of a neutron with an atom being expressed as a Fermi pseudopotential proportional to the scattering length. In neutron scattering the cross section is traditionally written as[30]

$$\frac{d^2\sigma}{d\omega d\Omega} = \frac{\sigma_T}{4\pi}\frac{k_i}{k_f}S(Q,\omega) \ , \tag{5.1}$$

where σ_T is the total scattering cross section for a single atom and k_i and k_f are the initial and final neutron wave vectors. The interaction of the neutron with the sample is described by the dynamic structure factor $S(Q,\omega)$, with energy transfer $\omega = (\hbar^2/2m_n)(k_i^2 - k_f^2)$ and momentum transfer $Q = |k_i - k_f|$.

The dynamic structure factor is the Fourier frequency transform of the density-density correlation function of the sample. It is convenient to separate the structure function into two components

$$S(Q,\omega) = \frac{\sigma_{coh}}{\sigma_T}S_{coh}(Q,\omega) + S_{inc}(Q,\omega) \ . \tag{5.2}$$

In the first term, S_{coh} is the coherent structure factor[31] and represents density fluctuations which involve *different* atoms. Thus

$$S_{coh}(Q,\omega) = \frac{1}{\pi N}Re \int_0^\infty dt \ e^{i\omega t} \sum_{i\neq j} < \Psi_0|e^{-i\vec{Q}\cdot\vec{x}_i(t)}e^{i\vec{Q}\cdot\vec{x}_j(0)}|\Psi_0 > \ , \tag{5.3}$$

where $\vec{x}_i(t)$ and $\vec{x}_j(t)$ are the time dependent positions of the two atoms, and Ψ_0 is the wave function for the initial state of the system before the scattering. This term describes the collective behavior of many atoms including collective excitations, crystalline order, etc. The second term S_{inc} is the incoherent structure factor, which only involves the motion of single atoms. Again, this may be written as the Fourier transform of a time correlation function

$$S_{inc}(Q,\omega) = \frac{1}{\pi N} Re \int_0^\infty dt \; e^{i\omega t} \sum_i < \Psi_0 | e^{-i\vec{Q}\cdot\vec{x}_i(t)} e^{i\vec{Q}\cdot\vec{x}_i(0)} | \Psi_0 > \; , \qquad (5.4)$$

which now only involves the scattering from a *single* atom at different times. Consequently, this structure factor will be sensitive to single-particle motions such as self diffusion and, as we will see, the momentum distribution.

At high energy and momentum transfers the scattering simplifies. Typical momenta involved in collective behavior correspond to neutron wavelengths comparable to the interatomic spacing. For Q in excess of these momenta, the exponentials in (5.3) will then oscillate rapidly from atom to atom and, on average, cancel out. Therefore, S_{coh} will not contribute in the high-Q limit and only S_{inc} will be left. This is the limit of importance to momentum distribution experiments, so we may focus on the scattering from individual atoms, commonly referred to as "Deep-Inelastic Neutron Scattering."[32]

If, in addition, the energy transferred to an atom by a neutron is large compared with the potential energies due to neighboring atoms, the final state of the struck atom will then be that of a free particle. Under these conditions, the incoherent scattering function reduces to the well-known "Impulse Approximation"[33] (IA), so called because the scattering particle is supposed to impart a momentum to the struck atom in a time so short that neighboring atoms are unable to respond. Then

$$S_{IA}(Q,\omega) = \frac{1}{\rho} \int \frac{d\vec{p}}{(2\pi)^3} \; n(|\vec{p}|) \; \delta\left(\omega - \frac{Q^2}{2M} + \frac{\vec{Q}\cdot\vec{p}}{M}\right) \; , \qquad (5.5)$$

where M is the mass of the struck atom.

The scattering law (5.5) exhibits characteristic features which have often been used as an indication that conditions for the validity of the IA have been reached. The scattering is centered at and symmetric about the recoil energy $\omega_r = Q^2/2M$. In neutron scattering the location of the scattering peak is determined by the mass of the struck particle and different constituents in the sample can be separated by their different recoil energies. In addition, the width of the observed scattering, at constant Q, is proportional to Q times the width of the momentum distribution. In the IA limit, the scattering is no longer a function of the energy and momentum transfer separately. For isotropic systems, where $n(p)$ depends only on the magnitude of p, the scattering becomes a function of a single variable

$$Y = \frac{M}{Q}\left(\omega - \frac{Q^2}{2M}\right) \; , \qquad (5.6)$$

which is just the longitudinal momentum, $p_{||}$. The scattering law may then be rewritten

$$J_{IA}(Y) = \frac{Q}{M} S_{IA}(Q,\omega) = \frac{1}{4\pi^2\rho} \int_{|Y|}^\infty dp \; p n(p) \; . \qquad (5.7)$$

The function $J_{IA}(Y)$ is known as the Compton profile. The asymptotic behavior expressed by (5.7), called Y scaling, was first emphasized by West[34] in the context of electron scattering from nuclei. It is of course more broadly applicable to scattering processes that meet the criteria for application of the impulse approximation. On the other hand, we must point out that Y-scaling is a necessary but not a sufficient condition for validity of the impulse approximation. For example, in systems with hard-core potentials, Y-scaling is predicted at high Q even though the impulse approximation does not hold.[35,36]

As an illustration of this kind of scaling, consider the scattering from liquid helium,[37] plotted in Fig. 4 as $J(Y)$, at Q's of 7, 12, and 23 Å^{-1} . When the experimental data for $S(Q,\omega)$ is plotted versus ω (not shown), the peak centers and widths vary greatly with Q. However, when converted to $J(Y)$ these results nicely demonstrate the predicted Y-scaling behavior. The scattering is symmetric and centered at $Y=0$, and the width of the scattering is independent of the momentum transfer Q of the measurement.

While Fig. 4 illustrates the expected Y-scaling behavior, deviations are in fact evident. The measurements at the highest Q (23 Å^{-1}) agree best with the impulse-approximation predictions. At the lower Q's one sees significant deviations from IA scattering (largest at 7 Å^{-1}) in the form of shifts of the peak center to the left. These and other deviations from the predictions of the impulse approximation are of great importance in experimental determinations of the momentum distribution in real experiments where Q, while large, is still finite. Such discrepancies have been termed "Final-State Effects," because they are thought to arise from the interactions of the struck atom with neighboring atoms. Another factor may be the binding of the target atom in the condensed phase, an 'initial state effect' which is ignored in deriving the impulse approximation. Deviations from the IA have received considerable theoretical attention and will be discussed more fully later.

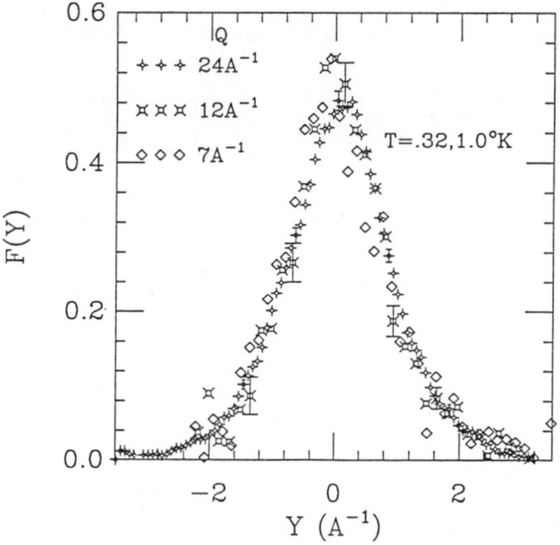

Fig. 4 Inelastic neutron scattering from liquid ^{4}He at momentum transfers of 7 Å^{-1}, 12 Å^{-1} (at 1.0 K) and 24 Å^{-1} (at 0.35 K). The results, plotted as $J(Y)$, all fall on approximately the same curve, illustrating the Y-scaling behavior (from ref 37).

Compton Scattering

An appropriate probe of electronic systems is the photon, whose coupling to the electron charge is also well described in first-order perturbation theory. The theoretical discussion proceeds in close analogy with that for neutron scattering from atoms. Accordingly, when the momentum transferred by the photon is large compared to the interelectronic spacing, the properties of individual electrons are probed. This condition, corresponding to what is termed Compton scattering, is met for photons at hard X-ray energies (tens of KeV). Compton scattering is the basis of the earliest of the momentum distribution experiments.[38] It has yielded considerable information on the electronic structure of systems in condensed phases.[1]

The cross section for the scattering of photons by a system of electrons in the high-Q regime is

$$\frac{d^2\sigma}{d\Omega d\omega} = \left(\frac{d\sigma}{d\Omega}\right)_0 \frac{\omega_2}{\omega_1} \frac{m_e}{Q} \int d\vec{p}\, n(|\vec{p}|)\; \delta\left(\omega - \frac{Q^2}{2m_e} + \frac{\vec{Q}\cdot\vec{p}}{m_e}\right) , \qquad (5.8)$$

where ω_1 and ω_2 are the frequencies of the incident and scattered photon, Q is the momentum transferred by the photon, and m_e is the mass of the electron. The factor $(d\sigma/d\Omega)_0$, specified below, is the elementary photon-electron scattering cross section. As in the impulse approximation to the neutron-scattering problem, the double-differential cross section in the Compton-scattering case may be expressed in terms of a single scaling variable, Y. It is given by some trivial factors, times the Compton profile

$$J(Y) = \int_{p_x} \int_{p_y} n(p_x, p_y, Y) dp_x dp_y . \qquad (5.9)$$

Here, we have written the Compton profile in a form which highlights its interpretation as the longitudinal momentum distribution, obtained by integrating $n(p)$ over the components of p transverse to Q.

The cross section for Compton scattering consists of a spin-independent and a spin-dependent component. The spin-independent component, which corresponds to the dominant mechanism for the scattering of unpolarized X-rays, is associated with the elementary Thompson cross section

$$\frac{d\sigma}{d\Omega}\bigg|_0 = \left(\frac{e^2}{m_e c^2}\right)^2 \qquad (5.10)$$

where e is the charge of the electron and c is the speed of light. In measuring this component of the scattering, one probes the full electronic $n(p)$, irrespective of the spin of the electron.

The spin-dependent component

$$\frac{d\sigma_s}{d\Omega}\bigg|_0 = \left(\frac{e^2}{m_e c^2}\right)^2 \left(\frac{\omega_2}{\omega_1}\right)^2 \left(\frac{1-\cos(\phi)}{m_e c}\right) \vec{S}\cdot(\vec{k_1}\cos(\phi) - \vec{k_2}) \qquad (5.11)$$

where $\vec{k_1}$ and $\vec{k_2}$ are the incident and final wavevector of the photon, ϕ is the scattering angle, and $\vec{S}$ is the spin of the electron. This component is only of interest for polarized radiation. Using polarized X-rays, the momentum distribution of one particular spin state can be measured. With the recent developments in synchrotron radiation sources, magnetic Compton scattering studies of the spin-dependent momentum distribution have become practical.

In Compton scattering, the struck particle is always an electron and the scattering peak is always centered at $Q^2/2m_e$. However, the structure of the scattering peak may be used to separate the contributions from different 'types of electrons,' such as conduction and core electrons. The width of a Compton profile is proportional to $Q\Delta p/m_e$, where Δp is the width of the momentum distribution. The core electrons, which are tightly bound, will have a broad profile, while the conduction electrons, which are in extended plane-wave states, will have a relatively narrow profile. Fig. 5 illustrates this distinction using the Compton profile of beryllium. The narrow contribution results from the conduction electrons, while the broad background is due to the tightly bound core electrons. The sharp change in slope of the scattering, which is due to the jump in $n(p)$ at the Fermi surface of the conduction electrons, is also clearly visible.

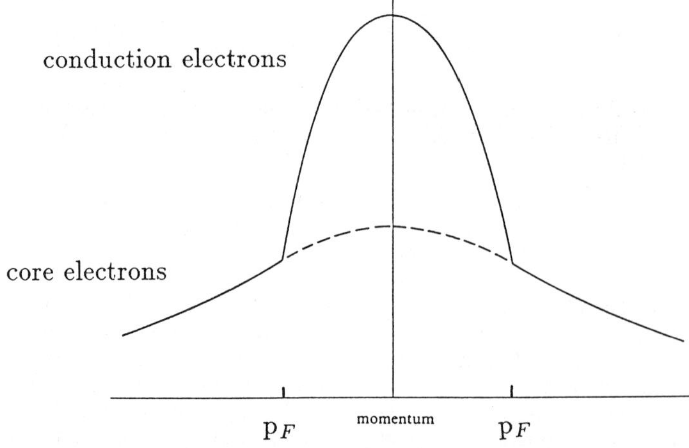

Fig. 5 Schematic illustration of the Compton profile from beryllium. The broad component, which has been extended through the central region by the dashed line, is due to the tightly bound core electrons. The narrow component is due to the nearly free conduction electrons. The sharp change in slope at the Fermi surface is clearly visible (from ref 28).

Quasielastic Electron-Nucleus Scattering

The appropriate probe for studying the momentum distribution of nucleons in nuclei is the electron, which couples electromagnetically to nucleons. The situation here is much more complicated than in electronic and atomic systems. It is important to use relativistic kinematics in describing the scattering. Moreover, the nucleon-nucleon interaction is incompletely characterized, and its complexity makes theoretical calculations difficult. Most important, internal degrees of freedom of the nucleons are easy to excite with the energy and momentum transfers required to approach conditions for validity of the impulse approximation.

The impulse-approximation result for the scattering of electrons by the nucleus is

$$S_{IA}(Q,\omega) = \frac{\nu}{\rho} \int_M^\infty \frac{d\sigma(Q)}{dM_E} \int \frac{d\vec{p}}{(2\pi)^3} n(|\vec{p}|)\delta(\omega - \sqrt{M_E^2 + (Q+p)^2} + \sqrt{M^2 + p^2})$$

(5.12)

where ν is the single-particle level degeneracy ($=4$), $n(p)$ is the momentum distribution of nucleons in the ground state of the target, and $d\sigma(Q)/dM_E$ is the cross section for electron scattering from a single nucleon leading to an excitation of the nucleon with rest mass M_E (e.g. excitation of the $\Delta(1238$ MeV$)$).

Y-scaling is predicted by the impulse approximation only if subnucleonic degrees of freedom are not excited, and the final mass of the recoiling constituent(s) (M_E) is the same as the nucleon mass. This condition, corresponding to what is known as "quasielastic electron-nucleus scattering" (QENS), places upper bounds on the applicable momentum and energy transfers. The associated scaling function is given by

$$F_{IA}(Y) = \frac{QS_{IA}(Q,\omega)}{(\omega + M)} \sigma_{el}(Q) = \frac{\nu}{4\pi^2\rho} \int_{|Y|}^{\infty} dp\; pn(p) \;, \qquad (5.13)$$

which is similar to the formulae arising in deep-inelastic neutron scattering and Compton scattering. With relativistic kinematics, the scaling variable Y now takes the somewhat different form

$$Y = p_{\parallel} = \frac{M}{Q}\left(\omega - \frac{Q^2}{2M} + \frac{\omega^2}{2M}\right) \;. \qquad (5.14)$$

Because in QENS the energy and momentum transfers are limited by the necessity to avoid the excitation of nucleonic internal degrees of freedom, it is difficult to achieve momentum transfers more than a few times larger than the characteristic momenta of nucleons in a nucleus, and Y-scaling is observed only for negative Y. This Y-scaling portion of the cross section corresponds to the high-p components in the tail of the momentum distribution. Of course, it is just these high-p components which provide the most information on short-range correlations in nuclei.

Another consequence of the limited ranges of energy and momentum transfers in QENS is that the off-shell character of the target particle in the many-body medium, prior to collision with the probe, can be far more important than in neutron or Compton scattering. Thus, while scaling behavior may still be observed, one needs to modify the scaling variable to take account of the binding of the constituent particles. In particular, Sick and coworkers[39,40] proposed using $Y_S = p_{\parallel}$ as determined from the kinematic relation

$$[(p + Q)^2 + M^2]^{1/2} - M \;=\; \omega - \mathrm{SE} \;, \qquad (5.15)$$

where the constant shift SE is a suitable average separation energy for removal of a nucleon from the nucleus, and irrelevant recoil effects have been omitted. Improved scaling plots of data are obtained in terms of Y_S. Other forms for the scaling variable have been proposed as well, notably in attempts to incorporate effects of "final-state interactions" into the analysis.[41]

As an example of quasielastic electron-nucleus scattering, analyzed in terms of the variable Y_S, consider the results for a carbon target[40] shown in Fig. 6. For Y_S below -0.1, the experimental results nicely illustrate the Y-scaling predictions. Around $Y_S = 0$ and above, the breakdown of Y scaling is seen as internal degrees of freedom of the nucleon are excited.

The behavior illustrated in Fig. 6 is all the more impressive when one notes that the scattering intensity changes by almost four orders of magnitude over the kinematic range in Y_S for which scaling appears to hold. In contrast, the Y-scaling plot for ^{4}He shown in Fig. 4 extends over less than two orders of magnitude in the scattering intensity. This disparity is due to the much higher backgrounds which prevail in current neutron-scattering experiments, as compared with QENS. In the nuclear work, the results found in the regime of large $-Y_S$ are suggestive of an exponential fall-off of the momentum distribution which might be a general feature of strongly-interacting many-body systems. It would be of great interest to test this conjecture in future neutron-scattering experiments on quantum fluids.

Before turning to other issues, we would like to address the unfortunate semantic confusion which can arise in trying to compare the three types of experiments: deep-inelastic neutron scattering, Compton scattering, and quasielastic electron-nucleus scattering. In condensed-matter physics, "Quasi-Elastic Neutron Scattering" (also QENS) refers to scattering measurements in which the energy transfer is small, as for example in experiments which study diffusive phenomena. Inelastic scattering refers to any scattering process in which the initial and final energies of the scattering (probe) particle are significantly different. "Deep-Inelastic Neutron Scattering" refers to higher-energy-transfer experiments which are aptly described in terms of scattering off single atoms in condensed-matter systems. In nuclear physics, inelastic scattering refers instead to cases in which there is excitation of nucleonic substructure, including any process which creates new particles. Thus, even though the energy of the scattering electron may change substantially in quasielastic electron-nucleus scattering, the process is termed "quasielastic" because the recoiling constituent particle has the same rest mass as it did before the collision. Finally, "Deep-Inelastic Electron-Nucleus Scattering" refers to experiments which are best described in terms of scattering of the probe electrons off the quark constituents of nucleons.

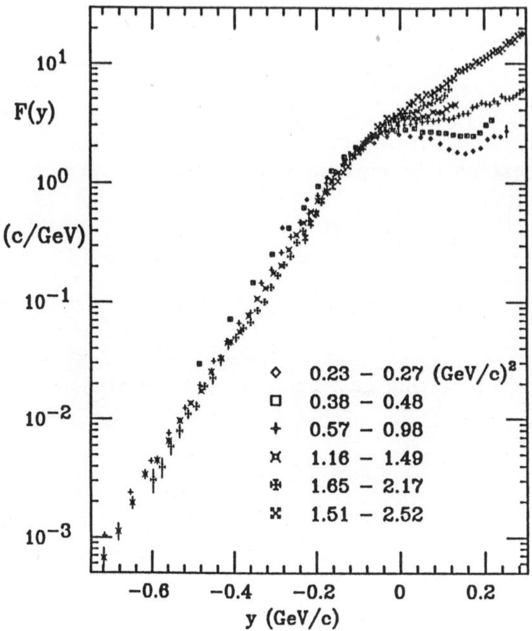

Fig. 6 Quasielastic Electron Nucleon Scattering (QENS) from ^{12}C at several different momentum transfers. For $-Y > 0.1$ GeV/c the results all fall on the same curve, illustrating the Y-scaling behavior. Y-scaling breaks down for larger values of Y due to the excitation of internal degrees of freedom of the nucleon (from ref 40).

Final State Effects

As described above, the measurements of the momentum distribution in atomic, nuclear, and electronic systems by scattering experiments all depend implicitly on the validity of the impulse approximation. The impulse approximation assumes that the particle probe (neutron, photon, electron) scatters off a single particle (atom, electron, nucleon) in the many-body system in a time so short that the particles neighboring the struck particle have no time to react to the perturbation caused by the probe. The response of the struck particle is determined entirely by its initial momentum distribution, and it recoils from the collision in a free-particle state of high momentum and energy. Obviously, this description becomes more accurate the higher the momentum and energy transferred in the scattering process (i.e. the shorter the probe/struck-particle interaction time), and the weaker the interactions between the struck particle and its neighbors.

In real experiments where the momentum and energy transfers are finite and interparticle interactions are never zero, the impulse approximation is always of limited validity. In atomic systems, deviations from the impulse approximation arise from atom-atom interactions; in nuclear problems, from nucleon-nucleon interactions; and in Compton scattering, from both electron-electron and electron-ion core interactions and orthogonalization. The important questions then are the size and form of the deviations from the impulse approximation and the extent to which they limit our ability to infer the single-particle momentum distribution from scattering data.

At high momentum and energy transfers, the most important deviations from the impulse approximation are due to collisions of the recoiling particle with its neighbors, which are called final-state effects. Hohenberg and Platzman[33] considered this problem for deep-inelastic neutron scattering from helium. Elementary arguments lead to the prediction that the impulse-approximation result will experience a Lorentzian broadening with width

$$\Delta Y_{FWHM} \approx \rho \sigma_{tot}(Q) \ , \tag{5.16}$$

where $\sigma_{tot}(Q)$ is the atom-atom scattering cross section. Heuristically, in terms of the uncertainty principle, this broadening is due to the finite 'lifetime' of the struck particle before it collides with its neighbors. The simple result (5.16) for the width of final-state broadening holds in several of the modern theories for FSE. However, there is much debate over the details of the lineshape. For one thing, the original Hohenberg-Platzman Lorentzian-broadening model cannot be rigorously correct since it violates the ω^2 sum rule on $S(Q, \omega)$.

Equation (5.16) suffices to indicate the properties which are most important in determining the FSE in real systems. FSE are larger for stronger interactions between particles and lower momentum transfers. Two idealized examples bracket the physical cases: In the noninteracting gas, FSE are entirely lacking, and Y-scaling holds trivially. At the other extreme of a gas of hard spheres, FSE are always present, even at infinite Q. Nevertheless, as we have already remarked, Y-scaling is predicted for the limiting case of hard-core interactions even though the impulse approximation is not valid. The repulsive cores for the real systems of interest to us are plotted in Fig. 3. (The attractive portion of the potential plays a negligible role in FSE.) Real systems may be ranked in *descending* order as follows, according to the strength of the core potential, appropriately scaled in terms of the characteristic momenta: atomic systems such as helium quantum fluids; nucleons in nuclei; electronic systems; and finally color-neutral systems of quarks and gluons. In *ascending* order, the ranking of the corresponding inclusive scattering experiments according to maximal momentum transfers, again scaled with characteristic momenta, is: quasielastic electron-nucleus scattering; deep-inelastic neutron scattering; Compton scattering; and deep-inelastic scattering of electrons from nucleons (which really refers to scattering off their constituent quarks and gluons). Thus, FSE are expected to be significant for both deep-inelastic neutron scattering (DINS) on quantum fluids and quasi-elastic electron-nucleus scattering (QENS), while they are more readily avoided in Compton scattering on electronic systems and are absent in particle physics.

An important consequence of (5.16) for deep-inelastic neutron scattering is that the approach to the impulse approximation will be very slow with increasing Q, because the $\sigma_{tot}(Q)$ decreases logarithmically with Q for He-He scattering. Therefore, helium is very close to a hard-sphere system where approximate Y-scaling behavior is obtained without the impulse approximation being valid. The final-state broadening interferes with observation of the features in the momentum distribution which are the most interesting, i.e. the Bose condensate peak in ^{4}He and the Fermi surface discontinuity in ^{3}He. While in QENS the $\sigma_{tot}(Q)$ falls more rapidly with increasing Q, the obtainable Q's are comparatively much smaller when scaled by characteristic momenta.

Any attempt to determine momentum distributions in DINS and QENS must somehow take into account the final-state effects. At present, we depend on theoretical calculations to provide the appropriate corrections. Unfortunately, this adds another layer of complication and uncertainty to the interpretation of the experimental results. The theory of final-state interactions is currently an area of very active research, fraught with numerous controversial issues and conflicting conclusions. The quantitative characterization of FSE presents a major challenge for many-body theorists. Apart from highly nontrivial dynamical considerations, this problem involves aspects of the strongly-interacting ground state (e.g., the two-body reduced density matrix) which are difficult to evaluate. However, if we assume that the many-body calculations of momentum distributions are correct, theories for FSE can be tested using scattering data.[42]

The recent non-Lorentzian broadening theory by Silver[43] for ^{4}He, based on earlier work by Gersch and Rodriguez[44] yields the broadening function shown in Fig. 7, which is to be convoluted with the impulse-approximation prediction, Eq. (5.7), to obtain the predicted scattering. While the FWHM of the function in Fig. 7 is comparable to that from the Hohenberg-Platzman Lorentzian-broadening theory, Eq. (5.16), Silver's theory has negative wings at large $|Y|$ so that it satisfies the ω^2 sum rule which requires that the second moment of the broadening function be zero. The additional physics which this theory takes into account is the pair-correlation function of the interacting ground state, which governs the collision rate as a function of recoil distance. The overall effect of final-state broadening predicted by Silver is much smaller than in the original Hohenberg-Platzman treatment, and it produces excellent agreement with the recent experiments of Sosnick et $al.$ as discussed further in the next section.[45]

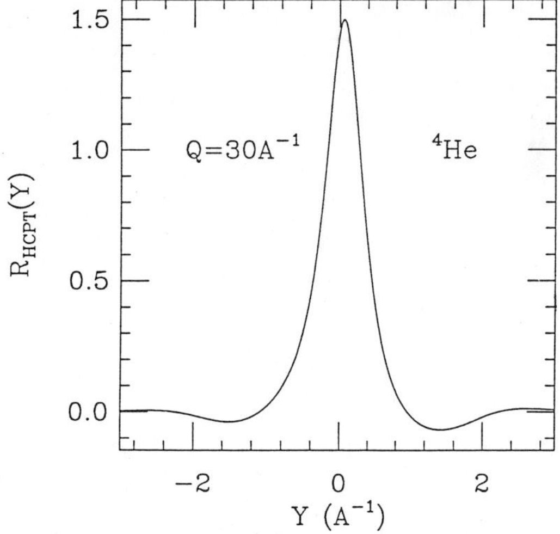

Fig. 7 Final State Effect broadening function $R(Y)$ for ^{4}He. The width of the central peak is approximately $\rho\sigma(Q)$, where $\sigma(Q)$ is the atom-atom scattering cross section, in agreement with simple 'lifetime' arguments. The negative wings are required to satisfy the ω^2 sum rule (from ref 43).

Scale Breaking

Deviations from Y-scaling may arise from final-state interactions, but a more fundamental cause is the excitation of internal degrees of freedom of the constituent particles of the sample. The latter effect is often referred to as "scale breaking." We have already seen that such scale breaking may occur in the nuclear context. It is also present in measurements of momentum distributions in other kinds of systems.

22

By way of illustration, we shall amplify upon two examples, one from 'atomic' and one from nuclear physics.

First, consider the scattering of neutrons by a collection of atoms. Fig. 8 shows an overview of the scattering function in the $Q - \omega$ plane for a simple atom with no internal excitations. At low Q and ω the scattering is due to collective excitations — phonons, rotons, diffusive modes, Bragg scattering, etc. At higher momentum transfers the collective excitations are damped out and the scattering is dominated by single-particle excitations. Finally, at large enough Q's, the scattering is described by the IA, and information on $n(p)$ can be obtained.

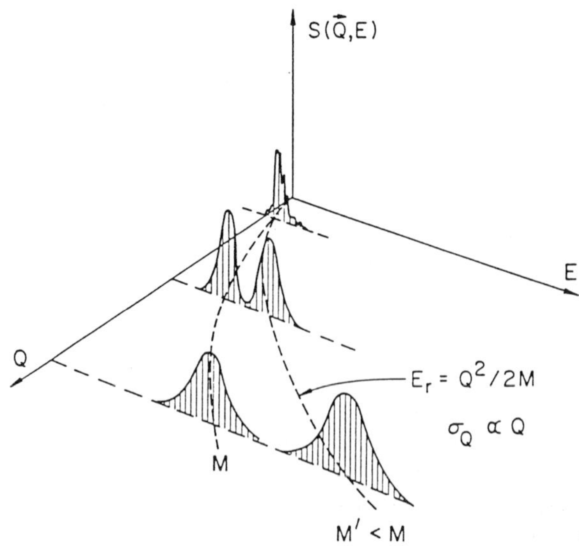

Fig. 8 Schematic of the dynamic structure factor $S(Q,\omega)$ as a function of energy and momentum transfer. The scattering at low Q and ω is dominated by collective excitations - phonons, rotons, diffusive modes, Bragg scattering, etc. At higher Q and ω the collective excitations are damped and the single particle properties determine the scattering. At very high Q and ω the scattering is described by the IA. In this region the peak is centered at the recoil energy $(Q^2/2m)$ and the width of the peak is proportional to Q. The scattering from particles with two different masses is shown to illustrate the ability to separate the scattering from different constituents of the sample is the IA regime (From ref 46).

Scale breaking can be illustrated by the scattering from a molecule which has both translational and internal degrees of freedom. Bulk molecular hydrogen furnishes an excellent example. The translational and internal modes (the latter consisting of vibrations and rotations of the molecule about its center of mass) are essentially decoupled. The IA is easily generalized to give

$$S_{IA}(Q,\omega) = \sum_n \frac{f_n}{\rho} \int \frac{d\vec{p}}{(2\pi)^3} \; n(|\vec{p}|) \; \delta\left(\omega - \omega_n - \frac{Q^2}{2M} + \frac{\vec{Q}\cdot\vec{p}}{M}\right) \,, \qquad (5.17)$$

where f_n and ω_n are respectively the structure factor and energy of the nth internal excitation. Fig. 9 shows the scattering from hydrogen[47] when two internal excitations, rotational excitations in this case, are excited. The rotational excitations shift the peak center from the recoil energy. The observed scattering is then the sum of the two shifted peaks, which can be resolved in this case. Y-scaling breaks down when multiple internal excitations are excited. At low energy transfers, only the first rotational excitation is excited and a unique scaling variable can be defined. However, at higher energy transfers many different excitations contribute to the scattering, each centered at a different energy. It is clear that no unique scaling variable can be defined in this case and that Y-scaling will break down.

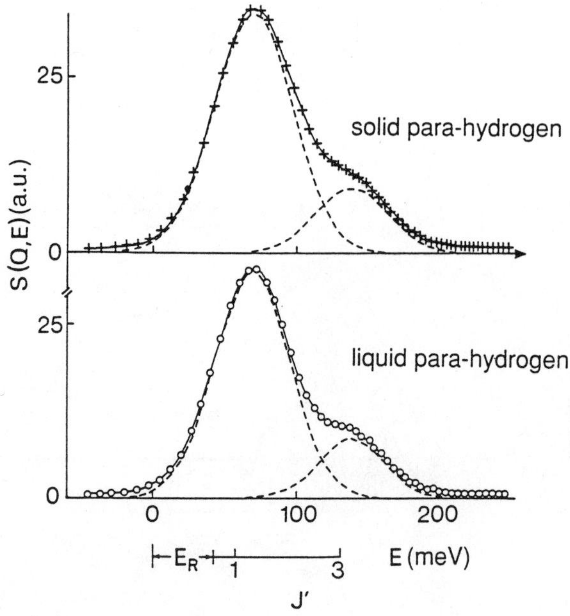

Fig. 9 The observed scattering from liquid and solid para-hydrogen. The crosses are the experimental results. The dashed lines are the predicted IA scattering from the hydrogen molecule when the J=0 to J=1 (14.7 meV) and J=0 to J=3 (88.2 meV) rotational transitions are excited (from ref 47).

The situation is similar in nuclear physics. Fig. 10 shows an overview of the nuclear response function. The resemblance to Fig. 8 for the atomic case is striking. At low Q and ω, nuclear response is dominated by elastic scattering, by inelastic scattering to low-lying states, and by collective excitations such as the giant resonances. At higher momentum and energy transfers the scattering is described essentially by the impulse approximation (region II). Y-scaling applies only in this region. At even higher momentum and energy transfers, nucleonic excitation and particle creation break the scaling behavior. Finally, at extremely high energies scaling behavior of a different sort, associated with electron scattering from the quark constituents of the nucleon, sets in. (For further details on deep-inelastic scattering from nucleons and "x-scaling," see the article by West in these proceedings.[29])

As an example of scale breaking in nuclear physics, refer once again to the measurements on ^{12}C plotted in Fig. 6. At low momentum transfers, the quasielastic scattering can be described by Y_S scaling up to $Y_S = -0.1$ GeV/c or so. However, as the incident energy is increased, more energy is available to create internal excitations and Y-scaling breaks down at lower and lower values of Y_S. This is exactly analogous to what is seen in neutron scattering from molecular hydrogen.

We emphasize that scale-breaking phenomena have nothing to do with the break-down of the impulse approximation, which is due to final-state effects. Rather, scale breaking (as the term is conventionally understood) is due to the excitation of internal degrees of freedom of constituent particles. In principle, momentum distributions could still be determined in the presence of scale breaking by inverting a more general impulse-approximation formula, such as eq. (5.12) or eq. (5.17). However, this has yet to be attempted in practice.

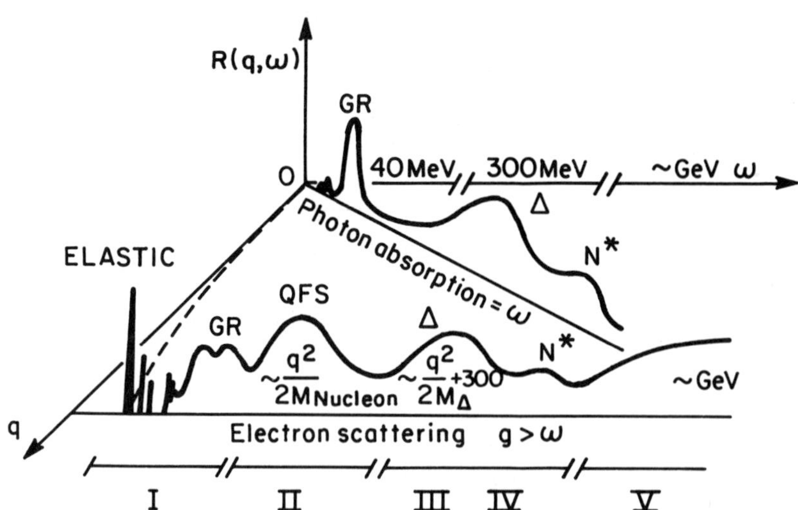

Fig. 10 Schematic of the nuclear response function $R(Q,\omega)$ as a function of energy and momentum transfer. The scattering at low energies (region I) is dominated by the elastic peak and inelastic scattering from low-lying states and the giant resonances (GR). At slightly higher energy transfers (region II) *quasifree scattering* (QFS) from nucleons initially bound in the nuclear medium is observed. This is the region where the momentum distribution of the nucleons is accessible using QENS. At larger energy transfers (region III-IV) sufficient energy is available to create pions and to excite the nucleon (Δ, N^*, resonances, etc.). Finally, at very large energy and momentum transfers (region V), electrons are scattered directly from the quark constituents of the nucleons (from ref 48).

Limitations on Determining $n(p)$

The impulse approximation provides a simple relationship between the single-particle momentum distribution, $n(p)$, and the observed scattering expressed in terms of a Compton profile, $J(Y)$. The experimental goal is to extract the momentum distribution from scattering measurements. Unfortunately, the extraction is hampered by the fact that prominent — and physically interesting — features in $n(p)$ may not be strongly reflected in $J(Y)$, due to the integral form of the impulse-approximation relationship (cf. eqs. (5.7), (5.8), (5.12), and (5.17)).

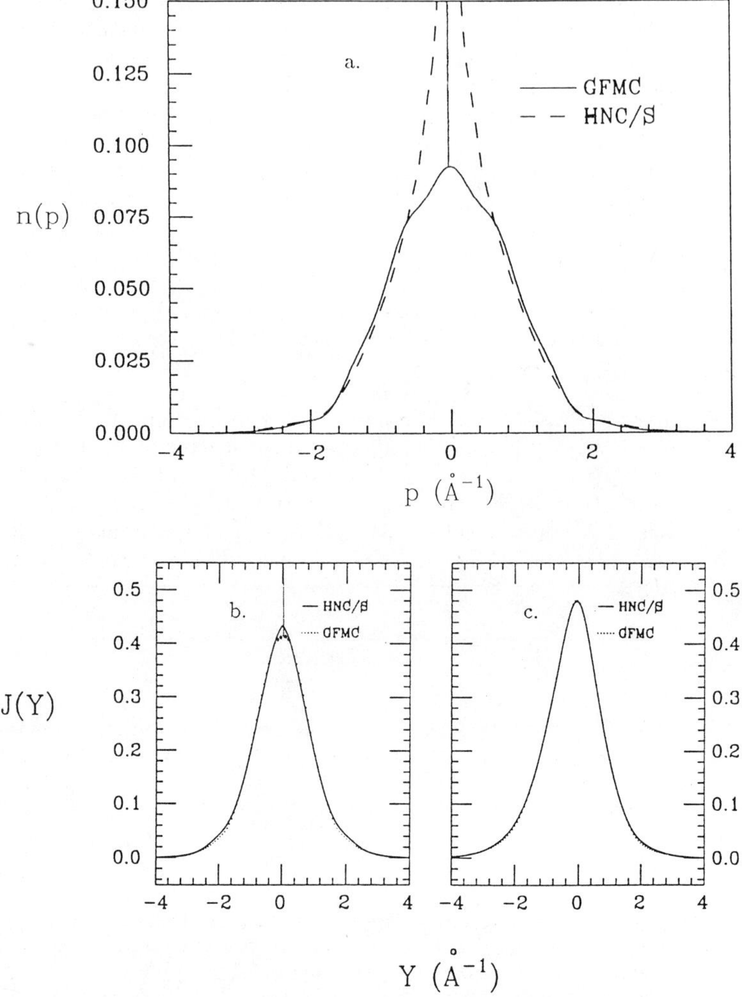

Fig. 11 Illustration of the relationship between $n(p)$ and $J(Y)$. a) shows the results of GFMC[24] (solid curve) and variational[49] (dashed curve) calculations of $n(p)$ in the ground state of liquid ^{4}He. Both calculations predict a condensate fraction of 10 % which gives a δ-function at $p = 0$ (not visible in the variational calculation). In addition, the variational calculation predicts a singular behavior at small p which is absent in the GFMC result. b) shows the effect of transforming $n(p)$ to $J(Y)$ using (5.7). The δ-function is still present. However, the singular behavior, which is the dominant feature in the variational $n(p)$ at small p, is now much less prominent. c) shows the results of including the final state broadening shown in Fig. 7. The two curves are now nearly indistinguishable.

To illustrate this fact, consider the ground-state momentum distribution of liquid ^{4}He. Fig. 11a shows the results of two different microscopic calculations, based on GFMC[24] and variational[49] (HNC/S) techniques. In both cases, $n(p)$ has a delta function at $p = 0$ representing the Bose condensate, with condensate fraction $n_0 = 9.2$ %. The results of the two calculations are also very similar at large p. However, they differ markedly at small but finite p. The variational calculation reproduces the p^{-1} singular behavior at small p which is due to coupling of long-wavelength density fluctuations to the condensate. On the other hand, the GFMC result for $n(p)$ shows no sign of this singularity, because of the finite size of the simulation box. Thus, while the two results for $n(p)$ are palpably different, the corresponding $J(Y)$ curves generated by the impulse approximation (Fig. 11b) differ only very slightly. When the respective impulse-approximation results for $J(Y)$ are in turn broadened by the final-state effects given by Silver's theory[43] (Fig. 11c), the predictions corresponding to the two microscopic calculations become indistinguishable. We conclude that the scattering experiment is insensitive to the singular behavior of $n(p)$ at small p which accompanies a proper treatment of long-range correlations. This insensitivity is obviously due to the fact that, in the isotropic case, $J(Y)$ reduces to an integral over $pn(p)$.

The above comparison documents an important limitation on the ability to extract information on the momentum distribution from scattering measurements. The physically measured quantity is the Compton profile $J(Y)$, not the momentum distribution $n(p)$. While, in principle, $n(p)$ could be obtained by numerical differentiation of $J(Y)$ data, it is clear that excellent statistics − perhaps unattainable in practice − would be required to distinguish experimentally between the two microscopic calculations for $n(p)$.

EXAMPLES

To round out this overview we present a few examples of momentum distributions in atomic, electronic, and nuclear systems. We give a sampling of both theoretical and experimental results, obtained both for classical systems, where statistics do not play a significant role, and for quantum systems, where they do.

Classical Systems

In atomic systems we encounter numerous examples in which the momentum distribution assumes the classical Maxwell-Boltzmann form. Except for the lighter elements, such as helium and hydrogen, the quantum nature of the constituent atoms plays a minor or insignificant role in the momentum distribution. In general, the interactions between the atoms are relatively strong, whereas zero-point motion is small. Accordingly, there is little overlap of the atomic wave functions in the condensed phases, which means that exchange effects are unimportant. Then it is an excellent approximation to treat the atomic 'particles' classically, with quantum effects included perturbatively as required.

Fig. 12 depicts the momentum distribution of liquid neon at 29.6 K, as inferred from inelastic neutron-scattering measurements.[50] The scattering results are well described in terms of a simple Gaussian form for $n(p)$, in accord with the classical prediction. For given particle mass, the width of the classical momentum distribution is determined by the temperature of the system. The average kinetic energy per atom, which is proportional to the width of the momentum distribution, is then expected to have the familiar ideal-gas value $3k_BT/2$. The kinetic energy extracted from the inferred momentum distribution for neon is 48.2 K, which is somewhat higher than the expected value of 44.4 K.

The differences between the measured and expected kinetic energies for neon can be attributed to quantum effects. These are small, although observable because neon is relatively light. The corrections are smaller still in the heavier rare-gas systems, such as argon, krypton, and xenon. The principal correction to the classical result comes from the small zero-point motion of the atoms. When this effect is small, corrections of increasing order may be made using the series

$$< K >= \frac{3}{2} k_B T \left(1 + \frac{1}{12}(\theta/T)^2 - \frac{1}{240}(\theta/T)^4 + \cdots \right) \ , \tag{6.1}$$

where $\theta^2 = (\hbar^2/3mk^2) < \Delta\Phi >$. The quantity $\Delta\Phi$ is the Laplacian of the potential energy. In the solid phase, the form of the corrections for zero-point motion is very similar, the role of θ being played by the Debye temperature, θ_D. Returning to the comparison of kinetic energies for liquid neon and incorporating the quantum zero-point correction, one obtains a theoretical prediction for the kinetic energy of 49.5 K, considerably improving the agreement with the measured kinetic energy.

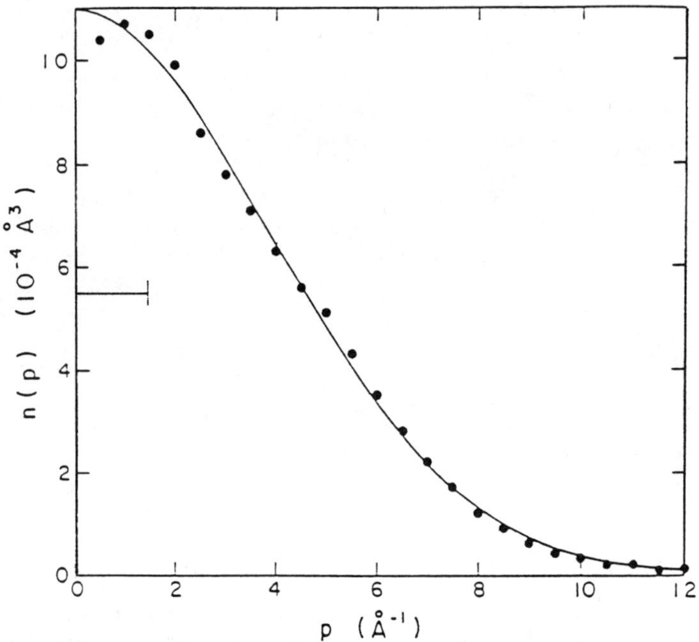

Fig. 12 The normalized momentum distribution for liquid neon at T=29.6 K. The dots are the experimental results. The solid line is a Gaussian, the classical form for $n(p)$, with the same root-mean-square momentum (from ref 50).

Bose Systems

Liquid ^{4}He provides the most celebrated example in which the Bose character of the constituent particles has a crucial influence on the properties of the many-body system, and especially on its momentum distribution. The interatomic potential in helium has a weak attractive portion and a condensed liquid phase does not form, at vapor pressure, until the temperature is decreased to 4.2 K. Indeed, the interactions are so weak that the system remains in a liquid phase down to zero temperature, when no external pressure is applied. Due to the weak attractive interactions and the light mass of the helium atom, the quantum zero-point motion is very large in the liquid. This implies a very large overlap of atomic wave functions naively proposed to describe the system. Correspondingly, exchange effects are expected to be very important.

28

The high temperature properties of liquid helium are similar to those of conventional liquids. However, when the temperature is lowered to about 2.2 K there occurs a phase transition from liquid helium I to a new phase, liquid helium II. The transition is signaled by a sharp feature in the specific heat — the famous λ-point anomaly. Liquid helium II, often called the superfluid phase, appears to contain a superfluid component which flows without viscosity and is responsible for an anomalously large thermal conductivity. Normal dissipative effects are attributed to a normal-fluid component. According to this "two-fluid model," the fraction of superfluid increases to unity at absolute zero. The success of the phenomenological two-fluid picture is thought to be indicative of the appearance of a Bose condensate in the momentum distribution,[51] a manifestation of Bose statistics with striking macroscopic repercussions. There has been a forty-year history of attempts to observe this Bose condensate by neutron scattering experiments.[52]

Fig. 13a shows the measured scattering, $J(Y)$, in the normal liquid phase (liquid He I) at 3.5 K, where no condensate is present. Theoretical predictions for the scattering are drawn as solid and dashed lines. The momentum distribution, calculated using the PIMC method,[26] is nearly Gaussian at helium II temperatures and has been converted to $J(Y)$ via the impulse-approximation (IA) formula. To allow direct comparison with experimental data, this prediction for $J(Y)$ has been artificially broadened using a model of the instrumental resolution, resulting in the dashed curve. The agreement is excellent. The significance of the solid curve will be explained below.

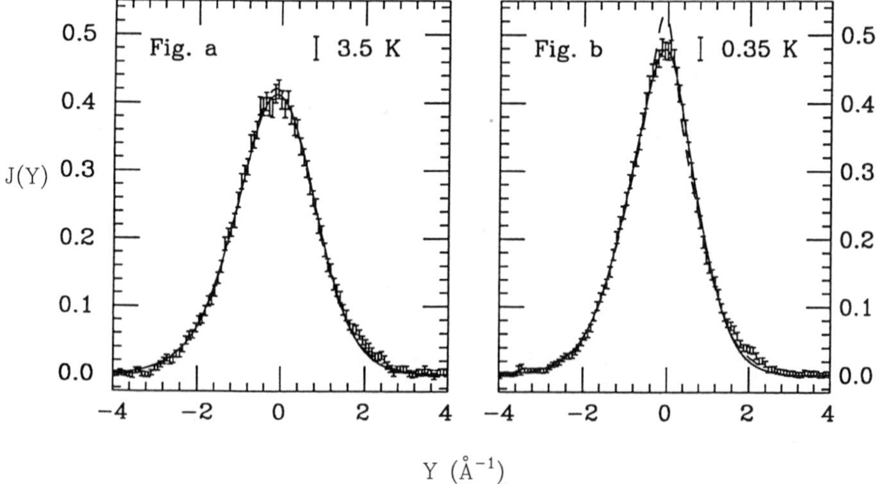

Fig. 13 The measured scattering in the normal liquid phase at 3.5 K (a) and the superfluid phase at 0.35 K (b) of liquid ^{4}He. The dashed curves are the theoretical predictions for $n(p)$ transformed to $J(Y)$ and convoluted with instrumental resolution. PIMC[26] calculations at 3.33 K have been used for comparison with the normal liquid and are in excellent agreement with the experimental results. GFMC[24] calculations have been used for comparison with the superfluid and large discrepancies exist near $Y = 0$. The solid curves are again the theoretical predictions, but including the FSE broadening shown in Fig. 7. The agreement between theory and experiment is now excellent in both the normal and superfluid phases (from ref 53).

Fig. 13b shows the measured scattering in the superfluid phase (liquid He II) at 0.35 K. The dashed line is again an instrumentally-broadened theoretical prediction based on the IA formula, this time using as input the $n(p)$ from a ground-state GFMC calculation[24] which yields a condensate fraction of 9.2 % . Due to the instrumental broadening, the IA prediction no longer exhibits a distinct condensate peak. However, it retains a sharper peaking around $Y = 0$ than in the normal liquid, which is symptomatic of an appreciable condensate. Unfortunately, the dashed curve gives too large an intensity around $Y = 0$ and is in generally poor agreement with the observed scattering.

The results of this comparison in the superfluid phase could be interpreted as an indication that the condensate fraction is far smaller than theoretically predicted. However, such an inference is unjustified without a consideration of the final-state broadening of the condensate peak. We may take final-state effects into account by convoluting the instrumentally-broadened IA result with a final-state broadening function furnished by the recent theory of Silver[43] (Fig. 7). In the normal-liquid case (Fig. 13a), this produces the solid curve, which is indistinguishable from the predicted scattering without final-state effects. Thus, final-state interactions have little effect in the presence of the broad momentum distribution of the normal liquid.

Turning to the superfluid case, the solid curve in Fig. 13b gives the theoretically predicted scattering when final-state broadening as well as instrumental resolution is incorporated. There is now striking agreement with the experimental data! This analysis provides a dramatic example of the importance of final-state broadening when a sharp feature is present in the momentum distribution. Moreover, the excellent agreement between theory and experiment furnishes strong support for the existence of a condensate in the superfluid and for the validity of the microscopic calculations of the momentum distribution.

Fermi Systems

Examples in which the Fermi character of the constituent particles is strongly reflected in $n(p)$ are quite widespread and can be found in atomic, electronic, and nuclear systems.

Fig. 14a displays the results of some representative microscopic calculations of the momentum distribution of the interacting electron gas (one-component plasma), at a density pertinent to sodium. The solid line is from a variational calculation based on an optimal Jastrow wave function,[54] which accounts, in an average way, for the effects of both short-range and long-range correlations. The long- and short-dashed lines correspond to older 'perturbative' calculations,[55] which invoke the random-phase approximation and thus concentrate (in a more detailed manner) on the long-range correlations. The momentum distribution of the noninteracting Fermi gas is also plotted for reference.

Fig. 14b demonstrates the dependence of the momentum distribution on the electronic density, smaller r_s corresponding to higher density. The two densities considered differ by a factor $2^{1/3}$. The curves were obtained by the same optimal Jastrow treatment which gave the solid line in Fig. 14a, the required expectation values being evaluated in a hypernetted-chain approximation which neglects elementary diagrams.[54] The dots show the results of comparable variational Monte Carlo calculations,[56] which attest to the validity of this approximation. We observe that the discontinuity Z_{k_F} of $n(p)$ at the Fermi surface narrows as the density decreases, which implies that the system is becoming more strongly coupled. This behavior is due to the fact that the screening of the long-range coulombic interaction between the electrons becomes less effective at lower density. The contrary behavior is seen in the atomic (viz. liquid ^{3}He) and nuclear cases, where the basic interactions are of short range and Z decreases as the density increases.

Fig. 14c collects a variety of theoretical results for the discontinuity Z, or "quasiparticle pole strength," of the homogeneous electron gas, and indicates that this system can display a range of behaviors between weak and strong coupling, in the density regime relevant to alkali metals. The result of an experimental determination of Z for metallic Na is included for comparison.[57]

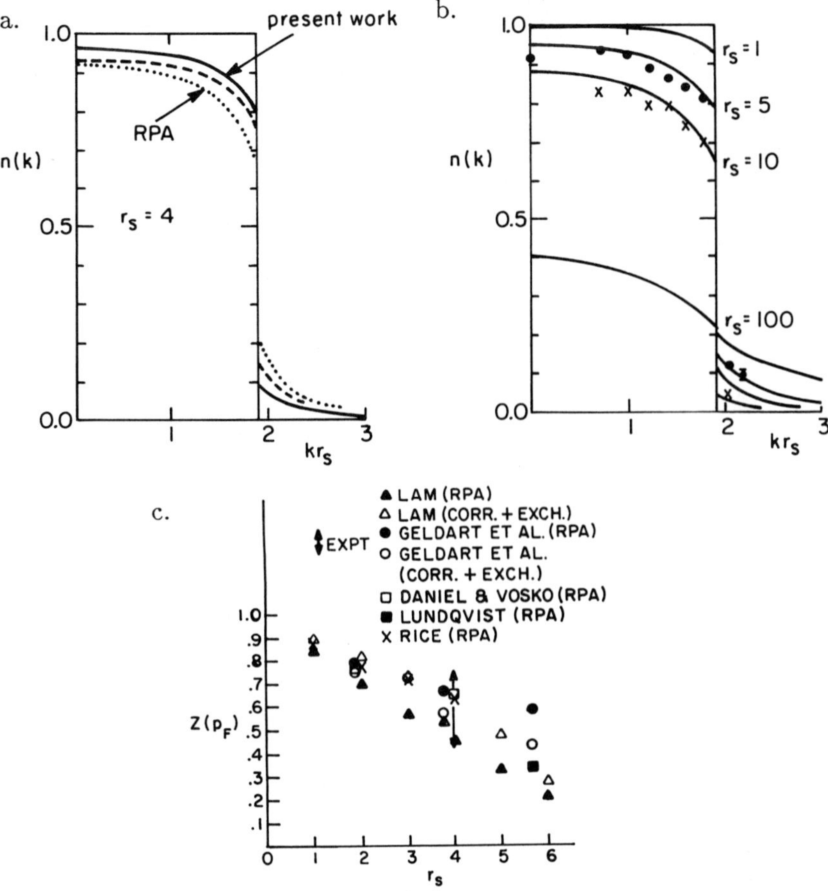

Fig. 14 Theoretical calculations of $n(p)$ for the interacting electron gas. a) $n(p)$ at a density comparable to Na. The solid line is a variational calculation and the dashed lines are perturbative calculations. $n(p)$ for the noninteracting Fermi gas is also shown for reference (from ref 54). b) Density dependence of $n(p)$. The curves are variational calculations of $n(p)$ and the open circles and crosses are Monte Carlo results. Smaller values of r_s correspond to higher density for the interacting electron gas (from ref 54). c) Theoretical results for the discontinuity Z of the homogenous electron gas as a function of density. The experimental results for metallic Na are included for comparison (from ref 57).

Fig. 15 sketches the momentum distributions obtained in a recent theoretical study of nuclear matter within the method of correlated basis functions.[58] These results document the importance of the tensor interaction in depleting the Fermi sea. With tensor forces present, the Fermi surface discontinuity is $Z \sim 0.7$, very close to that found in the optimal Jastrow calculation for the uniform electron gas at the density corresponding to metallic Na. In this sense, nuclear matter and the electron gas may be considered as comparably strongly interacting. Omitting the tensor component of the nucleon-nucleon potential, the nuclear-matter Z increases to about 0.85. In finite nuclei, the Fermi-surface discontinuity is broadened.

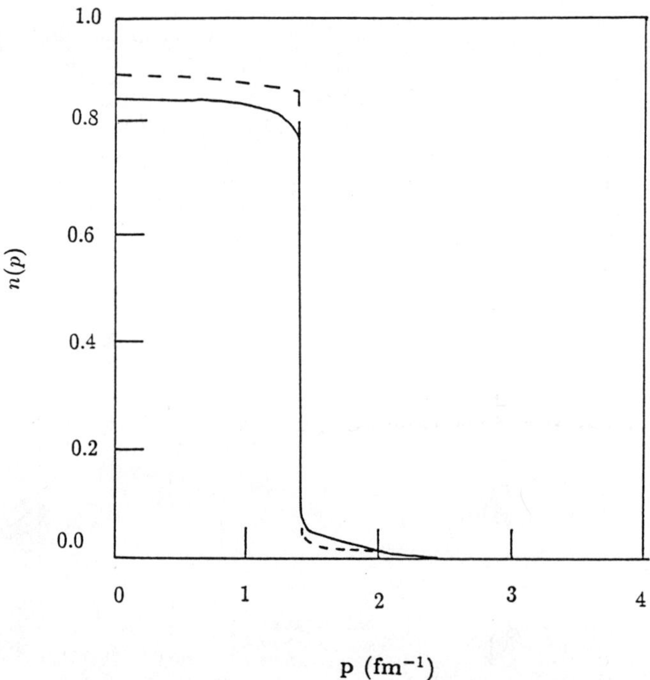

Fig. 15 Momentum distribution in nuclear matter using the method of correlated basis functions. The solid curve includes the tensor component of the nucleon-nucleon potential, while the dashed curve shows the effect of omitting this component (adapted from ref 58).

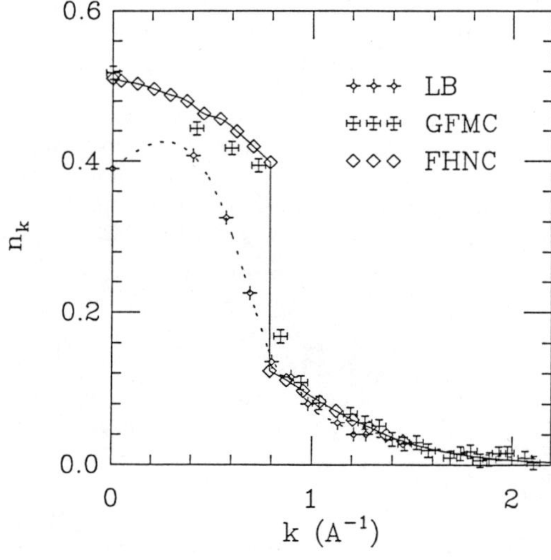

Fig. 16 Theoretical $n(p)$ for liquid ^{3}He. The FHNC[59] and GFMC[24] calculations yield very similar predictions for $n(p)$ with a Fermi-surface discontinuity of roughly 0.3. The Bouchaud and Lhuillier (BL) calculation[60], which uses a radically different trial wave function incorporating odd-wave pairing, predicts that the Fermi-surface disappears entirely.

In Fig. 16 we reproduce the results of three different theoretical evaluations of the momentum distribution in liquid ^{3}He at $T = 0$. The fixed-node GFMC calculation[24] and the variational treatment labeled FHNC[59] predict essentially the same results for $n(p)$. There is a Fermi-surface discontinuity of roughly 0.3 (not resolved in the GFMC evaluation). This small value supports the view that liquid ^{3}He is the most strongly interacting of the Fermi systems we have considered. The FHNC variational treatment is predicated on a conventional trial ground-state wave function including Jastrow two-body correlations, triplets, and backflow; it makes use of "scaling" and cluster-expansion procedures in addition to Fermi-hypernetted-chain (FHNC) resummation.

By contrast, the variational Monte Carlo calculation of Bouchaud and Lhuillier,[60] assumes a radically different trial wave function for the ^{3}He ground state, incorporating odd-wave pairing correlations as well as a Jastrow factor. This departure from tradition has a dramatic consequence for $n(p)$: the Fermi surface disappears entirely. The implications of the Bouchaud-Lhuillier work are yet to be fully explored, but it suggests that there is still room for improvement in our fundamental understanding of liquid ^{3}He, which serves as a prototype for testing approaches to other strongly-correlated Fermi systems.

CONCLUSION

We hope that this overview has convinced readers that momentum distributions are fundamental to our understanding of quantum many-body systems in most areas of physics. Even though the energy and length scales vary by more than ten orders of magnitude, we have identified common conceptual and methodological elements in theoretical and experimental studies of atomic, electronic, and nuclear systems. For lack of space, our overview has not developed upon the similar analogies which exist for particle physics,[29] electrons in atoms and molecules,[61] single atoms in potential wells such as metal hydrides,[62,63] and positron-annihilation studies of electronic systems.[7] Nor have we discussed momentum-distribution experiments in which more than one outgoing particle is detected.

Recently, there have been impressive advances in the theoretical calculation of momentum distributions. Numerical calculations, using GFMC and PIMC techniques, have benefited greatly from the accessibility of supercomputers and are beginning to provide extremely accurate descriptions of the momentum distribution. In addition, there have been significant conceptual improvements in the variational wave functions used to describe many-body systems. Moreover, the number of strongly-interacting many-body systems for which momentum distributions are of interest is expanding rapidly, particularly with the discoveries of strongly-correlated electronic systems such as the heavy-fermion materials, high-temperature superconductors, and structures displaying quantum Hall effects.[8] Accompanying these discoveries is an expanding variety of possibilities for variational wave functions of Fermi systems.

We have also discussed recent theoretical advances in the calculation of corrections to the impulse approximation due to final-state effects in atomic and nuclear systems. These effects can be quite important since they cannot be experimentally eliminated in feasible experiments on atomic and nuclear systems. We have shown that a quantitative theoretical characterization is essential to the detailed interpretation of the experimental results.

There have also been impressive advances on the experimental side. The development of new sources with increased energy and flux is expanding both the range of systems that can be studied and the amount of detailed information that can be obtained. Spallation neutron sources, such as IPNS, ISIS, and LANSCE, provide much larger fluxes at higher energies than were hitherto attainable. Synchrotron sources offer many orders of magnitude greater flux than available in laboratory X-ray facilities, and they have stimulated a revival of Compton-scattering experiments. New synchrotron sources, such as the Advanced Photon Source with special insertion devices including wigglers and undulators, will further enhance the capabilities for this class of measurement by several more orders of magnitude, and in particular they will facilitate the measurement of magnetic Compton profiles. New positron facilities and imaging capability should yield enormous improvements in studies of the Fermi surface of electronic systems. New facilities for nuclear physics research, such as CEBAF, will furnish much higher intensities for electron-scattering studies of nuclei and will facilitate the exploration of high-momentum components of the nuclear wave function. One can only imagine what momentum-distribution studies will become possible with the 20 TeV energies of the Superconducting Super Collider, or what constituent particles will form the basis for the impulse approximation (Higgs bosons?). Accompanying the development of these new, more intense and higher energy sources is the development of new and improved instrumentation with much better resolution. Most of these technical advances are discussed in more detail elsewhere in this volume.

In conclusion, we are now poised for an explosive growth of knowledge within the next few years about momentum distributions in many-body systems in all areas of physics. It is our conviction that this progress will benefit greatly from an interdisciplinary sharing of concepts and methodology, which has motivated the organization of this workshop. Our investigations of momentum distributions have convinced us, once again, of the essential unity of physics.

The following research support is acknowledged. PES: NSF Grant No. DMR-8704288; OBES/DMS support of the Intense Pulsed Neutron Source at Argonne National Laboratory under DOE grant W-31-109-ENG-38. RNS: OBES/DMS support of the Los Alamos Neutron Scattering Center at the Los Alamos National Laboratory. JWC: Condensed Matter Theory Program of the Division of Materials Research, National Science Foundation, under Grant No. DMR-8519077.

References

[1] B. Williams, **Compton Scattering**, (Mc Graw Hill, 1977).

[2] A. N. Antonov, P. E. Hodgson, I. Zh. Petkov, **Nucleon Momentum and Density Distributions in Nuclei**, (Clarendon Press - Oxford, 1988).

[3] J. Gavoret and P. Noziéres, Ann. Phys. (N. Y.) **B28**, 349 (1964).

[4] O. Penrose and L. Onsager, Phys. Rev. **B104**, 576 (1956).

[5] C. N. Yang, Rev. Mod. Phys. **B34**, 694 (1962).

[6] J. Clark and M. Ristig, contribution at this conference.

[7] S. Berko, contribution at this conference.

[8] S. Doniach, contribution at this conference.

[9] K. A. Brueckner, Phys. Rev. **97**, 1353 (1955); **100**, 36 (1957).

[10] J. Goldstone, Proc. Roy. Soc. (London) **A239**, 267 (1957).

[11] B. D. Day, Rev. Mod. Phys. **39**, 719 (1967).

[12] A. B. Migdal, Soviet Phys.–JETP **B5**, 333 (1957).

[13] J. M. Luttinger, Phys. Rev. **B119**, 1153 (1960).

[14] A. D. Jackson, A. Lande, and R. A. Smith, Physics Reports **B86**, 55 (1982).

[15] R. B. Laughlin, Phys. Rev. Lett. **50**, 1395 (1983); Surf. Sci. **142**, 163 (1984).

[16] J. M. Chen, J. W. Clark, and D. G. Sandler, Z. Phys. A **B305**, 223 (1982).

[17] E. Krotscheck, Phys. Rev. A **B26**, 3536 (1982).

[18] G. Senger, M. L. Ristig, K. E. Kurten, and C. E. Campbell, Phys. Rev. **B33**, 7562 (1986).

[19] Q. N. Usmani, B. Friedman, and V. R. Pandharipande, Phys. Rev. **B25**, 4502 (1982); Q. N. Usmani, S. Fantoni, and V. R. Pandharipande, Phys. Rev. **B26**, 6123 (1982); M. F. Flynn, Phys. Rev. **B33**, 91 (1986).

[20] A. Fabrocini and S. Rosati, Nuovo Cimento **BD1**, 567, 615 (1982); M. Viviani, E. Buendia, A. Fabrocini, and S. Rosati, Nuovo Cimento **BD8**, 561 (1986).

[21] S. Vitiello, K. Runge, and M. H. Kalos, Phys. Rev. Lett. **60**, 1970 (1988).

22 E. Feenberg, **Theory of Quantum Fluids** (Academic, New York, 1969).

23 J. W. Clark and P. Westhaus, Phys. Rev. **141**, 833 (1966).

24 P. Whitlock and R. M. Panoff, Can. J. Phys. **B65**, 1409 (1987).

25 R. M. Panoff and J. Carlson, Phys. Rev. Lett. **62**, 1130 (1989).

26 D. M. Ceperley and E. L. Pollock, Phys. Rev. Lett. **56**, 351 (1986).

27 A. H. Compton, Phys. Rev. **21**, 207 (1923); **21**, 483, (1923).

28 For a history, see the article by R. H. Stuewer and M. J. Cooper in reference 1.

29 G. West, contribution at this conference.

30 G. L. Squires **Introduction to the Theory of Thermal Neutron Scattering** (Cambridge University Press, 1978).

31 Note, that for the purposes of this discussion we have used a non-standard definition of the coherent structure factor.

32 G. Reiter and R. Silver, Phys. Rev. Lett. **54**, 1047 (1985).

33 P. C. Hohenberg and P. M. Platzman, Phys. Rev. **152**, 198 (1966).

34 G. B. West, Phys. Rep. **18C**, 263 (1975).

35 J. Weinstein and J. W. Negele, Phys. Rev. Lett. **49**, 1016 (1982).

36 R. N. Silver and G. Reiter, Phys. Rev. **B35**, 3647 (1987).

37 J. W. Clark and R. N. Silver, Proceedings of the Vth International Conference on Nuclear Reaction Mechanisms, Varenna, Italy 1988 from data taken by T. R. Sosnick, W. M. Snow, P. E. Sokol, R. N. Silver, to be published, and W. G. Stirling, E. F. Talbot, B. Tanatar, H. R. Glyde, to be published.

38 J. W. M. Du Mond, Phys. Rev. **33**, 643 (1929).

39 I. Sick, D. Day, and J. S. McCarthy, Phys. Rev. Lett. **45**, 871 (1980).

40 D. B. Day *et al.*, Phys. Rev. Lett. **59**, 427 (1987).

41 I. Sick, Comments Nucl. Part. Phys. **18**, 109 (1988).

42 P. E. Sokol, T. R. Sosnick, W. M. Snow, and R. N. Silver, contribution at this conference.

43 R. N. Silver, Phys. Rev. **B38**, 2283 (1988), and contribution at this conference.

44 H. A. Gersch and L. J. Rodriguez, Phys. Rev. **A8**, 905 (1973).

45 P. E. Sokol, contribution at this conference.

46 R. O. Simmons and P. E. Sokol, Physica **136B**, 156, 1986.

47 W. Langel, D. L. Price, R. O. Simmons, and P. E. Sokol, Phys. Rev. **B38**, 11275 (1988).

48 B. Frois in **Progress in Particle and Nuclear Physics** , Vol 13, ed. A. Faessler.

49 E. Manousakis, V. R. Pandharipande, and Q. N. Usmani, Phys. Rev. **B31**, 7022 (1985).

50 V. F. Sears, Can. J. Phys. **59**, 555 (1981).

51 F. London, Nature **141**, 643 (1938).

52 for reviews see e.g. E. C. Svensson, V. F. Sears, Physics **137B**, 126 (1986).

53 T. R. Sosnick, W. M. Snow, and P. E. Sokol, to be published.

54 L. J. Lantto, Phys. Rev. **B22**, 1380 (1980).

55 J. Lam, Phys. Rev. **B3** , 3243 (1971).

56 D. Ceperley, unpublished.

57 P. Eisenberger, L. Lam, P. M. Platzman, and P. Schmidt, Phys. Rev. **B6**, 3671 (1972); and references therein.

58 S. Fantoni and V. R. Pandharipande, Nucl. Phys. **BA427**, 473 (1984).

59 A. Fabrocini, V. R. Pandharipande, and Q. N. Usmani, to be published.

60 J. P. Bouchaud and C. Lhuillier, Europhys. Lett. **B3**, 1273 (1987).

61 I. McCarthy, in **High Energy Excitations in Condensed Matter.** LASL Report No. LA-10227-Z, Los Alamos National Laboratory, NM.

62 G. Reiter, contribution at this conference.

63 D. L. Price and R. Hempleman, contribution at this conference.

MOMENTUM DISTRIBUTIONS THEORY

OVERVIEW OF MOMENTUM DISTRIBUTION CALCULATIONS

John W. Clark

McDonnell Center for the Space Sciences
and Department of Physics
Washington University, St. Louis, Missouri 63130, USA

Manfred L. Ristig

Institut für Theoretische Physik
Universität zu Köln, 5000 Köln 41, BRD

INTRODUCTION

It is our purpose to summarize recent progress in the calculation of momentum distributions in helium, nuclear, and electronic systems based on advanced many-body theories. Beginning with elementary definitions and concepts, we compile the known properties (sum rules, limiting and singular behaviors, etc.) which one would like such calculations to reproduce. We then proceed to review the ideas and strategies underlying the varied microscopic methods − variational, perturbative, stochastic − which are available for realistic computation of $n(\mathbf{p})$. Without delving into the technical details, we hope to convey some understanding of the relative merits of these approaches in specific physical contexts, as well as a feeling for their quantitative reliability. An epitome of state-of-the-art calculations is followed by our assessment of the current status of the computational effort toward accurate prediction of $n(\mathbf{p})$ in strongly-interacting quantum systems.

WHAT DO WE WANT TO CALCULATE?

We begin with a statement of fundamental definitions.[1,2]

Def. In quantum mechanical terms, the *momentum distribution* $n(\mathbf{p})$ is the average number of particles with momentum $\mathbf{p}$, an expectation value

$$n(\mathbf{p}) = <\Psi \,|\, \sum_\sigma a_{\mathbf{p}\sigma}^\dagger \, a_{\mathbf{p}\sigma} \,|\, \Psi> \tag{1}$$

in the unit-normalized N-particle state $|\Psi>$.

Def. A closely related quantity is the *average occupancy of single-particle orbital* i:

$$n_i = <\Psi \,|\, a_i^\dagger a_i \,|\, \Psi> \quad . \tag{2}$$

The choice of single-particle representation, defined by an orthonormal basis $\{|i>\}$ of one-body states, is arbitrary. The single-particle nature of n_i is underscored by the fact that it may be regarded as the expectation value of a symmetric sum of one-body operators,[1] $\Sigma_{j=1}^{N} \mu_i(j)$, with $\mu_i(j)|i'>^{(j)} = \delta_{i'i}|i'>^{(j)}$.

In common usage (as seen in the literature) the term *momentum distribution* often refers to the average occupancy of a plane-wave orbital of momentum **p** and *given spin* σ.

For the most part, we shall forget about spins and consider a homogeneous, isotropic fluid, since this makes the writing simpler without missing any crucial points. The choice of plane-wave orbitals i then implies $n_i = n(\mathbf{p}) = n(p)$, where p is the magnitude of **p**.

Def. Another single-particle quantity, the *one-body density matrix*, is defined for the unit-normalized pure state $|\Psi>$ by

$$\rho_1(\mathbf{r}_1,\mathbf{r}_1') = N \int d\,\mathbf{r}_2 \cdots d\,\mathbf{r}_N \Psi^*(\mathbf{r}_1\mathbf{r}_2 \cdots \mathbf{r}_N)\Psi(\mathbf{r}_1'\mathbf{r}_2 \cdots \mathbf{r}_N) \quad . \tag{3}$$

The diagonal elements $\rho_1(\mathbf{r}_1,\mathbf{r}_1)$ of this matrix yield the local density $\rho(\mathbf{r}_1)$, which is just a constant ρ in the uniform fluid. Homogenity and isotropy of the system require that $\rho_1(\mathbf{r}_1,\mathbf{r}_1') = \rho_1(|\mathbf{r}_1 - \mathbf{r}'_1|) \equiv \rho_1(r)$.

When spins are present, we have to multiply the right-hand side of (3) by the single-particle level degeneracy ν and take the trace over all N spins.

Some useful generalizations are the following:

Def. The l-*body density matrix* ρ_l, $l = 1, 2, 3, \cdots N$, is formed by taking the trace of

$$\Psi^*(\mathbf{r}_1\mathbf{r}_2 \cdots \mathbf{r}_N)\Psi(\mathbf{r}_1'\mathbf{r}_2' \cdots \mathbf{r}_N') \tag{4}$$

over $N - l$ particles and attaching an appropriate normalization factor.

Generalization to finite T: Expectation values become averages over an equilibrium ensemble. The pure state $|\Psi>$ is replaced by an equilibrium-ensemble statistical mixture, defining an N-body density matrix of which we take partial traces.

WHAT DO WE KNOW EXACTLY? ('TRIVIAL' RESULTS)

Several important relations and limiting cases follow directly from these basic definitions and from the elementary quantum mechanics of Bose and Fermi systems.

First, consider the familiar example of a large number N of noninteracting particles in a cubic box with periodic boundary conditions. Evaluation of the momentum distribution and one-body density matrix for this problem is trivial. In the Bose ground state, all the particles reside in the single-particle state of zero momentum (Bose-Einstein condensate), i.e., $n(\mathbf{p}) = N\delta_{p0}$ and the condensate fraction n_o is unity. The corresponding one-body density matrix is just equal to the constant density ρ. In the Fermi ground state, $n(\mathbf{p})/\nu$ is unity for $p \leq k_F$ and zero for $p > k_F$, where $k_F = (6\pi^2\rho/\nu)^{1/3}$ is the Fermi momentum. The associated one-body density matrix is $\rho_1(|\mathbf{r}_1 - \mathbf{r}_1'|) = \rho l(k_F |\mathbf{r}_1 - \mathbf{r}_1'|)$, where $l(x) = 3x^{-3}(\sin x - x \cos x)$ is the Slater factor. (The results quoted for fermions account for spin through the level degeneracy parameter ν.) We notice in these simple examples that $n(\mathbf{p}) = n(p)$ is essentially the Fourier transform of $\rho_1(r)$ (a relation of obvious generality).

If the particles feel no mutual interactions but are subject to an external field, or if we adopt a mean-field picture of an interacting system, the states are built from a complete orthonormal set of orbitals $|i>$, in accordance with some independent-particle model. The average occupancy n_i of the single-particle states i is trivially evaluated, and it is straightforward to evaluate the momentum distribution, in terms of the amplitudes $<i|\mathbf{p}>$. The one-body density matrix is given by

$$\rho_1(\mathbf{r}_1,\mathbf{r}_1') = \Sigma_{j=1}^{A} <\mathbf{r}_1'|i_j>^* <\mathbf{r}_1|i_j> \quad , \tag{5}$$

where the sum is over occupied orbitals, and spins have been suppressed.

40

The effect of interactions will be to kick particles out of the zero-momentum condensate of a noninteracting Bose gas or out of the Fermi sea of a noninteracting Fermi gas, or, more generally, out of the occupied orbitals of the independent-particle model. One speaks of *depletion* of the condensate or the Fermi sea; such depletion is a convenient measure of the strength of the interactions.

Considering finite temperatures T, and supposing that T is high enough that $k_B T \equiv \beta^{-1}$ overwhelms the interaction energy per particle and any pseudo-attraction or repulsion due to statistics, we have the *ideal-gas limit* for the one-body density matrix, $\rho_1(\mathbf{r}_1,\mathbf{r}_1') \propto \exp[-m(\mathbf{r}_1 - \mathbf{r}_1')^2/2\hbar^2\beta]$, where m is the particle mass and the normalization has been left indefinite. Correspondingly, the momentum distribution is also a Gaussian function, namely $n(p) = (\beta/2\pi m)^{3/2}\exp[-\beta p^2/2m]$ if normalized to unity.

Let us now go on to list the general features of $n(p)$ and $\rho_1(r)$ as defined in (1) and (3).

Momentum distribution and one-body density matrix are related by Fourier transformation:

- For fermions having single-particle level degeneracy ν (e.g. spin $I = (\nu - 1)/2$),

$$n(p) = \nu^{-1}\!\int \rho_1(r)e^{i\mathbf{p}\cdot\mathbf{r}}\, d\mathbf{r} \quad . \tag{6}$$

- For bosons we need to allow for macroscopic occupation of the zero-momentum single-particle state, corresponding to (partial) Bose-Einstein condensation. Thus

$$n(p) = \langle\Psi|a_{\mathbf{p}}^{\dagger}a_{\mathbf{p}}|\Psi\rangle = Nn_o\delta_{p0} + n'(p) \quad , \tag{7}$$

where $n_o = \lim_{r \to \infty}\rho_1(r)/\rho$ is the condensate fraction and

$$n'(p) = \int [\rho_1(r) - \rho_1(\infty)]e^{i\mathbf{p}\cdot\mathbf{r}}\, d\mathbf{r} \tag{8}$$

is the continuous component of the momentum distribution.

Properties of $\rho_1(r)$. Behavior of $\rho_1(r)$ at small r.

- For the normalization chosen in definition (3),

$$\lim_{r \to 0}\rho_1(r) = N/\Omega = \rho \quad . \tag{9}$$

This limit expresses the conservation of particle number.

- The one-body density matrix is related to the kinetic energy expectation value according to

$$\frac{K}{N} = \frac{1}{2m}\langle p^2\rangle = -\frac{\hbar^2}{2m}\frac{1}{\rho}\frac{d^2\rho_1(r)}{dr^2}\Big|_{r=0} \quad . \tag{10}$$

Thus, the kinetic energy is proportional to the curvature of the one-body density matrix at the origin.

Properties of $\rho_1(r)$. Behavior of $\rho_1(r)$ at large r.

- *Bose particles*:

$$\lim_{r \to \infty}\rho_1(r)/\rho = n_o \quad . \tag{11}$$

This behavior, with finite n_o, is an expression of off-diagonal long-range order[3,4] (ODLRO). Macroscopic occupation of the zero-momentum single-particle state is manifested as a non-zero asymptotic value of $\rho_1(r)$.

- *Fermi particles*:

$$\lim_{r \to \infty} \rho_1(r) = 0 \quad .$$
(12)

Normal Fermi system:

At large r, $\rho_1(r)$ displays damped oscillations due to Pauli exchange correlations; upon Fourier transforming, these oscillations produce a discontinuity of $n(p)$ at the Fermi surface (see next section).

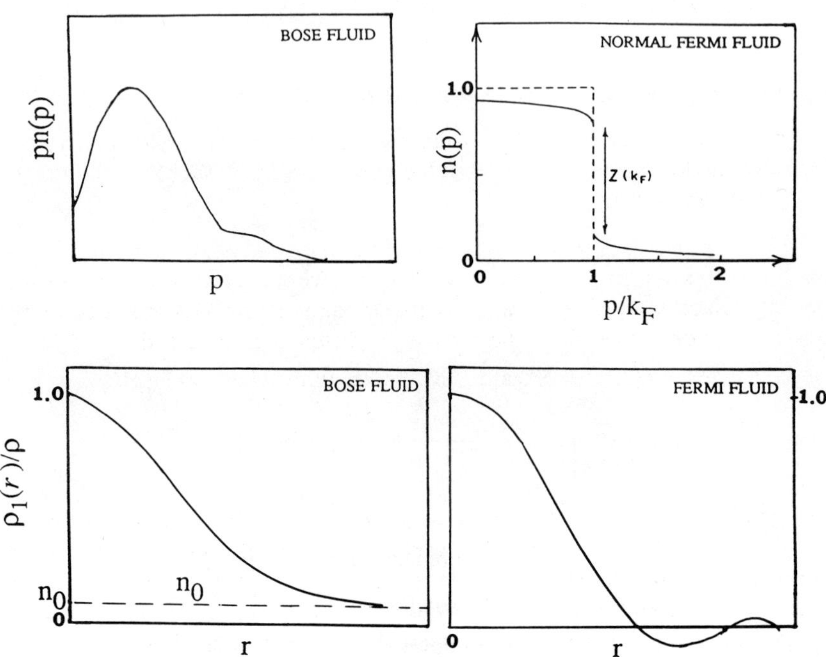

Fig. 1. Typical behaviors of $n(p)$ and $\rho_1(r)$ at zero temperature.

Properties of $n(p)$ – Corresponding to behaviors of $\rho_1(r)$ at small and large r.

(Small r):

- *Particle-number sum rule*

$$\sum_{\mathbf{p}} n(\mathbf{p}) = N \quad .$$
(13)

- *Kinetic-energy sum rule*

$$\frac{1}{2m}\sum_{\mathbf{p}} p^2 n(\mathbf{p}) = K \quad .$$
(14)

(Large r):

- For *Bose particles*

$$n(0) = n_o N \quad . \tag{15}$$

In general, there is macroscopic occupation of the $\mathbf{p} = 0$ state.

- For *Fermi particles*, no state is macroscopically occupied, and one ordinarily expects $n(\mathbf{p}) = O(1)$ for $p \le k_F$ and $n(\mathbf{p}) \ll 1$ for $p > k_F$.

Typical behaviors of $n(p)$ and $\rho_1(r)$ at $T = 0$ are illustrated in Fig. 1.

It is also useful to record some basic relations for the *two-body density matrix*. The diagonal elements of $\rho_2(\mathbf{r}_1\mathbf{r}_2,\mathbf{r}_1'\mathbf{r}_2')$ give the two-particle distribution function $p_2(\mathbf{r}_1\mathbf{r}_2)$. For a homogeneous, isotropic fluid, the latter becomes $\rho^2 g(|\mathbf{r}_1 - \mathbf{r}_2|)$, where $g(r)$ is the radial distribution function. In Fourier space, the two-body density matrix is given by the expectation value $n(\mathbf{k},\mathbf{p},\mathbf{q}) = \langle \Psi | a^\dagger_{\mathbf{k}+\mathbf{q}} a^\dagger_{\mathbf{p}-\mathbf{q}} a_{\mathbf{p}} a_{\mathbf{k}} | \Psi \rangle$.

WHAT DO WE KNOW EXACTLY? (NON-TRIVIAL RESULTS)

In this section, we collect, without derivation, some exact results which have been deduced for three-dimensional fluids from the general structure of many-body theories of interacting quantum systems.

Small-p behavior of $n(p)$ in Bose systems. The non-condensate portion $n'(p)$ of the momentum distribution displays a singular behavior which depends on whether the basic two-particle interaction is of short-range character (as in liquid He) or has the long-range character of a Coulomb potential. The behavior is different at finite T than it is at $T = 0$.

- *At zero temperature*, in the ground-state, the result for *short-range forces* is[5,6]

$$n'(p) \sim n_o \frac{mc}{2} \frac{1}{p} \quad (p \to 0) \quad . \tag{16}$$

This behavior is due to the presence of long-wavelength virtual phonons in the ground state. Gavoret and Noziéres[5] derived the result (16) by resumming the perturbation expansion for the one-particle Green's function $G(p,\Omega)$ in the limit $(p,\Omega) \to 0$. Chester and Reatto[6] obtained the same behavior for a Jastrow wave function incorporating the long-wavelength phonon modes. For *Coulombic forces* (charged Bose gas), the corresponding result is[7,2]

$$n'(p) \sim n_o \frac{m\hbar\omega_o}{2} \frac{1}{p^2} \quad (p \to 0) \quad , \tag{17}$$

where ω_o is the plasma frequency. The singularity goes from p^{-1} to p^{-2} as the appropriate large-r asymptotic behavior of the two-body correlation function $f(r)$ in the Jastrow wave function changes from $f - 1 \sim -(mc/\hbar)(1/2\pi^2\rho r^2)$ to $f - 1 \sim -(m\omega_o/\hbar)(1/2\pi\rho r)$, replacing phonon modes by plasmons.

- *At small finite T* (meaning $0 < T \ll T_\lambda$, where T_λ is the lambda-transition temperature), the singularity for short-range forces becomes[8-13]

$$n'(p) \sim \frac{n_o m k_B T}{p^2} \quad (p \to 0) \quad . \tag{18}$$

More generally, there exist exact small-T results for $n_o(T)/n_o(T=0)$ (Refs. 10,11), for $\rho_1(r,T)$ at large r (Ref. 10,11), and for $\rho_1(r,T) - \rho_1(r,T=0)$ at small r (Ref. 10). In addition, Schulz and Ristig[12] have studied the Bose problem with short-range interactions within the paired-phonon model and obtained analytic expressions for

$\rho_1(r,T)/\rho_1(r,T=0)$ and $n_o(T)/n_o(T=0)$ which conform to these results.

For recent and enlightening discussions of the small-p behavior of $n(p)$ in superfluid ^{4}He, and especially its implications for the extraction of the condensate fraction, see Ref. 13.

Behavior of $n(p)$ near k_F, in normal Fermi liquids. Under reasonable assumptions involving the validity of perturbation expansions, one has *Luttinger's theorem*[14,15]: In the interacting system the momentum distribution maintains a Fermi-surface discontinuity, the volume of momentum space within the Fermi surface being unchanged by the interactions. This result was derived by field-theoretic methods, based on the behavior of the one-particle Green's function in perturbation theory. In addition, under rather general conditions[16-18]

$$\lim_{p \to k_F^{\pm}} \frac{dn(p)}{dp} = -\infty \quad , \tag{19}$$

i.e., the slope of $n(p)$ at the Fermi surface is infinite, whether approaching it from below or above.

Behavior of $n(p)$ at large p. In order for the kinetic-energy integral in (14) to remain finite, the momentum distribution should decline faster than p^{-5}. Experimental[19] and computational[20] evidence, as well as simple models[21,22] point to an exponential fall-off of the momentum distribution at high momenta, in some of the systems under consideration. The large-p behavior should not depend on statistics, but should otherwise be governed by the two-body problem at short distances.

HOW DO WE CALCULATE $n(p)$ AND $\rho_1(r)$?

The quantum systems examined in this book include, prominently, (i) the helium liquids, (ii) nuclear matter and nuclei, and (iii) electronic systems. For the purposes of our discussion, electronic systems will be represented (and over-simplified!) by the electron gas, or one-component plasma: a system of electrons at uniform density, immersed in a positively charged uniform background with the same absolute charge density as the electrons. Traditionally, the helium systems are regarded as the most strongly-interacting, with an almost-hard-core interaction, and the electron gas as the most weakly interacting, the inner Coulomb repulsion being very soft by comparison. The nuclear case lies somewhere between. Historically, the helium problems have proved the hardest to solve, again because of the hard interaction. On the other hand, the nuclear problems present a singular frustration and a special challenge because of the strong state dependence (dependence on spin, isospin, angular momentum) and noncentral character of the bare two-particle interaction, and also because of ambiguities in the determination of this interaction from first principles. Moreover, judgments about strong vs. weak coupling depend on the density being studied. This fact is vividly illustrated by the electron gas, which is distinguished from the other examples by the long-range nature of the Coulomb interaction. Consequently, strong coupling prevails in the electron gas in the limit of *low* density, whereas helium and nuclear systems become more strongly interacting as the density increases. In all these cases the strength of the interaction may be gauged by the depletion of the "condensate," whether that be the Bose-Einstein condensate or the Fermi sea. Quantitatively, one examines the deviation of n_o or Z_{k_F} from unity, where Z_{k_F} is the discontinuity of $n(p)$ at k_F in a normal Fermi system.

In dealing with all these systems, we are necessarily driven to computational methods, since purely analytic treatment is hopeless. Computational many-body methods may be classified in several different ways: They may be based on wave functions or on field theory. They may be variational or perturbative. They may be stochastic or non-stochastic. In any case the singular, or near-singular character of the basic forces involved precludes a simple, stepwise perturbative calculation within an independent-particle (plane-wave) basis. For the helium and nuclear problems, we must, as a *minimal* requirement, account nontrivially for the strong short-range repulsions; for the electronic systems, it is the long range of the interaction which needs primary attention. In the language of conventional diagrammatic perturbation theory, we must sum the ladders in the helium and nuclear cases, and the rings for the electron gas. Actually, if we expect to do a quantitative job, we must do much more than that in the helium problems, and also rather more than the indicated minimum for nuclear and electronic systems.

What follows is a set of sketches of the many-body methods now in practice, with a bit about how they work and what their limitations are.

Variational Theory

Specification of Trial Functions. This approach is obviously wave-function based, and the evaluation of the corresponding expectation values may be non-stochastic or stochastic. The essential feature is that the basic physics of the problem must be put in by hand, through the choice of ground-state trial function. This is both an advantage and a disadvantage: the practitioner has a great deal of control over what he will get, but that (usually!) rules out surprises. The general form of trial wave function is

$$\Psi_{\text{trial}} = \Psi_o = F \Phi_o \ . \tag{20}$$

In this expression, Φ_o is a *model function*, adequate for the ground state when the interactions are weak (e.g. describing a noninteracting Fermi sea or Bose condensate), while F is a *correlation operator*, whose task it is to build in at least the short-range correlations due to the inner repulsion.

Jastrow pair correlations provide the most popular starting point:

$$F = F_J = \prod_{i<j} f(r_{ij}) = \prod_{i<j} \exp[-u(r_{ij})/2] \tag{21}$$

The pair correlation function $f(r)$ is assumed to approach unity for large r, so as to ensure the cluster-decomposition property.[23] The expectation value $<H>$ of the Hamiltonian with respect to Ψ_o is to be evaluated, and minimized with respect to the choice of $f(r)$. Ideally, this involves solution of the Euler equation $\delta<H>/\delta u(r) = 0$, which determines the *optimal* Jastrow correlations. A common alternative, less demanding, is to adopt a simple parameterized form for $f(r)$. The McMillan[24] or Schiff-Verlet[25] choice, $u(r) = (b/r)^5$, has been employed in many studies, the parameter b being fixed by numerical minimization of $<H>$. Solution of the Euler equation, e.g. in the framework of Bose or Fermi hypernetted-chain theory,[23,26–30] has a considerable advantage: it leads to a reasonable description of long-range as well as short-range correlations within the Jastrow setting. For optimal correlations, the asymptotic behavior of $f(r)$ is given by

$$f(r) - 1 = \text{const.}/r^n \quad (r \to \infty) \ , \tag{22}$$

where $n = 2$ applies for short-range interactions as in liquid ^{4}He and ^{3}He, while $n = 1$ applies for the long-range interaction of the electron gas. These behaviors properly reflect the presence of virtual phonons or plasmons, respectively.[6,7,28]

Using the language of conventional diagrammatic perturbation theory, the (optimal) Jastrow treatment sums rings and ladders in a self-consistent manner, within an average propagator approximation.[31] For problems involving very strong couplings (large depletions) [as in ^{4}He and ^{3}He liquids], *or* for problems in which the elementary excitations are not dominated by a collective mode of energy $\hbar\omega^c(q) = \hbar^2 q^2/2mS_o(q)$, where $S_o(q)$ is the static structure function of the noninteracting system [i.e. the average-propagator approximation breaks down] (cf. Ref. 32,33), *or* for problems in which the bare interaction displays strong state dependence [as in nuclear systems], the Jastrow model becomes inadequate. Thus, if we seek a quantitative description within variational theory, it is generally necessary to go beyond the Jastrow ansatz. This extension takes several forms, depending on the problem considered.

The *Feenberg function,*

$$F = F_F = \exp\frac{1}{2}\left[-\sum_{i<j} u_2(r_{ij}) - \sum_{i<j<k} u_3(ijk) - \sum_{i<j<k<l} u_4(ijkl) - \ldots - u_N(12\ldots N)\right] , \quad (23)$$

wherein a given u function is dependent only on spatial coordinates and cannot be decomposed additively into fewer-body functions, provides an exact representation of the ground state of a system of spinless bosons. It may also be used as a correlation factor in Fermi problems. In both liquid ^{4}He and liquid ^{3}He, current implementation involves the use of pair functions u_2 (corresponding to the Jastrow level), supplemented by "triplets" u_3. Physically, triplets may describe *backflow*[34,35,27] around a moving particle. A specific form adopted in the most recent variational Monte Carlo studies of liquid ^{3}He is[36]

$$F_{2,3} = \exp\frac{1}{2}\left[-\sum_{i<j}\hat{u}(r_{ij}) - \frac{\lambda_T}{2}\sum_{i,j}\sum_{l}\xi(r_{li})\xi(r_{lj})\mathbf{r}_{li}\cdot\mathbf{r}_{lj}\right] \quad (24)$$

with $\hat{u}(r) = u(r) - \lambda_T\,\xi^2(r)r^2$ and $\xi(r) = \exp\{-[(r-r_T)/w_T]^2\}$, the new variational parameters being λ_T, r_T, and w_T. In liquid ^{3}He the further introduction of *momentum-dependent backflow correlations* is found to produce a significant lowering of the energy.[36–38] Such correlations effect corrections to the average-propagator approximation embodied in the Jastrow model.[33] In practice, they may be included through a modification of the Slater determinant Φ_o:

$$\det(\exp i\,\mathbf{k}_i\cdot\mathbf{r}_j) \rightarrow \det(\exp\{i\,\mathbf{k}_i\cdot[\mathbf{r}_j + \sum_l\eta(r_{lj})\mathbf{r}_{lj}\,]\}) . \quad (25)$$

A specific choice for the function $\eta(r)$, used in the recent Monte Carlo calculations[36] on ^{3}He, is $\eta(r) = \lambda_B\exp\{-[(r-r_B)/w_B]^2\} + \lambda_B'/r^3$, the additional variational parameters being $\lambda_B, \lambda_B', r_B$, and w_B.

The complicated operatorial character of the nucleon-nucleon interaction necessitates the introduction of *state-dependent correlation operators* in attempts at quantitative description of nuclear systems.[39,40] For example, the Urbana and Argonne v_{14} interactions[41,42] have the form $\sum_{n=1,14} v^{(n)}(r_{ij})O_{ij}^{(n)}$, where $O_{ij}^{(n)} = 1, \tau_i\cdot\tau_j, \sigma_i\cdot\sigma_j, (\tau_i\cdot\tau_j)(\sigma_i\cdot\sigma_j), S_{ij}, S_{ij}(\tau_i\cdot\tau_j), \mathbf{L}\cdot\mathbf{S}, (\mathbf{L}\cdot\mathbf{S})(\tau_i\cdot\tau_j), (\mathbf{L}\cdot\mathbf{S})^2,$ $(\mathbf{L}\cdot\mathbf{S})^2(\tau_i\cdot\tau_j), L^2, L^2(\tau_i\cdot\tau_j), L^2(\sigma_i\cdot\sigma_j), L^2(\sigma_i\cdot\sigma_j)(\tau_i\cdot\tau_j)$. The correlation operators in current use[40,43] reflect this dependence at a tractable level in which two-body correlations quadratic in L are disregarded:

$$\hat{F} = \mathcal{S}[\prod_{i<j} F_{ij}] , \qquad F_{ij} = \sum_{n=1,8} f^{(n)}(r_{ij})O_{ij}^{(n)} , \quad (26)$$

where $\mathcal{S}$ is a symmetrizer. Spin-dependent correlations of type $\sigma_i\cdot\sigma_j$ have also been invoked in treating liquid ^{3}He (Ref. 38).

Some new initiatives in variational theory – based on new proposals for correlated trial wave functions – should be mentioned.

Shadow wave functions are based on the idea that a fictitious set of "shadow" particles may be introduced in such a way as to preserve some symmetry of the real system. In particular, Vitiello, Runge, and Kalos[44] (see also Ref. 45) have studied liquid and solid ^{4}He using the wave function

$$\Psi(R) = \int \exp\left[-\frac{1}{2}\sum_{i<j} u(r_{ij}) - \sum_{k} \zeta(\mathbf{r}_k - \mathbf{s}_k) - \frac{1}{2}\sum_{l<m} u_s(s_{lm})\right] dS \qquad (27)$$

with the choices $\zeta(\mathbf{r}_k - \mathbf{s}_k) = C(\mathbf{r}_k - \mathbf{s}_k)^2$, $u(r) = (b/r)^5$, and $u_s(s) = (b_s/s)^5$. The configurations $S = (s_k)$ of the shadow particles are integrated out, leaving a symmetrical, translationally invariant trial wave function which allows a unified description of liquid and solid phases. *A priori* introduction of a crystal lattice is not required. Moreover, such a wave function does a measurably better job for the liquid phase than the (also symmetrical) Jastrow ansatz.

Another new proposal, which has generated some controversy, concerns the ground state of liquid ^{3}He and involves a modification of the Jastrow wave function through a revised choice of the model function Φ_o in (20) rather than a more elaborate correlation operator F. Let N_σ, $\sigma = \uparrow$ or $\downarrow$, denote the number of ^{3}He atoms with spin up or down, assumed to be equal and even. Bouchaud and Lhuillier[46] envision a strong pairing within up-spin and down-spin subsystems. Their trial wave function takes the form $\Psi_o = F_J \Phi_\uparrow \Phi_\downarrow$, where the model function

$$\Phi_\sigma(1,2,\cdots N_\sigma) = \mathcal{A}[\phi(12)\phi(34)\cdots\phi(N_\sigma{-}1, N_\sigma)] \quad , \qquad (28)$$

with $\mathcal{A}$ an antisymmetrizer, is the projection onto N-space of a BCS-type superstate. Specializing to the odd-state pairing function $\phi(ij) = \mathbf{n}\cdot\mathbf{r}_{ij}e^{-\gamma^2 r_{ij}^2}$, these authors have performed variational Monte Carlo calculations (see below) which yield an excellent ground-state energy and optimal $\gamma^{-1} \sim 8$–13 Å, corresponding to strong coupling of pairs. The momentum distribution no longer has a discontinuity at the Fermi momentum, and the Landau Fermi-liquid picture, so dear to physicists working on liquid ^{3}He, breaks down.

Evaluation of Expectation Values – Non-stochastic. Once a wave function has been specified, the physics which can be described by the model has been decided, and thereafter the only approximations involved are computational: we have to face the highly non-trivial problem of *calculating* the desired expectation values (energy, momentum distribution, ...). The correlations prevent exact reduction to few-body integrals, as is done in Hartree-Fock theory, but various cluster-expansion schemes are available[47] for systematic approximate evaluation in terms of cluster integrals. A preferable alternative is to set up a diagrammatic representation of the terms in such cluster expansions, and attempt to sum up subseries of diagrams to all orders, guided by the classical theory of imperfect gases.[48] For Jastrow (and Feenberg) correlations, this approach has led to Bose and Fermi hypernetted-chain procedures[23,26,28] (denoted respectively HNC and FHNC). In HNC or FHNC, one evaluates a given expectation value (e.g. the radial distribution function $g(r)$ or momentum distribution $n(p)$) by solving a set of coupled, nonlinear integral equations which resums cluster expansions in $f^2 - 1$ or $f - 1$. The Fermi hypernetted-chain equations are more numerous and more complicated than the Bose equations because of the presence of an extra diagrammatic bond corresponding to exchange. For the radial distribution function, there is just one HNC integral equation, whereas there are four to solve in FHNC. For the momentum distribution, the structure of the Jastrow $n(p)$ in terms of sums of irreducible diagrams was first established by Ristig and Clark[1]; HNC and FHNC equations for the evaluation of these diagram sums were subsequently derived by Fantoni.[49]

Unfortunately, HNC-FHNC methods are not closed, since the summations are performed for an input set of *elementary diagrams* (classified as simple but non-nodal[48,23]) representing conditions of compact, irreducible correlation among 0, 4, 5, ... particles. (Thus, one speaks of (F)HNC/n approximations, where $n = 0$ means that all elementary diagrams are dropped and $n = 4, 5, \cdots$ means that elementary diagrams with up to 4, up to 5, ... particles are included.) The problem is that the elementary diagrams involve multidimensional integrals, e.g., four-body integrals for (F)HNC/4, etc. Hence their sum cannot be evaluated in closed form. The simplest approximation, (F)HNC/0, already leads to reasonable results for some properties of the helium systems and is in fact quite good for model nuclear and electronic problems. It is especially noteworthy that even at the (F)HNC/0 level, optimal Jastrow theory achieves a self-consistent summation of ring, ladder, and self-energy diagrams in the average-propagator approximation intrinsic to that theory.[31] On the other hand, truly quantitative prediction, most emphatically in liquid ^{4}He and liquid ^{3}He, may call for something better. To this end, various *scaling approximations*[50–54,37,38] have been introduced to simulate the effects of elementary diagrams. Typically, such an approach might involve explicit calculation of the simplest elementary diagrams, and scaling of their contributions with one or more adjustable parameters, so as to fulfill certain exact properties or to match certain predictions of an independent calculation. The domain of validity of these approximations has not been established, although significant agreement with Monte Carlo results can be obtained.

For trial wave functions $\Psi_o = F\Psi_o$ going beyond the Feenberg F of Eq. (23), or beyond the choice of a translationally-invariant independent-particle Φ_o, the non-stochastic computational options are in a less satisfactory state. At present, one must resort to straightforward cluster expansion,[47] or highly restricted resummation of higher cluster diagrams (e.g. single-operator chains, etc.[39,55,38,40]). Convergence of these procedures is problematic at high densities. Agreement with Monte Carlo stochastic evaluation is not always good.

Evaluation of Expectation Values – Stochastic. Monte Carlo computation of correlated expectation values is now the method of choice, but it also has some limitations. Implementing the Metropolis algorithm,[56] it can handle Jastrow, triplet, and momentum-dependent backflow correlations. The Metropolis algorithm generates a set of configurations $R = (\mathbf{r}_1 \cdots \mathbf{r}_N)$ characterized by an equilibrium probability distribution proportional to $|\Psi|^2$, where Ψ is the given wave function. In basic Metropolis, you throw the dice, and move a particle. The move is accepted with probability unity if $|\Psi(R_{\text{new}})| > |\Psi(R_{\text{old}})|$; otherwise the move is accepted with probability $|\Psi(R_{\text{new}})|/|\Psi(R_{\text{old}})|$. This is done many times, until the system has equilibrated. The variational energy is computed as the average of the local energy $H\Psi/\Psi$, over configurations so generated.

Evaluation of the one-body density matrix proceeds from the observation[57,58] that $\rho_1(\mathbf{s})$ measures the change in the wave function $\Psi(\mathbf{r}_1 \cdots \mathbf{r}_N)$ for a given displacement $\mathbf{r}_i \to \mathbf{r}_i + \mathbf{s}$. Move particle i, chosen randomly or systematically, by an amount $\mathbf{s}$ and average the ratio of the new value of the wave function to the old one, over an ensemble of configurations generated by the Metropolis routine:

$$\rho_1(\mathbf{s}) = \left\langle \frac{\Psi(\mathbf{r}_1 \cdots \mathbf{r}_i + \mathbf{s} \cdots \mathbf{r}_N)}{\Psi(\mathbf{r}_1 \cdots \mathbf{r}_i \cdots \mathbf{r}_N)} \right\rangle . \tag{29}$$

There are some obvious limitations of Monte Carlo techniques for evaluating correlated expectation values.

- The routine must be run long enough to reach a condition of equilibrium, and long enough to generate a large sample of equilibrium configurations, so that statistical errors are small.

- The number N of particles in the simulation sample must be taken large enough that *box-size effects* are unimportant, or can be adequately estimated. For ^{4}He, box-size effects on the accuracy of $\rho_1(r)$ at large distance are discernible in a comparison of condensate-fraction results for variational trial functions referring to $N = 64$ and 512 particles,[58] a discrepancy ~0.005 being noted. However, if the 64-particle results are corrected for the omission of long-wavelength phonon contributions, essential agreement is obtained for $\rho_1(r)$ at the two particle numbers. In a corresponding variational Monte Carlo study[58,36] of liquid ^{3}He, including optimal Jastrow correlations, plus triplets and momentum-dependent backflow with optimized parameters (cf. (24), (25)), $\rho_1(r)$ and $n(p)$ have been calculated at $N = 54$ (with checks at 66). In this case one has to worry about how reliably one can predict the oscillations of $\rho_1(r)$ at large r which produce the Fermi-surface discontinuity and give Z_{k_F}. The issue of finite-size effects in liquid ^{3}He requires further study, with tests at a substantially larger N. Another consideration is that the functions $n(\mathbf{p})$ and $\rho_2(\mathbf{r})$ are isotropic in the bulk liquid, whether it be ^{4}He, ^{3}He, nuclear matter, ... , whereas these functions show a directional dependence for a finite sample. Hence *sphericalizing* of MC results is necessary at some stage. Problems may arise in particular from the fact that sphericalizing and Fourier transforming do not commute.

- State-dependent (spin-, isospin-dependent) interactions and correlations − as in nuclear physics − are much more demanding and stochastic techniques for dealing with them are still under development (see, for example, Ref. 59).

Field Theoretic and Perturbative Approaches

Traditional many-body theory started with the exploitation of field-theoretic techniques, ordinarily in the context of perturbation expansions of quantities such as the single-particle Green's function, or more directly the ground-state energy. This species of approach, which is documented by a voluminous literature,[60] has been most fruitful in application to model problems, to limiting cases of weak interactions and low or high density, and to asymptotic dynamical properties. The perturbation expansions are carried out in the strength of the bare interaction, and are predicated on plane-wave basis states (or shell-model states in the case of finite nuclei). Hence infinite resummations of selected classes of diagrams are necessary before sensible results can be obtained for the strongly-interacting systems of concern to us.

Approaches Based on One-Particle Green's Function. Historically, these have been most useful for the one-component plasma and other electronic problems. The *long-range* Coulombic forces require summation of random-phase-approximation (RPA) or *ring diagrams* in the perturbation expansion of the single-particle propagator. This is the basic approximation, multiply refined.[61,62]

The one-particle Green's function $G(\mathbf{p},t) = -i <\Psi | \mathrm{T}[a_{\mathbf{p}}(t)a_{\mathbf{p}}^{\dagger}(0)] | \Psi>$, where T here denotes the Dyson time-ordering operator, describes propagation of a hole or a particle in the interacting medium. In terms of this quantity and its Fourier frequency transform $G(\mathbf{p},\Omega)$, the momentum distribution is given by $n(p) = \lim_{t \to 0^-} G(\mathbf{p},t)$ and $n(p) = -i(2\pi)^{-1} \int_C dE\, G(\mathbf{p},E)$, where the integration contour C runs along the real axis and is closed by a semi-circle in the upper-half plane. The interacting Green's function $G(\mathbf{p},t)$ is in turn given in terms of the noninteracting Green's function $G_o(\mathbf{p},t)$ by a Dyson equation involving the self-energy $\Sigma(\mathbf{p},E)$; in momentum space this equation has the exact solution $G(\mathbf{p},\Omega) = (p^2/2m + \Sigma(\mathbf{p},\Omega) - \Omega)^{-1}$.

Approaches Based on Brueckner-Bethe-Goldstone Expansion. This approach, entailing massive rearrangement of the perturbation expansion for the ground-state energy and other quantities (e.g. $n(p)$) has been very popular in nuclear problems. The strong *short-range* repulsion demands summation of *ladder diagrams*. This is accomplished by rearrangement of diagrams according to the number of hole lines (producing a hole-line expansion) and by introduction of Brueckner's g-matrix.[63] The g-matrix formalism permits convenient treatment of noncentral, spin(-isospin-) dependent interactions. Calculations normally stop at the two-hole-line level, and this is indeed the case for current calculations of the momentum distribution in extended systems. At high nuclear densities or for very strongly interacting systems like the helium liquids, this approach ceases to be useful, due to the intractability of terms with three or more hole lines and problematic convergence.

Modern reincarnations. The evolution of more powerful schemes from conventional field-theoretic ideas is seen in self-consistent Green's function theory[64] and the parquet formalism.[31] These new approaches are yet to be applied to quantitative evaluation of $n(p)$ for the systems of interest. Another relative of the original field-theoretic perturbative procedures is the coupled-cluster method.[65] This method is highly developed for nuclear systems, the electron gas, and quantum chemistry problems, but it does not appear well suited to strongly-coupled quantum fluids like the helium liquids.

Method of Correlated Basis Functions

We may combine the variational option of introducing explicit correlations with the techniques of perturbation theory and resummation of Goldstone or Feynman diagrams. This leads to a highly versatile set of approaches called correlated-basis theory[66,67,27,33] (CBF). The single configuration $F|\Phi_o>$ of the variational description is extended to a basis $\{F|\Phi_m>/<\Phi_m|F^{\dagger}F|\Phi_m>^{1/2}\}$ of normalized but nonorthogonal correlated states, where the $|\Phi_m>$ form a complete orthonormal set of model functions, suitable as a basis for a weakly interacting system. Again, F builds in the most important correlations. Cutting through the rather dense formalism that grows out of this idea, one can say that CBF transforms the problem of bare particles interacting via strong forces to a problem of dressed particles interacting through much weakened effective forces. The analogs of the (RPA) ring sum and the (g-matrix) ladder sum may be performed.[47,68,32,69] The necessity for such resummations is reduced by the fact that rings and ladders are already included approximately in the zeroth-order, variational description; on the other hand, explicit propagator corrections now account for subtle energy- and momentum-dependent effects which are elusive in the purely variational approach.[33] Moreover, inadequate spin and isospin dependence of the chosen correlation operator F may be repaired by low-order perturbation corrections. Thus, CBF combines the best features of variational and perturbative strategies. It also shares some of their drawbacks: (i) In principle, there remain questions of convergence of the perturbation series or its rearrangements. At a more technical level, the relations between the two options of working directly with a nonorthogonal basis, or transforming first to an orthogonal correlated basis, need further clarification. (ii) Evaluation of correlated matrix elements, both nondiagonal and diagonal, is far from trivial, requiring, as in variational theory, the introduction of cluster-expansion, integral-equation (e.g. HNC), or stochastic (Monte Carlo) techniques. In practice, CBF methods have been remarkably successful across the range of interaction strengths and densities of interest to us; thus they can produce quantitative results for the helium liquids and nuclei and have worked quite well in the electronic problems for which they have been tried. As yet, however, realistic numerical application to momentum distributions — beyond the variational level — has been limited to nuclear matter.[43]

Exact Stochastic Methods

Green's Function Monte Carlo (GFMC). The many-body problem for a pure quantum state is attacked head-on. Computation is facilitated by transforming the Schrödinger equation to an integral equation with a shifted energy scale,

$$\Psi^{(n+1)}(R) = (E_T + E_C)(H + E_C)^{-1}\Psi^{(n)}(R) \quad , \tag{30}$$

where E_T is a trial energy near the ground-state energy and E_C is chosen so that the spectrum of $H + E_C$ is positive. Repeated iteration of (30) is accomplished by random walks, beginning with an optimized trial function Ψ_T. An asymptotically unbiased estimator for the energy is provided by $E_M^{(n)} = \int \Psi^{(n)}(R)H\Psi_T(R)dR / \int \Psi^{(n)}(R)\Psi_T(R)dR$, which is in fact an upper bound to the exact ground-state energy E_o for all n, with $\lim_{n \to \infty} E_M^{(n)} = E_o$. Evaluation of the one-body density matrix $\rho_1(\mathbf{r})$ proceeds via (29). Details may be found in Refs. 70,71.

The GFMC method just outlined assumes that the desired wave function can be interpreted as a probability density. This requirement is obviously met for the Bose ground state, where the method has been spectacularly successful. However, for fermions the wave function must change sign because of antisymmetry, presenting a serious obstacle to the application of GFMC in Fermi problems. The following strategies are representative of the current state of the art:

- *Transient estimation.* The Fermi wave function is obtained as a difference of two positive functions, $\Psi = \Psi^+ - \Psi^-$. One is confronted by exponential decay of the Fermi signal in a background of symmetric noise. Thus one must hope for a rapid approach to the ground-state solution, which means that a good starting point (trial function) is essential.

- *Fixed-node approximation.* Alternatively, one may decide to live with the nodes of a good variational trial function. The GFMC algorithm is then implemented within the prescribed regions where the wave function is of one sign, to obtain the best wave function with the given nodal surfaces.

Besides the problem of dealing with wave functions that change sign, what other blemishes of GFMC are worth noting? As in any Monte Carlo procedure, the issues of equilibration and finite simulation size must be dealt with. In GFMC, an 'importance function,' taken from an optimized variational calculation, is used to initiate the evolution. Conceivably, the true ground state could be missed, if the relaxation time from the initial state is very long. Again, in evaluation of $\rho_1(r)$ at large r, finite-box-size effects may be significant.

Path-Integral Monte Carlo (PIMC). The equilibrium properties of a many-body system at finite temperature $T = 1/k_B\beta$ are determined by integrals over the N-particle density matrix $\rho_N(R,R';\beta) = \langle R | \exp(-\beta H) | R' \rangle$. The path-integral Monte Carlo method is based on the semi-group property that a configuration-space convolution of two density matrices at temperature T is again a density matrix, at temperature $T/2$. Repeated application of this property allows the computation of a low-temperature density matrix (e.g. at 2 K) from a path integral over a product of density matrices (corresponding e.g. to a path with 20 steps in configuration space) all referring to a temperature in the classical regime (e.g. 40 K). Thus, the quantum system is mapped onto a classical system of higher dimension. A well-understood high-temperature expansion is available for the classical density matrix. To accommodate Bose statistics, a sum over all permutations of particle labels in R must be evaluated. The integral over paths and the sum over permutations are carried out by a generalization of the Metropolis Monte Carlo scheme. To improve convergence, a special algorithm has been implemented in which moves of particle coordinates are coupled with moves in permutation space. For detailed exposition, see Refs. 72-75.

This approach has the profound advantages that it is potentially exact and that thermodynamic properties of the quantum system can be quantitatively evaluated over the whole range of temperatures from the classical limit to the ground state, allowing, for the first time, a microscopic study of the lambda transition in liquid ^{4}He. Disadvantages include path-discretization errors, which can be estimated by performing calculations at several step sizes, and the difficulty of including exchange effects, which becomes formidable for Fermi systems. The greater precision in the prediction of some properties, relative to analytically-based methods for treating finite temperatures (see e.g. Ref. 76), is won at the expense of a vast increase in computer time. The attendant practical considerations limit simulations to a small number of particles ($\sim$ 100), with periodic boundary conditions imposed to minimize edge effects. The small simulation size presents a problem for accurate description of phase transitions, and in particular the (second-order) lambda transition in ^{4}He, which is sensitive to boundary effects. Technical difficulties arise in the computation of $\rho_1(r)$ at large r, where the condensate fraction is determined, the situation worsening near the lambda point.[74] Nevertheless, a coherent and essentially quantitative picture of the temperature dependence of the condensation fraction and the momentum distribution of liquid ^{4}He emerges, and the PIMC method shows great promise for further quantitative exploration of strongly-interacting quantum systems.

STATE-OF-THE ART CALCULATIONS

In this section we shall pay our respects to the best of the current round of microscopic calculations of momentum distributions for helium, nuclear, and electronic systems. The results of several of these calculations (and improvements and extensions of them) will be discussed at length in other contributions to this volume, by the authors themselves. We shall avoid up-staging these reports by restricting our comments to the obvious. In each case the calculated condensate fraction n_o or Fermi-surface discontinuity Z_{k_F} will be quoted to indicate the strength of the coupling, as appropriate.

Superfluid ^{4}He *Ground State.* State-of-the-art variational[53] and Green's function Monte Carlo[58] calculations of $n(p)$ and $\rho_1(r)$ have been carried out for the Aziz HFDHE2 interaction.[77,78] The variational calculations of Manousakis, Pandharipande, and Usmani[53] are based on a trial wave function including Jastrow pair correlations and triplets, the required expectation values being computed within the HNC method, with a scaling prescription for including effects of elementary diagrams (HNC/S). The best such calculation (OJTV) employs optimized pair correlations, and, as a consequence, the predicted momentum distribution does manifest the p^{-1} singularity of Eq. (16). This singularity does not appear in the GFMC results of Whitlock and Panoff,[58] which refer to simulations of $N = 64$ atoms with periodic boundary conditions (supplemented by variational Monte Carlo studies at $N = 64$ and 512). At the experimental equilibrium density of 0.0218 Å^{-3}, OJTV and GFMC calculations agree upon the value 0.092 for the condensate fraction. However, they differ with regard to the density dependence of n_o, somewhat higher values for n_o being predicted by OJTV at the higher-density points.

Liquid ^{4}He *at Finite T.* The path-integral Monte Carlo computations of Ceperley and Pollock[74] (updated by Ceperley in his contribution to this volume) provide the current microscopic benchmarks for the temperature dependence of the condensate fraction, and more broadly for the momentum distribution, at temperatures ranging from 1.18 K across the lambda point to 3.33 K. Over this temperature range the kinetic energy per particle at $\rho = 0.0218$ Å^{-3} rises from 14.2 K to 16.0 K; the GFMC value for the ground state[58] is 14.5 K. The calculations are based on the Aziz HFDHE2 potential, and most runs involved in Ref. 74 were performed for 64 atoms.

Normal Ground State of Liquid ^{3}He. Again we only cite recent calculations employing the Aziz interaction. Applying an FHNC/S scaling approximation, Fabrocini, Pandharipande, and Usmani[54] have determined $n(p)$ and $\rho_1(r)$ for a trial function containing semi-optimized Jastrow correlations, along with spatially dependent triplet correlations and momentum-dependent backflow. Whitlock and Panoff[58,36] have carried out variational Monte Carlo and fixed-node GFMC computations of these quantities for 54 (and in some cases 66) particles in a box with periodic boundary conditions. At the experimental equilibrium density (near 0.0163 Å^{-3}), Ref. 54 quotes a value $Z_{k_F} = 0.275$ for the Fermi-surface discontinuity; the value for the fixed-node GFMC calculation[58] appears to be substantially smaller (perhaps 0.22), although in this case the sharp Fermi surface is compromised by finite-size effects. A critical comparison of these two calculations is needed. It is interesting to note that the GFMC results for $n(p)$ in the (large-p) tail region are consistent with an exponential fall-off, but not a Gaussian.[20] The variational results of Fabrocini *et al.* do not show a divergent slope of $n(p)$ as $p \rightarrow k_F^{\pm}$.

Nuclear Matter. The momentum distribution of symmetrical nuclear matter has been evaluated for "realistic" two-nucleon interactions by Fantoni and Pandharipande,[43] within the state-dependent variational approach and the CBF method, and by Butler and Koonin,[79] using a Brueckner-Bethe-Goldstone (BBG) hole-line expansion. The former study is based on the Urbana v_{14} two-nucleon potential,[41] supplemented by a phenomenological density-dependent three-nucleon interaction (TNI); the latter assumes the familiar Reid-soft-core[80] and Paris[81] potentials. All three choices of interactions have some basis in meson-exchange theory and yield good fits to the two-nucleon data. In the variational evaluation of Ref. 43, a correlation operator of type (26) is adopted, with correlation functions $f^{(p)}(r)$ determined by constrained variation as in Ref. 82. In a CBF calculation using the same correlation operator, the first-order wave function constructed by nonorthogonal CBF perturbation theory is inserted into (1) [with $\sigma \rightarrow \sigma\tau$ to account for isospin] and terms of higher than second order in the resulting expression are dropped. Non-central correlations are found to dominate the correction to the Fermi-gas $n(p)$. The variational treatment does not give a singular slope of $n(p)$ as $p \rightarrow k_F^{\pm}$; however, there is a CBF term which produces the expected behavior (19). At the experimental equilibrium density, corresponding to $k_F = 1.33$ fm^{-1}, the CBF calculation gives $Z_{k_F} = 0.7$. The BBG calculation of Ref. 79, carried through two-hole-line terms, yields an $n(p)$ very similar to the CBF prediction, but with a somewhat smaller discontinuity at k_F. The population in the tail region is strongly influenced by the tensor force, being reduced by an order of magnitude when the tensor component of the interaction is omitted. We should mention that $n(p)$ has been studied by the Liége group[17,83] for a variety of interactions, using various perturbation-based techniques.

Finite Nuclei. Variational calculations of the momentum distributions of neutrons, protons, and nucleon clusters in $A = 3,4$ nuclei have been performed for the Urbana and Argonne two-nucleon interactions, supplemented by the model-VII three-body interaction.[84] Pair-product (Jastrow-like) correlation operators (26) containing central (1), spin $(\sigma_i \cdot \sigma_j)$, and tensor $(S_{ij}\tau_i \cdot \tau_j)$ two-body components are assumed. For ^{3}He, a five-channel Faddeev wave function has also been studied. Evaluation of multi-dimensional integrals was aided by the Metropolis Monte Carlo algorithm. Comparison of the nucleon momentum distributions $n(p)$ in $A = 2,3,4$ nuclei with that for nuclear matter shows a strong dependence on A at small p and a distinct suggestion of saturation in the large p-regime, the finite-A curves approaching the nuclear matter curve from below as A increases. At large p, the slope of $\log n(p)$ vs. p is roughly the same for all four A values, a feature which may point to some kind of universality associated with the two-body problem. For the Reid interaction, there exist variational[85] and Faddeev[86] treatments in ^{3}He and a coupled-cluster calculation[87] in ^{4}He.

At low energies in the finite nuclear context it is more natural to consider the average occupancy of the relevant shell-model orbitals. Evidence of significant departures from the mean-field picture, ascribed to strong short-range correlations,[88,89] is found in recent (e,e′p) experiments[90,91] which indicate that the depletion of such orbitals can be on the order of 15% or more for single-particle states below the Fermi energy.

Electronic Systems – the Electron Gas. The starting point for most determinations of $n(p)$ for the electron gas is the RPA self-energy (see Ref. 62 for a concise survey). The running variable conventionally used to characterize the density in this system is the radius of the volume per electron, measured in units of the Bohr radius a_o and denoted r_s. The random-phase approximation is accurate in the small-r_s regime, but must be subjected to various modifications and corrections to give quantitative results for r_s values in the range 1.8-5.6 corresponding to conduction-electron densities in "free-electron-type" metals. Sodium ($r_s = 3.97$) would appear to be a good test case in that the interaction between conduction and ion-core electrons is weak and the Fermi surface is very nearly spherical. At r_s near 4, various implementations of RPA[92] (differing for example in the treatment of the Dyson equation or in the inclusion or neglect of exchange corrections) yield values of Z_{k_F} scattered from 0.45 to 0.66. X-ray Compton scattering indicates $Z_{k_F}(\mathrm{Na}) = 0.55 \pm 0.15$.

More recent electron-gas calculations exploit the power of variational and CBF approaches. The efficacy of the latter is demonstrated by the fact that the exact high-density limit for the ground-state energy is reproduced already in second-order CBF perturbation theory if an optimal Jastrow correlation operator F is adopted.[93] Lantto[94] has presented results for the momentum distribution at the optimal Jastrow variational level. At $r_s = 4$, he finds $Z_{k_F} = 0.71$. It is noteworthy that in this treatment there is a logarithmic singularity at the Fermi surface (cf. (19)), by virtue of the long-range $1/r$ correlations appropriate to the Coulomb problem.[7,2,28] Krotscheck[95] has estimated Z_{k_F} in terms of an ω-mass (see Refs. 83,16,17) derived from a correlated RPA calculation of the self-energy,[32,33] the results showing excellent agreement with Lantto's determination of this quantity from the $n(p)$ discontinuity. These Jastrow-variational and Jastrow-CBF studies used FHNC techniques to evaluate expectation values and matrix elements; comparison with available variational Monte Carlo data[96] on $n(p)$ indicates that the neglected elementary-diagram effects are small. Even the uniform limit of the Jastrow model, which gives a good account of the effects of virtual plasmons, yields quantitatively useful results.[7,2]

The most accurate microscopic predictions for the ground state of the electron gas have been obtained by Ceperley and Alder[97] using the GFMC method. The N dependence of the simulations was determined empirically for systems with 38 to 246 particles, allowing extrapolation to $N = \infty$. Comparison of existing Jastrow and CBF results with the benchmarks provided by this calculation suggests that an accurate description of such properties as $n(p)$ can also be attained, with much less computational effort, using refined variational wave functions or the CBF approach.

WHERE DO WE STAND?

Current experiments demand quantitative microscopic predictions for the momentum distributions of the helium liquids, nuclei, and electronic systems, under realistic conditions of density and interactions. We conclude with a brief assessment of how close we are to achieving that goal.

The *helium systems*, which have consistently rebuffed conventional field-theoretic, perturbation-based approaches, have yielded to modern methods which begin with a reasonable picture of the strong correlations. Presumably we now have an excellent description of the ground state of liquid ^{4}He, from variational and Green's function Monte Carlo treatments, and finite temperatures are being adequately mapped out with path-integral Monte Carlo techniques. For the ground state of liquid ^{3}He, the consensus is that the required computational solution is nearly in hand, although the precision available for Bose systems is much harder to attain. Updated CBF calculations for this system are expected to refine our understanding based on variational and GFMC calculations, and especially to provide a better description of the normal-state $n(p)$ near k_F. The work of Bouchaud and Lhuillier[46] challenges this complacent view of the $T = 0$ ^{3}He problem.

Finite-temperature calculations for liquid ^{3}He are still in their infancy.

Many-body techniques for studying *nuclear matter* are highly developed, but their predictive power is hampered by the complications and ambiguities inherent to nuclear forces. The finite geometry of real *nuclei* introduces further complications which greatly reduce the tractability of viable microscopic approaches. There is a prevailing climate of uncertainty regarding the range of validity of the conventional picture of nuclei in terms of nucleons.

Our account of *electronic systems* has been limited in the extreme, being confined to the uniform electron gas as an idealization of the conduction electrons in a metal. In dealing with mobile electrons in real metals and real solids, lattice and finite-temperature effects inhibit but do not preclude the application of the many-body approaches which have proved so successful for the strongly-interacting systems considered in this survey. Calculation of momentum distributions of the electrons in atoms and molecules presents technical obstacles similar to those for finite nuclei. Significant progress in the extension of variational, CBF, and stochastic methods to nonuniform and/or finite systems (see, e.g. Refs. 98,59) supports the view that these modern many-body methods will prove effective in describing a variety of correlation phenomena in the rich domain of electronic systems. As further evidence we may cite the recent work of Vollhardt and coworkers.[99] These authors apply cluster-diagrammatic techniques to achieve an incisive understanding of Hubbard-type models within the context of a Gutzwiller variational wave function[100] (a specialized form of the Jastrow *Ansatz*). In particular, their approach yields an approximation-free evaluation of the momentum distribution and energy in one dimension, for arbitrary density and interaction strength.

We should not close without making the point that "Man cannot live by momentum distributions alone." A recent analysis of final-state effects in deep-inelastic scattering[101] has reminded us that, for strongly-interacting systems, the spatial correlation structure can have a vital influence on the scattering law even at very large momentum transfers. Indeed, there is a growing realization that it may be important to understand nontrivial aspects of the two-body density matrix $\rho_2(\mathbf{r}_1,\mathbf{r}_2,\mathbf{r}_1'\mathbf{r}_2')$ in order to extract unambiguous information on $n(\mathbf{p})$ (i.e. on $\rho_1(\mathbf{r}_1,\mathbf{r}_1')$) from inclusive inelastic experiments. Accordingly, microscopic evaluation of ρ_2 should be high on the agenda of the many-body theorist. Some initial steps in this direction are taken in another contribution.[102]

ACKNOWLEDGMENT

This research has been supported in part by the Condensed Matter Theory Program, Division of Materials Research, U. S. National Science Foundation, under Grant No. DMR-8519077 (JWC) and by the Deutsche Forschungsgemeinschaft under Grant Nr. Ri 267 (MLR). MLR enjoyed the hospitality of the Department of Physics, Washington University, during a sabbatical leave from the University of Köln. We thank R. N. Silver for many stimulating discussions.

REFERENCES

1.   M. L. Ristig and J. W. Clark, Phys. Rev. B **14**, 2875 (1976).

2.   M. L. Ristig, in *From Nuclei to Particles*, ed. A. Molinari (North-Holland, Amsterdam, 1981), p. 340.

3.   O. Penrose and L. Onsager, Phys. Rev. **104**, 576 (1956).

4.   C. N. Yang, Rev. Mod. Phys. **34**, 694 (1962).

5.   J. Gavoret and P. Noziéres, Ann. Phys. (N. Y.) **28**, 349 (1964).

6.   G. V. Chester and L. Reatto, Phys. Lett. **22**, 276 (1966).

7. P. M. Lam and M. L. Ristig, Nucl. Phys. **A328**, 267 (1979); Kinam **1**, 407 (1979).

8. N. N. Bogoliubov (unpublished); N. N. Bogoliubov, *Lectures on Quantum Statistics* (Gordon and Breach, New York, 1970), Vol. 2, pp. 58 and 185.

9. P. C. Hohenberg and P. C. Martin, Phys. Rev. Lett. **22**, 69 (1963); P. C. Hohenberg and P. C. Martin, Ann. Phys. (N. Y.) **34**, 291 (1965).

10. R. A. Ferrell, N. Menyhard, H. Schmidt, F. Schwabl, and P. Szepfalusy, Ann. Phys. (N. Y.) **47**, 565 (1968).

11. L. Reatto and G. V. Chester, Phys. Rev. **155**, 88 (1967).

12. N. Schultz and M. L. Ristig, Z. Physik **38**, 293 (1980).

13. A. Griffin, Phys. Rev. B **30**, 5057 (1984); **32**, 3289 (1985).

14. A. B. Migdal, Soviet Phys.−JETP **5**, 333 (1957).

15. J. M. Luttinger, Phys. Rev. **119**, 1153 (1960).

16. J. P. Jeukenne, A. Lejeune, and C. Mahaux, Phys. Reports **25**, 83 (1976).

17. R. Sartor and C. Mahaux, Phys. Rev. C **21**, 1546 (1980).

18. V. A. Belyakov, Sov. Phys. JETP **13**, 850 (1961).

19. I. Sick, D. Day, and J. S. McCarthy, Phys. Rev. Lett. **45**, 871 (1980); I. Sick, private communication.

20. R. M. Panoff, private communication, and contribution at this conference; J. Carlson, R. M. Panoff, K. E. Schmidt, P. A. Whitlock, and M. H. Kalos, Phys. Rev. Lett. **55**, 2367(C) (1985).

21. R. D. Amado, Phys. Rev. C **14**, 1264 (1976); R. D. Amado and R. M. Woloshyn, Phys. Rev. C **16**, 1255 (1977).

22. J. Heatherington, private communication, and contribution at this conference.

23. J. W. Clark, Prog. Part. Nucl. Phys. **2**, 89 (1979); and references therein.

24. W. L. McMillan, Phys. Rev. **138**, A442 (1965).

25. D. Schiff and L. Verlet, Phys. Rev. **160**, 208 (1967).

26. S. Rosati and S. Fantoni, Lecture Notes in Physics **138**, 1 (1981); and references therein.

27. C. E. Campbell, *Progress in Liquid Physics*, ed. C. A. Croxton (Wiley, New York, 1978), Chap.6; C. C. Chang and C. E. Campbell, Phys. Rev. B **15**, 4238 (1977); C. E. Campbell and F. J. Pinski, Nucl. Phys. **A328**, 210 (1979).

28. E. Krotscheck, Phys. Rev. A **15**, 397 (1977).

29. L. J. Lantto and P. J. Siemens, Nucl. Phys. **A317**, 55 (1979).

30. E. Krotscheck, R. A. Smith, J. W. Clark, and R. M. Panoff, Phys. Rev. B **24**, 6383 (1981).

31. A. D. Jackson, A. Lande, and R. A. Smith, Physics Reports **86**, 55 (1982).

32. E. Krotscheck, Phys. Rev. A **26**, 3536 (1982).

33. E. Krotscheck, in *Quantum Fluids and Solids*, Proceedings of the Symposium on Quantum Fluids and Solids, Sanibel, Florida, 1983, ed. E. D. Adams and G. G. Ihas (AIP, New York, 1983), p. 132; J. W. Clark and E. Krotscheck, Lecture Notes in Physics **198**, 127 (1984).

34. R. P. Feynman and M. Cohen, Phys. Rev. **102**, 1189 (1956).

35. V. R. Pandharipande, Phys. Rev. B **18**, 218 (1978); K. E. Schmidt and V. R. Pandharipande, Phys. Rev. B **19**, 2504 (1979).

36. R. M. Panoff, in *Condensed Matter Theories*, Vol. 2, ed. P. Vashishta, R. K. Kalia, and R. F. Bishop (Plenum, New York, 1987); J. Carlson and R. M. Panoff, to be published; R. M. Panoff, private communication.

37. E. Manousakis, S. Fantoni, V. R. Pandharipande, and Q. N. Usmani, Phys. Rev. B **28**, 3770 (1983).

38. M. Viviani, E. Buendia, S. Fantoni, and S. Rosati, Phys. Rev. B **38**, 4523 (1988).

39. V. R. Pandharipande and R. B. Wiringa, Rev. Mod. Phys. **51**, 821 (1979).

40. R. B. Wiringa, V. Fiks, and A. Fabrocini, Phys. Rev. C **38**, 1010 (1988).

41. I. E. Lagaris and V. R. Pandharipande, Nucl. Phys. **A359**, 331 (1981).

42. R. B. Wiringa, R. A. Smith, and T. L. Ainsworth, Phys. Rev. C **19**, 1207 (1984).

43. S. Fantoni and V. R. Pandharipande, Nucl. Phys. **A427**, 473 (1984).

44. S. Vitiello, K. Runge, and M. H. Kalos, Phys. Rev. Lett. **60**, 1970 (1988).

45. L. Reatto and G. L. Masserini, Phys. Rev. B **38**, 4516 (1988).

46. J. P. Bouchaud and C. Lhuillier, Europhys. Lett. **3**, 1273 (1987); J. P. Bouchaud, A. Georges, and C. Lhuillier, J. Phys. France **49**, 553 (1988).

47. J. W. Clark, Lecture Notes in Physics **138**, 184 (1981).

48. J. M. J. van Leeuwen, J. Groeneveld, and J. de Boer, Physica **25**, 792 (1959).

49. S. Fantoni, Nuovo Cim. **A44**, 191 (1978).

50. Q. N. Usmani, B. Friedman, and V. R. Pandharipande, Phys. Rev. B **25**, 4502 (1982); Q. N. Usmani, S. Fantoni, and V. R. Pandharipande, Phys. Rev. B **26**, 6123 (1982); M. F. Flynn, Phys. Rev. B **33**, 91 (1986).

51. A. Fabrocini and S. Rosati, Nuovo Cimento **D1**, 567, 615 (1982); M. Viviani, E. Buendia, A. Fabrocini, and S. Rosati, Nuovo Cimento **D8**, 561 (1986).

52. M. Puoskari and A. Kalos, Phys. Rev. B **30**, 152 (1984).

53. E. Manousakis, V. R. Pandharipande, and Q. N. Usmani, Phys. Rev. B **31**, 7022 (1985).

54. A. Fabrocini, V. R. Pandharipande, and Q. N. Usmani, to be published.

55. I. Lagaris, Anales de Fisica **81** (Num. 1 Especial), 39 (1985).

56. N. Metropolis, A. Rosenbluth, M. Rosenbluth, A. H. Teller, and E. Teller, J. Chem. Phys. **21**, 1087 (1953).

57. D. Ceperley, G. V. Chester, and M. H. Kalos, Phys. Rev. B **16**, 3081 (1977).

58. P. Whitlock and R. M. Panoff, Can. J. Phys. **65**, 1409 (1987).

59. J. Carlson and M. H. Kalos, Phys. Rev. C **36**, 27 (1987); V. R. Pandharipande, S. C. Pieper, and R. B. Wiringa, to be published.

60. D. J. Thouless, *The Quantum Mechanics of Many-Body Systems* (Academic, New York, 1961); D. Pines, *The Many-Body Problem* (Benjamin, New York, 1961); A. A. Abrikosov, L. P. Gorkov, and I. E. Dzyaloshinksi, *Methods of Quantum Field Theory in Statistical Physics* (Prentice-Hall, Englewood Cliffs, N. J., 1963); A. L. Fetter and J. D. Walecka, *Quantum Theory of Many-Particle Systems* (McGraw-Hill, New York, 1971); G. E. Brown, *Many Body Problems* (North-Holland, Amsterdam, 1972).

61. M. Gell-Mann and K. A. Brueckner, Phys. Rev. **106**, 364 (1957).

62. P. Eisenberger, L. Lam, P. M. Platzman, and P. Schmidt, Phys. Rev. B **6**, 3671 (1972); and references therein.

63. H. A. Bethe, Ann. Rev. Nucl. Sci. **21**, 93 (1971).

64. W. H. Dickhoff, Phys. Lett. **210B**, 15 (1988). A. Ramos, A. Polls, and W. H. Dickhoff, Phys. Lett., in press.

65. H. Kümmel, K. H. Lührmann, and J. G. Zabolitzky, Physics Reports **36**, 1 (1978); R. F. Bishop and H. G. Kümmel, Physics Today **52** (March 1987).

66. J. W. Clark and E. Feenberg, Phys. Rev. **113**, 388 (1959).

67. E. Feenberg, *Theory of Quantum Fluids* (Academic, New York, 1969).

68. J. M. Chen, J. W. Clark, and D. G. Sandler, Z. Phys. A **305**, 223 (1982).

69. E. Krotscheck and J. W. Clark, Lecture Notes in Physics **138**, 356 (1981).

70. D. M. Ceperley and M. H. Kalos, in *Monte Carlo Methods in Statistical Physics*, ed. K. Binder (Springer, Berlin, 1979), Vol. 7 of Topics in Current Physics; K. E. Schmidt and M. H. Kalos, in *Monte Carlo Methods in Statistical Physics*, ed. K. Binder (Springer, Berlin, 1984); Vol. 36 of Topics in Current Physics.

71. K. E. Schmidt and J. W. Moskowitz, J. Stat. Phys. **43**, 1027 (1986).

72. E. L. Pollock and D. M. Ceperley, Phys. Rev. B **30**, 2555 (1984).

73. D. M. Ceperley and E. L. Pollock, Phys. Rev. Lett. **56**, 351 (1986).

74. D. M. Ceperley and E. L. Pollock, Can. J. Phys. **65**, 1416 (1987).

75. D. M. Ceperley, contribution to this volume.

76. G. Senger, M. L. Ristig, K. E. Kürten, and C. E. Campbell, Phys. Rev. B **33**, 7562 (1986).

77. R. A. Aziz, V. P. S. Nain, J. S. Carley, W. L. Taylor, and G. T. McConville, J. Chem. Phys. **70**, 4330 (1979).

78. M. H. Kalos, M. A. Lee. P. A. Whitlock, and G. V. Chester, Phys. Rev. B **24**, 115 (1981).

79. M. N. Butler and S. E. Koonin, Phys. Lett. **205B**, 123 (1988); M. N. Butler, Ph.D. Thesis, California Institute of Technology (1987).

80. R. V. Reid, Ann. Phys. **50**, 411 (1968).

81. M. Lacombe *et al.*, Phys. Rev. C **21**, 861 (1980).

82. I. E. Lagaris and V. R. Pandharipande, Nucl. Phys. **A359**, 349 (1981).

83. C. Mahaux, Lecture Notes in Physics **138**, 50 (1981); C. Mahaux, P. F. Bortingnon, R. A. Broglia, and C. H. Dasso, Physics Reports **120**, 1 (1985); and references therein.

84. R. Schiavilla, V. R. Pandharipande, and R. B. Wiringa, Nucl. Phys. **A449**, 219 (1986).

85. C. Ciofi degli Atti, E. Pace. and G. Salme, in *Perspectives in Nuclear Physics at Intermediate Energies*, ed. S. Boffi, C. Ciofi degli Atti, and N. M. Giannini (World Scientific, Singapore, 1984).

86. A. E. L. Dieperink, T. de Forest, I. Sick, and R. A. Brandenberg, Phys. Lett. **63B**, 261 (1976).

87. J. G. Zabolitzky and W. Ey, Phys. Lett. **76B**, 527 (1978).

88. V. R. Pandharipande, C. N. Papanicolas, and J. Wambach, Phys. Rev. Lett. **53**, 1133 (1984).

89. M. Jaminon, C. Mahaux, and H. Ngô, Nucl. Phys. **A440**, 228 (1985).

90. E. N. M. Quint *et al.*, Phys. Rev. Lett. **58**, 1088 (1987).

91. J. W. A. den Herder *et al.*, Phys. Lett. **184B**, 11 (1987).

92. E. Daniel and S. H. Vosko, Phys. Rev. **120**, 2041 (1960); D. J. W. Geldart, A. Houghton, and S. H. Vosko, Can. J. Phys. **42**, 1938 (1964); T. M. Rice, Ann. Phys. **31**, 100 (1965); B. I. Lundqvist, Physik Kondensierten Materie **7**, 117 (1968); J. Lam, Phys. Rev. B **3**, 3243 (1971).

93. L. J. Lantto, E. Krotscheck, and R. A. Smith, Lecture Notes in Physics **142**, 287 (1981).

94. L. J. Lantto, Phys. Rev. B **22**, 1380 (1980).

95. E. Krotscheck, Ann. of Phys. **155**, 1 (1984).

96. D. Ceperley, Phys. Rev. B **18**, 3126 (1978); and unpublished.

97. D. M. Ceperley and B. J. Alder, Phys. Rev. Lett. **45**, 566 (1980).

98. E. Krotscheck, M. Saarela, and J. L. Epstein, Phys. Rev. B **38**, 111 (1988), and references therein.

99. W. Metzner and D. Vollhardt, Phys. Rev. Lett. **59**, 121 (1987); F. Gebhard and D. Vollhardt, Phys. Rev. Lett. **59**, 1472 (1987); W. Metzner and D. Vollhardt, Phys. Rev. B **37**, 7382 (1988).

100. M. C. Gutzwiller, Phys. Rev. A **134**, 923 (1964); **137**, 1726 (1965).

101. R. N. Silver, Phys. Rev. B **38**, 2283 (1988); and to be published.

102. M. L. Ristig and J. W. Clark, contribution to this volume.

MOMENTUM DISTRIBUTIONS IN QUANTUM LIQUIDS
FROM GREEN'S FUNCTION MONTE CARLO CALCULATIONS

R. M. Panoff

Kinard Laboratory of Physics
Clemson University
Clemson, SC 29634-1911

and

P. A. Whitlock

Courant Institute of Mathematical Sciences
New York University
New York, NY 10012

I. INTRODUCTION

As evidenced by the broad scope of topics covered in this Workshop, significant progress has been made in recent years both in experimental techniques and the analysis of results from neutron-diffraction studies of momentum distributions in quantum many-body systems. Comparison with theory requires both a proper consideration of final-state effects, as explained by other contributors to this volume[1], as well as a basic understanding of the computational procedures employed to generate the momentum distribution from quantum simulations.

In our approach, we attempt to solve the Schroedinger Equation

$$H\psi_0 = E\psi_0, \qquad (1)$$

to obtain the ground-state wave function ψ_0 for a given Hamiltonian H. However, in many cases the Hamiltonian is approximate and/or ψ_0 is replaced by some trial function, ψ_T. Uncertainties in the numerical evaluation of the energy or other ground-state properties such as the momentum distribution may therefore be so large as to prevent a meaningful comparison of experiment and theory.

Green's function Monte Carlo[2-5] (GFMC) is a powerful method which, for a specified H, stochastically solves the Schroedinger equation exactly for ψ_0, within statistical sampling errors and finite-size effects. The direct output from GFMC calculations are configurations – lists of particle coordinates distributed with a probability distribution $p(R) \propto \psi_0(R)$. Ensemble averages over these configurations yield accurate information about the ground-state properties of the simulated system.

The methods discussed here are generally applicable to all quantum fluids and solids, however, for purposes of comparison we report here results of calculations on the bulk atomic liquids Helium-4 and Helium-3, the paradigm boson and fermion systems, as well as an idealized two-dimensional system of ^{4}He atoms. Rather than dwell on particular details of the calculations, we gather here a selection of available results – some previously reported – to demonstrate the strengths and limitations of the GFMC method. The confidence we have in our work comes, not primarily from our gratifying agreement with new experimental analyses, but from the efforts we have made to control or eliminate approximations in our calculations, achieving an internal consistency in the numerical evaluation of distinct quantities which are formally equivalent.

The rest of this paper is organized as follows: Section II outlines the GFMC method and the basic equations used to evaluate the momentum distribution and related quantities. In Section III, we discuss the reliability of the extrapolation procedure used to evaluate ground-state properties other than the energy. Finite size effects are examined in Section IV, and we conclude with a consideration of model dependence.

II. BASIC EQUATIONS OF THE GFMC METHOD

The simulation of quantum many-body systems by the Green's function Monte Carlo (GFMC) algorithm has been described in detail elsewhere.[2-6] Briefly, the Schroedinger equation (1) is transformed into an integral equation with a shifted energy scale,

$$\psi^{n+1}(R) = \left(\frac{E_T + E_C}{H + E_C} \right) \psi^n(R), \tag{2}$$

where H is the Hamiltonian, R is a $3N$-dimensional vector and E_T and E_C are constants. Iterated many times, Eq. (2) converges exponentially to the lowest state of the Hamiltonian not orthogonal to the starting wave function. The value of the trial energy, E_T, is chosen to be a good approximation to the ground-state energy and the value of E_C is chosen so that the spectrum of $H + E_C$ is positive. The iterations of Eq. (2) are carried out by random walks, starting out with an optimized trial wave function. An asymptotically unbiased estimator for the energy is given by

$$E_M = \frac{\int \psi^n(R)H\psi_T(R)dR}{\int \psi^n(R)\psi_T(R)dR}. \tag{3}$$

If the initial iterate $\psi^0 = \psi_T$, E_M is an upper bound to the ground-state energy E_0 for all n, and as $n \rightarrow \infty$, $E_M = E_0$. Application of the GFMC method to bosonic systems such as many-body ^{4}He is straightforward since the wave function ψ is positive definite. Complications in applying GFMC to fermionic systems such as ^{3}He, due to the antisymmetric nature of the fermion wave function, are well known and widely discussed.[6-10] The fermion problem manifests itself as an exponential increase in statistical error as the calculation proceeds. A detailed comparison of methods designed to reduce or eliminate this difficulty may be found elsewhere.[11,12]

Importance sampling incorporated into the algorithm controls the variance of the GFMC calculation for both bosons and fermions. A trial wave function guides the random walks into areas of configuration space where the eigenfunction is greatest.

Good trial wave functions are obtained for the most part from variational calculations. In variational calculations, we calculate the expectation value of the Hamiltonian which is an upper bound to the ground-state energy, E_0,

$$
\begin{aligned}
E_0 &\leq \frac{\int \psi_T^*(R) H \psi_T(R) dR}{\int \psi_T^*(R)\psi_T(R) dR} \\
&= \int \frac{H\psi_T^*(R)}{\psi_T^*(R)} \frac{\psi_T^*(R)\psi_T(R)}{\int \psi_T^*(R')\psi_T(R')dR'} dR
\end{aligned}
\tag{4}
$$

The second form of the integral is easily calculated using the method of Metropolis et al. [13], usually with a low variance answer since $H\psi_T^*(R)/\psi_T^*(R)$ is nearly constant if $\psi_T(R)$ is close to the true ground-state wave function. A good trial wave function minimizes Eq. (4).

The trial function used in our calculations is generally of the form[14]

$$
\psi_T(R) = \exp\left(-\frac{1}{2}\sum_{i<j} u_2(r_{ij}) - \frac{1}{2}\sum_{i<j<k} u_3(r_i, r_j, r_k) + \cdots \right) \times D
\tag{5}
$$

where u_2 and u_3 are two- and three-body functions, respectively. D is 1 for boson systems and is a product of Slater determinants of orbitals, one for each spin state, in fermion systems. State-dependent correlations, including those which incorporate the effects of backflow, may be included in Eq. (5), for instance, by modifying the orbitals in D. Functional forms for u_2 and u_3 along with optimal parameters entering therein are presented elsewhere[12,14−16].

As stated above, the output from both variational and GFMC calculations are configurations composed of particle coordinates. These may be analyzed to determine other properties of the systems. We have confidence in the reliability of our GFMC calculations in both ^{4}He and ^{3}He. For both systems the HFDEH2 potential of Aziz et al. [17] serves well to describe the pair interaction. In Figure 1 are shown the equations of state[15,10] of liquid ^{4}He and ^{3}He from the GFMC calculations (points with error bars), where the fixed-node approximation was adopted for ^{3}He. Also shown are the experimental results (solid lines) with which we are in excellent agreement.

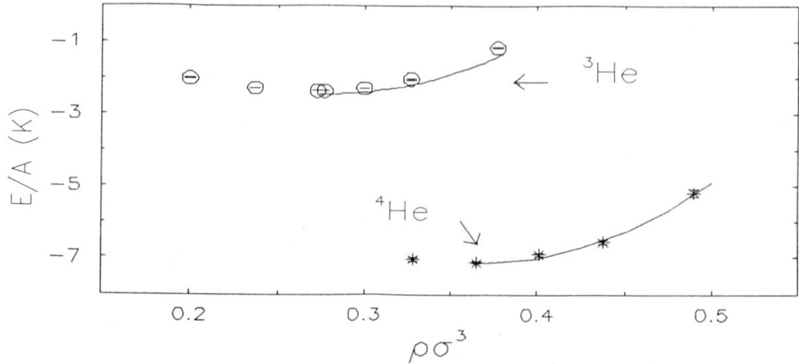

Figure 1. Equation of state of liquid ^{3}He and liquid ^{4}He.

Configurations resulting from the GFMC calculation of the ground state of ^{4}He and ^{3}He may be analyzed[18] to give $n(\vec{r})$ and $n(\vec{k})$ at k-vectors contained within the simulation box. The one-body density matrix, $n(\vec{r})$, is the Fourier transform of the momentum distribution,

$$n(\vec{r}) = \int e^{i\vec{k}\vec{r}} n(\vec{k}) d\vec{k}$$
$$= \left\langle \frac{\psi(r_1, r_2, \cdots, r_i + \vec{r}, \cdots, r_n)}{\psi(r_1, r_2, \cdots, r_i, \cdots, r_n)} \right\rangle . \qquad (6)$$

and is a measure of the change in the wave function for a given displacement r. The expectation value in Eq. (6) is evaluated by moving particle i by an amount r and averaging the ratio of the wave functions over an ensemble of configurations. Two different methods are generally used in this analysis. One method computes $n(r)$ by moving particle i in the configurations a random amount, giving n_0 accurately and the best values for $n(k)$ at specific k-vectors. A second way determines $n(r)$ by moving particle i increments of dr along a prescribed direction yielding precise values of $n(r)$ at small r and therefore better values for the kinetic energy.

The fraction of particles that have condensed into the zero-momentum state is determined from the asymptotic limit of $n(r)$,

$$n_0 = \lim_{r \to \infty} n(r). \qquad (7)$$

In our calculations, we obtain n_0 from the large-r behavior of $n(r)$, and hope that the asymptotic value has been attained within our simulation box. No condensate fraction is to be found in ^{3}He since it is a fermion system. This results from the Pauli exclusion principle and that $\lim_{r \to \infty} n(r)$ is identically equal to zero. In a fermion system, $n(r)$ fluctuates about zero for large r, reflecting the antisymmetry of the wave function.

As the calculation of $n(r)$ proceeds using the first of the two methods above, we determine $n(\vec{k})$ directly for those wave vectors contained in our simulation box. That is,

$$n(\vec{k}) = \left\langle \frac{e^{i\vec{k}\cdot(\vec{r}_i + \vec{r})} \psi(\vec{r}_i + \vec{r})}{\psi(\vec{r}_i)} \right\rangle . \qquad (8)$$

The spherical $n(k)$ can be obtained as the Fourier transform of the calculated $n(r)$,

$$n(k) = n_0 \delta(k) + \frac{1}{2\pi^2} \int e^{-ikr} [n(r) - n_0] r^2 dr. \qquad (9)$$

Given a reliable $n(k)$ the average kinetic energy per particle can be calculated from its second moment:

$$\langle KE \rangle = \frac{\hbar^2}{2m} \langle k^2 \rangle = \frac{\int_0^\infty k^4 n(k) dk}{\int_0^\infty k^2 n(k) dk}. \qquad (10)$$

To carry out the integrals in Eq. (10), the large-k behavior of $n(k)$ must be known. For this purpose, an exponential tail can be fit to $n(k)$. While there is no proof that the exact large-k behavior is exponential, the near-tail behavior of $n(k)$ suggests that this is at least a reasonable approximation. No great effort is required to produce a fit which is continuous, with a continuous first derivative, and for which the integral in Eq. (10) reproduces within statistics the value for the kinetic energy from the eigenvalue calculation.

Alternatively, the kinetic energy may be determined from the curvature of $n(r)$ as $r \to 0$, using

$$\left\langle k^2 \right\rangle = -\frac{3}{n(r)} \frac{d^2 n(r)}{dr^2}\bigg|_{r=0}. \tag{11}$$

Again, as a numerical check on the internal consistency of our calculations, the value of the kinetic energy obtained by this second method agrees within statistics with that from the eigenvalue calculation.

III. EXTRAPOLATION OF GROUND-STATE PROPERTIES

Because importance sampling has been incorporated into the algorithm to control the variance, GFMC simulation actually yields configurations – lists of particle coordinates – chosen from the asymptotic distribution $\psi_T \psi_0$. We may then define a "mixed estimation" for any operator F by

$$\langle F \rangle_M = \frac{\int \psi_0(R) F \psi_T(R) dR}{\int \psi_0(R) \psi_T(R) dR}. \tag{12}$$

The calculation of the total energy in the previous section is precisely such a mixed estimation (hence the subscript M in Eq. (3) above). In that case, ψ_0 is an eigenfunction of the Hamiltonian operator and so the mixed estimator is exact, i.e., the total energy calculated with Eq. (3) asymptotically approaches[5] E_0. On the other hand, to form ground-state expectation values for other operators which do not commute with the Hamiltonian, corrections must be introduced. Assuming that the trial function is close to the ground-state wave function, $\psi_T = \psi_0 + \delta\psi$, a linear extrapolation is made to obtain $\langle F \rangle_0$,

$$\langle F \rangle_0 = 2 \langle F \rangle_M - \langle F \rangle_V \tag{13}$$

where $\langle F \rangle_V$ is the expectation value from the variational calculation. The extrapolated expectation value gives $\langle F \rangle_0$ to order δ^2. While other methods[3,5] exist for obtaining $\langle F \rangle_0$, they have higher variance for small δ.

The linear extrapolation, Eq. (13), was extensively tested[19] in a simulation of ^{4}He with the Lennard-Jones potential. In that study, results for two very different Jastrow-type wave functions, incorporating two-body correlations only, were compared for a number of properties such as the kinetic energy and the radial distribution function, g(r). All of the quantities extrapolated to the same values within statistical errors.

Here we demonstrate the reliability of the extrapolation procedure by considering results with and without three-body correlations, and by considering different systems. In Figure 2 we compare results for $n(k)$ calculated by Eqs. (8) and (9) for two very different importance functions. The solid line includes both Jastrow (two-body)

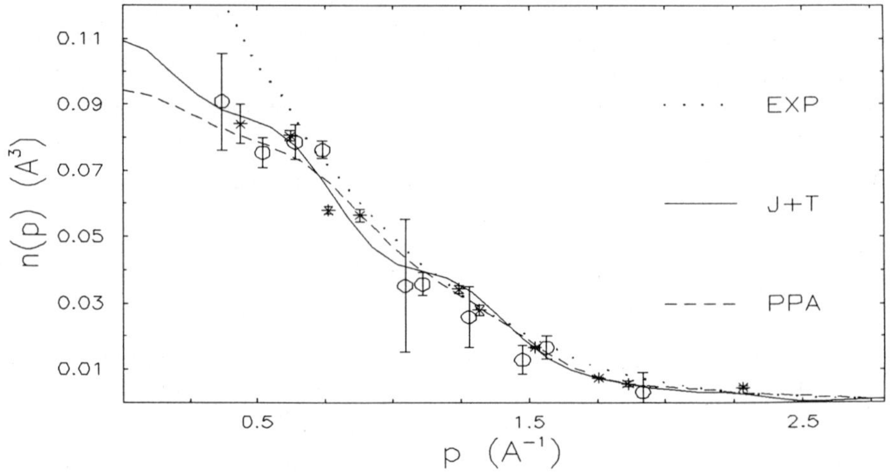

Figure 2. Momentum distribution $n(k)$ for liquid ^{4}He for optimized Jastrow (PPA) and Jastrow + Triplets importance functions.

and three-body correlations, where a simple McMillan form was used for the Jastrow factor. The dashed line comes from a calculation with two-body correlations only, where the functional optimization of the Jastrow wave function was carried out by the paired-phonon analysis (PPA) method as described by Pinski and Campbell[20]. Within statistics, it is hard to differentiate the two extrapolated curves, and both are in quite close agreement with the experimental data (the dotted line) of Woods and Sears[21].

Consider now the extrapolation of the momentum condensate fraction n_0. Table 1 displays results for n_0 in liquid ^{4}He, and Table 2 collects results for a model realization of 2-dimensional ^{4}He. Comparing importance functions with and without three-body correlations, for the two physically different systems similar conclusions can be made. In each case, while the two variational calculations are far apart – differing by as much as a factor of 3 – the extrapolated values are quite close and are indistinguishable within their respective statistical errors.

Table 1. Fraction of atoms condensed in the zero-momentum state from variational, mixed, and extrapolated GFMC calculations in liquid ^{4}He at equilibrium density for two choices of importance function.

ψ_T	n_0^{var}	n_0^{mixed}	n_0^{GFMC}
Jastrow (Optimized PPA)	0.1069 ± 0.0002	0.1002 ± 0.0003	0.0935 ± 0.0005
Jastrow + Triplet (McMillan J)	0.0562 ± 0.0005	0.0680 ± 0.0004	0.0798 ± 0.0008

Table 2. Fraction of atoms condensed in the zero-momentum state from variational, mixed, and extrapolated GFMC calculations for 2-dimensional ^{4}He at density $\rho = 0.0612\text{Å}^{-2}$ for two choices of importance function.

ψ_T	n_0^{var}	n_0^{mixed}	n_0^{GFMC}
Jastrow (McMillan J)	0.23±0.01	0.15±0.01	0.07±0.02
Jastrow + Triplet (McMillan J)	0.0563±0.0005	0.0624±0.0003	0.0595±0.0006

The wide spread of the variational results must be taken as a firm *caveat* for theorists trying to describe properties other than the energy with simple variational wave functions[22-24]: while the variational energy may be quite close to the ground-state energy expectation value, other properties may yet be indeterminately far away and may vary from system to system. For example, in both 3-d liquid ^{4}He and in the 2-d system, the Jastrow + triplet (J+T) wave function gave an energy that was much closer to the ground-state expectation value than with the Jastrow (J) wave function alone. The variational and mixed estimators for n_0, in contrast, were closer for the J+T wave function in 2-d ^{4}He and closer for the optimized Jastrow alone in 3-d ^{4}He.

Besides comparing the energies alone, then, an improved indicator of how "close" ψ_T actually is to ψ_0 may be found in the extrapolation procedure itself. The closer that ψ_T is to the true ground state, the smaller the extrapolation and the closer $\langle F \rangle_V$ is to $\langle F \rangle_M$. We conclude, therefore, that in this sense at least, the optimized (PPA) Jastrow wave function has a greater overlap with the ground-state wave function in the bulk liquid than the particular choice of J+T used in these studies. For the two-dimensional ^{4}He system, however, the J+T is clearly better.

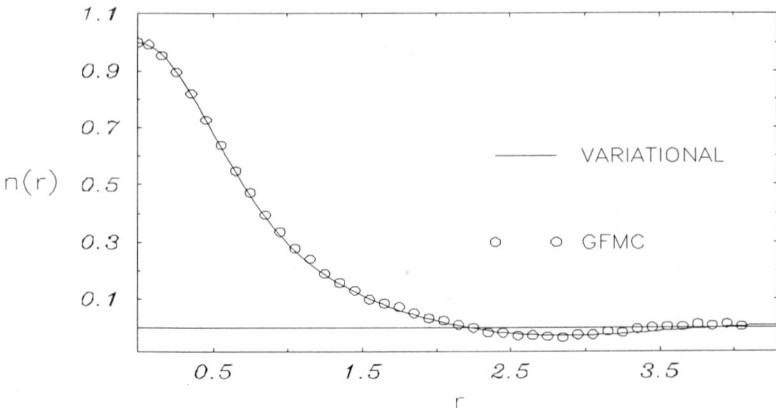

Figure 3. One-body density matrix $n(r)$ for liquid ^{3}He at density $\rho = 0.0163\text{Å}^{-3}$.

In the fermion system, we have also found[10,12,22,24] that the particular choices for the functional forms of Jastrow or triplet or backflow correlations, along with the values for the parameters contained therein, may be quite different while yielding very similar variational energies. However, other properties of the system, at the variational level may be very different. Nevertheless, for our choice of importance function and employing the fixed-node approximation of GFMC for fermions[10,12], the extrapolated ground-state properties ought to be quite reliable. As with its boson counterparts, we associate high reliability in our calculations with small extrapolations. Figure 3, for example, shows how close the variational and extrapolated values are for $n(r)$ in a 54-particle ^{3}He system at equilibrium density.

IV. FINITE-SIZE EFFECTS

Even with access to significant supercomputer resources, GFMC simulations carried out to date have been limited to, at most, several hundred particles subject to periodic boundary conditions. The question arises, therefore, to what extent properties calculated for these systems may be applied to an infinitely-extended liquid. In the brief discussion which follows, it may be helpful to state at the outset that we have found that finite-size effects are among the smaller sources of uncertainty in our calculations.

There are basically three different kinds of finite-size effects. First, in simulating systems at a given density, the particle number determines the size of the simulation cell. Small particle numbers limit information gained for ground-state properties to small-r behavior only; this in turn limits the reliability of numerical Fourier transforms used in calculating related properties. For instance, for too small a simulation volume, $n(r)$ may not yet have reached its asymptotic value, thereby affecting the estimate for the condensate fraction n_0; transforming a truncated $n(r)$ would also yield a poor evaluation of $n(k)$ for small-k. Second, since it is convenient for correlations among the particles to vanish at the sides of the periodic box, the size of the box determines the range of correlations. It is known that for the helium systems, the optimal Jastrow correlations go asymptotically as $1/r^2$. As the particle number and therefore the size of the periodic box is increased, longer range correlations may be taken into account.

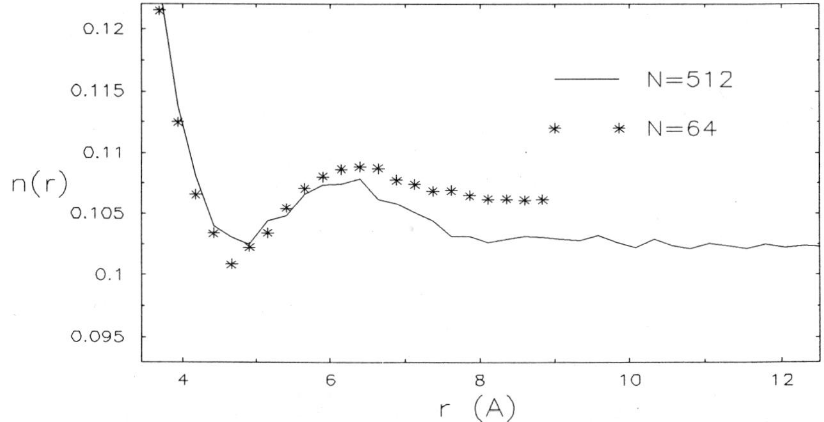

Figure 4. The one-body density matrix $n(r)$ calculated from variational configurations for liquid ^{4}He at $\rho = 0.0218\text{Å}^{-3}$.

Instead of considering the effects of having too few particles, we present results that show that these first two finite-size effects are small for the numbers of particles which are typically used in our detailed studies. In Figure 4, we plot the large-r behavior of $n(r)$ from variational calculations[18] with 64 and 512 ^{4}He atoms; while we do not have similar results for a GFMC comparison, similar behavior is expected. Noting the *very expanded energy scale*, it is clear that, while discernable, the difference between the 64- and 512-particles results in a few percent at most, and is much less than the uncertainty in the extrapolation discussed in the previous section. Table 3 compares calculations of the condensate fraction in 2-d ^{4}He for 64- and 100-particle simulations. Again, while discernable, the difference between these two calculations is much less than the differences in the 2-d extrapolation in Table 2 for two different importance functions.

Table 3. Fraction of atoms condensed in the zero-momentum state from variational, mixed, and extrapolated GFMC calculations for 2-dimensional ^{4}He at density $\rho = 0.0421\text{Å}^{-2}$ for $N = 64$ and $N = 100$ with Jastrow importance function.

N	n_0^{var}	n_0^{mixed}	n_0^{GFMC}
64	0.44±0.05	0.40±0.01	0.36±0.05
100	0.46±0.01	0.40±0.02	0.34±0.02

A third finite-size effect concerns the calculation of k-dependent properties. With the periodic boundary conditions, only certain discrete values of k consistent with the box size can be calculated. Nevertheless, for simulations carried out with different particle numbers, different values of k can be realized. Figure 5 shows calculations with 54 and 66 ^{3}He atoms with and without backflow. For each wave function separately, the discrete values of $n(k)$ for different N appear to lie on the same curve. Information about $n(k)$ can be "filled in" by increasing the particle count further.

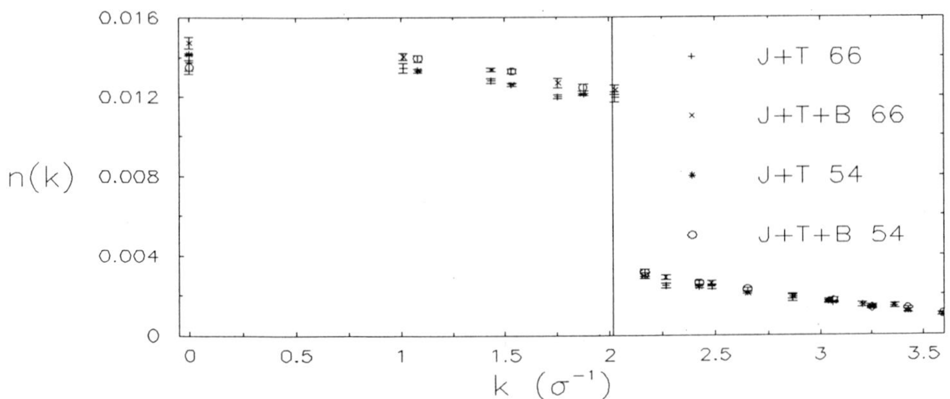

Figure 5. Momentum distribution $n(k)$ calculated from variational configurations for liquid ^{3}He at $\rho = 0.0163\text{Å}^{-3}$ for two different particle numbers and two different wave functions.

V. MODEL DEPENDENCE AND CONCLUSIONS

As the "technology" of GFMC has improved, the reduction of the numerical uncertainties allows us to focus on the question of model dependence. In particular, we are interested in finding out how dependent various ground-state properties are on the choice of the Hamiltonian H which enters Eq. (1). The choice of H is essentially a choice of the two- or more-body potentials $V_n(R)$ which are assumed to describe the particle interactions. Extensive tests with the energy expectation value[15] led to the selection of the HFDHE2 interaction as the two-body potential which best reproduces the equation of state for liquid ^{4}He. This same potential was then adopted for studies of ^{3}He and in 2-d ^{4}He, as well. While the total energy eigenvalue is quite sensitive to the details of H, it turns out that other quantities – e.g., the kinetic energy, the potential energy, and $g(r)$ – are much less dependent on the choice of V. In Figure 6 we demonstrate this by comparing $n(k)$ for ^{4}He for the Lennard-Jones and HFDHE2 interactions. There is little difference between the two curves, and both can be said to be consistent with the experimental curve. We conclude that refinements of the interparticle potentials[25] would change $n(k)$ very little and would not significantly affect our estimates of the magnitude of the condensate fraction or the kinetic energy per particle, for instance. The effects of inclusion of a 3-body interaction directly in H are believed to be small, based on perturbative estimates, but a firm conclusion will have to await further study.

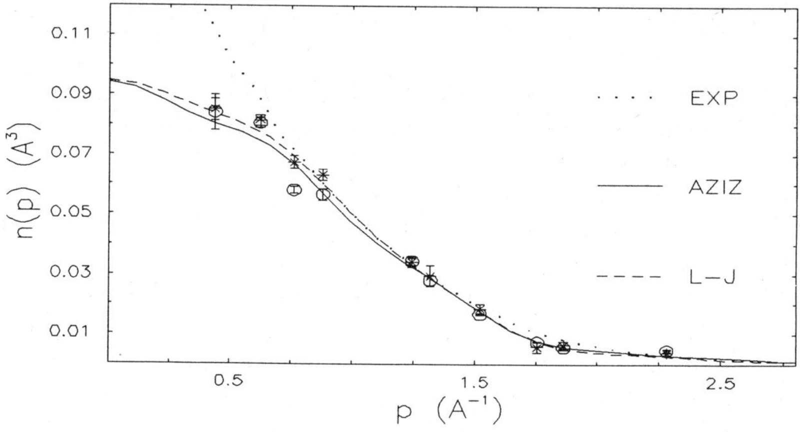

Figure 6. Momentum distribution $n(k)$ calculated from GFMC configurations for liquid ^{4}He at $\rho = 0.0218$Å^{-3} for two different potentials.

Green's function Monte Carlo, in conclusion, yields reliable estimates of ground-state properties of quantum systems, provided that sufficient care is taken to start with a good importance function, to monitor the extrapolation procedure, and to include enough particles in the simulations to reduce finite-size effects. Internal consistency of our calculations gives us confidence that for a given Hamiltonian, we can reliably investigate the ground state. As with the energy expectation value calculations, the agreement with experiment for the momentum distribution is impressive and gives confidence that the two-body HFDHE2 potential serves well as an effective many-body interaction for the helium systems. More detailed comparisons with experiment are given by other authors in this volume.[1]

We would like to thank our colleagues at the Courant Institute for many helpful discussions, and in particular J. M. C. Chen and M. H. Kalos for a careful reading of the manuscript. This work was supported by the Applied Mathematical subprogram of the Office of Energy Research, U.S. Department of Energy under Contract No. DE-AC02-76ER03077. The calculations have been performed on the Cray X-MP available at the National Center of Supercomputing Applications located at the University of Illinois through a grant of computer time from the National Science Foundation under Grant No. DMR-8419083.

REFERENCES

1. See especially contributions by R. N. Silver; T. R. Sosnik, M. Snow, and P. E. Sokol; and S. Koonin.

2. M. H. Kalos, D. Levesque, and L. Verlet, Phys. Rev. **A9**, 2178 (1974).

3. D. M. Ceperley and M. H. Kalos, in *Monte Carlo Methods in Statistical Physics*, K. Binder, ed; Topics in Current Physics, vol. 7 (Springer, Berlin, Heidelberg, New York, 1979) Chap. 4.

4. K. E. Schmidt and M. H. Kalos, in *Applications of the Monte Carlo Method in Statistical Physics*, K. Binder, ed; Topics in Current Physics, vol. 36 (Springer, Berlin, Heidelberg, New York, 1984).

5. K. E. Schmidt, in *Models and Methods in Few-Body Physics*, Lecture Notes in Physics, Vol. 273, (Springer, Berlin, 1987), p. 363.

6. J. Zabolitzky, in *Progress in Particle and Nuclear Physics*, A. Faessler, ed; vol. 16 (Pergamon, Oxford, 1986), and *Proceedings of the Conference on Frontiers of Quantum Monte Carlo*, J. Stat. Phys. 43 (1986).

7. D. M. Ceperley and B. J. Alder, Phys. Rev. Lett. **45**, 566 (1980).

8. D. M. Arnow, M. H. Kalos, Michael A. Lee and K. E. Schmidt, J. Chem. Phys., **77**, 5562 (1982).

9. K. E. Schmidt and J. W. Moskowitz, J. Stat. Phys. **43**, 1027 (1986).

10. R. M. Panoff, in *Condensed Matter Theories, Vol. II*, P. Vashista, R. Kalia, and R. Bishop, eds. (Plenum 1987).

11. J. Carlson and M. H. Kalos, Phys. Rev. **C32**, 1735 (1985).

12. R. M. Panoff and J. Carlson, submitted to Phys. Rev. Lett., 1989.

13. N. Metropolis, A. W. Rosenbluth, M. N. Rosenbluth, A. M. Teller, and E. Teller, J. Chem. Phys. **21**, 1087 (1953).

14. K. E. Schmidt, M. H. Kalos, Michael A. Lee and G. V. Chester, Phys. Rev. Lett. **45**, 573 (1980).

15. M. H. Kalos, Michael A. Lee, P. A. Whitlock and G. V. Chester, Phys. Rev. **B24**, 115 (1981).

16. P. A. Whitlock, G. V. Chester and M. H. Kalos, Phys. Rev. **B38**, 2418 (1988).

17. R. A. Aziz, V. P. S. Nain, J. S. Cerley, W. L. Taylor, and G. T. McConville, J. Chem. Phys. **70**, 4330 (1979).

18. P. A. Whitlock and R. M. Panoff, Can. J. Phys. **65**, 1409 (1987).

19. P. A. Whitlock, D. M. Ceperley, G. V. Chester and M. H. Kalos, Phys. Rev. **B19**, 5598 (1979).

20. F. J. Pinski and C. E. Campbell, Phys. Lett. **79B**, 23 (1978); F. J. Pinski (private communication); L. Whitney and C. E. Campbell (private communication).

21. A. D. B. Woods and V. F. Sears, Phys. Rev. Lett. **39**,415 (1977).

22. E. Manousakis, V. R. Pandharipande, and Q. N. Usmani, Phys. Rev. **B31**, 7022 (1985).

23. J. P. Bouchard and C. Lhuillier, Europhysics Lett. **3**, 1273 (1987).

24. M. Viviani, E. Buendia, S. Fantoni, and S. Rosati, Phys. Rev. **B38**, 4523 (1988).

25. Ronald A. Aziz, Frederick R. W. McCourt, and Clement C. K. Wong, Mol. Phys. **61**, 1487 (1987).

THE MOMENTUM DISTRIBUTION OF ^{4}HE AT NON-ZERO TEMPERATURE

D. M. Ceperley

National Center for Supercomputing Applications
Dept. of Physics
University of Illinois at Urbana-Champaign
Champaign, Il. 61820, USA

Path Integral Monte Carlo calculations provide a rigorous way of calculating equilibrium properties of boson systems at finite temperature. The only significant limitation is that the simulations are of a finite system of bosons, typically on the order of 100 atoms. Most results[1,2] have concerned liquid and solid ^{4}He. The method is based on the Feynman-Kac path integral expression[3] for the many body density matrix:

$$\rho(R,R';\beta) = \frac{1}{N!} \sum_P \int dR(t) \, \exp\left[\int_0^\beta -dt' V(R(t'))\right] \qquad (1)$$

where the integral is over all Brownian paths from R to PR', $\beta=1/kT$, R and R' are two sets of the 3N atomic positions, V(R) is potential energy function and PR' is a permutation (relabeling) of the N atoms. To use the above expression on a computer the "path" is made discrete and accurate approximations[4] are made to the high temperature density matrix. The discrete path and permutation is then sampled with a generalization of the Markov chain algorithm (or Metropolis) Monte Carlo method. A sufficiently long sample of the coordinates will then allow one to compute exact properties of the density matrix.

Fig. (1) shows a schematic picture of a typical configuration. Each circle represents an atom, the dark line connecting the dots is the path. The atoms not connected with any other atoms are permuting with themselves and thus "distinguishable". Figure (1) shows a multiply connected region similar to the interior of a torus. For such a region, it is possible to define the superfluid density[5] in terms of the average number of paths which wind around the torus; one is shown in this picture. Others are shown which do not wind and thus do not contribute to superfluidity. The probability of permuting with a nearby atom becomes large at low temperature. The superfluid transition is thus a kind of percolation transition. Comparison with experiment shows good agreement on the computed superfluid density.

The momentum distribution is the Fourier transform[6] of the single particle off-diagonal density matrix, n(r). All this means is that the path of one particle is not required to return to its starting point. One such path is shown in Fig. (1). At high temperatures, where exchange is not important, the two ends remain close to each other; with an average square distance of 1.5 $\hbar^2$/mT. In the superfluid phase the two ends can get arbitrarily far from each other by hooking onto a macroscopically long path. The condensate fraction (probability that a given atom has exactly zero momentum) equals the probability density that the two ends are far apart divided by

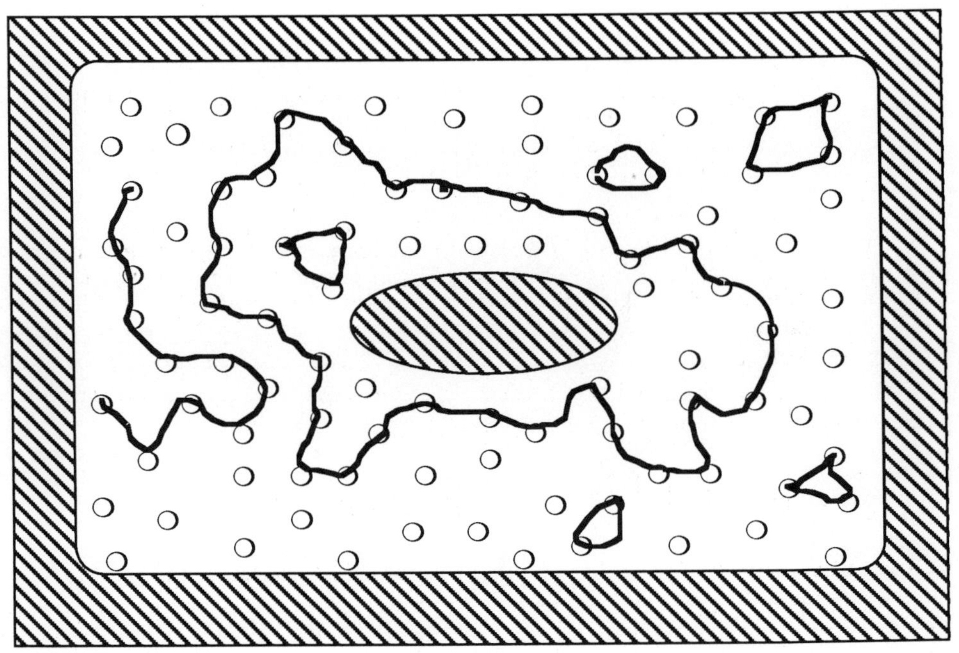

Figure 1. Schematic of a path in liquid helium. The shaded region represents the walls of the container, the circles the helium atoms, and the lines the exchanging paths. The paths of atoms not exchanging are not shown. The path encircling the hole in the container is responsible for superfluidity. The single particle density matrix and the momentum distribution are calculated by determining the end-to-end distribution of the path with two disconnected ends.

the density that they are close together. Thus the condensate fraction and the superfluid density are intimately related. However, they are not the same. In three dimensional liquid helium the condensate fraction is non-zero only when the superfluid fraction is non-zero. Helium films never have a condensate but can become superfluid.

The detailed procedures for the calculation of the momentum distribution are discussed in Ref. (2). Since that article was written we have done additional calculations of the momentum distribution of ^{4}He which we will now summarize.

CALCULATION OF J(y)

In Ref. (2) the momentum distribution of liquid helium at SVP from 1.2K to 3.3K is tabulated. However, what is measured by neutron scattering at high momentum transfers, assuming y-scaling holds, is closely related to the one dimensional momentum distribution, J(y), defined as the probability that an atom has a component of momentum in one direction equal to y. It is easily seen that this is the cosine transform of the single particle density matrix.

$$J(y) = 1/\pi \int_o^\infty dr \cos (ry) \, n(r) \qquad\qquad (2)$$

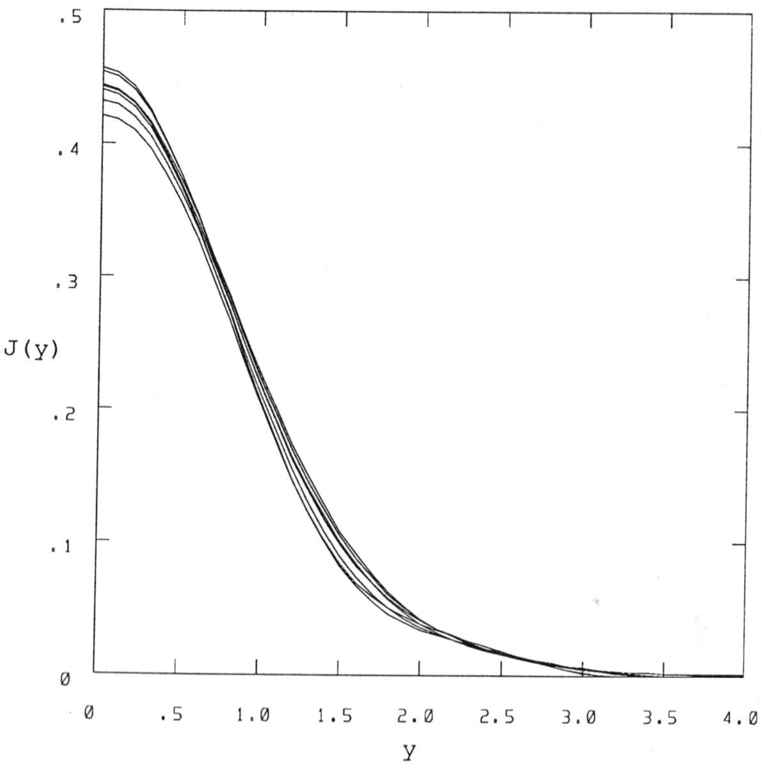

Figure 2.   J (in A) versus y (in A^{-1}) in liquid helium along SVP.   The density is about 0.022/A^3 and the temperatures range from 1.2K to 4K.

It is desirable to compare theory and experiment for J(y) rather than n(k) because the statistical error is smaller for J(y) than n(k) for results obtained both with computer simulations and from neutron scattering. To calculate n(k) from n(r) involves a 3 dimensional fourier transform which has a phase space factor r^2dr. This accentuates the large r statistical fluctuations. In principle, J(y) is obtainable from the n(k)'s in ref.(2) by the use of Fourier transforms and some extra assumptions about the large k behavior of n(k). It is preferable to go back to the original computer simulation data for n(r) and extract J(y) with Eq.(2).

Those results are given in Table 1 along with the computed condensate fraction and kinetic energy (in K). Of course, J(y) should be non-negative for all values of y. The occasional negative entry in Table 1 is a consequence of statistical fluctuations in n(r). This indicates that the results are not very reliable for y greater than about 3 A^{-1}, thus one cannot say much about the exponential behavior of the momentum distribution at large momenta.

Table 1. J(y) in liquid and solid helium.

state	bcc	fcc	nf	nf	sf	sf	sf	sf	sf	nf	nf	sf	sf	nf
T (K)	1.60	1.67	2.35	4.00	1.18	1.54	1.82	2.22	2.50	3.33	4.00	1.18	2.00	4.00
dens.	0.0288	0.0288	0.0288	0.0288	0.0218	0.0218	0.0219	0.0220	0.0218	0.0207	0.0207	0.0260	0.0260	0.0260
cond.	0.0000	0.0000	0.0002	0.0000	0.0687	0.0873	0.0634	0.0272	0.0131	0.0040	0.0017	0.0262	0.0120	0.0003
kin. energy	24.40	24.08	24.75	25.66	14.17	14.35	14.71	15.92	15.91	16.00	16.93	19.11	20.11	21.81
y(1/A)	J(y)													
0.0	0.3544	0.3579	0.3577	0.3502	0.4400	0.4212	0.4319	0.4428	0.4571	0.4542	0.4436	0.3913	0.3954	0.3837
0.1	0.3529	0.3564	0.3561	0.3487	0.4368	0.4184	0.4289	0.4395	0.4535	0.4507	0.4405	0.3893	0.3933	0.3817
0.2	0.3485	0.3519	0.3515	0.3444	0.4275	0.4101	0.4201	0.4299	0.4428	0.4404	0.4312	0.3834	0.3870	0.3757
0.3	0.3413	0.3444	0.3439	0.3373	0.4122	0.3965	0.4058	0.4145	0.4256	0.4238	0.4162	0.3738	0.3767	0.3660
0.4	0.3314	0.3343	0.3335	0.3276	0.3916	0.3780	0.3864	0.3939	0.4029	0.4020	0.3964	0.3607	0.3628	0.3529
0.5	0.3193	0.3217	0.3208	0.3156	0.3665	0.3551	0.3628	0.3691	0.3759	0.3760	0.3725	0.3444	0.3456	0.3371
0.6	0.3050	0.3070	0.3060	0.3016	0.3376	0.3284	0.3357	0.3412	0.3459	0.3471	0.3456	0.3253	0.3257	0.3189
0.7	0.2891	0.2906	0.2895	0.2860	0.3061	0.2990	0.3061	0.3112	0.3140	0.3165	0.3167	0.3040	0.3037	0.2990
0.8	0.2718	0.2728	0.2717	0.2690	0.2730	0.2678	0.2750	0.2804	0.2816	0.2852	0.2869	0.2810	0.2802	0.2779
0.9	0.2535	0.2540	0.2530	0.2511	0.2396	0.2358	0.2434	0.2495	0.2495	0.2542	0.2570	0.2569	0.2557	0.2561
1.0	0.2346	0.2347	0.2338	0.2327	0.2069	0.2043	0.2123	0.2196	0.2186	0.2243	0.2278	0.2323	0.2310	0.2340
1.1	0.2155	0.2151	0.2144	0.2141	0.1758	0.1742	0.1825	0.1911	0.1895	0.1960	0.1998	0.2077	0.2066	0.2121
1.2	0.1965	0.1957	0.1952	0.1955	0.1473	0.1465	0.1548	0.1647	0.1627	0.1696	0.1736	0.1838	0.1830	0.1906
1.3	0.1779	0.1767	0.1764	0.1774	0.1219	0.1219	0.1298	0.1406	0.1384	0.1455	0.1495	0.1609	0.1607	0.1699
1.4	0.1599	0.1584	0.1582	0.1598	0.1000	0.1008	0.1077	0.1189	0.1169	0.1237	0.1275	0.1394	0.1401	0.1502
1.5	0.1428	0.1410	0.1409	0.1431	0.0817	0.0833	0.0888	0.0998	0.0981	0.1045	0.1079	0.1198	0.1213	0.1318
1.6	0.1266	0.1247	0.1246	0.1273	0.0668	0.0694	0.0731	0.0833	0.0821	0.0876	0.0906	0.1021	0.1044	0.1148
1.7	0.1116	0.1095	0.1094	0.1125	0.0552	0.0585	0.0603	0.0690	0.0687	0.0732	0.0755	0.0865	0.0896	0.0993
1.8	0.0977	0.0956	0.0954	0.0989	0.0462	0.0502	0.0502	0.0570	0.0576	0.0610	0.0627	0.0729	0.0766	0.0855
1.9	0.0851	0.0830	0.0827	0.0865	0.0394	0.0439	0.0422	0.0471	0.0486	0.0508	0.0518	0.0612	0.0654	0.0732
2.0	0.0736	0.0716	0.0712	0.0752	0.0342	0.0388	0.0360	0.0388	0.0413	0.0423	0.0428	0.0512	0.0558	0.0624
2.1	0.0634	0.0615	0.0610	0.0650	0.0300	0.0344	0.0310	0.0321	0.0353	0.0354	0.0353	0.0427	0.0474	0.0531
2.2	0.0543	0.0526	0.0520	0.0560	0.0266	0.0303	0.0268	0.0267	0.0303	0.0296	0.0292	0.0355	0.0402	0.0449
2.3	0.0462	0.0448	0.0442	0.0480	0.0235	0.0263	0.0231	0.0222	0.0259	0.0247	0.0242	0.0292	0.0340	0.0379
2.4	0.0392	0.0380	0.0374	0.0409	0.0206	0.0222	0.0196	0.0186	0.0220	0.0205	0.0201	0.0238	0.0285	0.0319
2.5	0.0331	0.0321	0.0316	0.0348	0.0178	0.0180	0.0162	0.0155	0.0185	0.0168	0.0167	0.0189	0.0237	0.0267
2.6	0.0278	0.0270	0.0267	0.0295	0.0150	0.0140	0.0130	0.0129	0.0153	0.0136	0.0139	0.0147	0.0196	0.0223
2.7	0.0232	0.0227	0.0226	0.0248	0.0124	0.0103	0.0099	0.0107	0.0123	0.0108	0.0115	0.0110	0.0160	0.0185
2.8	0.0193	0.0190	0.0191	0.0209	0.0098	0.0070	0.0071	0.0087	0.0097	0.0084	0.0094	0.0079	0.0129	0.0154
2.9	0.0160	0.0159	0.0161	0.0175	0.0075	0.0044	0.0046	0.0070	0.0075	0.0064	0.0076	0.0055	0.0104	0.0127
3.0	0.0132	0.0133	0.0137	0.0146	0.0055	0.0023	0.0026	0.0055	0.0056	0.0048	0.0062	0.0037	0.0082	0.0105
3.1	0.0108	0.0111	0.0117	0.0122	0.0037	0.0009	0.0012	0.0044	0.0041	0.0034	0.0049	0.0027	0.0065	0.0087
3.2	0.0088	0.0092	0.0100	0.0101	0.0021	0.0000	0.0002	0.0035	0.0028	0.0024	0.0039	0.0022	0.0051	0.0071
3.3	0.0071	0.0077	0.0085	0.0084	0.0008	-0.0005	-0.0003	0.0028	0.0018	0.0016	0.0032	0.0022	0.0040	0.0059
3.4	0.0058	0.0063	0.0073	0.0069	-0.0003	-0.0008	-0.0004	0.0025	0.0010	0.0009	0.0025	0.0026	0.0032	0.0048
3.5	0.0046	0.0052	0.0063	0.0057	-0.0013	-0.0009	-0.0002	0.0023	0.0003	0.0003	0.0021	0.0030	0.0025	0.0039
3.6	0.0036	0.0043	0.0054	0.0047	-0.0020	-0.0010	0.0000	0.0022	-0.0003	-0.0002	0.0017	0.0035	0.0019	0.0031
3.7	0.0029	0.0035	0.0047	0.0039	-0.0026	-0.0011	0.0002	0.0022	-0.0008	-0.0006	0.0014	0.0039	0.0014	0.0025
3.8	0.0022	0.0029	0.0040	0.0032	-0.0030	-0.0013	0.0003	0.0022	-0.0011	-0.0009	0.0011	0.0040	0.0011	0.0020
3.9	0.0017	0.0023	0.0034	0.0026	-0.0031	-0.0014	0.0004	0.0020	-0.0012	-0.0011	0.0009	0.0040	0.0008	0.0016
4.0	0.0013	0.0018	0.0028	0.0021	-0.0030	-0.0014	0.0003	0.0018	-0.0012	-0.0012	0.0007	0.0039	0.0006	0.0012

Fig. (2) shows the computed J(y) along the SVP line from 1.18K to 4K. The assumption often made, that the non-condensed part of the momentum distribution is independent of temperature, is approximately correct. Most of the differences between the various J(y)'s seen in Fig. (2) are the result of statistical fluctuations. This is because in performing the cosine transform in Eq. (2) for a superfluid we assume that n(r) equals its asymptotic value, n_O, for $r>R_C$, where R_C is taken to be 5.5 A. Because the assumed value of n_O has statistical fluctuations of the order of $\delta n_O = 0.005$, this procedure will introduce noise in J(y) of the form $\delta n_O \sin(yR_C)/(\pi y)$. Thus, the statistical error in J(y) at small y can be as large as 0.01A, about what is observed in Fig. (2).

DENSITY DEPENDANCE OF THE MOMENTUM DISTRIBUTION

The data in Table 1 are at three different densities in units of atoms/A^3. The lowest density is very close to SVP where most theoretical and experimental studies have been carried out. The intermediate density, 0.02596/A is near the liquid-solid transition at zero temperature. Fig.(3) shows J(y) at the higher density. It is seen that the curves at low and high density are quite similar. The main effect is that the

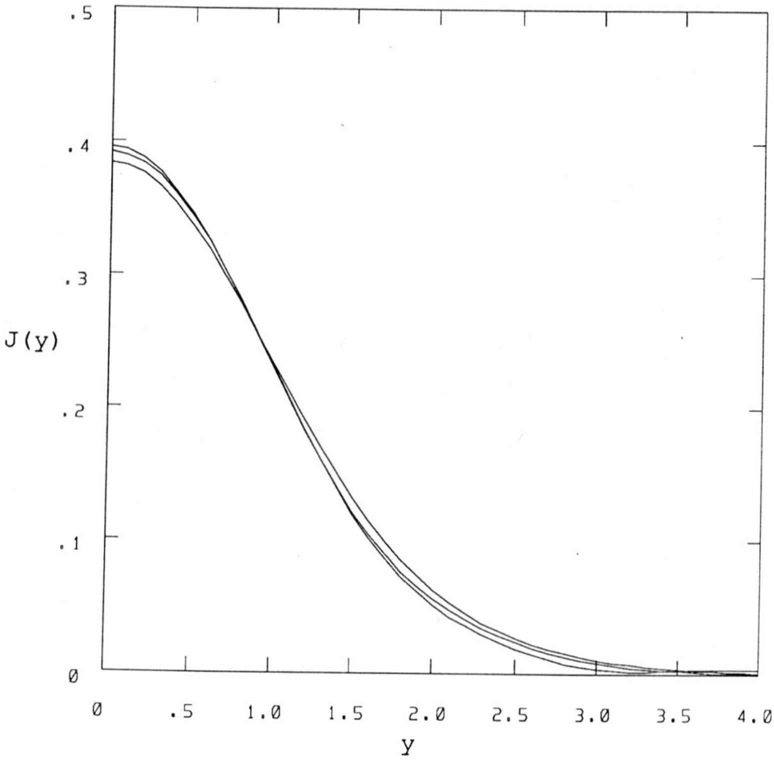

Figure 3. J (in A) versus Y (in A^{-1}) for liquid helium at a density of 0.026/A^3 and temperature from 1.2K to 4K.

zero-temperature kinetic energy has increased from 14.2K to 19.1K. The condensate fraction drops from 8% at SVP to 3%. This is in agreement with GFMC calculations[7] but not with neutron scattering experiments of Sokol.[8] Finally, Table 1 contains J(y) for an even higher density which cuts through the liquid-solid line.

VARIATION OF THE MOMENTUM DISTRIBUTION FROM LIQUID TO SOLID

At a higher density, we have computed the momentum distribution in both the liquid and solid phase (both a fcc structure and a bcc structure). The results are given in Table 1. Fig. (4) compares the J(y)'s from these calculations It is seen that the momentum distribution is very insensitive to whether the system is in a liquid or solid phase, less sensitive than the pair correlation function, for example. Neutron scattering measurements at the same density and temperature also show this independence with respect to the phase[9]. The momentum distribution is controlled by the cage surrounding an atom and the local environment does not change much when going from a normal liquid to a solid. That is controlled mainly by the density. We have not looked for angular dependance of the momentum distribution in the solid phase.

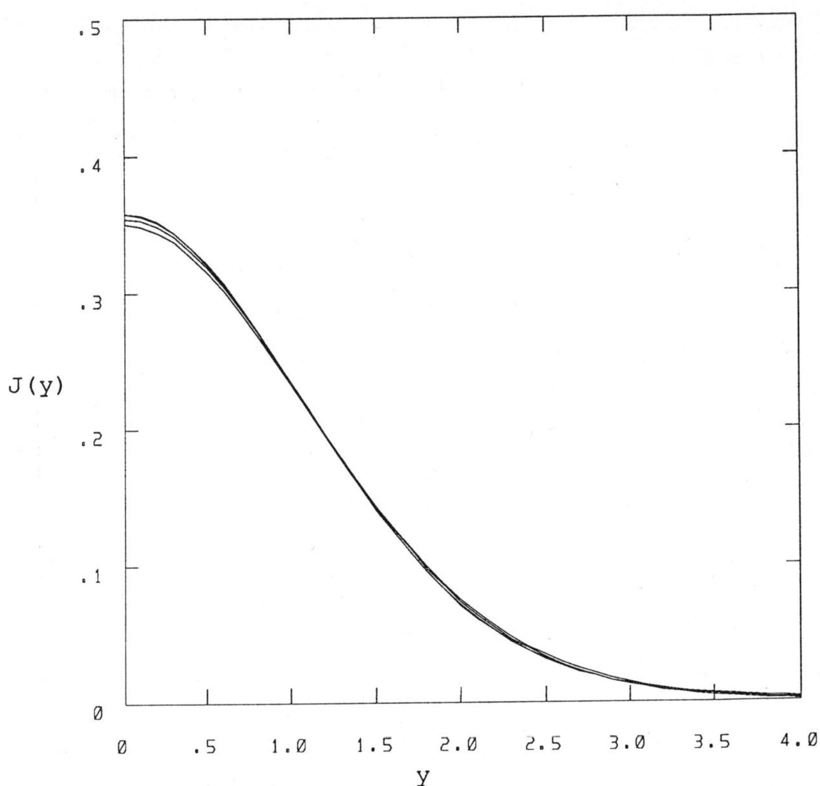

Figure 4. J (in A) versus y (in A^{-1}) for liquid and solid helium at a density of 0.0288/A^3 and temperatures from 1.6K to 4K.

EFFECT OF DIMENSIONALITY ON THE MOMENTUM DISTRIBUTION

The theoretical description of superfluidity in two dimensional He films is given by the theory of Kosterlitz and Thouless, rather than by the macroscopic occupation of the zero momentum state as it is bulk (3 dimensional) He. According to this theory the end-to-end distribution of the polymer in Fig. (1), $n(r)$, decays algebraically to zero at all finite temperatures, if the system is superfluid:

$$n(r) = n_O (d/r)^{-\eta} \quad \text{where} \quad \eta = mkT/2\pi h^2\rho \qquad (3)$$

The exponent η goes to zero linearly with the temperature; thus n_O is the 2 dimensional analogue of the condensate fraction. Ref. (10) contains the results of path integral calculations in the zero pressure liquid. In that paper, it was determined that 2 dimensional He films will become superfluid below 0.72K, and that $n_O = 0.22$ and $d = 3.7$ A. This value of n_O is similar to the results of Green's Function Monte Carlo calculations[11]. The parameter d has the interpretation in the Kosterlitz-Thouless theory of the vortex core size.

In Fourier transforming to get $J(y)$ one arrives at a singular behavior at small momentum, not a delta function:

$$J(y) \approx y^{\eta-1} \qquad (4)$$

The difference in the path integral simulations between the behavior in two and three dimensions is very noticeable. However experimentally it is not clear whether there will be such a pronounced effect of the dimensionality after the distribution function is broadened by instrumental and final state effects.

TWO BODY OFF-DIAGONAL DENSITY MATRIX

The experimental search by neutron scattering for the momentum condensate actually measures the density-density response function, $S(k,\omega)$. At very high momentum transfers $S(k,\omega)$ can be calculated with the aid of the impulse approximation and becomes proportional to the momentum distribution. However, $S(k,\omega)$ converges exceedingly slowly to this limit. Agreement between experiment and theory is only obtained with the use of more accurate expressions for these final state effects. For a detailed discussion see other contributions in this volume and their references. One of the major inputs to the final state effect calculations is the probability that a spectator atom carries off some of the momentum of the neutron which is proportional to the expectation value of

$$\sum_n a^\dagger_{k+q} a^\dagger_{n-q} a_n a_k \qquad (5)$$

where a and $a^\dagger$ are destruction and creation operators in momentum space. The Fourier transform of this operator can be written in coordinate space as the probability distribution that the two ends of the polymer are at positions $\mathbf{r}_1$ and $\mathbf{r}_1'$ and one other atom is at position r_2. Let us define $n_2(r_1, r_1', r_2; \beta)$ as that distribution function, normalized so that the integral over r_2 is the single particle off-diagonal density matrix, $n_1(r_1, r_1'; \beta)$.

We have calculated n_2 with Path Integral Monte Carlo. The full analysis of this function and its consequences for analysis of final state effects in neutron scattering will appear elsewhere. Here we will simply compare to a simple statistical approximation, similar in spirit to the superposition approximation of classical liquids. Let $g(r)$ be the radial distribution function, the average density a distance r from one atom divided by the total density. Then at small and large r_1-r_1' it is reasonable to write:

$$n_2^A(r_1, r_1', r_2; \beta) = n_1(r_1, r_1'; \beta)\,[g(r_2-r_1)g(r_2-r_1')]^{1/2} \quad (6)$$

To compare n_2 with n_2^A we have fixed the end-to-end distance of the polymer, r_1-r_1', at 4 A and averaged over the angle between r_1-r_1' and r_1-r_2. Shown in Fig. (5) are the approximate result

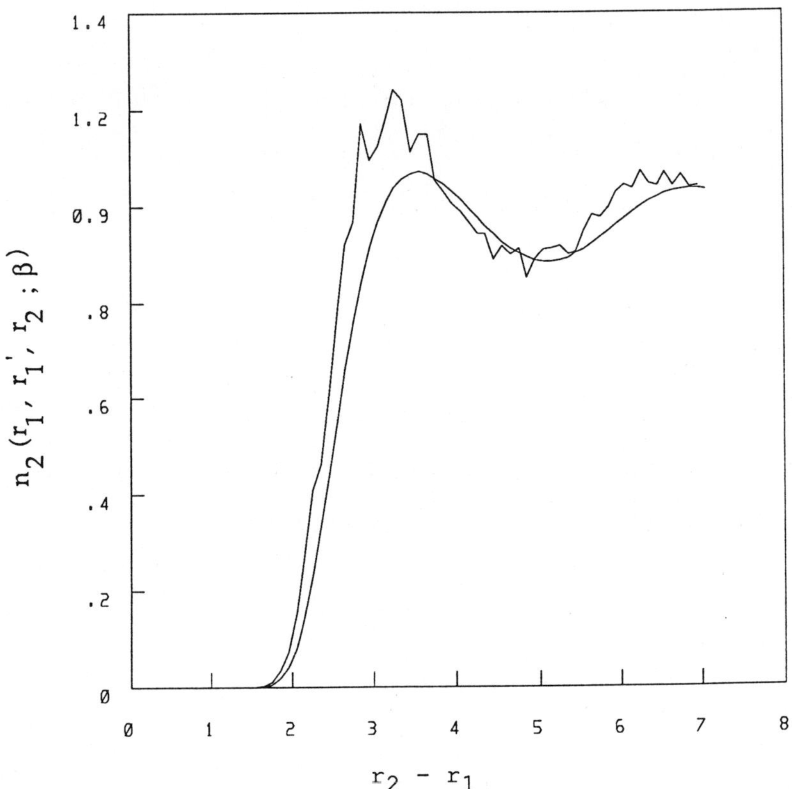

Figure 5. The sphericalized two partide density matrix at a density of $0.022/A^3$ and temperature 1.2K with $r_1 - r_1' = 4A$. The upper curve with noise is the result of PIMC. The smooth curve is the result of Eq. (6).

78

(the curve without noise) and the exact result (with statistical noise) for the spherical average of n_2 around $\mathbf{r}_1$ at a temperature of 1.2K and at SVP. The approximation is fairly accurate except at the first peak where it is too small by about 10%. Similar results are obtained at other values of $\mathbf{r}_1-\mathbf{r}_1'$ and at other temperatures.

CONCLUSIONS

The path integral method is able to calculate quite accurate momentum distributions for bose systems at finite temperature. At the present time there are no significant discrepancies with respect to experiment.

ACKNOWLEDGEMENTS

The simulations of helium films were done in collaboration with E. L. Pollock. The calculations of n_2 and the momentum distribution at high density were supported by a grant of computer time on the CRAY-X/MP at the National Resource for Supercomputing Applications, University of Illinois Urbana-Champaign.

REFERENCES

1. D. M. Ceperley and E. L. Pollock, Phys. Rev. Letts. **56**,351 (1986),
2. D. M. Ceperley and E. L. Pollock Can. J. of Physics,**65**,1416 (1987).
3. R. P. Feynman, Statistical Mechanics (Benjamin, Reading, Mass. 1972).
4. E. L. Pollock and D. M. Ceperley, Phys. Rev. **B30**,2555(1984).
5. E. L. Pollock and D. M. Ceperley, Phys. Rev. **B36**,8343(1987).
6. O. Penrose and L. Onsager, Phys. Rev. **104**,576(1956).
7. P. Whitlock, and R. M. Panoff, Can. J. of Phys.**65**,1409(1987).
8. P. Sokol, Can. J. of Phys.**65**,1393(1987).
9. R. O. Simmons, Can. J. of Phys. **65**,1401 (1987).
10. D. M. Ceperley and E. L. Pollock, Phys. Rev.**B**, in press,(1988).
11. R. B. Panoff and P. Whitlock, this volume.

MOMENTUM DISTRIBUTION IN LIQUID ^{4}HE AT LOW TEMPERATURES: VARIATIONAL METHODS

Efstratios Manousakis

Department of Physics and
Center for Materials Research and Technology
Florida State University, Tallahassee, Florida 32306

ABSTRACT

We review variational calculations of the momentum distribution and condensate fraction of atoms in liquid ^4He at zero and low temperatures. We use ground state and excited state wave functions which include Jastrow, three-body and backflow correlations and are obtained from variational calculations that give accurate energies and pair distribution functions over a wide density range. We calculate the change of the ground state momentum distribution due to creation of an elementary excitation and use it to calculate the change of the momentum distribution and condensate fraction at low temperatures ($T < 1^\circ K$). This calculation brings out the collective and single-particle character of the excitations in the long and short wavelength limit respectively and the interplay between the two at intermediate momenta. The expectation values are calculated by means of cluster expansions and making use of the hypernetted-chain equation and the scaling approximation to include the contribution of the elementary diagrams. We discuss the singularities of the momentum distribution at low momenta and low temperatures. We compare our results with momentum distributions obtained from Green's function Monte Carlo calculations and neutron scattering data.

1. INTRODUCTION

A great deal of theoretical effort has been invested towards a microscopic understanding of the ground state of dense quantum fluids in terms of the bare interaction between the constituent particles. Liquid helium is unique for it remains liquid even at absolute zero where the quantum coherence has fundamental consequences and for the simplicity of the atom-atom interaction. Five decades ago, London[1] proposed that the underlying mechanism, which drives the superfluid transition in the isotope liquid ^4He, is the Bose-Einstein condensation. It was already sixties when, after the suggestion of Miller, Pines and Nozières[2] that the condensate could be probed by neutron scattering experiments, Hohenberg and Platzman[3] proposed a definite deep inelastic neutron scattering experiment. The neutrons at high momentum transfers are expected to see the individual helium atoms distributed according to the micro-scopic momentum distribution. The fraction of atoms occupying the zero momentum state if any, would be a signature of the order parameter characterizing the superfluid transition. During the more recent years, various experimental techniques, which are reviewed in this workshop, have been employed to measure the momentum distribution in helium liquids. In this paper, we review the variational results for the momentum distribution obtained with the most accurate potentials and the best variational wave functions available.

Extensive calculations for liquid ^4He by Kalos et al.,[4] using the HFDHE2 in-teratomic potential of Aziz et al.,[5] and the Green's-function Monte Carlo (GFMC) method show that this is the most realistic potential available to study properties of condensed phases of ^4He.

The initial variational calculations used the Jastrow ansatz for the ground state wave function

$$\psi_0 = \prod_{i<j}^{N} f(r_{ij}), \tag{1.1}$$

where N is the number of particles; the pair correlation function $f(r)$ takes into account the short-ranged and long-ranged correlations between the atoms. The best variational results[6,7,8] obtained with (1.1) for the binding energy and equilibrium density of Bose liquid ^4He are $\sim 20\%$ off the presumably exact GFMC results. A more accurate wave function which includes optimized Jastrow and three-body correlations [Jastrow plus triplet $(J+T)$] is the following

$$\psi_0 = \prod_{i<j}^{N} f(r_{ij}) \prod_{i<j<k}^{N} f_3(r_{ij}, r_{jk}, r_{ik}). \tag{1.2}$$

The Jastrow part of the wave function is an optimized pair correlation found by solving the Euler-Lagrange equations with the asymptotic behavior[7]

$$f(r \to \infty) = 1 - \frac{mc}{2\pi^2 \hbar \rho} \frac{1}{r^2}, \tag{1.3}$$

which is a consequence of the long-wavelength phonons[9]. Here c is the velocity of sound. The three body correlation is written as

$$f_3(r_{ij}, r_{jk}, r_{ik}) = exp\left(-\frac{1}{2}\sum_{cyc}\sum_{l=0,1,2} \xi_l(r_{ij})\xi_l(r_{ik})P_l(\hat{r}_{ij}.\hat{r}_{ik})\right). \tag{1.4}$$

Here $\sum_{cyc}$ represents a sum of the three terms obtained by replacing ijk with jki and kij. P_l represents Legendre polynomials of lth order and $\hat{r}_{ij}$ are unit vectors. The ξ_l's are functions of the distance between the pairs of particles in the triplet and are determined variationally. The $l=1$ describes Feynman-Cohen[10] backflow correlations in the ground state, as suggested in Ref. 11, and gives the dominant contribution to the energy; the $l=0$ gives a small contribution and the $l=2$ has a negligible effect[8]. The ground state energy as a function of the density of liquid ^4He is shown in Fig.1. The squares with the error bars are GFMC results with the Aziz

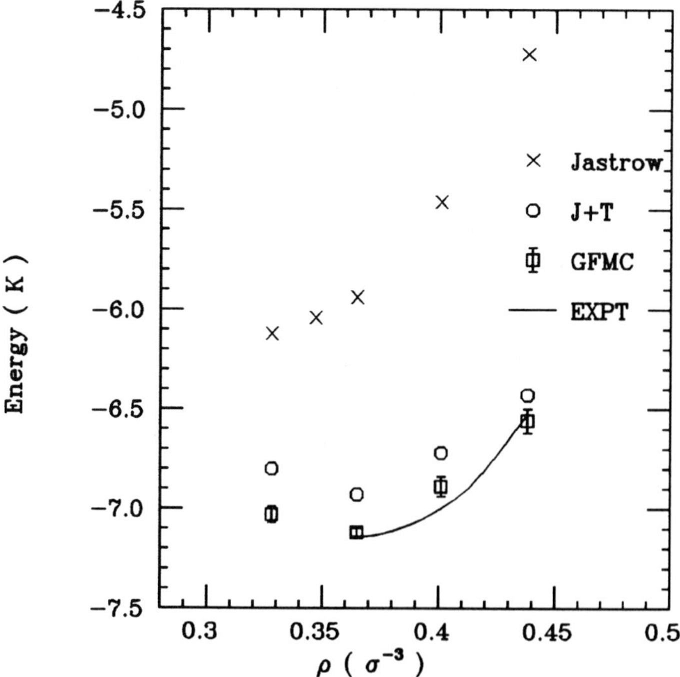

Figure 1.   The ground state energy for several densities calculated with Jastrow (crosses), Jastrow plus triplets (open circles) is compared with the GFMC (open squares) results and experiment (solid line).

interaction and they are very close to the experimental curve (solid line). The crosses represent the results obtained with the Jastrow wave-function (1.1).

The open circles obtained with the $J + T$ wave function are within 4% of the GFMC energies. The same wave function gives ground state energies for droplets of liquid ^4He which are also within $\sim 4\%$ of the presumably exact GFMC energies. The structure function obtained with the $J + T$ wave function is in excellent agreement with the GFMC results and experiment. We conclude that the Aziz potential and the approximate wave function (1.2) give good quantitative description of the ground state of liquid ^4He.

The nontrivial task in the variational calculations is the accurate evaluation of the expectation values with the wave functions (1.1) and (1.2). In Ref. 7 and 8 the Hypernetted-Chain-Scaling (HNC/S) approximation was introduced where scaling techniques were employed to include the contribution of the elementary diagrams. The results obtained with this technique are within $< 1\%$ agreement with the Monte Carlo integration techniques.

Wave functions and energies of elementary excitations in liquid ^4He have been studied by many authors. The first variational wave function of an elementary excitation with momentum $\vec{k}$ was the Bijl-Feynman ansatz

$$\psi_{\vec{k}} = \sum_{i=1}^{N} e^{i\vec{k}.\vec{r}_i}\psi_0. \tag{1.5}$$

83

An improved wave function

$$\psi_{\vec{k}} = \rho_B(\vec{k})\psi_0, \tag{1.6}$$

with

$$\rho_B(\vec{k}) = \sum_{i=1}^{N} e^{i\vec{k}.\vec{r}_i}\left(1 + i\sum_{j\neq i}^{N} \vec{k}.\vec{r}_{ij}\eta(r_{ij})\right) \tag{1.7}$$

was proposed by Feynman-Cohen[9] to take into account the backflow current. The backflow function $\eta(r)$ is determined by minimizing the excitation energy $e(k)$. We have carried out detailed variational calculations[12] of the spectrum of elementary excitations with the $\rho_B(\vec{k})$ and the ψ_0 which includes $J + T$. The best variational energies obtained with these wave functions are accurate in the phonon region up to excitation energies $e(k) \sim 8°K$, but they are $\sim 20\%$ above the experimental values in the maxon-roton region. We developed a perturbation expansion using correlated basis functions (CBF) generated by the $\rho_B(\vec{k})$ operators. The second order corrections to the excitation spectrum improve the agreement with the experiment significantly in the roton-maxon region. The same method was extended to calculate the dynamic structure function[13] at low and intermediate momentum transfers. We found peaks corresponding to one-quasiparticle and two-quasiparticle excitations. The location of the peaks and their strength are in semiquantitative agreement with those known from neutron inelastic scattering experiments.

In the next section of this paper, we outline the important features of the calculation of the ground state momentum distribution $n_0(p)$ and condensate fraction of atoms in liquid ^4He as obtained[14] with the variational wave function (1.2) and the HNC/S technique. In section 3, we review calculations[15] of the change $\delta n_{\vec{k}}(\vec{p})$ in the momentum distribution of the ground state due to creation of a single elementary excitation in the liquid. The collective and single-particle character of the excitations in the long and short wavelength limit as well as the interplay between the two at intermediate momenta is revealed. In section 5, we use $\delta n_{\vec{k}}(\vec{p})$ to calculate the momentum distribution at low temperatures due to thermally created elementary excitations in the liquid. The calculated correction is expected to be accurate at temperatures less than $1°K$. In section 5 we present our results at $T = 0$ obtained with the realistic variational wave function (1.2) and the finite temperature corrections. We discuss the singularities of the momentum distribution at low momenta which must be taken into account in the analysis of neutron scattering experiments to obtain accurate estimates of the condensate fraction. Our results are compared with momentum distributions obtained from GFMC calculations and neutron scattering data.

2. MOMENTUM DISTRIBUTION AT ZERO TEMPERATURE

The momentum distribution of helium atoms when the liquid is in the state ψ_x is defined as

$$n_{\vec{x}}(p) \equiv\, <\psi_{\vec{x}}|a^\dagger_{\vec{p}}a_{\vec{p}}|\psi_{\vec{x}}> \tag{2.1}$$

where $a^\dagger_{\vec{p}}$ creates a helium atom in the momentum state $\vec{p}$. $\psi_{\vec{x}}$ may be the ground state ψ_0 or the phonon-maxon-roton excitation $\psi_{\vec{k}}$. $n_{\vec{x}}(p)$ may be obtained from the Fourier transform of the one-particle density matrix

$$\rho_{\vec{x}}(\vec{r}_{11'}) \equiv N\frac{\int \psi^*_{\vec{x}}(\vec{r}_1{}',\vec{r}_2,...,\vec{r}_N)\psi_{\vec{x}}(\vec{r}_1,\vec{r}_2,...,\vec{r}_N)d^3r_2...d^3r_N}{\int |\psi_{\vec{x}}(\vec{r}_1,\vec{r}_2,...,\vec{r}_N)|^2d^3r_1...d^3r_N}. \tag{2.2}$$

Namely

$$n_{\vec{x}}(\vec{p}) = \int \rho_{\vec{x}}(\vec{r}_{11'})e^{i\vec{p}\cdot\vec{r}_{11'}}d^3r_{11'}. \tag{2.3}$$

Using the wave function (1.2) for the ground state $n_0(p)$ is obtained as

$$n_0(p) = n_c N \delta_{p,0} + n_0'(p), \tag{2.4a}$$

$$n_0'(p) \equiv \int \left(\rho_0(r) - \rho_0(r \to \infty) \right) e^{i\vec{p}\cdot\vec{r}}d^3r \tag{2.4b}$$

where n_c is the condensate fraction and $n_0'(p)$ represents the momentum distribution at finite p. The Kronecker-δ, which represents the $p = 0$ condensate is a result of the off-diagonal long-range order in the one-body density matrix

$$\rho_0(r_{11'} \to \infty) = n_c\rho. \tag{2.5}$$

Here ρ denotes the particle density.

TABLE I

$n_0(p)$ at the equilibrium density $\rho = 0.365\sigma^{-3}$.

P	$n_0(p)$	P	$n_0(p)$
0.015	5.01	0.65	0.38
0.035	2.36	0.85	0.29
0.050	1.81	1.05	0.22
0.055	1.66	1.25	0.15
0.075	1.33	1.45	0.10
0.095	1.14	1.65	0.058
0.115	1.02	1.85	0.033
0.135	0.93	2.05	0.024
0.155	0.86	2.25	0.020
0.175	0.81	2.45	0.015
0.195	0.76	2.65	0.009
0.215	0.73	2.85	0.006
0.235	0.70	3.05	0.003
0.250	0.69	3.25	0.002
0.450	0.50	3.35	0.001

The one-particle density matrix has been calculated in Ref. 14 using the $J + T$ wave function. Both n_c and $\rho_0(r_{11'})$ are functions of the usual and auxiliary distribution and nodal functions as well as elementary diagrams which are summed using the HNC/S technique. The accuracy of this approximation in calculating $\rho_0(r_{11'})$ was demonstrated[14] in the simpler case of Jastrow wave functions where excellent agreement was found with existing results of MC integration. Here, skipping the technical details of the calculation, we review some of the main results with the realistic $J + T$ wave function.

It may be verified that the asymptotic behavior (1.3) reflects the following singularity in $n_0(p)$

$$n_0(p \to 0) = n_c \frac{mc}{2\hbar} \frac{1}{p}. \tag{2.6}$$

In Table I we give the calculated $n'_0(p)$ at the equilibrium density $\rho = 0.365\sigma^{-3}$. We find that the value of the condensate fraction at this density is $n_c = 0.092$. For the density dependence of $n'_0(p)$ see Ref. 14. The results are compared with GFMC and neutron scattering data in section 5.

In the next section, we discuss the change in $n_0(p)$ due to creation of an elementary excitation in the liquid and we use it to calculate the changes in $n_0(p)$ due to thermally created excitation at low T.

3. STRUCTURE OF THE ELEMENTARY EXCITATIONS

We define the change in the momentum distribution caused by the creation of a single elementary excitation of momentum $\vec{k}$ in liquid ^4He as

$$\delta n_{\vec{k}}(\vec{p}) \equiv n_{\vec{k}}(\vec{p}) - n_0(p) \tag{3.1}$$

where both $n_0(p)$ and $n_{\vec{k}}(\vec{p})$ are defined by eq.(2.1) using ψ_0 and $\psi_{\vec{k}}$ for $\psi_{\vec{x}}$ respectively. In Ref. 15 $\delta n_{\vec{k}}(\vec{p})$ is calculated using the Feynman-Cohen wave function and the approximation (1.2) for ψ_0. Note that both $n_{\vec{k}}(\vec{p})$ and $n_0(p)$ are of order 1 while their difference is of order of $1/N$. A cluster expansion of the density matrices (2.2) in terms of the correlation functions is performed and the following irreducible form is obtained[15] for $\delta n_{\vec{k}}$:

$$\delta n_{\vec{k}}(\vec{p}) = \frac{1}{N}\left(-t_0(k)n_0(p) + t_+(k)n_0(|\vec{p} - \vec{k}|) + t_-(k)n_0(|\vec{p} + \vec{k}|) + t'(\vec{k}, \vec{p}) \right). \tag{3.2}$$

The $t_0(k)$, $t_+(k)$ and $t_-(k)$ are functions of the Fourier transforms of the usual and the auxiliary distribution and nodal functions obtained by the hypernetted-Chain-Scaling technique. $t'(\vec{k}, \vec{p})$ is the sum of diagrams which depend both on k and p and the angle between them in a non-trivial way. The definition of these functions is given in Ref. 15 in detail. The function $n_0(x)$ entering in the above expression is the ground state momentum distribution. There are three distributions of particles centered around $\vec{p} = 0$, $\vec{k}$ and $-\vec{k}$.

The asymptotic tail (1.3) of $f(r)$ gives the following long-wavelength limits:

$$t_0(k \to 0) = \frac{1}{2S(k)} + O(1) \tag{3.3a}$$

$$t_+(k \to 0) = \frac{1}{4S(k)} + O(1) \tag{3.3b}$$

$$t_-(k \to 0) = \frac{1}{4S(k)} + O(1) \tag{3.3c}$$

where the static structure function $S(k)$ becomes

$$S(k \to 0) = \frac{\hbar}{2mc}k. \tag{3.3d}$$

Hence in the long-wavelength limit $t_0(k)$, $t_+(k)$ and $t_-(k)$ diverge whereas t' remains finite. For finite k, there are correction terms of order 1. A long-wavelength phonon removes $1/2S(k)$ particles from the ground state $n_0(p)$ and divides them equally into two distributions $n_0(|\vec{p} - \vec{k}|)$ and $n_0(|\vec{p} + \vec{k}|)$ centered at $\vec{p} = \vec{k}$ and $\vec{p} = -\vec{k}$. The $n_0(p)$ has a singular term $Nn_c\delta_{p,0}$ and hence a phonon has $n_c/4S(k)$ particles each in states with $\vec{p} = \pm\vec{k}$. The change in the kinetic energy $\delta(KE)$ of the liquid on exciting a phonon can be obtained from $\delta n_{\vec{k}}(\vec{p})$ in the $k \to 0$ limit as

$$\delta(KE)(k \to 0) = \frac{\hbar^2 k^2}{2m}\frac{1}{2S(k)} \tag{3.4}$$

which is half of the total phonon energy obtained with both the Bijl-Feynman and Feynman-Cohen wave functions in the $k \rightarrow 0$ limit. Thus, a phonon in Bose liquids is, indeed, a pure harmonic vibration with half of its energy from kinetic and the other half from potential terms.

In the short-wavelength limit $(k \rightarrow \infty)$

$$t_0(k \rightarrow \infty) = t_+(k \rightarrow \infty) = 1 \qquad (3.5a)$$
$$t_-(k \rightarrow \infty) = t'(k \rightarrow \infty) = 0. \qquad (3.5b)$$

i.e, a single particle is removed from the ground state distribution and is put in the distribution $n_0(|\vec{p} - \vec{k}|)$ centered at $\vec{p} = \vec{k}$. In this limit the energy of the Feynman excitation is $\hbar^2 k^2/2m$ and equals the change in the kinetic energy.

In Fig. 2a, we give $t_0(k)$, $t_+(k)$ and $t_-(k)$ as calculated for finite k using Eq. (1.6) and (1.2) for $\psi_{\vec{k}}$ and ψ_0 respectively and in the HNC/S approximation.

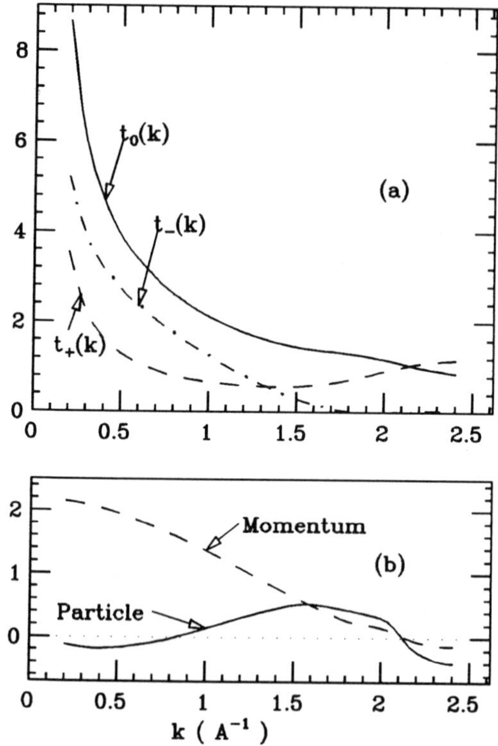

Figure 2. (a) The functions $t_0(k)$, $t_+(k)$ and $t_-(k)$. (b) The contributions $t'_n(k)$ and $t'_m(k)$ of the t' term to the particle and momentum conservations. k is the momentum of the excitation.

If we neglect the term t' for a moment, the creation of a single elementary excitation of momentum $\vec{k}$ removes $t_0(k)$ particles from the ground state momentum distribution;

$t_+(k)$ of them are distributed around $\vec{k}$ and $t_-(k)$ of them around $-\vec{k}$ with the ground state $n_0(x)$ with $x = |\vec{p} - \vec{k}|$ and $x = |\vec{p} + \vec{k}|$ respectively. This is not an accurate statement, however, for the term t' plays important role in the particle and momentum conservation

$$V \int \frac{d^3 p}{(2\pi)^3} \delta n_{\vec{k}}(\vec{p}) = 0, \qquad (3.6a)$$

$$V \int \frac{d^3 p}{(2\pi)^3} \delta n_{\vec{k}}(\vec{p}) \vec{p} = \vec{k} \qquad (3.6b)$$

where V is the volume of the system. These conservations imply the following sum rules:

$$t_0(k) = t_+(k) + t_-(k) + t'_n(k) \qquad (3.7a)$$
$$t_+(k) - t_-(k) + t'_m(k) = 1 \qquad (3.7b)$$

where

$$t'_n(k) = \int \frac{d^3 p}{(2\pi)^3 \rho} t'(k, p, \theta_p) \qquad (3.8a)$$

$$t'_m(k) = \frac{1}{k} \int \frac{d^3 p}{(2\pi)^3 \rho} t'(k, p, \theta_p) p \cos\theta_p. \qquad (3.8b)$$

The calculated values of $t'_n(k)$ and $t'_m(k)$ are plotted in Fig.2b. We see that the "background" term t' does not give significant contribution to the particle conservation (i.e, $t'_n(k)$ term); its contribution to the momentum conservation, however, is very important at small k. Since $t_0(k)$ measures the particles removed from the $n_0(p)$ due to creation of the excitation, we may say that for $k < 2\text{Å}^{-1}$, i.e for phonon-maxon states, more than one bare particles are involved in the excitation. In the creation of a maxon, for example, approximately $t_0(k \sim 1\text{Å}^{-1}) \sim 2$ bare particles participate. In the long-wavelength limit the number of participating particles diverges as $1/k$. For $k >> 2\text{Å}^{-1}$ a Feynman-Cohen excitation is constructed by only a single particle.

4. MOMENTUM DISTRIBUTION AT LOW TEMPERATURES

The expression (3.2) may be used to calculate the change $\delta n(T, p)$ of the ground state momentum distribution $n_0(p)$ due to thermally created excitations at low temperatures. At low temperatures ($T < 1°K$) the liquid can be thought of as a dilute gas of noninteracting elementary excitations, therefore the momentum distribution $n(T, p)$ is approximately given as

$$n(T, p) = n_0(p) + \delta n(T, p), \qquad (4.1a)$$

$$\delta n(T, p) = V \int \frac{d^3 k}{(2\pi)^3} \frac{1}{\exp(\beta e(k)) - 1} \delta n_{\vec{k}}(\vec{p}) \qquad (4.1b)$$

where β is the inverse temperature and $e(k)$ is the spectrum of the elementary excitations calculated with $\psi_{\vec{k}}$ in Ref. 12 and 13. Using Eq. (3.2) we may write

$$\delta n(T, p) = \delta n_1(T, p) + \delta n_2(T, p) + \delta n_3(T, p), \qquad (4.2a)$$

$$\delta n_1(T, p) = n_0(p) \int \frac{d^3 k}{(2\pi)^3 \rho} \frac{1}{\exp(\beta e(k)) - 1} t_0(k), \qquad (4.2b)$$

$$\delta n_2(T, p) = \int \frac{d^3 k}{(2\pi)^3 \rho} \frac{1}{\exp(\beta e(k)) - 1} (t_+(k) + t_-(k)) n_0(|\vec{p} - \vec{k}|), \qquad (4.2c)$$

$$\delta n_3(T, p) = \int \frac{d^3 k}{(2\pi)^3 \rho} \frac{1}{\exp(\beta e(k)) - 1} t'(\vec{k}, \vec{p}). \qquad (4.2d)$$

In this temperature range ($T < 1°K$) only long-wavelength phonons can be excited, i.e excitations with $k < 0.2\mathring{A}^{-1}$. In Ref. 12 and 13, we found that the wave function (1.6) is accurate for such excitations i.e with energies $e(k < 0.2\mathring{A}^{-1}) < 4°K$. In this temperature range, we may neglect the δn_3 term because the integral of t' is small for $k < 0.2\mathring{A}^{-1}$. In this range of k and T, we can further approximate $e(k \to 0) = \hbar c k$ and $t_0(k)$, $t_+(k)$ and $t_-(k)$ by the expressions (3.3). Therefore, we obtain

$$\delta n_1(T,p) \simeq -\frac{m}{12\hbar^3 \rho c} T^2 n_0(p) = -\left(\frac{T}{T_0}\right)^2 n_0(p), \tag{4.3a}$$

$$\delta n_2(T,p) = \int \frac{d^3 k}{(2\pi)^3 \rho} \frac{1}{\exp(\beta e(k)) - 1} \frac{mc}{\hbar c} n_0(\vec{p} - \vec{k}), \tag{4.3b}$$

$$n(T,p) = \left(1 - \left(\frac{T}{T_0}\right)^2\right) n_0(p) + \delta n_2(T,p). \tag{4.3c}$$

where $T_0 \sim 7.6K$. Thus, the temperature dependence of the condensate fraction at low T is given by

$$n_c(T) = n_c(0)\left(1 - \left(\frac{T}{T_0}\right)^2\right). \tag{4.4}$$

The last equation has also been derived by a phenomenological approach[16] and from the structure of perturbation theory at finite T[17].

The term $N n_c \delta_{p,0}$ in $n_0(p)$ gives rise to terms in δn_2 which have singular behavior in the $p \to 0$ limit. These singular terms are

$$\delta n_s(T,p) = \frac{n_c(0)mc}{\hbar p} \frac{1}{\exp(\beta e(k)) - 1} \simeq \frac{n_c(0)m}{\hbar^2 \beta} \frac{1}{p^2} - \frac{n_c(0)mc}{2\hbar} \frac{1}{p} + \dots \tag{4.5}$$

$n_0(p)$ has exactly the same $1/p$ singularity with opposite sign (see Eq. (2.6)). Thus, for $\beta \hbar c p << 1$, the $1/p$ term in $n_0(p)$ is canceled by that in $\delta n(T,p)$. This cancelation has also been pointed out by Griffin[18]. The present method can provide $\delta n(T,p)$ for small T but for any p. In Ref. 15 we give tables of the change $\delta n(T,p)$ for several values of T and p obtained with the full $\delta n_{\vec{k}}(\vec{p})$. It appears that the expression

$$n(T,p) \simeq \left(1 - \left(\frac{T}{T_0}\right)^2\right) n_0(p) + \frac{n_c(0)mc}{\hbar p} \frac{1}{\exp(\beta \hbar c k) - 1} \tag{4.6}$$

is a good approximation for low T and it may be used in neutron scattering experiments to find the contribution from the singular terms. The $\delta n(T, p \neq 0)$ is positive in this temperature range; thus, it appears that the atoms are removed from the $p = 0$ condensate and placed in states with $\hbar c p \lesssim \pi T$.

5. RESULTS AND COMPARISONS

In Fig. 3 we compare the results of our calculations of the momentum distribution at the equilibrium density with the GFMC[19] and the neutron scattering data[20]. The solid line is our $n_0(p \neq 0)$ obtained with the wave function (1.2) which includes optimized Jastrow and three-body correlations. In these results the $n(p)$ is normalized such that

$$\int n(p)d^3 p = 1. \tag{5.1}$$

The dashed-dotted curve is the result of the GFMC calculation obtained from a simulation of 64 particles in a periodic box. The differences at low p and the small differences at intermediate and higher p between the variational $n_0(p)$ and that obtained by the GFMC calculation may be attributed either a) to the approximate nature of the

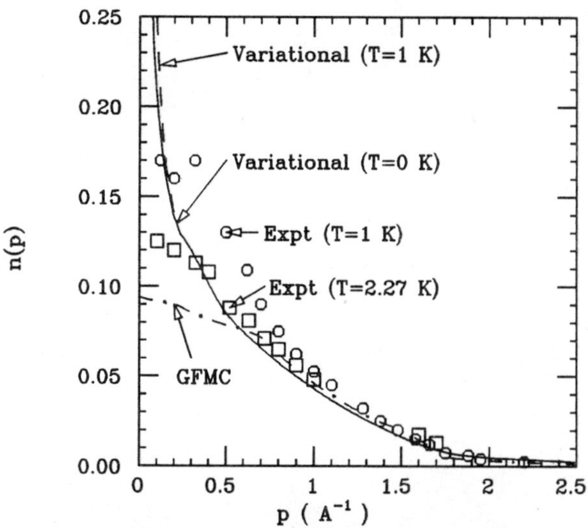

Figure 3.   The momentum distribution at $p \neq 0$ obtained from the variational calculation at $T = 0$ (solid line) is compared with the GFMC and experimental results. The dashed line is obtained from the variational calculation at $T = 1°K$.

variational wave function or to the HNC/S approximation or b) to finite size-effects in the GFMC calculation which does not produce the correct $1/p$ singularity at low p. Because in the GFMC calculation the particles are forced to remain in a finite box, they are pushed at higher momenta. The dashed line in Fig. 3 shows the variational $n(p, T = 1°K)$ and gives the magnitude of the change in $n_0(p)$ due to finite and low temperature effects. The $n(T, p)$ for any low temperature $(T < 1°K)$ can be approximately obtained from the Eq. (4.6) using $n_0(p)$ from table I. More accurately, one can use the tables given in Ref. 15. Eq. (4.6) may still be useful in neutron scattering experiments, for its simplicity. The open circles and open squares are the experimental $n(p)$ obtained at $T = 1°K$ and just above the λ transition at $T = 2.27°K$ respectively. The difference between the $n(p)$ at $T = 1°K$ and at $T = 2.27°K$ is attributed[20] to scattering from the condensate and the singularities in $n(p)$ in the $p \to 0$ limit. The theoretical $n(p)$ has a δ-function singularity at $p = 0$ due to the condensate and the $1/p$ and $1/p^2$ singularities in the $p \to 0$ limit. The difference at low p, between the theoretical and the experimental $n(p)$ at $T = 1°K$ is mainly due to the broadening in the experimental $n(p)$ of all these singularities. When $n(p)$ is extracted from the measured $S(q, \omega)$ using the impulse approximation the final state interactions have been neglected. Furthermore, the validity of the impulse approximation for systems with hard core interactions has been criticized[21].

In Fig. 4 the variational n_c at zero temperature as a function of density ρ is given by the open circles (solid line). The open squares are the GFMC results[19]. At the equilibrium density $\rho = 0.365\sigma^{-3}$, the variational $(J + T)$ and GFMC results are the same. The GFMC value of n_c at the intermediate density is about 25% below the variational, but they are close at the highest density $\rho = 0.438\sigma^{-3}$. The triangle is the experimental result of Ref. 22 available only at the equilibrium density. The crosses are obtained from neutron scattering measurements of Sokol el al.[23]. All the lines are drawn as guides to the eye. It is interesting that the density dependence of the variational results and that of the experimental results of Sokol et al., is linear and they are in crude agreement. However, the density dependence of Mook's data[24], approximately represented here by the dashed-dotted line, is much stronger.

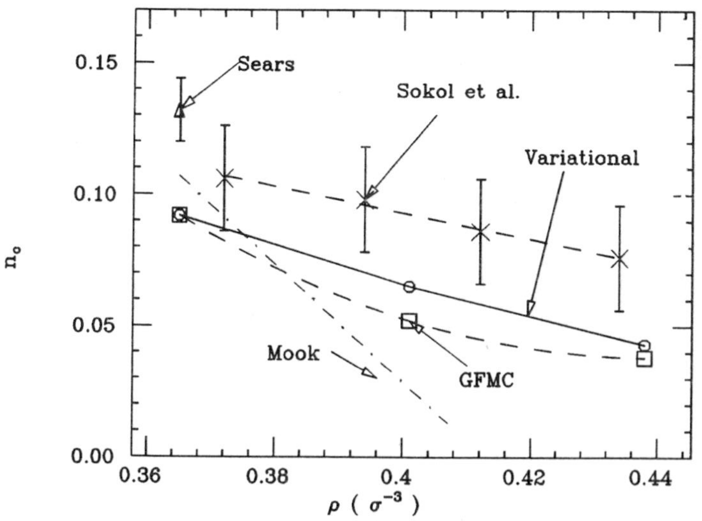

Figure 4. The condensate fraction as a function of density as obtained from the variational calculation (solid line) compared with GFMC and experimental results. The lines are guides to the eye.

At $T < 1°K$, the condensate fraction depends weakly on temperature. Using Eq.(4.4) and taking $n_c(0) = 0.092$ we find $n_c(T = 1°K) = 0.090$, which is a negligible difference.

6. ACKNOWLEDGEMENTS

This work was supported in part by the Florida State University Supercomputer Computations Research Institute which is partially funded by U.S Department of Energy through contract No DE-FC05-85ER250000.

REFERENCES

1. F. London, Nature (London) **141**, 643 (1938); Phys. Rev. **54**, 947 (1938).
2. A. Miller, D. Pines, and P. Nozières, Phys. Rev. **127**, 1452(1962).
3. P. C. Hohenberg and P. M. Platzman, Phys. Rev. **152**, 198(1966).
4. M. H. Kalos, M. A. Lee, P.A. Whitlock and G. V. Chester, Phys. Rev. **B 24**, 115 (1981).
5. R. A. Aziz, V. P. S. Nain, J. S. Carley, W. L. Taylor and D. T. Mcconvill, J. Chem. Phys. **70**, 4330 (1979).
6. K. Schmidt, M. H. Kalos, M.A. Lee and G. V. Chester, Phys. Rev. Lett. **45**, 573 (1980).
7. Q. N. Usmani, B. Friedman and V. R. Pandharipande, Phys. Rev. **B 25**, 6123(1983).
8. Q. N. Usmani, S. Fantoni, and V. R. Pandharipande, Phys. Rev. **B 26**, 6123(1983).
9. L. Reatto and G. V. Chester, Phys. Lett. **22**, 276(1966).
10. R. P. Feynman and M. Cohen, Phys. Rev. **102**, 1189(1956).
11. V. R. Pandharipande, Phys. Rev. **B 18**, 218(1978).
12. E. Manousakis and V. R. Pandharipande, Phys. Rev. **B 30**, 5062(1984).
13. E. Manousakis and V. R. Pandharipande, Phys. Rev. **B 33**, 150(1986).

14. E. Manousakis and V. R. Pandharipande and Q. N. Usmani, Phys. Rev. **B 31**, 7022(1985).
15. E. Manousakis and V. R. Pandharipande, Phys. Rev. **B 31**, 7029(1985).
16. R. A. Ferrell , N. Menyhard, H. Schmidt, F. Schwabl and P. Szepfalusy, Ann. Phys. (N.Y) **47**, 365(1968).
17. K. Kehr, Z. Phys. **221**, 291(1969).
18. A. Griffin, Phys. Rev. **B 30**, 5057(1984).
19. P. Whitlock and R. M. Panoff, Can. J. Phys. **65**, 1409(1987).
20. V. F. Sears, E. C. Svensson, P. Martel and A. D. B. Woods, Phys. Rev. Lett. **49**, 279(1982).
21. J.J. Weinstein and J. W. Negele, Phys. Rev. lett. **49**, 1016(1982).
22. V. F. Sears, Phys. Rev. **B 28**, 5109(1983).
23. P. E. Sokol, R. O. Simmons, R. O. Hilleke, and D. L. Price, unpublished.
24. H. A. Mook, Phys. Rev. Lett. **51**, 1454(1983) and as quoted in Table 4 of Ref. 19.

IMPULSE APPROXIMATION AND CORRECTIONS

X AND Y SCALING

Geoffrey B. West

Theoretical Division, T-8, MS B285
Los Alamos National Laboratory
Los Alamos, NM 87545

I INTRODUCTION

Until the discovery almost twenty years ago that the structure functions of nucleons, as measured in deep-inelastic electron scattering exhibited a point-like scaling behavior, the idea that hadrons really were bound states of quarks was not generally taken seriously.[1] Quarks somehow were thought of as fictitious objects carrying certain quantum numbers which could explain the general features of the hadronic spectrum.[2] In spite of the great success of this "naïve" quark model, quarks as real hard objects which could be "seen" in electron scattering were not given much credence by the cognoscenti. Indeed the "real" nucleon was still thought of as a "bare" nucleon surrounded by a mesonic cloud and so was expected to respond as a soft spongy object when probed by a hard scattering. Thus the observation of a point-like scaling behavior in deep-inelastic electron scattering played a crucial rôle in establishing quarks as "real" objects from which all hadronic matter is constructed.[1,3] Indeed these experiments completely re-oriented our thinking about strong interactions and opened the way to re-establishing quantum field theory as the paradigm for describing all fundamental interactions. Since that time enormous progress has been made and it is fair to say that we now have an almost universally accepted realistic quantum field theory describing all the fundamental interactions with the possible exception of gravity.[4] Quarks enter as a fundamental fermionic field with strong interactions mediated by massless vector bosons, called gluons. This strong interaction part of the theory is called quantum chromodynamics (QCD) because it closely mimics QED. The crucial difference is that the gluons, as well as the quarks, carry the fundamental "charge" of the theory, here called color. In QED, the photons are not, of course, electrically charged (unlike the electrons) and so do not directly interact with each other. It is generally believed that it is this curious complexity, namely that gluons are colored (and therefore self-interact) that ultimately leads to the confinement of quarks (and more generally of color itself).

For, the great paradox surrounding the quark model is that even though quarks are undoubtedly the fundamental constituents of hadrons and, as already emphasized, have in fact been "seen" in deep inelastic experiments, it seems impossible to isolate them in the laboratory. Over the years many arguments based on QCD have been proposed for explaining this apparent paradox. Although none are truly convincing and we await the ultimate rigorous proof that "infrared slavery" can co-exist with "asymptotic freedom" most believe that such a proof is indeed a consequence of QCD.[4]

Prior to the development of QCD there was no serious framework allowing one to address the sorts of questions raised by confinement and scaling. In the naïve quark model a nucleon was typically thought of much like a nucleus, namely as non- relativistic constituents bound by some potential. This actually gives a remarkably good description of the hadronic spectrum, as well as other low energy phenomena such as decay rates.[5] It is therefore not unreasonable to address the scaling vs. confinement paradox within this theoretical lab. of non-relativistic many-body theory. It is as if we were to suppose that the nucleons in a nucleus were bound to each other by a confining potential such as an harmonic oscillator, for example, and were to ask whether the non- relativistic structure functions, as measured in deep inelastic electron scattering, scale; and, if so, how would the confinement mechanism manifest itself? It was out of considerations such as these that the phenomenon of y-scaling in non-relativistic systems such as nuclei and liquids arose.[3]

It is ironic that, although much of the intuition for interpreting the high energy data as scattering from structureless constituents came from nuclear physics (and to a lesser extent atomic physics) virtually no data existed for nuclear targets in the non-relativistic regime until relatively recently. It is therefore not so surprising that, in spite of the fact that the basic nuclear physics has been well understood for a very long time, the corresponding non-relativistic scaling law was not written down until <u>after</u> the relativistic one, relevant to particle physics, had been explored.[3] Of course, to the extent that these scaling laws simply reflect quasi-elastic scattering of the probe from the constituents, they contain little new physics once the nature of the constituents is known and understood. On the other hand, <u>deviations from scaling</u> represent corrections to the impulse approximation and can reflect important dynamical and coherent features of the system. Furthermore, as will be discussed in detail below, the scaling curve itself represents the single particle momentum distribution of constituents inside the target. **It is therefore prudent to plot the data in terms of a suitable scaling variable since this immediately focuses attention on the dominant physics.** Extraneous physics, such as Rutherford scattering in the case of electrons, or magnetic scattering in the case of thermal neutrons is factored out and the use of a scaling variable (such as y) automatically takes into account the fact that the target is a bound state of well-defined constituents.

In this talk I shall concentrate almost entirely on non-relativistic systems. Although the formalism applies equally well to both electron scattering from nuclei and thermal neutron scattering from liquids, I shall, because of my own background, usually be thinking of the former. On the other hand I shall completely ignore spin considerations (as well as relativistic corrections) so, ironically, the results actually apply more to the latter case!

I shall first show how the classic sum rules of Bethe and Heisenberg already suggest the possibility of scaling. In order for scaling to be a useful tool, one must clearly have some idea as to how it is approached as one increases $\underline{q}^2$. To this end we carry out a dynamical calculation to show how corrections can be evaluated in terms of the basic inter-nucleon potential. Indeed, as a byproduct of this I shall show that scaling follows even if the potential is confining; that is, the system behaves in a free manner even though the constituents are forever bound. This is not, of course, relevant for nuclei but may have some bearing on the relativistic hadronic problem. Apart from these potential corrections to scaling there is also the very physical effect of correlations; i.e., the system will not behave in a quasi-free fashion until $|q|$ exceeds the coherence length in the system. Finally, there is a further constraint to scaling from nuclear systems and that is the eventual effect of meson production. By the time q is large enough for scaling to dominate, it may also be large enough for pions to be produced. Since we shall consider only nucleon degrees of freedom and remain in the non-relativistic domain our calculations ignore this contribution. Thus our "predictions" are only useful provided one remains in a region where pion production is negligible. It should be remarked that in particle physics, the analog to pion production is gluon production (radiative corrections) and in QCD these are responsible for the phenomenon of asymptotic freedom which (up to log log corrections!) justifies the simple impulse approximation (the "quark-parton" model), where gluons can be ignored![4,6]

I   Cross-Section Formulae

We begin by considering spinless non-relativistic scattering from a target composed of Z scattering centers such as is the case of a nucleus or of a macroscopic liquid. The formalism that I shall review applies in fact almost precisely to the case of thermal neutron scattering from liquids. In general, the process to be discussed is illustrated in Fig. 1: the scattered probe particle (an electron, say) is detected without regard to the fate of the target final states. In terms of the energy loss (ν) and momentum transfer (q) it is convenient to introduce the structure function (appropriate to Coulomb scattering).

$$W(\nu, q^2) \equiv \frac{(d^2\sigma/d\Omega\, dE)}{(d\sigma/d\Omega)_{Ruth}} \tag{1}$$

$(d\sigma/d\Omega)_{Ruth}$ is the classical Rutherford scattering cross-section for structureless particles. [For liquids one factors out the cross-section for the magnetic interaction of the neutron probe with the constituent atoms; for such cases the symbol S is used for W and ω for ν]. From the Fermi golden rule W is given by

$$W(\nu, q^2) = \sum_f |< \Psi_f| \sum_{i=1}^{z} Q_i e^{iq\cdot r_i} |\Psi_0 > |^2 \delta(E_f - E_o + \nu) \tag{2}$$

where Q_i is the charge of the i'th constituent whose position is r_i. $\Psi_{0(f)}$ is the initial (final) state of the target. [For the purpose of this talk I shall assume that the constituent is structureless; to include its structure, one simply replaces Q_i by the relevant elastic form factor]. Using the Heisenberg equations of motion together with the completeness of the set of final states f (i.e. the conservation of probability) one can express (2) as a ground state expectation value

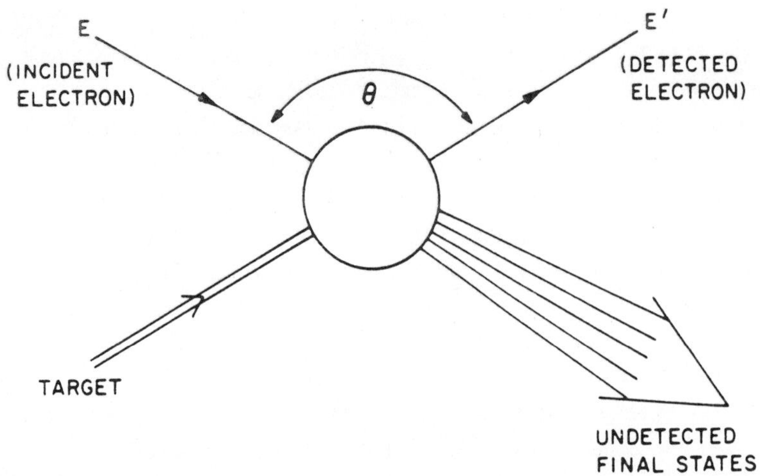

Figure 1. General graph illustrating inclusive scattering from an arbitrary target.

$$W(\nu, q^2) = \mathrm{Im}$$

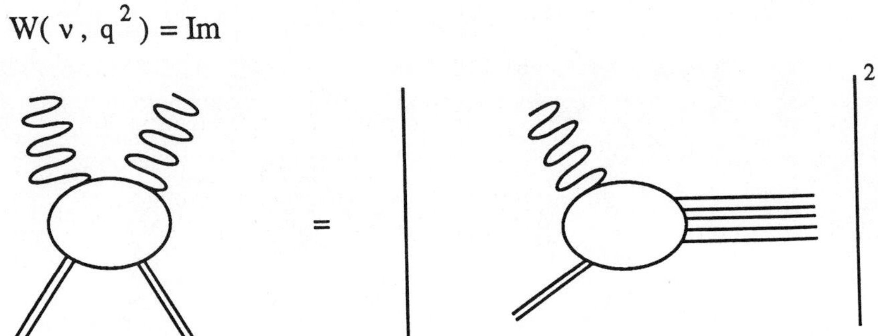

Figure 2. The Optical Theorem relating W to the imaginary past of the forward virtual Compton amplitude.

$$W(\nu, q^2) = \int_{-\infty}^{\infty} \frac{dt}{2\pi} e^{i\nu t} < \Psi_0 | \sum_{i,j} Q_i Q_j e^{i q \cdot r(t)} e^{-i q \cdot r(o)} | \Psi_0 > \tag{3}$$

The price paid for eliminating the sum over final states is the need for knowledge of the time development of $r_i(t)$. This, of course, is governed by the Hamiltonian of the system whose general structure is taken to be

$$H = -\sum_i \frac{\nabla_i^2}{2\mu} + V(r_1, \cdots\cdots, r_z) \tag{4}$$

where μ is the mass of the constituents. Although we will not need to do this in what follows, it is usually assumed that the potential V can be expressed as a sum of 2-body potentials.

$$V(\underline{r}_1 \cdots \underline{r}_z) = \sum_{i<j} v(\underline{r}_i - \underline{r}_j) \tag{5}$$

Indeed this usually leads to a 2nd quantized many-body description in terms of creation-destruction operators $a_{\underline{k}}$:

$$W(\nu, q^2) = \int_{-\infty}^{\infty} \frac{dt}{2\pi} e^{i\nu t} < \Psi_0 |[\rho_q(t), \rho_q(o)]|\Psi_0 > \tag{6}$$

The density operator is given by

$$\rho_q \equiv \sum_{\underline{k}} a_{\underline{k}+q}^{+} a_{\underline{k}} \tag{7}$$

Its time development is controlled by the Hamiltonian, eq. (4) which, in this formalism, can be expressed as

$$H = \sum_{\underline{k}} \frac{k^2}{2\mu} a_k^{+} a_k + \frac{1}{2} \sum_{k} v(\underline{k}) \rho_{\underline{k}}^{+} \rho_{\underline{k}} \tag{8}$$

$v(\underline{k})$ is just the Fourier transform of the 2-body potential $v(r)$ defined through eq. (5). This field theoretic description of eqs. (2) and (3) allows one to think of W as the imaginary part of the corresponding (virtual) photon, forward Compton scattering amplitude as illustrated in Fig. 2. The question we wish to address is what is the behaviour of W when q becomes very large?

II   Sum Rules and x-Scaling

Historically, sum rules have played an important rôle in the development of scaling results. Let us briefly review how they are derived and how they can be used to motivate scaling behaviour. Returning to the representation (3) it is clear that

$$I(q^2) \equiv \int_{-\infty}^{\infty} d\nu W(\nu, q^2) = \sum_{ij=1}^{z} \langle \Psi_o | Q_i Q_j e^{iq \cdot (\underline{r}_i - \underline{r}_j)} | \Psi_o \rangle$$

$$= \sum_{i=1}^{z} Q_i^2 + \sum_{ij=1}^{z} Q_i Q_j \langle \Psi_o | e^{iq \cdot (\underline{r}_i - \underline{r}_j)} | \Psi_0 \rangle \tag{9}$$

Notice that by integrating over ν we now only require the coordinates r_i and r_j at the same time so that, as operators, they now commute. Effectively, this means that by forming the sum rule, the explicit dynamics is integrated out. It is convenient to separate the double sum over i and j into incoherent and coherent contributions; thus

$$I(q^2) = \sum_{i=1}^{z} Q_i^2 + \sum_{ij=1}^{z} Q_i Q_j \langle \Psi_0 | e^{iq\cdot(r_i - r_j)} | \Psi_0 \rangle \tag{10}$$

Notice that for identical particles with $Q_i = 1$,

$$I(q^2) \rightarrow \begin{cases} z & \text{when } q^2 \rightarrow \infty \\ z^2 & \text{when } q^2 = 0 \end{cases} \tag{11}$$

showing how the two extreme regimes pick out the incoherent from the coherent. In general, the **sum rule has integrated out the explicit dependence on dynamics so that the approach to scaling for $I(q^2)$ is completely governed by correlations alone.**

Now, below elastic threshold where the target stays intact and recoils (in the Lab. frame) with energy $q^2/2M$, W necessarily vanishes; thus, for $\nu < q^2/2M$, $W = 0$. The asymptotic form of the sum rule therefore reads

$$\lim_{q^2 \rightarrow \infty} \int_{q^2/2M}^{\infty} d\nu\, W(\nu, q^2) \approx \sum_{i=1}^{z} Q_i^2 \tag{12}$$

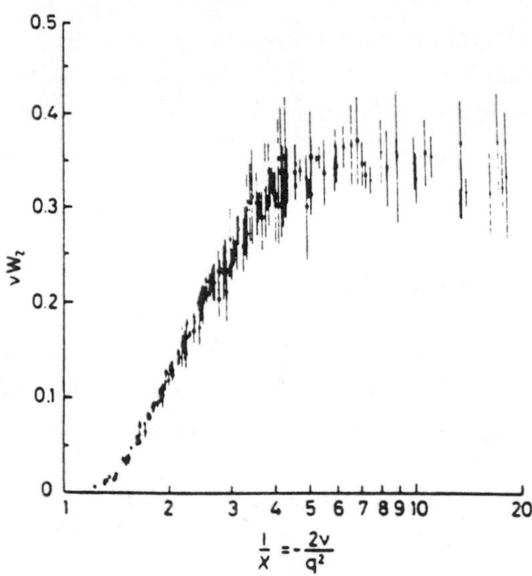

Figure 3. Plot of μW vs. $1/x$ for various values of q^2 taken from early SLAC data. The data lies on a universal scaling curve.

It is obviously convenient to scale q^2 out of the limits on the integral by introducing the non-relativistic version of the Bjorken scaling variable [7]

$$x \equiv \frac{q^2}{2M\nu} \tag{13}$$

In that case, eq. (12) becomes

$$\lim_{q^2 \to \infty} I(q^2) = \lim_{q^2 \to \infty} \int_o^1 \frac{dx}{x} \nu W(\nu, q^2) = \sum_{i=1}^{z} Q_i^2 \tag{14}$$

i.e. $I(q^2)$ must eventually be independent of q^2. Since, by definition $W > 0$, this suggests (but certainly does not prove) that the integrand, namely

$$F(q^2, x) \equiv \nu W(q^2, \nu) \tag{15}$$

itself becomes independent of q^2 and a function of x alone. This is Bjorken x-scaling and is illustrated in Fig. 3. Note that the data shown there is taken in a relativistic domain with q^2 being the square of the 4-momentum transferred.

III   Dynamical Origin of Non-Relativistic x-Scaling

The "derivation" of x-scaling from the sum rule (14) is no more than heuristic since to prove that $F(q^2, x)$ scales would require showing that <u>every</u> moment sum rule of F exists and is independent of q^2. In fact, in relativistic quantum field theory based on the QCD Lagrangian, scaling is proven via a consideration of all such moments. [4,6] In that case one actually show that all the moments of F deviate from constants by small, but calculable, powers of $\ln q^2$: thus

$$M_n(q^2) \equiv \int_0^1 dx\, x^{n-2}\, F(x, q^2) \tag{16}$$

$$\approx \frac{C_n}{(\ln q^2)^{\gamma_n}} \tag{17}$$

where the γ_n are known. Thus exact Bjorken scaling is actually violated logarithmically although in a predicted pattern. Fig. 4 shows a plot of M_n^{1/γ_n} vs. $\ln q^2$ for $n = 3, 4, 5$ and 6, together with the theoretical predictions which are (unambiguously) calculated from QCD. The agreement is remarkably good and is one of the main reasons why QCD is believed as <u>the</u> theory of the strong interactions. The case $n = 2$ is special in that the sum rule reflects energy-momentum conservation which forces $\gamma_2 \equiv 0$. Furthermore, unlike the general case, C_2 is also calculable; in fact, one finds

$$C_2 = \left(\frac{N_q}{N_q + N_g} \right) \sum_{i=1}^{N_f} \frac{Q_i^2}{N_f} \tag{18}$$

where $N_{q(g)}$ is the number of quarks (gluons) in the theory and N_f the number of flavors. With $N_g = 8$ [i.e. the colour group is SU(3)] and with 6 flavors of quarks (up, down, strange, charm, bottom and top) one predicts $C_2 = 5/34 \approx 0.147$. Fig. 5 shows the data for n=2 again showing remarkably good agreement with the prediction from QCD. Actually, the agreement is even more satisfying when one realizes that QCD also predicts that the scaling constant C_2 should be approached logarithmically from <u>above</u>.

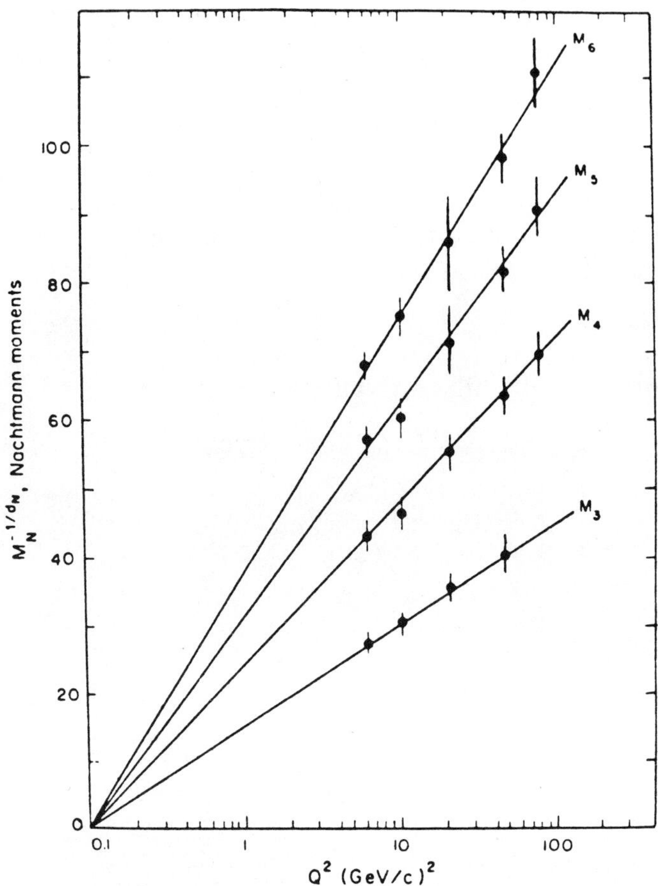

Figure 4. Graph showing agreement of QCD predictions for $M_n(q^2)$ with the experimental data [see Eqs. (16) and (17)].

The general pattern of scale-violation therefore is that each of the moments vanishes with increasing q^2 subject to the constraint that the area under the scaling curve (i.e. the $n = 2$ moment) remains fixed. For F itself this therefore translates into a pattern illustrated in Fig. 6.

Returning to the non-relativistic systems under discussion it would be nice to derive scaling directly for the structure function itself. To do so, let us examine the first-quantized expression, Eq. (3). It is worth remarking that although much formal and phenomenological work has been based upon the 2nd quantized representation, Eq. (6), (especially for liquids) it turns out to be much more convenient for our purposes to stay with the first quantized version.

From a judicious use of operator identities one can re-express Eq. (3) in the form [8]

$$W(\nu, q^2) = \int_{-\infty}^{\infty} \frac{dt}{2\pi} e^{i[\nu - q^2/2\mu]t} \langle \Psi_0 | \sum_{ij=1}^{z} Q_i Q_j e^{iq \cdot (\underline{r}_j - \underline{r}_i)} T e^{iq/\mu \cdot \int_0^t \underline{p}_i(t')\,dt'} | \Psi_0 \rangle \qquad (19)$$

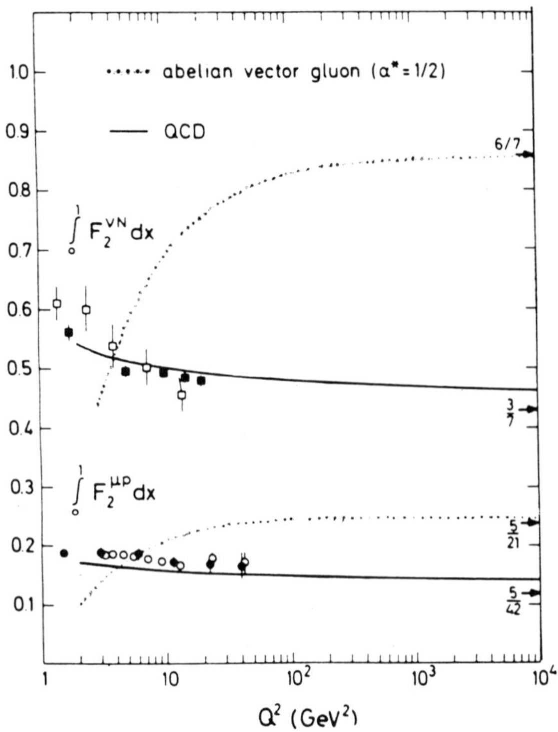

Figure 5. The behavior of the n=2 moment vs. q^2 for neutrino and muon scattering showing approach to a constant from above.

Here, p_i is the momentum operator for the ith constituent and the coordinates $\underline{r}_j$ and $\underline{r}_i$ are both evaluated at the same time $t = 0$; T stands for the usual time-ordering operator for the phase integral. If we express this as a function of q^2 and x rather than q^2 and ν by using the definition (13), then the first phase-factor in (19) becomes

$$i\left(\nu - \frac{q^2}{2\mu}\right) t = -i \frac{q^2 t}{2\mu x}\left(x - \frac{\mu}{M}\right) \qquad (20)$$

The second phase factor $e^{iq \cdot (\underline{r}_j - \underline{r}_i)}$ distinguishes between the incoherent ($i = j$, where it ceases to oscillate) and the coherent ($i \neq j$); this was discussed in the context of the sum rule following Eq. (9). For large q^2 we need only keep the incoherent piece where $\sum Q_i Q_j \rightarrow$

$\sum Q_i^2$. The final phase factor (which involves the momentum operator p_i) remains $O(q)$ when $q^2 \to \infty$ at fixed x. In this limit the first factor, Eq. (20) therefore dominates since it is $O(q^2)$. **Thus, as with sum rule, all explicit dependence on dynamical operators eventually disappears in the Bjorken limit** and we are left with

$$F(q^2, x) \;\equiv\; \nu W(q^2, \nu)$$

$$\approx \; \left(\sum_i Q_i^2\right) \delta\left(x - \frac{\mu}{M}\right) \tag{21}$$

In other words νW does indeed scale to a function of x only but the result is independent both of dynamics and the structure of the target! The sum rule (14) is thereby satisfied in an essentially trivial fashion: by choosing this set of variables, one is effectively making the scattering appear as if from completely static constituents. This is therefore not a particularly useful way of presenting the data for non-relativistic systems since one learns little directly about the dynamics or momentum distribution of the target. As already indicated above, such is not the case for the relativistic situation.

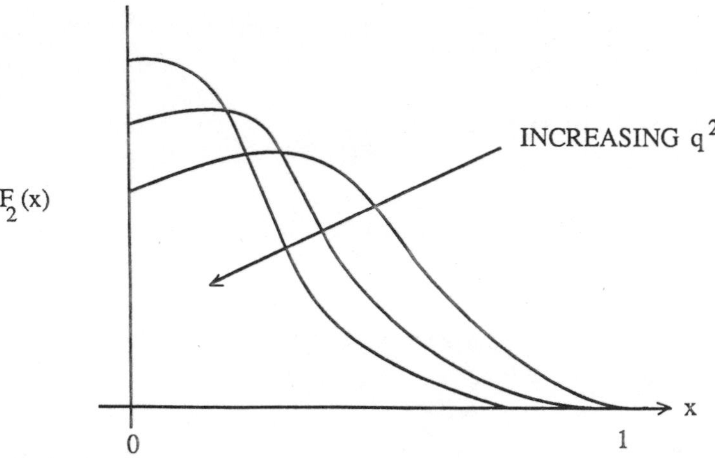

Figure 6. Pattern of scaling violation predicted by QCD.

IV y-Scaling

We saw in the previous section that, although νW can be expected to scale to a function of x only, the result is relatively uninteresting as far as extracting useful information about the target is concerned. As already intimated above, the reason is that in the Bjorken limit the quasielastic piece in (19) grows like q^2 [as explicitly shown in (20)] whereas the dynamical piece only grows like q. To extract dynamics we need to make both of these pieces of comparable magnitude when $q \to \infty$. This is the rationale for y-scaling.

To further motivate y-scaling it is useful to use the equation of motion:

$$\frac{dp_i}{dt} = i[H, p_i] = - \nabla_i V(r_1, \cdots r_z) \equiv F_i \tag{22}$$

where F_i is the force felt by the ith constituent due to the presence of all other constituents. This can be formally solved to read

$$p_i(t) = p_i(o) + \int_o^t dt' F_i(t') . \tag{23}$$

When substituted into (19) this leads to the following exact expression:

$$W(\nu, q^2) = \langle \Psi_0 | \sum_{i,j=1}^{z} Q_i Q_j e^{iq \cdot (r_i - r_j)} T \int_{-\infty}^{\infty} \frac{dt}{2\pi} e^{i\frac{qt}{\mu}[\mu y - p_{iz} - \int_o^t (1 - t'/t) F_{iz}(t') dt']} | \Psi_0 \rangle \tag{24}$$

Here we have defined the z-direction as that of q and introduced the dimensionless variable

$$y \equiv \frac{2\mu\nu - q^2}{2\mu q} \tag{25}$$

If we now think in terms of $q \to \infty$ at fixed y (rather than x) then the static quasielastic ($q^2 = 2\mu\nu$) and dynamical contributions have equal weight. To see how this leads to y-scaling introduce $\beta \equiv qt$ (and $\beta' \equiv qt'$) then the incoherent part of Eq. (24) can be re-expressed as

$$qW(\nu, q^2) = \langle \Psi_0 | \sum_{i=1}^{z} Q_i^2 T \int_{-\infty}^{\infty} \frac{d\beta}{2\pi} e^{i\beta[y - \frac{p_{iz}}{\mu} - \frac{1}{q} \int_o^\beta d\beta'(1 - \frac{\beta'}{\beta}) \frac{F_i(\beta'/q)}{\mu}]} | \Psi_0 \rangle \tag{26}$$

Apart from suppressing the coherent contribution which vanishes rapidly with q^2, this expression is exact. Now, if we take $q \to \infty$ at fixed y, it is clear that the term in the exponent containing F_i also eventually vanishes and we are left with

$$\mathcal{F}(y, q^2) \equiv qW(\nu, q^2) \to < \Psi_0 | \sum_{i=1}^{z} Q_i^2 \delta(y - \frac{p_{iz}}{\mu}) | \Psi_0 > \tag{27}$$

Thus F becomes a function of y only; its form clearly depends on the internal structure of the targets through the momentum operator p_i.

Like the result of x-scaling, y-scaling reflects incoherent quasi-elastic scattering from the constituents. However, whereas the former corresponds to a "static" snapshot of the target, the latter allows for a "dynamic" picture. To see this more explicitly let us express (27) in a momentum representation (in which k_i is the eigenvalue of p_i):

$$\mathcal{F}(y, q^2) \approx \sum_i Q_i^2 \int \frac{d^3 k_1}{(2\pi)^3} \cdots \frac{d^3 k_{\mathcal{Z}}}{(2\pi)^3} | < \Psi_0 | \underline{k}_1 \cdots \underline{k}_{\mathcal{Z}} > |^2 \delta(k_{iz} - \mu y). \tag{28}$$

If $|f(\underline{k}_i)|^2$ is a single-particle momentum distribution defined by

$$|f(k_i)|^2 \equiv \left[\int \frac{d^3 k}{(2\pi)^3} \right]_i | < \Psi_0 | \underline{k}_1 \cdots \underline{k}_i \cdots \underline{k}_{\mathcal{Z}} > |^2 \tag{29}$$

where the integration symbol means integrate over the momenta of <u>all</u> the constituents <u>except</u> the i'th, then (in the symmetric case), (28) reduces to

$$\mathcal{F}(y, q^2) \approx \mathcal{Z} \int \frac{d^2 k_\perp}{(2\pi)^3} \int_{-\infty}^{\infty} dk_z |f(k_\perp, k_z)|^2 \delta(k_z - \mu y) \tag{30}$$

Thus for large q^2, $\mathcal{F}(y, q^2) \equiv qW(\nu, q^2)$ scales to a function of y which measures the longitudinal momentum distribution of constituents inside the target.

It is clear from this discussion that the approach to $y-$ scaling is governed by correlations as well as explicit dynamics. The expression given in Eq. (24) allows for a systematic expansion in powers of $1/q$. [Actually, with some reasonable approximations, one can translate this into an expansion in power of $e^{1/q}$]. Thus the scaling phenomenon simply reflects the fact that the target can be well described by $\mathcal{Z}$ scattering centers. In this sense, it is the correction and the approach to scaling that contain the really interesting physics. On the other hand, in the high energy case discussed above where it was <u>not</u> known that hadrons were definitely composed of quarks, the scaling phenomena itself was the clearest evidence for the ultimate establishment of the quark model. It should be pointed out that the relativistic analogue of y-scaling is, in fact, x-scaling (even though x-scaling is "trivial" non-relativistically!) Thus in the relativistic Bjorken limit $F(q^2, x) \equiv \nu W(q^2, \nu)$ "measures" the momentum distribution of quarks inside hadrons much like $\mathcal{F}(q^2, y) \equiv qW(q^2, \nu)$ measures that of nucleons inside nuclei or atoms inside liquids.

Typical scaling curves for electron scattering from nuclear targets ($\lesssim GeV$ range) and for neutron scattering from liquids ($(\lesssim KeV$ range) are shown in Fig. 7. [9] When developed, the theoretical discussion above leads to many interesting results which are in agreement with these data. Some of these are the following:

V   Remarks and Conclusions

- Scaling results whether F_i is a confining force or not; this is clear from Eq. (26). Thus even for interparticle potentials $V(r) \sim r^n$ for $r \to \infty$, the system behaves as if the constituents were essentially free.

- Corrections arising from dynamical effects (i.e. from finite F_i/q effects) dominate those due to correlations. An expansion in this parameter leads to the conclusion that, for (non-relativistic nuclei), scaling should be approached from above as $q^2 \to \infty$; see Fig. 8. Furthermore, this correction should vanish at $y = 0$. These systematics are confirmed by the data - see Fig. 7.

- For a symmetric system one can derive the exact result that [3]

$$\mathcal{F}(0 , q^2) \equiv \langle \frac{\mu}{2k} \rangle \tag{31}$$

Taking $u(r) \sim e^{-\alpha r}/r$ as a single-particle wave function for a nucleus (where $\alpha \approx 170\,MeV$ is solely determined by the binding energy) leads to $\mathcal{F}(0) \sim 2$. This is again in reasonable agreement with the data showing how the overall normalization of the data can be understood.

- In terms of $\mathcal{F}$, the sum rule (12) reads

$$\int_{-\infty}^{\infty} dy \mathcal{F}(y , q^2) \approx \sum Q_i^2 \tag{32}$$

which, because $|\Psi_0\rangle$ is normalized to unity, is in agreement with Eq. (27).

- It is straightforward to derive other sum rules; for example

$$\int_{-\infty}^{\infty} dy y^2 \mathcal{F}(y , q^2) \approx (\sum Q_i^2) \frac{2}{3} \langle \frac{T}{\mu} \rangle \tag{33}$$

where $T \equiv \bar{p}^2/2\mu$ is the kinetic energy operator. Thus this second moment of $\mathcal{F}$ measures the mean kinetic energy of the constituents. More generally one can derive an infinite sequence of sum rules which relate moments of $\mathcal{F}$ to matrix elements of operators:

$$\int_{-\infty}^{\infty} dy y^{2n} \mathcal{F}(y , q^2) \approx (\sum_i Q_i^2) \langle \Psi_0| \sum_{m=0}^{\infty} a_m p^{2(n-m)} \left(\frac{F}{\mu q}\right)^m |\Psi_0\rangle \tag{34}$$

Although this is not particularly useful in non-relativistic systems where one can work directly with the original expression Eq. (24), its analogue in relativistic field theory is the key to progress, as briefly discussed in Section IV.

- In liquids (and even possibly in nuclei) the interparticle potential contains a hard core which means that F_i cannot be smoothly defined over all space. This requires special treatment—indeed, as discussed elsewhere in these proceedings, Silver [9] has suggested that, although scaling is not spoiled by the presence of hard-core potentials, the naïve interpretation in terms of momentum distribution needs to be revised. He has suggested a convolution structure which incorporates the hard-core as a final-state effect.

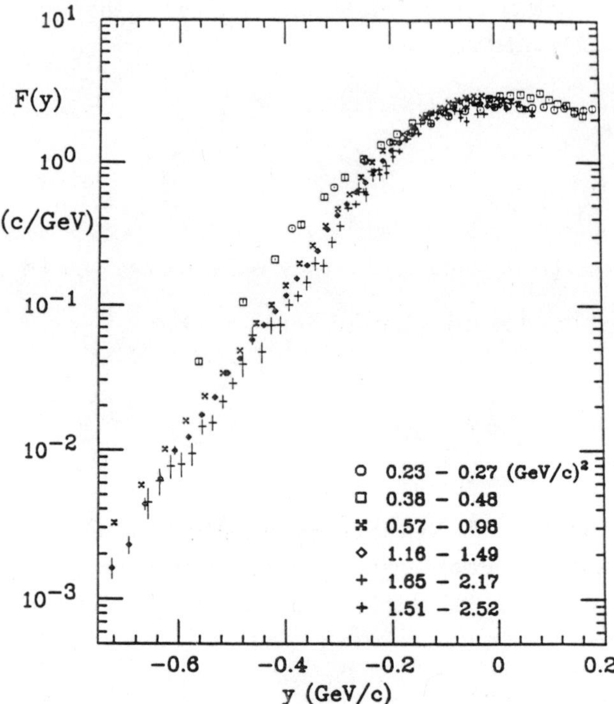

Figure 7. F(y) vs. y from electron scattering from nuclei.

- Perhaps one of the most peculiar aspects of the nuclear data is that, over a considerable range of y, $\mathcal{F}(y) \sim e^{-a|y|}$. There seems to be no straightforward reason why $\mathcal{F}(y)$ should exhibit such a simple structure.

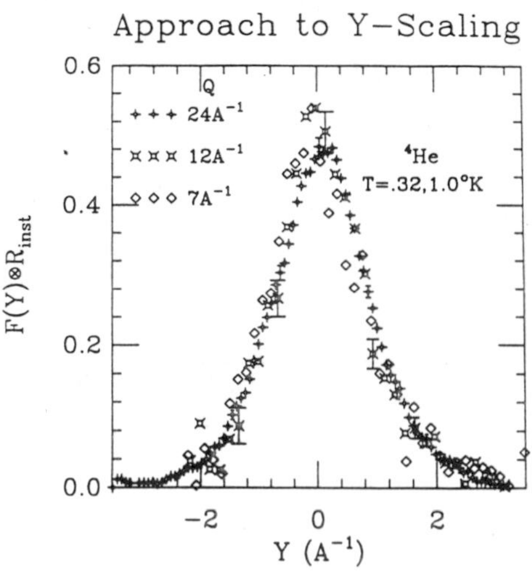

Figure 8. F(y) vs. y for scattering from liquid He⁴.

- Finally we should emphasize that relativistic corrections (to y-scaling) have been completely ignored . These are important for the nuclear case because much of the data is in a kinematic range where $|q| \gtrsim |GeV$ i.e. where $q^2/\mu^2 \gtrsim 0(1)$. Estimating these corrections in the region below which relativistic Bjorken scaling relevant to a quark-gluon description is a difficult business. Kinematic attempts in which relativistic kinematics is substituted in the essentially non-relativistic formulae are clearly not entirely justifiable even though they probably account for some important effects.[8] Unfortunately this clouds the issue of how one extracts meaningful information from the data. It is important to note that an added complication in the relativistic regime is that mesons can be produced. Since these are new degrees of freedom beyond the nucleonic ones, scaling is thereby broken, as is clearly seen in Fig. 8. Meson production is the origin of the scale violations seen for $y > 0$. It is amusing that the same problem occurs in the ultra- relativistic case where gluons are ultimately produced. However, it is the magic of QCD and its attendant property of asymptotic freedom that limits such scaling violations to be only logarithmic in nature. Indeed, as emphasized in Section IV, the precise nature of the logarithms are an "exact" prediction of the theory and their brilliant experimental confirmation is one of the main reasons that QCD is accepted today as <u>the</u> theory of the strong interactions!

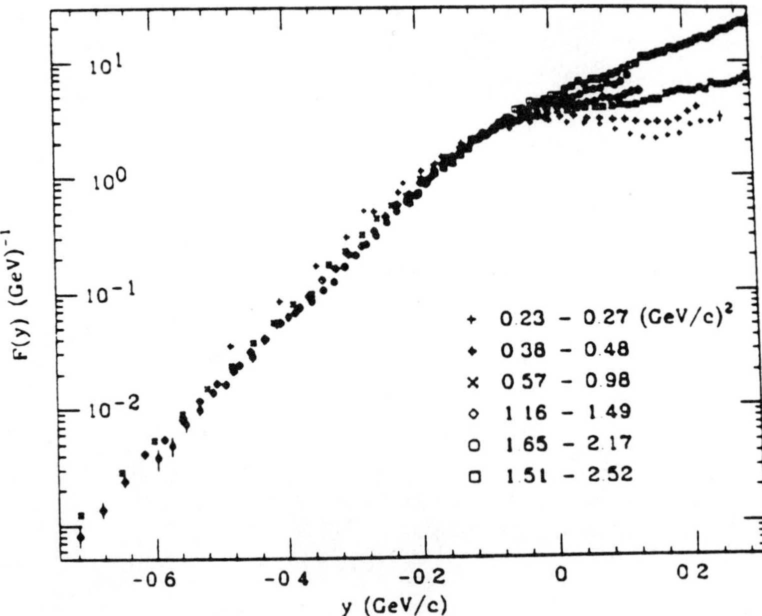

Figure 9. F(y) vs. y for a nuclear target showing the breaking of scaling for $y > 0$ arising from the production of new degrees of freedom (pions). In QCD, the analogous process (productions of gluons) induces only the mild logarithmic breaking of scaling indicated in Figs. 4 and 5; see Eqs. (16) and (17).

References

[1] For early reviews of the phenomenology see, e.g., F. Gilman, Phys. Rep. **4c**, 94 (1972) or R. P. Feynman "Photon-Hadron Interactions," (W.A. Benjamin, Reading, MA 1972). See also refs. [3] and [4].

[2] This is generally the attitude expressed in "The Eightfold Way," by M. Gell-Mann and Y. Ne'eman. (W.A. Benjamin, NY 1964).

[3] G. B. West, Phys. Rep. **18c**, 264 (1975).

[4] T-P. Cheng and L-F. Li, "Gauge Theory of Elementary Particle Physics Particle Physics," (Oxford Univ. Press, NY 1984).

[5] J. J. J. Kokkedee "The Quark Model," (W.A. Benjamin, NY 1969).

[6] H. D. Politzer, Phys. Rep. **14c**, 129 (1974).

[7] J. D. Bjorken and E. Paschos, Phys. Rev. **185**, 1975 (1969).

[8] G. B. West in "Electron and Pion Interactions with Nuclei at Intermediate Energies," (Eds. W. Bertozzi, S. Costa and C. Schaerf, Harwood Acad. Press, NY 1980) p. 417.

[9] See other contributions to this volume from R. Arnold, I. Sick and R. Silver.

[10] I. Sick, this volume.

FINAL STATE EFFECTS IN QUANTUM FLUIDS

Richard N. Silver

Theoretical Division and Los Alamos Neutron Scattering Center
MS B262 Los Alamos National Laboratory
Los Alamos, New Mexico 87545

INTRODUCTION

The extraction of momentum distributions from high energy scattering experiments depends on the validity of the impulse approximation (IA), which has been extensively discussed in the workshop overview [1]. In the IA

$$F_{IA}(Y) = \frac{Q}{M} S_{IA}(Q, \omega) = n_0 \delta(Y) + \frac{1}{4\pi^2 \varrho} \int_{|Y|}^{\infty} k\, n_k\, dk \quad . \qquad (1)$$

Here $Y = M\omega/Q - Q/2$ is the West [2] scaling variable, n_k is the momentum distribution, n_0 is the Bose condensate fraction in the case of superfluid 4He, and ϱ is the density. The IA asssumes that in the scattering process the kinetic energy imparted to a recoiling particle is large compared to the potential energy due to neighboring particles. While this may be true for x–ray Compton scattering from electronic systems, it is less true for quasielastic electron–nucleus scattering, and it is invalid for deep inelastic neutron scattering (DINS) from quantum fluids and solids such as helium. In the latter case, the interatomic potential [3] has a steeply repulsive core which is never negligible compared to the kinetic energies which can be imparted in feasible neutron scattering experiments. The corrections to the IA due to interactions of the recoiling atom with neighboring atoms are termed "final state effects" (FSE). This paper presents the theory of FSE for the case of deep inelastic neutron scattering from 4He quantum fluids. The lessons should also be applicable to momentum distribution experiments in many other systems.

The simplest theory for FSE was presented in the original application of the IA to quantum fluids by Hohenberg and Platzman [4] (HP) in 1966. They postulated that the IA was lifetime broadened due to collisions of the recoiling atom with neighboring atoms at rate $1/\tau = \varrho \sigma_{tot}(Q)Q/M$, where $\sigma_{tot}(Q)$ is the He–He total cross section. Then the neutron scattering law is a convolution

$$F(Y) \equiv \frac{Q}{M} S(Q, \omega) = \int_{-\infty}^{\infty} dY' \; R(Y - Y') \; F_{IA}(Y') \; , \qquad (2)$$

where the broadening predicted by HP is Lorentzian

$$R_{HP}(Y) = \frac{1}{\pi} \frac{\Gamma}{Y^2 + \Gamma^2} \; ; \qquad \Gamma(Q) \equiv \varrho \sigma_{tot}(Q) \; . \qquad (3)$$

The FSE broadening satisfies $\Delta Y_{FWHM} = 2\varrho\sigma_{tot}(Q)$ and Y–scales to the extent that $\sigma_{tot}(Q)$ is independent of Q. In fact, $\sigma_{tot}(Q)$ for 4He [2] decreases as $log\ Q$ with "glory" oscillations due to Bose statistics which decrease as $1/Q$. The observation of oscillations in $S(Q, \omega)$ for $Q \le 10A^{-1}$ [5] has been thought to provide confirmation of the HP theory. However, the ω^2–sum rule on $S(Q, \omega)$ requires

$$\int_{-\infty}^{\infty} F(Y) \; Y^2 \; dY = \frac{2M}{3} < K.E. > \quad \Rightarrow \quad \int_{-\infty}^{\infty} R(Y) \; Y^2 \; dY = 0 \; , \qquad (4)$$

which is violated by the Lorentzian tails predicted by the HP theory.

The solution of this problem was first discussed by Gersch and Rodriguez [6] (GR) in 1973. They developed a many–body theory for FSE using novel time–ordered cumulant expansions for $S(Q, \omega)$. Due to a difference in the description of FSE as dephasing rather than lifetime broadening, $R(Y)$ at small $|Y|$ should be positive but satisfy $\Delta Y_{FWHM} \approx \varrho\sigma_{tot}(Q)$, which is a factor of 2 smaller than the HP theory. $R(Y)$ should be negative at large $|Y|$ in order to satisfy the ω^2–sum rule. This is achieved by incorporating the physics of the real space correlations in the ground state wave function, which implies that the scattering rate at short recoil distances should be zero. In hindsight, we believe that the GR solution was qualitatively correct, and it is compared to the latest experiments in ref. [7]. However, it was largely ignored in the more than twenty subsequent papers on FSE [8], which can be traced to several problems: the cumulant methods are novel and have not been widely applied elsewhere; the paper predates the concept of Y–scaling and reliable many–body calculations of n_k; and, the numerical results were contained in an experimental paper [9] which is most notable for reporting an erroneously low 2% Bose condensate fraction in superfluid 4He.

In the past two years, the Quasiclassical Approximation (QCA) of Silver and Reiter [10] and the Hard Core Perturbation Theory (HCPT) of Silver [11] have provided new approaches to FSE which confirm and extend the original GR theory. The HCPT predictions for FSE have been successfully compared to recent pulsed neutron scattering experiments [12] in normal and superfluid 4He, which is further discussed by Sosnick, et al. [13]. The present paper outlines the physical ideas, the new theoretical approaches and the numerical predictions for FSE, while refering the reader to other papers for more detail.

QUASICLASSICAL APPROXIMATION

The simplest theory for FSE which includes the effect of real space correlations in the ground state is the QCA theory [10]. We begin with the standard expression

$$S(Q, \omega) = \frac{1}{\pi} \int_0^\infty dt\, e^{i\omega t} \; < e^{-i\vec{Q}\cdot\vec{x}(t)} e^{+i\vec{Q}\cdot\vec{x}(0)} > \quad . \tag{5}$$

for the incoherent part of the neutron scattering law, assuming the coherent part is negligible at high Q. Here, $<>$ denotes the ground state expectation value. Since the wavelengths are short, one can hope to apply semiclassical methods to evaluate Eq. (5). A quasiclassical approximation based on the Wigner distribution function [14] suggests that at high Q we can replace the quantum mechanical operator $\vec{x}(t)$ by a classical trajectory for particles of momentum $\vec{Q}/2 + \vec{p}$

$$\vec{x}(t) \to \frac{\vec{Q}t}{2M} + \frac{\vec{p}t}{M} + \delta\vec{x}(t) \quad , \tag{6}$$

where $\delta\vec{x}(t)$ is the difference between collisional and collisionless classical trajectories. Then

$$S(Q, \omega) \approx \frac{1}{\pi}\mathrm{Re}\int_0^\infty dt \int \frac{d^3\vec{p}}{(2\pi)^3} n(\vec{p})\, e^{it\left(\omega - \frac{Q^2}{2M} - \frac{\vec{Q}\cdot\vec{p}}{M}\right)} \ll e^{i\delta\vec{x}(t)\cdot\vec{Q}} \gg \quad . \tag{7}$$

Here $\ll \gg$ denotes an average over all such trajectories. If we ignore collisions, the IA, Eq. (1), is recovered from the $\delta\vec{x}(t) \to 0$ limit of Eq. (7). The QCA is also exact for neutron scattering from an harmonic oscillator. Further applications of the QCA are discussed by Reiter [15].

Consider then the effect of collisions for a single trajectory at $\vec{Q}$'s large compared to $\vec{p}$. In this case the phase factor, $\delta\vec{x}(t) \cdot \vec{Q}$, is zero for times less than the collision time, t_c, and it is very large for times greater than the collision time. Then,

$$e^{i\delta\vec{x}(t)\cdot\vec{Q}} \simeq \Theta(t_c - t) \quad , \text{ which implies } \quad \ll e^{i\delta\vec{x}(t)\cdot\vec{Q}} \gg \; \simeq P_{NC}(t) \tag{8}$$

when one averages over trajectories. Here, $\Theta(t)$ is a step function and $P_{NC}(t)$ is the probability of no collisions in time t. It is convenient to use scaled variables, $x=Qt/M$, which is the twice the distance along the trajectory for a particle of momentum $Q/2$. $P_{NC}(x)$ can be calculated from the collisions per unit distance along the trajectory, $\Gamma(x)$, which depends on the spatial distribution of particles as expressed by the radial distribution function, $g(r)$. Thus,

$$P_{NC}(x) = \exp\left(-\int_0^{x/2} \Gamma(x')dx'\right) \; ; \quad \Gamma(x) = \varrho\pi \int_0^{r_0^2} db^2\, g(r(x,b)) \quad . \tag{9}$$

Here, r_o is the effective hard sphere radius of a helium atom for the collision determined by the turning point of the classical trajectory, and b is the impact parameter for the collision. The $r(x,b)$, given by $r(x,b)^2 = b^2 + [x + \sqrt{r_o^2 - b^2}]^2$, is appropriate when x is defined as the distance between the edges of the particles, as illustrated in Fig. 1. Substituting Eq. (9) into Eq. (7) results in a physically appealing interpretation of $R(Y)$ as the Fourier transform of the probability of no collisions as a function of recoil distance, with the West scaling variable, Y, as the conjugate momentum, i.e.

$$R_{QCA}(Y) = \frac{\text{Re}}{\pi} \int_0^\infty dx\, e^{iYx}\, P_{NC}(x) \quad . \tag{10}$$

To the extent that r_o is independent of Q, QCA predicts a Y–scaling FSE broadening. In the limit of ignoring real space correlations, i.e. $g(r) \to 1$, the HP theory, Eq. (3), is recovered with $\Delta Y_{FWHM} \approx Q\sigma_{tot}$ and $\sigma_{tot} = \pi r_o^2$ for classical collisions. The inclusion of real space correlations has a dramatic effect, as illustrated in Fig. 2. When a particle recoiling from a neutron collision begins its classical trajectory, its initial position is far away from the hard cores of the potential due to neighboring atoms which are responsible for final state effects. It must travel some distance before it encounters a hard core, so there are no collisions at small x. Thus $P_{NC}(x)$ is one for small x, and all of its derivatives are zero as $x \to 0$. This implies that it satisfies the ω^2–sum rule, Eq. (4), by letting $R_{QCA}(Y)$ be positive at small $|Y|$ and negative at larger $|Y|$. The QCA provides a very simple way to derive the essential qualitative physics of FSE.

However, the QCA can not be expected to be quantitatively correct. It appears to be plagued by a number of factors of 2. One difficulty can be traced to the neglect of forward diffractive scattering in the QCA. The quantum mechanical σ_{tot} for a hard sphere is twice the classical value of πr_o^2 due to forward diffractive scattering. For example, the QCA prediction, $\Delta Y_{FWHM} \approx Q\pi r_o^2$, is a factor of 2 smaller than expected. QCA is also incapable of describing glory oscillations which are due to interference between forward and backward diffractive scattering for identical Bose particles. Another obvious factor of 2 comes from giving the classical trajectory a momentum of $Q/2$ rather than Q. Further, one expects that x in $r(x,b)$ should be the distance between the centers of the particles rather than the distance between the edges as in QCA. These difficulties will be overcome in the fully quantum theory to be discussed in the next section, but the mathematical form and physical consequences of the final answer will resemble Eqs. (9–10).

THE QUANTUM THEORY OF FSE

Perturbation theory about the non–interacting state is notoriously slowly convergent for almost hard core systems such as helium. However, excellent variational and Monte Carlo methods exist to determine ground state properties such as n_k and $g(r)$. The n_k determines the IA prediction, and the QCA leads to the expectation that $g(r)$ is essential to the correct physics of FSE. This motivates an approach to calculate $S(Q,\omega)$ by perturbative expansion about the strongly interacting ground state wave function using projection techniques. In the following, I shall outline the steps in the derivation of a quantum theory of FSE. I shall emphasize the approach and identify where approximations have been made. I shall refer the reader to other papers where details may be found.

The neutron scattering law, $S(Q,\omega)$, is related to the dynamic susceptibility, $\chi(Q,\omega)$, by

$$\chi(Q,\omega) = \frac{1}{\pi N} < [\hat{S}^Q(\omega), \hat{\varrho}_{-Q}] > \quad ; \quad \text{Im } \chi(Q,\omega) = S(Q,\omega) - S(Q,-\omega) \quad , \tag{11}$$

where

$$\hat{S}^Q(\omega) \equiv i \int_0^\infty dt e^{i\omega t} e^{i\hat{H}t} \hat{\varrho}_Q e^{-i\hat{H}t} \quad ; \quad \hat{\varrho}_Q \equiv \sum_k \hat{a}_{k+Q}^+ \hat{a}_k \quad . \tag{12}$$

The kth component of the dynamic susceptibility, χ_k, may be defined by

$$\chi_k \equiv \frac{1}{\pi N} < [\hat{S}^Q(\omega), \hat{a}_{k-Q}^+ \hat{a}_k] > \quad . \tag{13}$$

Then diagrammatic perturbation theory can prove that the χ_k satisfy an exact Dyson equation

$$\chi_k = \chi_k^o \left[1 + \sum_{k'} I(k,k') \chi_{k'} \right] \quad ; \quad \chi_k^o = \frac{1}{\pi N} \frac{n_{k-q} - n_k}{\omega - e_{k-Q} + e_k + i\epsilon} \quad . \tag{14}$$

Ignoring collisions corresponds to setting the kernel, $I(k,k')$, to zero, in which case the IA is recovered from Eq. (14). The kernel is calculated [11,16] by truncating a projection superoperator expansion of $\chi(Q,\omega)$ at lowest order in the two–body t–matrix and keeping only the non–zero terms in the "asymptotic limit" of high Q and hard core interactions:

$$I(k,k') = -\frac{1}{n_k n_{k'}} \sum_{k_1} T_{k_1, k-Q, k-k'}^B \Phi(k_1, k, k-k') \quad . \tag{15}$$

A diagrammatic representation of the Dyson equation for FSE is shown in Fig. 3. The dashed line in Fig. 3 represents the Bose symmetrized two–body t–matrix

$$T_{k_1,k_2,q}^B = -\frac{4\pi}{2\mu i k_{CM}} \sum_{l \; even} (2l+1)(e^{2i\delta_l} - 1) P_l(\cos\theta) \Rightarrow t^B(Q,q) \quad . \tag{16}$$

The shaded box in Fig. 3 represents the two–particle density matrix

$$\Phi(k_1, k_2, q) = \; < \hat{a}_{k_1+q}^+ \hat{a}_{k_2-q}^+ \hat{a}_{k_2} \hat{a}_{k_1} > \quad . \tag{17}$$

In the limit of keeping only the $q=0$ part of Φ, the Lorentzian broadening theory of Hohenberg and Platzman, and more specifically Platzman and Tzoar [17], is recovered. The more general expression, Eq. (17), retains all the information about the spatial correlations in the ground state wave function which are required to get the correct physics of FSE. In particular, Φ is related by a sum rule to $g(r)$:

$$\frac{1}{N} \sum_{k_1, k_2} \Phi(k_1, k_2, q) = \varrho \int d^3 r \, e^{iq \cdot r} g(r) \quad . \tag{18}$$

Also,

$$\lim_{q \to 0} \frac{1}{N} \sum_{k_2} \Phi(k_1, k_2, q) \to n_{k_1} \quad . \tag{19}$$

I take the two–body t–matrix to be on–shell and I make semiclassical approximations [18] for Eq. (16), which should certainly be valid at high Q. I also approximate Φ to satsify Eqs. (18–19), i.e.

$$\sum_{k_1} \Phi(k_1, k, q) \approx n_k \varrho \int d^3 r \, e^{iq \cdot r} g(r) \quad . \tag{20}$$

Equation (20) amounts to a separable kernel approximation for the Dyson equation for χ_k, which results in the form, Eq. (2), for the final state broadening as a convolution of an $R(Y)$ with the IA. Then an exact analytic solution of the Dyson equation for

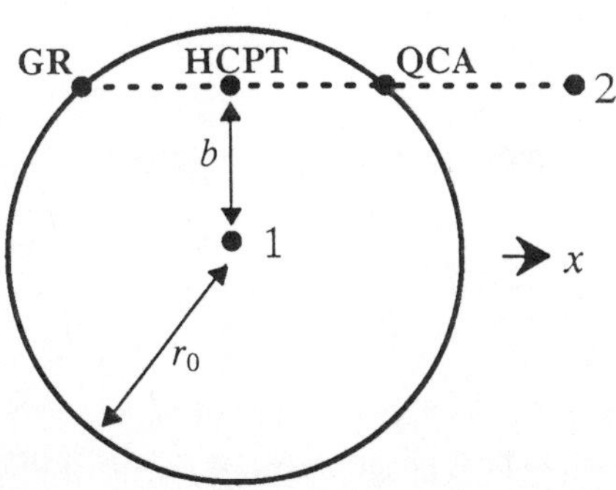

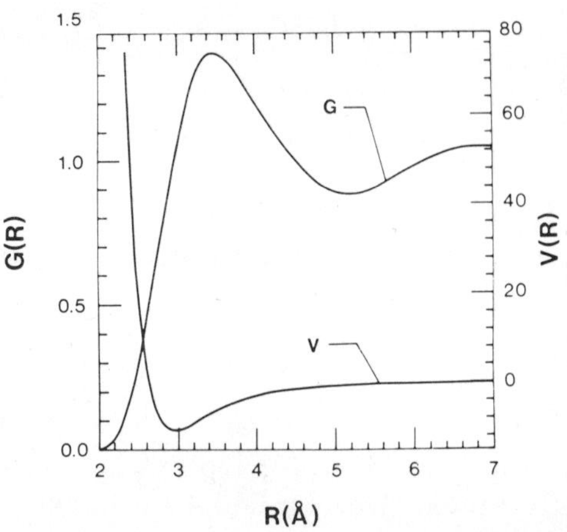

FIG. 1 – Schematic representation of classical trajectories which contribute to the GR, QCA, and HCPT theories for FSE. Shown is the initial configuration of particles 1 and 2 immediately after a neutron imparts momentum Q to 1, which moves to the right to collide with 2 with impact parameter b. The circle represents the range of the steeply repulsive core about 1. In each theory, x is the distance between particle 2 and the point with the corresponding theory label.

FIG. 2 – Plot of the radial distribution function, $g(r)$, for ^{4}He and the He–He potential, $V(r)$. After a neutron collision, the recoiling particle starts its trajectory in the attractive well of the potential at some distance from the steeply repulsive core responsible for final state effects. The scaling variable, Y, is conjugate momentum to distance, x, along the classical trajectory.

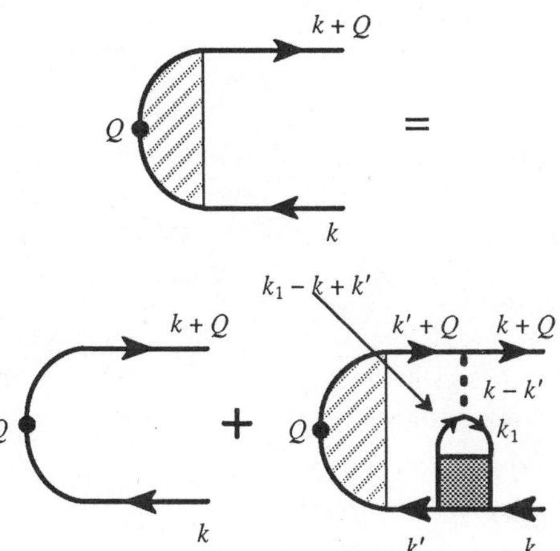

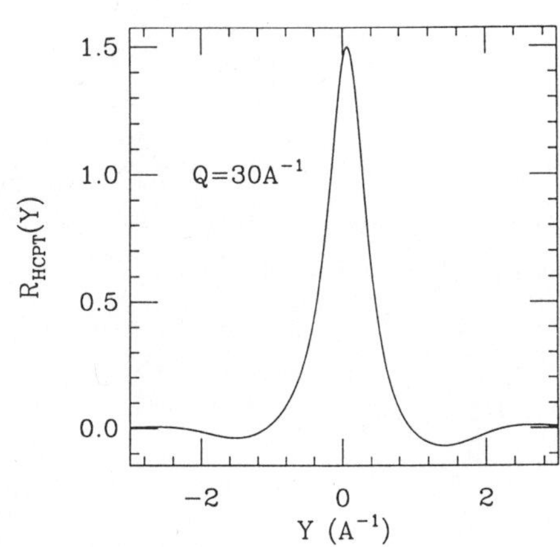

FIG. 3 – Diagrammatic representation of the Dyson equation, Eqs. (14–15), for the HCPT of FSE. The shaded box is the two–particle density matrix, Φ. The dashed line is the two–particle t–matrix. Arrows denote the direction of the flow of momentum. Only the lines with Q carry high momentum.

FIG. 4 – The final state effects (FSE) broadening function, $R(Y)$, predicted by HCPT for ^{4}He.

the χ_k exists, details of which may be found in ref. [11].

The final result for $R(Y)$ is identical to Eq. (10) obtained by the QCA, with

$$P_{NC}(x) = \exp\left(-\int_0^x \Gamma(x')dx'\right) \; ; \quad \Gamma(x) = \varrho\pi \int_0^\infty db^2 \, f_b \, g(r(x,b)) \quad . \tag{21}$$

This is similar to Eq. (9) but differs in three details. First, there is no factor of 2 in the limit of the integral. Second,

$$f_b = e^{2i\delta_b} - 1 + e^{2i\delta_b - \frac{i\pi Qb}{2}} \quad . \tag{22}$$

Here, δ_b is the semiclassical (JWKB) phase shift for He–He scattering at impact parameter b. The third term in Eq. (22) is due to Bose symmetrization and gives rise to the hard sphere glories in the He–He total cross section. The real part of $\Gamma(x)$ is again the collisions per unit distance. With $\mathrm{Re}\,\Gamma(\infty) = \varrho\sigma_{tot}/2$ by the optical theorem, there is an interesting cancellation of the 2 in $\sigma_{tot} = 2\pi r_o^2$, so that $\mathrm{Re}\,\Gamma(\infty)$ is simply the classical result, $\varrho\pi r_o^2$. Finally, $r(x,b) = \sqrt{x^2 + b^2}$ for HCPT. The x's appropriate for the HCPT, QCA and GR theories are shown in Fig. 1.

HCPT PREDICTIONS FOR EXPERIMENT

The HCPT predictions [19] for deep inelastic neutron scattering on 4He can be evaluated using as input the variational and Monte Carlo n_k, the experimental $g(r)$ measured by neutron diffraction [20], and the semiclassical predictions for the He–He phase shifts. The semiclassical methods are accurate at high Q, especially since the He–He potential at short distances is inferred from measured σ_{tot} by inverting the semiclassical formulae [3].

Figure 4 shows the $R(Y)$ predicted at $Q = 30$ A^{-1} which is achievable at pulsed neutron sources, but is much larger than reactor experiments for which $Q \leq 12$ A^{-1}. While $\Delta Y_{FWHM} \approx \varrho\sigma_{tot}(Q)$ as expected, $R(Y)$ is negative at larger $|Y|$ as required to satisfy the ω^2-sum rule. Figure 5 shows specifically how this is accomplished by displaying the integrand of the sum rule, $Y2R(Y)$. The sum rule requires the area under this curve to be zero. The scale of the oscillations in $Y2R(Y)$ is determined by the scale of the spatial structure in $g(r)$, i.e. the period is approximately $2\pi/2.5$ A^{-1}.

The n_k predicted for superfluid 4He by Greens' Function Monte Carlo (GFMC) [21] is shown as the solid line in Fig. 6. GFMC predicts a 9.2% Bose condensate fraction, n_0. The n_k predicted for normal fluid 4He by Path Integral Monte Carlo (PIMC) [22] at T=3.3 K is shown as the dashed line. The normal fluid n_k is smooth and approximately Gaussian.

The predictions for $F(Y)$ for normal fluid 4He are shown in Fig. 7. The pluses are the IA prediction for $F(Y)$ calculated by inserting the PIMC n_k into Eq. (1). The solid line is the HCPT prediction including FSE, which is calculated using the $R(Y)$ in Fig. 4 assuming that $g(r)$ is relatively temperature independent. FSE are predicted to

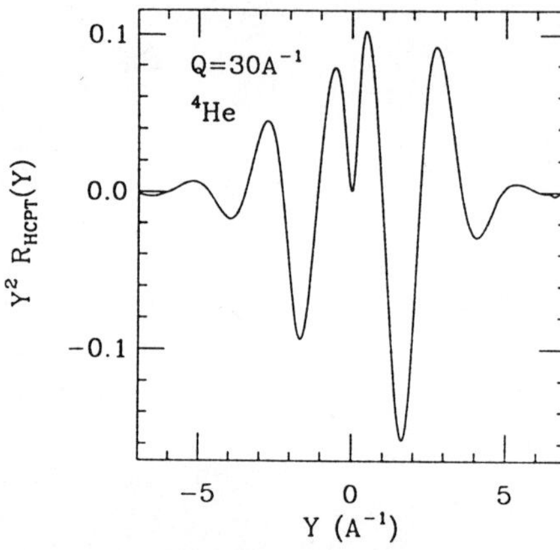

FIG. 5 – $Y^2 R(Y)$ predicted by HCPT. The ω^2 sum rule requires the area under this curve to be zero. The scale of the structure is governed by Fourier transform of the radial distribution function, $g(r)$.

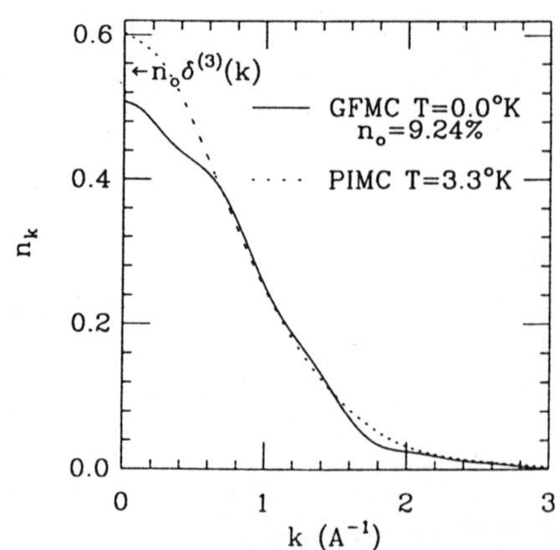

FIG. 6 – Momentum distributions for liquid ^{4}He. The solid line is the Greens' Function Monte Carlo (GFMC) prediction for the superfluid at T=0K, which has a delta function at $k=0$ with a 9.2 % Bose condensate fraction. The dashed line is the Path Integral Monte Carlo (PIMC) prediction for the normal fluid at T=3.33K.

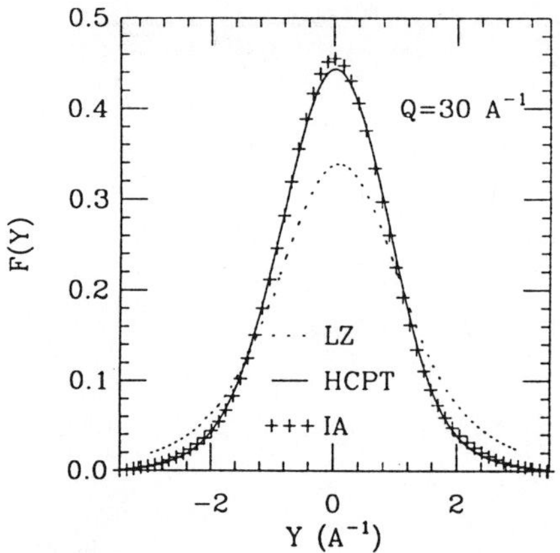

FIG. 7 – The neutron scattering law, $F(Y)$, for normal ^{4}He at T=3.33K using the PIMC momentum distribution. The pluses are the impulse approximation (IA) prediction. The solid line is the HCPT prediction which includes broadening due to final state effects (FSE). The dashed line is the Lorentzian (LZ) broadening prediction obtained by ignoring ground state spatial correlations and setting $g(r)$ to one.

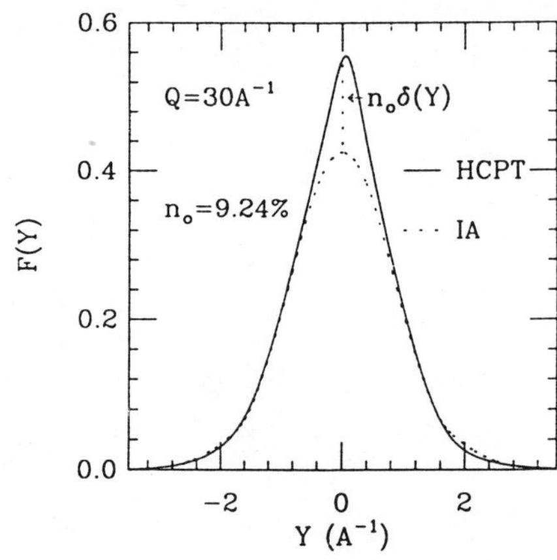

FIG. 8 – The neutron scattering law, $F(Y)$, for superfluid ^{4}He at T=0K using the GFMC momentum distribution. The dashed line is the impulse approximation (IA) prediction, which has a delta function at $Y=0$ due to the Bose condensate with 9.2% of the weight. The solid line is the prediction of HCPT which includes broadening due to final state effects (FSE).

be almost negligible for the normal fluid at $Q = 30$ A^{-1}, and the IA should be an excellent description. For comparison, the dashed line is the prediction of the Lorentzian broadening theory obtained by setting $g(r)$ to one. The FSE predicted by HCPT are much smaller than those predicted by theories which ignore spatial correlations. This is expected since the effect of spatial correlations is to reduce collisions at short recoil distances.

The corresponding IA and HCPT predictions of $F(Y)$ for the superfluid are shown in Fig. 8. The IA is calculated by inserting the GFMC n_k in Fig. 6 into Eq. (1). The dashed line is the IA prediction which has 9.2 % of the weight in a delta function at $Y=0$. FSE are included in the HCPT prediction shown by the solid line, which is calculated by convoluting the IA prediction with the $R(Y)$ in Fig. 4, according to Eq. (2). The Bose condensate delta function in the IA is completely wiped out by FSE broadening, so that it is no longer a distinct peak. However, the Gaussian width of the overall distribution is unchanged by FSE as is expected from the ω^2 –sum rule, Eq. (4). Intensity is robbed from high $|Y|$ in order to preserve the sum rule.

Thus, the Bose condensate peak will not become resolvable by going to Q's much larger than reactor experiments. However, such Q's permit a theory of FSE, which allows tests of calculations of n_k using DINS data. This prediction does not imply that values of n_0 can not be inferred directly from DINS data, but the problem has become ill–conditioned as is discussed by Sivia and Silver [23] in these proceedings.

Since this theory of FSE predicts that the Bose condensate is unobservable as a distinct peak at $Q = 30$ A^{-1}, a critical question becomes at what Q's the Bose condensate may become observable as a distinct peak. As in the original HP theory, the broadening scales with $\sigma_{tot}(Q)$ for 4He, which is very slowly decreasing with increasing Q. Figure 9 shows the classical turning point, r_o , of the He–He potential [3] as a function of Q on a linear–log scale, which shows that the decrease is approximately logarithmic. Figure 10 shows the slow approach to the IA predicted by HCPT by plotting $F(Y)$ at several Q. HCPT FSE vary as $O(Q^0)$ for hard core potentials, and approximately as $O(\log Q)$ for the He–He potential. Experimentalists should certainly try to test this prediction by experiments at much larger Q. However, Q's greater than 100 A^{-1} with adequate resolution are unfeasible with present spectrometers.

The behavior of FSE should be similar for DINS experiments on 3He Fermi liquid, where a primary goal is to observe the Fermi surface discontinuity predicted in n_k. The solid line in Fig. 11 is the Fermi HyperNetted Chain (FHNC) prediction [24] for n_k which has a discontinuity, and the dashed line is the prediction of the Lhullier and Bouchaud wave function [25] which does not. The FSE predicted by HCPT and the $F(Y)$ at $Q = 30$ A^{-1} for 3He using the FHNC n_k as input have been calculated. An apparent n_k could be inferred from a measured $F(Y)$ by assuming the IA was correct and differentiating the data with respect to Y according to Eq. (1). The apparent n_k is shown as the dot–dashed line in Fig. 11, in which the Fermi surface discontinuity is no longer visible due to FSE. It resembles the Lhullier and Bouchaud prediction. Again, the Q dependence of the FSE broadening would be similar to 4He, so that the Fermi surface discontinuity in 3He should not be expected to become directly observable by performing experiments at much higher Q.

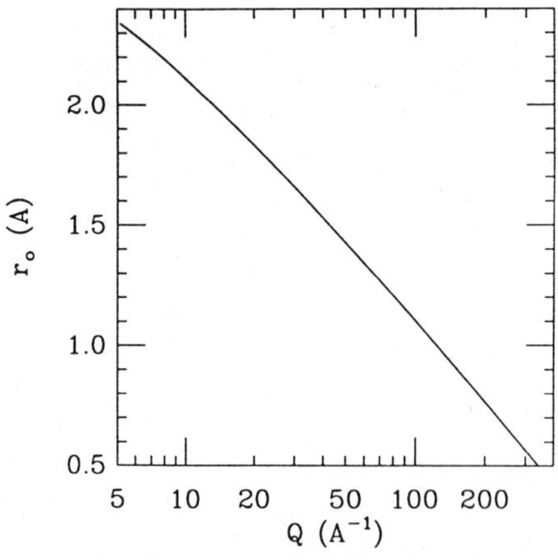

FIG. 9 – Classical turning points of the He–He potential versus Q, plotted on linear–log scale. This shows that the width of the FSE broadening of the IA, $R(Y)$, should decrease approximately as $O(\log Q)$ with increasing Q.

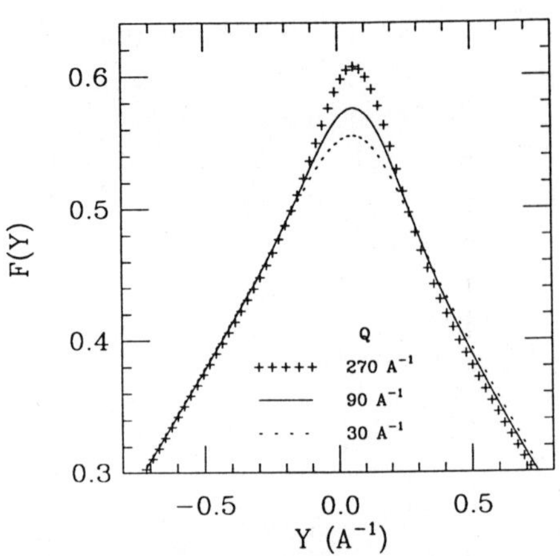

FIG. 10 – Approach to the impulse approximation (IA) with increasing Q of the neutron scattering law, $F(Y)$, for superfluid ⁴He at T=0K as predicted by HCPT. Only the region near $Y=0$ relevant to the Bose condensate is shown. The Bose condensate does not become a distinct peak at any experimentally feasible Q.

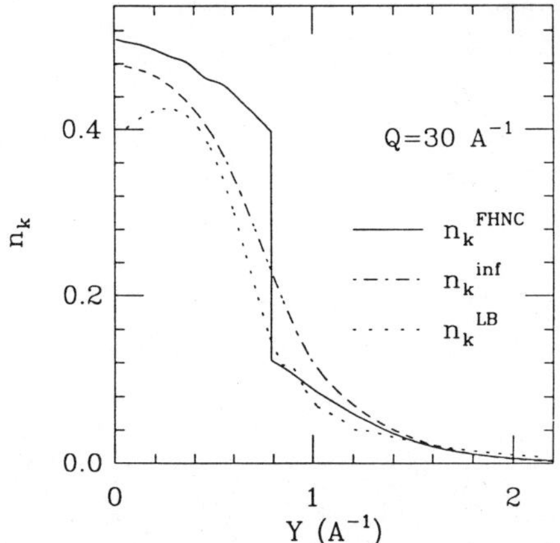

FIG. 11 – Momentum distributions for normal ³He Fermi liquid at T=0K. The solid line is the Fermi Hypernetted Chain (FHNC) prediction. The dashed line is the prediction using the Lhullier & Bouchaud (LB) wave function. The dot–dash line is the apparent momentum distribution after broadening the FHNC–IA neutron scattering law by the HCPT prediction for FSE. The Fermi surface discontinuity is smoothed out by FSE.

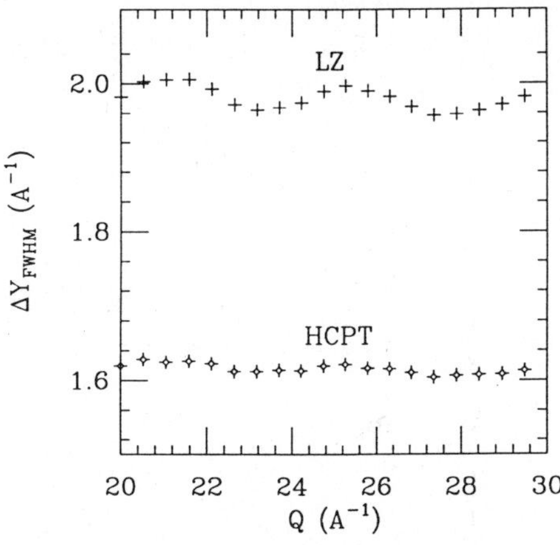

FIG. 12 – Full Width Half Maximum (FWHM) of the neutron scattering law, $F(Y)$, for superfluid ⁴He at T=0K including FSE broadening. The HCPT is at the bottom. The Lorentzian broadening prediction (LZ) at the top is obtained by setting $g(r)$ to one. The FWHM and the glory oscillations are much smaller for HCPT than for LZ. They are close to the impulse approximation (IA) prediction.

DISCUSSION AND EXTENSIONS

The direct comparison of the HCPT predictions for $F(Y)$ with pulsed neutron experiments at $Q = 23$ A^{-1} on 4He is discussed in ref. [12], and in the paper by Sosnick, et al. [13] in these proceedings. The agreement is excellent with no adjustable parameters. A discussion of the formal relation of HCPT to other theories may be found in ref. [8], and a comparison between the data and these theories is presented in ref. [7]. HCPT has the best agreement with experiment. Given this success, one may question the need for improvements. However, in addition to further experimental tests and applications to other systems, there remain many outstanding problems in the theory of FSE.

First, the approximation of the two–particle density matrix (Φ) by Eq. (20) satisfies Eqs. (18–19), but it does not satisfy other known properties of Φ. Recently Ristig and Clark [26] have calculated a general structure for Φ which remedies this defect. By substituting their expression for Φ into Eqs. (14–15), one can formally show that the convolution form of the FSE, Eq. (2), and the relative temperature independence of FSE should be regarded as approximations. The critical issue remaining is whether the excellent agreement between theory and experiment [12] becomes fortuitous after the corrections to the IA using better Φ are calculated.

The approximations of HCPT, such as the semiclassical treatment of the t–matrix, begin to fail at the lower Q's of reactor experiments. While HCPT has kept the leading $O(\log Q)$ FSE, Sears [27] has proposed that the $O(Q^{-1})$ FSE corrections to the IA are approximately antisymmetric in Y. They can be the dominant corrections in cases where the HCPT FSE are small, e.g. the almost Gaussian momentum distributions such as for normal 4He as shown in Fig. 7. There appears such a slight asymmetric component even in the $Q = 23$ A^{-1} experiments [12], and such an asymmetry is pronounced in reactor experiments on 4He and other noble gases. The glory oscillations are also an $O(Q^{-1})$ correction, which may be estimated by retaining the third term in Eq. (22). Figure 12 shows that the glory oscillations should be much smaller than predicted by the Lorentzian broadening theories which ignore ground state spatial correlations. However, a complete theory for FSE which includes all the $O(Q^{-1})$ terms remains to be developed and compared to experiment, and the connection to the Sears [27] expansion of FSE in powers of Q needs to be clarified.

The theory should be extended to the problem of FSE in quasielastic electron–nucleus scattering (QENS). As is discussed in more detail in ref. [28], QENS can be compared to DINS on 4He by appropriate scaling of the physical variables. The nuclear repulsive core is softer by three orders of magnitude and the nuclear density is comparatively low, which suggest that FSE might be smaller in QENS. However, QENS requires much smaller Q's in order to avoid inelastic processes and much larger Y's in order to obtain information on short–range correlations in nuclei, which suggests that FSE should be more important. Some naive applications of HCPT to QENS are discussed in ref. [28], but the HCPT approximations are less valid for QENS. For example, the semiclassical methods and neglect of the self–energy of the initial particle can hardly be justified at low Q's, and the on–shell approximation for the two–body t–matrix becomes less valid at large $|Y|$. Hopefully, the superoperator projection methods will provide a starting point for a more complete theory of FSE in QENS.

I believe there remains little doubt about the general properties of the FSE for DINS in helium. However, the quasiclassical approximation and the superoperator expansions I have used in their derivation are no more familiar to the condensed matter physics community than the time–ordered cumulant expansions originally used by Gersch and Rodriguez [6]. All of these methods deserve further development, as well as applications to other problems involving the dynamical response of systems with strong ground state correlations. In addition, a treatment of FSE using diagrammatic perturbation theory starting with the non–interacting ground state remains to be provided.

ACKNOWLEDGEMENT

Research supported by the Office of Basic Energy Sciences of the U.S. Dept. of Energy. I thank K. Bedell, J. W. Clark, G. Rieter, and P. E. Sokol for many helpful discussions.

REFERENCES

[1] J. W. Clark, R. N. Silver, P. E. Sokol, this volume.
[2] G. B. West, Physics Reports $\underline{18}$, 263 (1975).
[3] R. Feltgen, H. Kirst, K. A. Koehler, F. Torello, J. Chem. Phys. $\underline{26}$, 2360 (1982).
[4] P. C. Hohenberg, P. M. Platzman, Phys. Rev. $\underline{152}$, 198 (1966).
[5] P. Martel, E. C. Svensson, A. D. B. Woods, V. F. Sears, R. A. Cowley, J. Low Temp. Phys. $\underline{23}$, 285 (1976).
[6] H. A. Gersch, L. J. Rodriguez, Phys. Rev. A$\underline{8}$, 905 (1973).
[7] T. R. Sosnick, W. M. Snow, P. E. Sokol, R. N. Silver, this volume.
[8] See R. N. Silver, Phys. Rev. B$\underline{38}$, 2283 (1988) for a summary of prior theories.
[9] L. J. Rodgriguez, H. A. Gersch, H. A. Mook, Phys. Rev. A$\underline{9}$, 2085 (1974).
[10] R. N. Silver, G. Reiter, Rap. Comm. Phys. Rev. B1 $\underline{35}$, 3647 (1987).
[11] R. N. Silver, Rap. Comm. Phys. Rev. B$\underline{37}$, 3794 (1988), ibid. B$\underline{38}$, 2283 (1988).
[12] T. R. Sosnick, W. M. Snow, P. E. Sokol, R. N. Silver, to be published.
[13] T. R. Sosnick, W. M. Snow, P. E. Sokol, this volume.
[14] G. Reiter, R. N. Silver, Phys. Rev. Lett. $\underline{54}$, 1047 (1985).
[15] G. Reiter, this volume.
[16] R. N. Silver, *Condensed Matter Theories*, V. $\underline{4}$, Plenum Press (1989), to be published.
[17] P. M. Platzman, N. Tzoar, Phys. Rev. B$\underline{30}$, 6397 (1984).
[18] R. K. B. Helbing, J. Chem. Phys. $\underline{50}$, 493 (1969).
[19] R. N. Silver, to be published.
[20] E. C. Svensson, V. F. Sears, A. D. Woods, P. Martel, Phys. Rev. B$\underline{21}$, 3638 (1980).
[21] P. Whitlock, R. Panoff, Can. J. of Phys. $\underline{65}$, 1409 (1987).
[22] D. M. Ceperley, E. L. Pollock, Can. J. of Phys. $\underline{65}$, 1416 (1987).
[23] D. S. Sivia, R. N. Silver, this volume.
[24] E. Fabrocini, S. Fantoni, V. R. Pandharipande, to be published.
[25] J. P. Bouchaud, C. Lhullier, Europhys. Lett. $\underline{3}$, 1273 (1987)
[26] M. L. Ristig, J. W. Clark, this volume.
[27] V. F. Sears, Phys. Rev. B$\underline{30}$, 44 (1984)
[28] J. W. Clark, R. N. Silver, *Proceedings of the Third International Conference on Nuclear Reaction Mechanisms*, Varenna, Italy, June 13–18, 1988, E. Gadioli, ed. Supplemento $\underline{66}$, p. 531–540 (1988).

APPROACH TO THE IMPULSE APPROXIMATION IN QUANTUM SOLIDS AND FLUIDS

H. R. Glyde

Department of Physics and Astronomy
University of Delaware
Newark, Delaware 19716 USA

and

W. G. Stirling

Department of Physics
University of Keele
Keele, Staffs ST5 5BG U.K.
and S.E.R.C. Daresbury Laboratory, Warrington U.K.

ABSTRACT

We discuss calculations and observed values of $S(Q,\omega)$ in liquid and solid helium in the wave vector transfer range $3 \lesssim Q \lesssim 20$ Å^{-1}. The aim is to display deviations of $S(Q,\omega)$ from the Impulse Approximation, $S_{IA}(Q,\omega)$. In a calculation based on a T-matrix in solid helium we find that the deviation, $\delta S = S(Q,\omega) - S_{IA}(Q,\omega)$, is well accounted for by the leading correction term δS_1 to the IA for $10 \leq Q \leq 20$ Å^{-1}. An analytic expression for δS_1 is given. In liquid ^{4}He we present new measurements of $S(Q,\omega)$. In the range $3 \leq Q \leq 10$ Å^{-1} the width and peak position of $S(Q,\omega)$ oscillate with Q, as noted by Martel et al., due to final state interactions. These oscillations constitute substantial deviations from the IA and probably exist up to $Q = 15$-20 Å^{-1}. Similar oscillations in liquid ^{3}He are calculated to be much smaller.

1. INTRODUCTION

In most solids and fluids, the dynamic form factor $S(Q,\omega)$ is expected to go over to the impulse approximation form as $Q \to \infty$. Exceptions are systems of particles interacting via potentials having a perfectly hard core[1] or in which there remains finite transfer of momentum between atoms on short time scales,[2] $t \to 0$. Typical atomic time scales in solids are $\tau_A \sim \nu_D^{-1} \sim 10^{-12}$ sec, where $\omega_D = 2\pi\nu_D$ is a Debye frequency.

In this paper, we examine the approach to the impulse approximation (IA). We consider ranges of momentum transfer, $\hbar Q$, where we do not expect the IA to be even approximately valid. Our aim is to display

deviations from the IA. We also take this opportunity to present some
new experimental results[3] for liquid ^{4}He and discuss new calculations[3,4]
in liquid ^{3}He and ^{4}He.

2. THE IMPULSE APPROXIMATION

The IA to $S(Q,\omega)$ is[5,6]

$$S_{IA}(Q,\omega) = \int d\vec{p}\, n(\vec{p})\, \delta\left(\omega - \omega_R - \frac{\vec{Q}\cdot\vec{p}}{M}\right) \qquad (1)$$

where $n(\vec{p})$ is the momentum distribution of the helium atoms normalized
to $\int d\vec{p}\,n(p) = 1$, $\omega_R = \hbar Q^2/2M$ is the free atom recoil frequency and M
is the helium atom mass. The IA may be derived by (a) approximating the
coherent $S(Q,\omega)$ by the incoherent $S_i(Q,\omega)$ and (b) neglecting the impulse
on the scattering atom due to its neighbors compared to the impulse $\hbar Q$
transferred to the scattering atom from the neutron. Criterion (a)
requires $Q \gtrsim 8$ Å^{-1} in helium at svp. The second criterion is

$$\hbar Q \gg F\tau_S \qquad (2)$$

where τ_S is the scattering time. The scattering time may be clearly
defined as a typical time over which $S(Q,t)$ contributes to $S(Q,\omega)$ in the
Fourier transform $2\pi S(Q,\omega) = \int dt\, e^{i\omega t} S(Q,t)$. This definition gives τ_S
$\sim M/Q\langle p^2\rangle^{\frac{1}{2}}$ where $\langle p^2\rangle^{\frac{1}{2}}$ is the RMS value of the momentum. In liquid ^{4}He,
Green Function Monte Carlo[17] (GFMC) calculations find a kinetic energy
of $\langle KE\rangle = 14.47 \pm 0.09$ K. This corresponds to $\langle (p/\hbar)^2\rangle^{\frac{1}{2}} \approx 1.6$ Å^{-1}
giving $\tau_S = 0.02\times10^{-12}$ sec at $Q = 20$ Å^{-1}, which is approximately $\tau_A/100$.
In a solid, criterion (b) may be re-stated as $\tau_A \gg \tau_S$ or $\omega_R \gg \omega_D$. The
IA is thus basically a short scattering time approximation.[7]

Quite independently of the IA, the second moment of $S_i(Q,\omega)$ is
related to the kinetic energy of an atom, $\langle KE\rangle$, by[8]

$$\sigma^2 \equiv \int d\omega\, S_i(Q,\omega)(\omega - \omega_R)^2 = \frac{4}{3}\,\omega_R\langle KE\rangle/\hbar. \qquad (3)$$

In ^{4}He the scattering is purely coherent. Since $S(Q,\omega) \to S_i(Q,\omega)$ at Q
$\gtrsim 8$ Å^{-1} in helium, the second second moment σ^2 of the observed $S(Q,\omega)$
may be used to determine $\langle KE\rangle$, provided the second moment in (3) can be
calculated with confidence.

If the momentum distribution $n(\vec{p})$ is a Gaussian function, $n(\vec{p})$
$= n(p_x)n(p_y)n(p_z)$

$$n(p_z) = (2\pi\langle p_z^2\rangle^{\frac{1}{2}})\, e^{-\frac{1}{2}\frac{p^2}{\langle p_z^2\rangle}} \qquad (4)$$

the $S_{IA}(Q,\omega)$ in (1) also reduces to a Gaussian,

$$S_{IA}(Q,\omega) = (2\pi\sigma^2)^{\frac{1}{2}} e^{-\frac{1}{2}(\omega-\omega_R)^2/\sigma^2} \qquad (5)$$

In this special case, we may obtain the second moment $\sigma^2 = \langle(\omega-\omega_R)^2\rangle$ from the width of the Gaussian. The FWHM of the Gaussian (5), is $W = (8\ell n2)^{\frac{1}{2}}\sigma = 2.35\sigma$. In this special case the $\langle KE\rangle$ may be obtained from the observed W as

$$\langle KE\rangle = \left(\frac{3M}{2\hbar^2}\right)\left(\frac{\sigma}{Q}\right)^2 = \left(\frac{3M}{2\hbar^2}\right)(8\ell n2)^{-1}\left(\frac{W}{Q}\right)^2. \qquad (6)$$

For ^{4}He, $\hbar^2/M = 12.12$ K$\mathrm{\AA}^{-1}$. From (6) we therefore expect (W/Q) to be a constant when the IA is valid.

3. SOLID HELIUM

The dynamics of solid helium may be evaluated from first principles more accurately than that of the liquid, essentially because a solid is easier to treat than a liquid. Also the exchange interaction is small in the solid ($\sim 10^{-3}$ the direct term) so that the nuclear dynamics of solid ^{3}He and ^{4}He are quite similar. To display deviations from the IA we present[9] calculations of $S_i(Q,\omega)$ based on the self consistent phonon (SCP) theory coupled with a T-matrix treatment of the repulsive core of the potential. The latter is critical and describes the short range

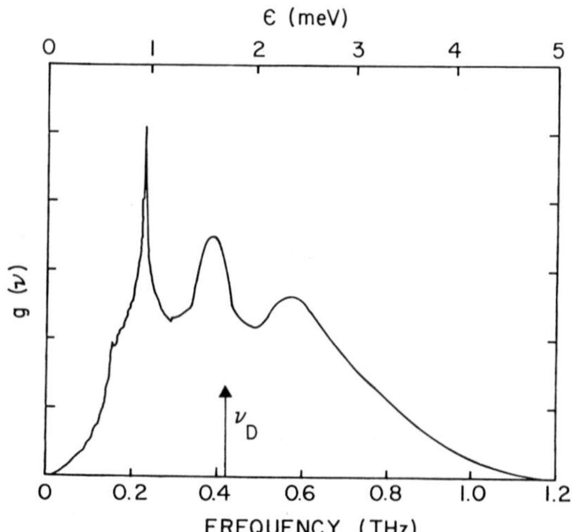

Fig. 1. Phonon density of states, $g(\nu)$, for bcc ^{3}He at $V = 24$ cm^3/mole calculated using the SCP theory. The high frequency tail arises from the anharmonic response functions. The observed Debye temperature $\theta_D = 20$ K (Ref. 10) gives $\nu_D = 0.416$ THz.

correlations in the atomic motion due to the steeply repulsive core in
$v(r)$. The _anharmonic_ response of the crystal can be evaluated and, from
this, an anharmonic density of phonon frequencies $g(\omega)$ can be
constructed. An example of $g(\nu)$ calculated in this way for bcc ^{3}He at V
= 24.00 cm^3/mole (just above the melting line) is shown in Fig. 1. We
note that, this $g(\nu)$ has high frequency tails reaching well above the
emipirically determined Debye frequency. These high frequency tails
account for approximately 40% of the kinetic energy, $\langle KE \rangle \approx 22$ K, in bcc
^{3}He at this volume[9] and tails arise from the high energy components in
the dynamics induced by the high energy, repulsive core of $v(r)$.

The $g(\nu)$ or $g(\omega)$, once calculated, may be taken to describe the
dynamics needed for $S_i(Q,\omega)$ and other properties such as θ_D and $\langle KE \rangle$.
For example, at T = 0 K and ignoring interference terms,

$$S_i(Q,\omega) = \frac{1}{2\pi} \int_{-\infty}^{\infty} dt \; e^{i\omega t} \; e^{-2W} \exp\left[\omega_R \int_0^{\infty} d\omega \; g(\omega) \; \frac{1}{\omega} \; e^{-i\omega t} \right] \qquad (7)$$

where
$$2W = \omega_R \left(\frac{3\hbar}{2k\theta_D} \right) = \omega_R \langle \omega^{-1} \rangle \quad \text{and}$$

$$\langle KE \rangle = \frac{3}{4} \hbar \langle \omega \rangle \; ; \; \sigma^2 = \omega_R \langle \omega \rangle \qquad (8)$$

where $\langle \omega^n \rangle = \int d\omega g(\omega) \omega^n$ are moments of $g(\omega)$.

In Fig. 2 we show $S_i(Q,\omega)$ in bcc ^{3}He calculated from (7) using the
$g(\omega)$ of Fig. 1. Also shown in Fig. 2 is the corresponding scattering
function S_{IA} and the deviation, $\delta S = S_i - S_{IA}$, from the IA. S_{IA} may be
calculated from (7) in the short time limit by expanding the last
exponential in (7) and retaining terms up to t^2; $e^{-i\omega t} \approx 1 - i\omega t - \omega^2 t^2$.
In this case we recover (5) with σ^2 given by (8). We may also calculate
the leading correlation, $\delta S_1(Q,\omega)$ to the IA by keeping the term in t^3 in
the expansion of $e^{-i\omega t}$. In this way we obtain

$$S_i(Q,\omega) \approx S_{IA}(Q,\omega) + \delta S_1(Q,\omega) \qquad (9)$$

where $\delta S_1(Q,\omega)$, arising from the t^3 term, is

$$\delta S_1(Q,\omega) = -\frac{\langle \omega^2 \rangle}{\langle \omega \rangle^2} \frac{(\omega-\omega_R)}{2\omega_R} \left[1 - \frac{(\omega-\omega_R)^2}{2\sigma^2} \right] S_{IA}(Q,\omega) \qquad (10a)$$

or

$$\delta S_1(y) = -\frac{\langle \omega^2 \rangle}{\langle \omega \rangle^2} \frac{y}{\hbar Q} \left[1 - \frac{E_y}{\langle KE \rangle} \right] S_{IA}(y). \qquad (10b)$$

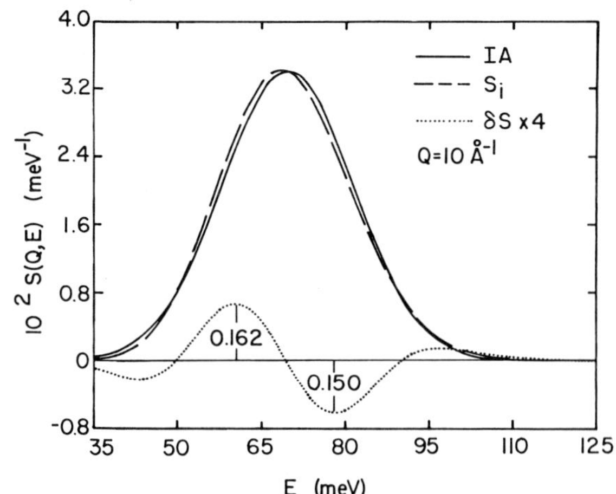

Fig. 2. $S_i(Q,\omega)$ (-----) and $S_{IA}(Q,\omega)$ (———)
calculated using $g(\nu)$ in Fig. 1 for bcc
^{3}He at V = 24 cm^3/mole. Also shown in
$\delta S = S_i - S_{IA}$ (·····) enlarged by a
factor of 4.

In (10b) $y = M/Q(\omega-\omega_R)$ is the "y scaling" variable introduced by West.[13]
The function y has dimensions of momentum and in practice is restricted
to values comparable to the atomic momentum $|\vec{p}|$. $E_y = y^2/2M$ is then
the kinetic energy associated with y.

Notice that Eqns. (7), (9) and (10) are a simple example of an
expansion[13] of $S(Q,\omega)$ about $S_{IA}(Q,\omega)$ in powers of Q^{-1}.

The $\delta S(Q,\omega)$ shown in Fig. 2 and the $\delta S_1(Q,\omega)$ in eqn. (10) show
several interesting features. Firstly, $\delta S(Q,\omega)$ is a predominantly
<u>antisymmetric</u> function about ω_R. $\delta S_1(Q,\omega)$ in (10a) is clearly an
<u>antisymmetric</u> function of $(\omega-\omega_R)$ and vanishes at $\omega = \omega_R$. For ω near ω_R
(in the region of the peak) $\delta S_1(Q,\omega)$ is negative for $\omega > \omega_R$ and positive
for $\omega < \omega_R$. Thus $\delta S_1(Q,\omega)$ displaces the peak of $S_i(Q,\omega)$ to lower ω.
Secondly, $\delta S_1(Q,\omega)$ has a second zero at $\omega = \omega_R \pm \sqrt{3}\ \sigma$. Thus the
correction δS_1 becomes positive at high ω $(\omega \gtrsim \omega_R + \sqrt{3}\ \sigma)$ which
contributes toward the high frequency tail of $S_i(Q,\omega)$. The position of
the second zero depends on the kinetic energy i.e. $\delta S_1 = 0$ when $E_y =$
$\langle KE \rangle$. (In an ideal world this might be used to measure the $\langle KE \rangle$).
Finally, δS_1 decreases as Q^{-1} compared to S_{IA}. In Fig. 3 we show
calculations of $S_i(Q,\omega)$, $S_{IA}(Q,\omega)$ and $\delta S(q,\omega) \equiv S_i(Q,\omega) -S_{IA}(Q,\omega)$ for
solid ^{4}He at three Q values. We may see that the full δS is decreasing
as Q^{-1} compared to $S_{IA}(Q,\omega)$. For the calculations in Fig. 3 we used a
model $g(\omega)$ having the same general shape as of Fig. 1, but was
constrained to reproduce the observed[11] $\theta_D = 50$ K and the MC value[12] of
$\langle KE \rangle = 36.6$ K.

From Figs. 2 and 3, we see that the leading term δS_1 appears to
describe the total δS quite well. Sears[13] has shown that $\delta S_1(Q,\omega)$ is
antisymmetric and decreases as $Q^{-1}S_{IA}$ quite generally, independent of
the interaction. These features suggest that much of the deviation from
$S_{IA}(Q,\omega)$ could be removed by identifying and eliminating the
antisymmetric contribution to $S_i(Q,\omega)$. This has certainly been
effective[13,14] in liquid ^{4}He and is expected[15] to be effective in solid
He.

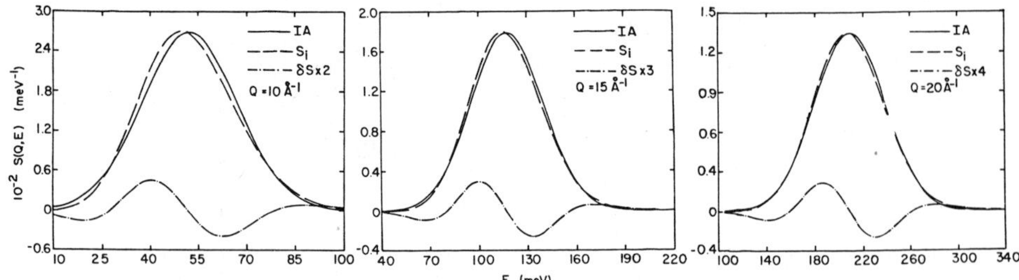

Fig. 3. $S_i(Q,\omega)$, IA = $S_{IA}(Q,\omega)$ and $\delta S = S_i - S_{IA}$ calculated using a model $g(\nu)$ constructed to reproduce the Monte Carlo $\langle KE \rangle = 36.6$ K (Ref. 12) and observed $\theta_D = 50$ K (Ref. 11) for bcc ^{4}He at V = 16 cm^3/mole. Comparison of δS at Q = 10, 15 and 20 shows δS scales as $Q^{-1}S_{IA}$, as suggested by eqn. (10).

The present model may be criticized as being too close to a simple harmonic model. However, in the present case, the high frequency tail component of $g(\omega)$ depends on anharmonic contributions via the hard core of $v(r)$ and this is the feature that makes it realistic. The magnitude of the high frequency tail determines how large corrections to the IA will be. For example, δS_1 in (10) depends on $\langle \omega^2 \rangle$. The higher order terms in δS depend on $\langle \omega^n \rangle$, $n \geq 2$. If $v(r)$ had an "untreated" perfectly hard core so that infinitely large ω components were introduced, then clearly the IA would never be reached. In the present model the magnitude of the high frequency tail depends sensitively on the treatment of the hard core, here using a T-matrix. A formulation which cuts the hard core off more severely leads to a significantly smaller values of $\langle \omega^n \rangle$.[9]

Eqn. (3) has been used successfully to obtain atomic kinetic energies in solid helium.[16] As noted, (3) is independent of the IA. In practice the chief assumption is the shape of $S(Q,\omega)$ so that the second moment σ^2 can be inferred from the FWHM of $S(Q,\omega)$. For Q = 20 Å^{-1}, $S(Q,\omega)$ in Fig. 3 appears to be quite well approximated by a Gaussian.

4. LIQUID ^{4}He

To consider final state interactions in liquid ^{4}He we discuss measurements and calculations of $S(Q,\omega)$ in the range $3 \leq Q \leq 12$ Å^{-1}. These data show important contributions to $S(Q,\omega)$ arising from interactions of the scattered atom with the remainder of the fluid.

In Fig. 4 we show $S(Q,\omega)$ at Q = 7, 10 and 12 Å^{-1}. The observed[3] $S(Q,\omega)$ show important deviations from a Gaussian centered at ω_R. The peak of $S(Q,\omega)$ lies below ω_R and $S(Q,\omega)$ has a longer tail on the high ω side than on the low ω side. So either corrections to the IA or a non-Gaussian form for $n(p)$ have skewed $S(Q,\omega)$ to low ω and introduced a high frequency tail. From the discussion above, this is expected from the leading correction term (10) to the IA. The lines in Fig. 4 are simple first principles calculations[4] of $S(Q,\omega)$. The input is the potential $v(r)$ and liquid volume. $S(Q,\omega)$ is calculated in the Random

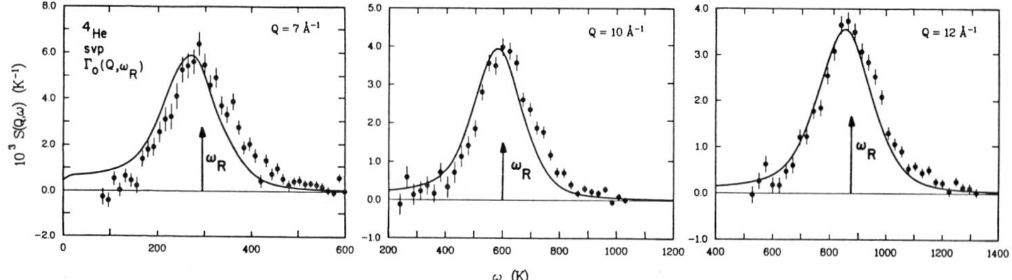

Fig. 4.   Observed (●) and calculated (solid line) $S(Q,\omega)$ including instrumental resolution for superfluid ^{4}He at svp and T = 1.2 K.   The observed intensity is converted to $S(Q,\omega)$ by requiring $\int d\omega S(Q,\omega) = 1$.   The calculated $S(Q,\omega)$ has been folded with a Gaussian of FWHM = 37.1 K (Q = 7 Å^{-1}), 82.1 K (Q = 10 Å^{-1}) and 92.1 K (Q = 12 Å^{-1}) to simulate the instrument resolution width.

Phase Approximation (RPA) which is valid when $Q \gg \langle p^2 \rangle^{\frac{1}{2}}$ where $\langle p^2 \rangle^{\frac{1}{2}}$ is a typical momentum of an atom in the fluid.   In ^{4}He $\langle (p/\hbar)^2 \rangle^{\frac{1}{2}} \approx 1.6$ Å^{-1} and in ^{3}He $\langle (p/\hbar)^2 \rangle \approx 1$ Å^{-1}.   The momentum distribution is given by the free atom Bose function $n(p) = (e^{\beta \varepsilon(p)} - 1)^{-1}$ with free atom energies $\varepsilon(p) = p^2/2m$ and temperature T = 3.2 K just above T_λ = 2.17 K.   The interaction between atoms is calculated in a T-matrix approximation valid at high Q where the liquid no longer supports collective excitations.   In Fig. 5 we present further the data at Q = 10 Å^{-1} fitted to a Gaussian.   Although the fit looks quite good there are deviations from a Gaussian at all Q which we will see are important.

In Fig. 6 we show the measured FWHM (W) of $S(Q,\omega)$ divided by Q.   In the upper half of Fig. 6, the data is from new measurements[3] of $S(Q,\omega)$ taken at the Institut Laue Langevin (ILL) while the lower half is from earlier measurements[14,17] of $S(Q,\omega)$ at Chalk River (CR).   The solid line

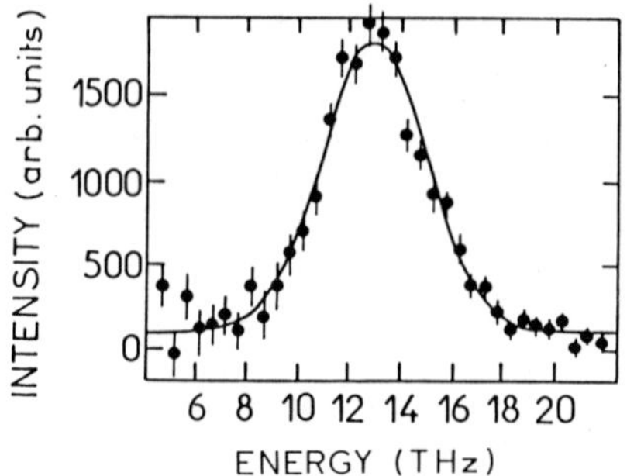

Fig. 5.   Observed intensity in superfluid ^{4}He at Q = 10 Å^{-1} (T = 1.2 K). The full line is a fitted Gaussian.

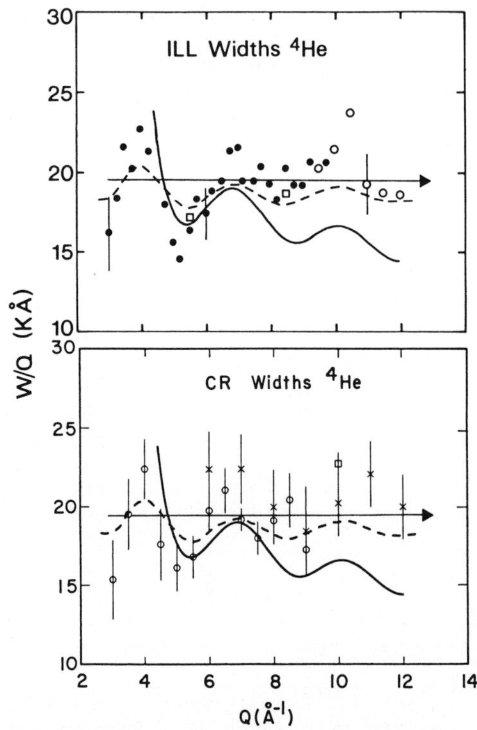

Fig. 6.   Full Width at Half Maximum, W, of $S(Q,\omega))$, divided by Q, superfluid. The data have been corrected for instrumental resolution effects. Upper half shows data from Stirling et al. (Ref. 3) taken at ILL and the lower half data from Martel et al. (Ref. 14) and Cowley and Woods (Ref. 16) taken at Chalk River. The solid line is the RPA, T-matrix calculation by Tanatar et al. (Ref. 4) and the dashed line a model calculation by Martel et al. (Ref. 14). The straight line with arrow is an estimated average of W/Q based on the data (W/Q = 19.5 K Å).

is the calculated[4] W/Q while the dashed line is a model calculation[14] based on the total ^{4}He-^{4}He atom cross-section fitted to give the correct mean value of W/Q.

Fig. 6 contains several interesting points. Firstly, in the impulse approximation, W/Q is a constant independent of Q. Indeed, eqn. (6) relates this constant value of W/Q to the atomic ⟨KE⟩ for the special case that n(p) is a Gaussian. The data shows clear deviations

from a constant W/Q due to final state interactions. These deviations
lie within the statistical error of the data for $Q \gtrsim 10$ Å^{-1} but are
almost certainly still there at higher Q. At $Q = 10$ Å^{-1}, Martel et
al.[14] were able to identify an antisymmetric component to $S(Q,\omega)$ of the
form (10). In addition, the oscillations in W extracted from the
calculated[4] $S(Q,\omega)$ continue to higher Q.

Secondly, the calculated W/Q (solid line) reproduce the observed
W/Q well in absolute value, in amplitude of the oscillation and in the
period of the oscillation. However, the calculated W/Q appears to
decrease with increasing Q. This is an artifact resulting from the
sharp peak in the calculated $S(Q,\omega)$ which makes it difficult to extract
the absolute W/Q precisely. If the calculated $S(Q,\omega)$ is folded with a
Gaussian to simulate instrument resolution and the folded width is
compared with experiment, then agreement is excellent. This is seen in
Fig. 4 for $Q = 12$ Å^{-1}, for example, where the calculated and observed
width of $S(Q,\omega)$ agrees well. For small Q ($Q \lesssim 5$ Å^{-1}), the calculated
W/Q continues to increase due to use of a frequency <u>independent</u>
T-matrix. Using a frequency <u>dependent</u> T-matrix the calculated W/Q turns
down again below $Q = 4$ Å^{-1} in agreement with experiment.[3] In the
calculations the oscillations in W/Q originate from oscillations in the
T-matrix interaction with Q; the T-matrix represents the interaction
between the scattered atom and the remainder of the fluid in the final
state. In the model of Martel et al.[14], denoted by the dashed line in
Fig. 6, the oscillations originate from oscillations in the ^{4}He-^{4}He atom
scattering cross-section.

In Fig. 7, we plot the deviation of the peak position of $S(Q,\omega)$,
$E(Q)$, from the recoil energy, $E_R = \hbar^2 Q^2/2M$, versus Q. In the IA, $E(Q)$ –
$E_R = 0$ (given by the dashed line in Fig. 6). The deviation oscillates
with Q and this oscillation can be reproduced in the calculations (solid
line). As with the width at small Q, the calculated $E(Q)-E_R$ is not
valid for $Q \lesssim 5$ Å^{-1} because the frequency dependence of the T-matrix
interaction is ignored. It is particularly interesting is that the
observed deviation does not appear to be going to zero at high Q.
However, the deviation is actually a very small percentage of E_R at
higher Q, e.g. at $Q = 10$ Å^{-1}, $E(Q)-E_R = -10$ K corresponds to
$(E(Q)-E_R)/E_R \approx 1.5\%$.

If care is taken to remove the leading correction to the IA and to
take account of the non-Gaussian $n(p)$, the atomic kinetic energy can be
obtained[19,20] from $S(Q,\omega)$. This yields kinetic energy values between 13
and 15 K which agree with the GFMC value of 14.47 ± 0.09 K quoted by
Whitlock and Panoff[17] and by Kalos et al.[18]

To investigate deviations from the IA further, accurate
measurements of the shape of $S(Q,\omega)$ at higher Q would be most
interesting. Such measurements are planned in the near future.

5. LIQUID ^{3}He

In Fig. 8 we compare $S(Q,\omega)$ in liquid ^{3}He calculated[4] in the RPA
and using a T-matrix interaction with the scattering intensity observed
by Mook[21] at $Q = 5.5$ Å^{-1} and with that observed by Sokol et al.[22] at $Q =$
13.85 Å^{-1}. In this case a <u>frequency dependent</u> T-matrix $\Gamma(Q,\omega)$ and a ^{3}He
atom self energy was included in the calculation. While the agreement
with Mook's data is only approximate, the agreement is better at higher
Q, but there $S(Q,\omega)$ is dominated by the kinetic or Doppler broadening.

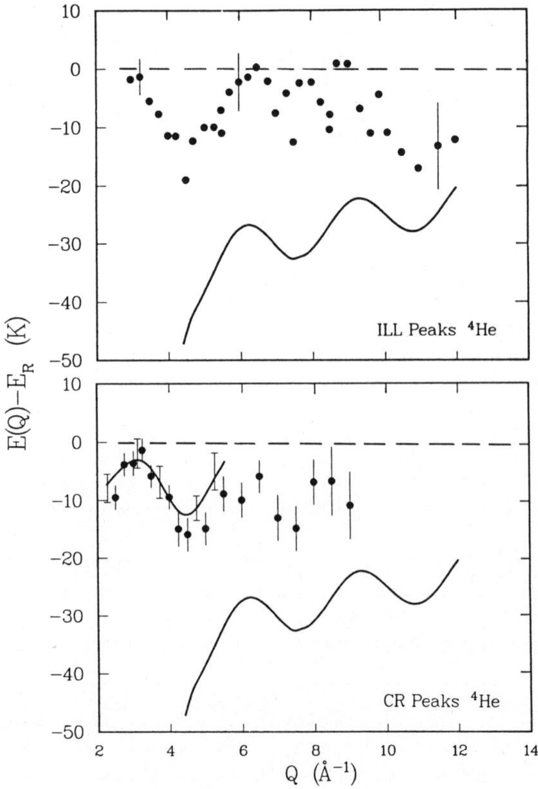

Fig. 7. The displacement of the peak
position of $S(Q,\omega)$ $(E(Q))$
below the recoil energy E_R.
Upper half shows data from
Stirling et al. (Ref. 3) and
lower half from Cowley and
Woods (Ref. 23). Solid line
is the RPA, T-matrix calculation
of Tanatar et al. (Ref. 4).

In Fig. 9 we show W/Q obtained from the calculated[4] $S(Q,\omega)$ in three
approximations. The simplest approximation is the dashed line which is
simply the Lindhard function in which the single particle energies $\varepsilon(k)$
have a self energy due to their interaction with the rest of the fluid.
This self energy oscillates with Q but these oscillations do not appear
in W/Q to any significant degree. Model 1 is the full $S(Q,\omega)$ using <u>free
particle energies</u> $\varepsilon^0(k) = \hbar^2 k^2/2M$ in the Lindhard function and a
T-matrix for ^{3}He atoms scattering in free space. Model 2 is the full
$S(Q,\omega)$ using full <u>interacting energies</u> in the Lindhard function and a
full frequency dependent T-matrix for particles scattering in the fluid.
Models 1 and 2 give similar results —especially for $Q \gtrsim 5$ Å^{-1}. These
models predict only small oscillations in W(Q)/Q - too small to be
observed. The oscillations in W/Q are small in liquid ^{3}He because the
kinetic or Doppler width is larger in ^{3}He than in ^{4}He and the width due
to the interaction is buried or lost in this constant Doppler width. In

132

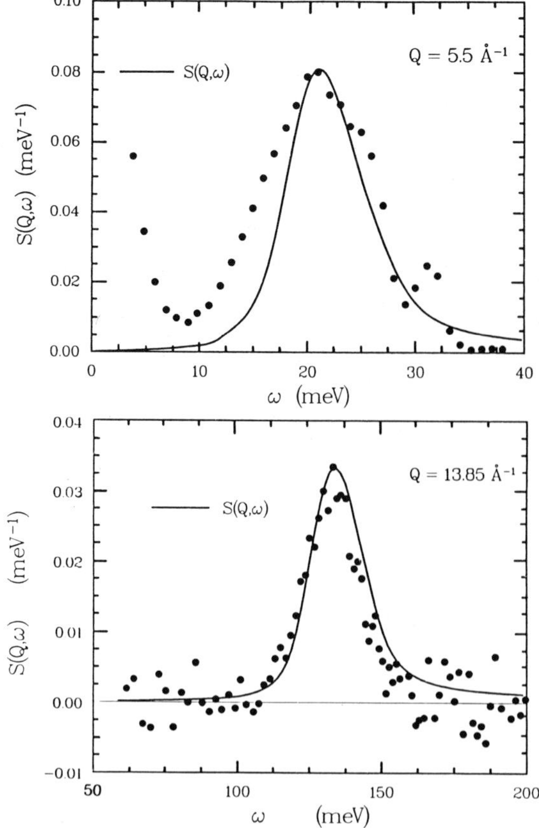

Fig. 8. RPA, T-matrix calculation of
S(Q,ω) compared with the
scattered intensity observed
by Mook (Ref. 21) at scattering
angle φ = 81.6° (upper half)
and with S(Q,ω) observed by
Sokol et al. (Ref. 22) at
Q = 13.85 Å⁻¹.

addition the amplitude of the oscillations in the interaction are
smaller in ^{3}He.

In Fig. 9 we show also the W/Q observed by Mook[21] and by Sokol et
al.[22]. The calculated widths agree well with the values of Sokol et al.
However, if a Gaussian S(Q,ω) is assumed and the second moment
calculated from the observed width above, then this second moment (or
kinetic energy) does not agree well with GFMC values of the kinetic
energy.[22] Clearly more refined techniques along the lines developed by
Silver[24] are required to verify n(p) and to extract kinetic energies in
liquid ^{3}He.

As in liquid ^{4}He, it would be interesting to have accurate line
shape measurements of S(Q,ω) for ^{3}He and to compare these directly with
calculated S(Q,ω).

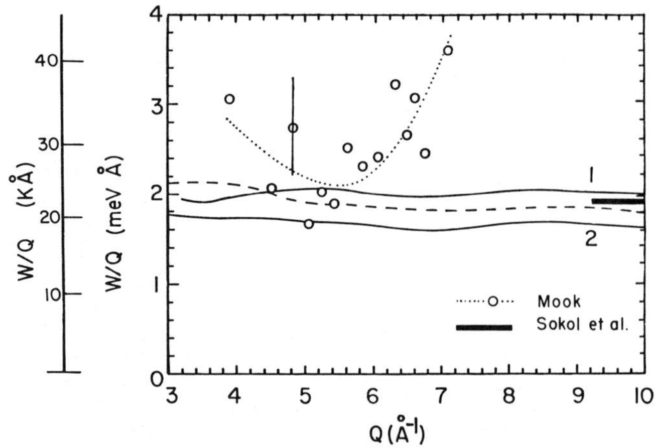

Fig. 9. FWHM divided by Q, W/Q, obtained from Lindhard function calculations of $S(Q,\omega)$ (-----), and from models 1 and 2 (———). The dots with a guide to the eye ($\cdots$) are Mook's data (Ref. 21) and (====) the widths of Sokol et al. observed at $12 \leq Q \leq 15$ Å^{-1} (Ref. 16).

ACKNOWLEDGEMENTS

Support of NATO through its collaborative grant program and the US Department of Energy, Office of Basic Energy Sciences under contract No. DE-FG02-84ER45082 is gratefully acknowledged.

REFERENCES

1. J. J. Weinstein and J. W. Negele, Phys. Rev. Lett. 49, 1016 (1982).
2. G. Reiter and R. Becher, Phys. Rev. B 32, 4492 (1985).
3. W. G. Stirling, E. F. Talbot, B. Tanatar and H. R. Glyde, J. Low Temp. Phys. 72 (in press) (1988).
4. B. Tanatar, E. F. Talbot and H. R. Glyde, Phys. Rev. B 36, 2425, 8376 (1987).
5. A. Miller, D. Pines and P. Nozières, Phys. Rev. 127, (1962) 1452.
6. P. C. Hohenberg and P. M. Platzman, Phys. Rev. 152, (1966) 198.
7. H. A. Gersch and L. J. Rodriguez, Phys. Rev. A 8, 905 (1974).
8. A. Rahman, K. S. Singwi, and A. Sjölander, Phys. Rev. 126, 986 (1962).
9. H. R. Glyde, J. Low Temp. Phys. 59, 561 (1985); L. K. Moleko and H. R. Glyde, Phys. Rev. Lett. 54, 901 (1985); Phys. Rev. B (in press).
10. D. S. Greywall, Phys. Rev. B 15, 2604 (1977); 16, 5129 (1977).
11. R. A. Reese, S. K. Sinha, T. O. Brun, and C. R. Tilford, Phys. Rev. A 3, 1688 (1971).
12. P. A. Whitlock, D. M. Ceperley, G. V. Chester, and M. H. Kalos, Phys. Rev. B 19, 5598 (1979).

13. G. B. West, Phys. Rep. 18C, 263 (1975); V. F. Sears, Phys. Rev. B 30, 44 (1984); 185, 200 (1969); A. S. Rinat (preprint).

14. P. Martel, E. C. Svensson, A. D. B. Woods, V. F. Sears, and R. A. Cowley, J. Low Temp. Phys. 23, (1976) 285.

15. B. Tanatar, G. E. Lefever, and H. R. Glyde, J. Low Temp. Phys. 62, 489 (1986).

16. P. E. Sokol, D. A. Peek, R. O. Simmons, D. L. Price and R. O. Hilleke, Phys. Rev. B 33, 7787 (1986) and references therein.

17. P. A. Whitlock and R. M. Panoff, Can. J. Phys. 65, 1409 (1987).

18. M. H. Kalos, M. A. Lee, P. A. Whitlock, and G. V. Chester, Phys. Rev. B 24, 115 (1981).

19. V. F. Sears, Phys. Rev. B 28, 5109 (1983).

20. A. D. B. Woods and V. F. Sears, J. Phys. C 10, L341 (1977).

21. H. A. Mook, Phys. Rev. Lett. 55, 2452 (1985).

22. P. E. Sokol, K. Sköld, D. L. Price and R. Kleb, Phys. Rev. Lett. 54, 909 (1985); J. Carlson et al, Phys. Rev. Lett. 55, 2367 (1985).

23. R. A. Cowley and A. D. B. Woods, Can. J. Phys. 49, 177 (1971).

24. R. N. Silver, Phys. Rev. B 37, 3794 (1988).

QUANTUM FLUIDS AND SOLIDS

MOMENTUM DISTRIBUTIONS IN LIQUID HELIUM

P.E. Sokol

Department of Physics
The Pennsylvania State University
University Park, PA 16802

T.R. Sosnick† and W.M. Snow‡

Intense Pulsed Neutron Source and Materials Science Division
Argonne National Laboratory
Argonne, IL 60439

INTRODUCTION

Helium is unique among the elements in that it remains liquid even at $T=0$ due to its light mass and weak attractive interactions. Consequently, the zero point motion is very large and the atomic wavefunctions of the helium atoms have a large overlap. This leads to many interesting properties in the liquid that depend on the quantum mechanical statistics of the particles. For example, the superfluid properties of liquid ^{4}He are a consequence of the Bose statistics obeyed by ^{4}He atoms. Alternately, the Fermi liquid behavior and the low temperature superfluid phase of liquid ^{3}He are a direct consequence of the Fermi statistics obeyed by ^{3}He atoms.

Liquid ^{4}He has been extensively studied due to its unique properties and we shall concentrate on it in this review. The liquid does not form, at atmospheric pressure, until nearly 4 K, illustrating the weakness of the attractive interactions. The properties of the liquid at high temperature, near liquification, are similar to conventional liquids. However, as the temperature is lowered a transition from the high temperature phase (Helium I) to a new phase (Helium II) occurs at $\approx$ 2.2 K. This transition is marked by a sharp feature in the specific heat — the famous λ transition — and by a change in the thermal and mass transport properties. In this new phase the liquid appears to develop a component that flows without viscosity and has infinite thermal conductivity — the superfluid — and the phase is often called the superfluid phase. Phenomenologically, this new phase can be described by the 'two-fluid' model, where the liquid is composed of two interpenetrating liquids, a superfluid component that approaches unity at $T=0$ and a normal component that is responsible for dissipation.

---

† Present address:Biophysics Division, Los Alamos National Laboratory, Los Alamos, NM 87545

‡ Permanent address: Lyman Laboratory of Physics, Harvard University, Cambridge, MA 02138

The appearance of the superfluid phase, with its unique macroscopic properties, and the success of the 'two-fluid' model in describing the transport properties are intimately linked to the Bose statistics obeyed by ^{4}He atoms. London[1] first suggested that the underlying microscopic cause for the spectacular macroscopic effects observed in the superfluid phase was due to the appearance of a Bose condensate, a macroscopic occupation of a single quantum state. Since this insightful suggestion there has been extensive theoretical work and the idea of a Bose condensate, or a broken Bose symmetry, has remained central to our understanding of the superfluid phase.

The most direct signature of the Bose condensate is in the single particle momentum distribution $n(p)$. In the normal liquid the momentum distribution has a broad Gaussian shape, as predicted for classical systems, with a width determined by the quantum zero point motion. In the superfluid phase a new feature appears, the Bose condensate. The condensate appears as a δ-function singularity in $n(p)$ with an intensity proportional to n_0, the condensate fraction.

There has been extensive experimental work attempting to verify the existence of the condensate. Hohenberg and Platzman[2] originally suggested that inelastic neutron scattering at high momentum transfers Q, where the Impulse Approximation IA can be used to directly relate the observed scattering to $n(p)$, could provide a means of directly observing the condensate. There is now a long history of attempts to directly verify the existence of the Bose condensate.[3] Unfortunately, despite many years of experimental work, a direct observation of the condensate has still not occurred.

Most of the previous experimental studies aimed at determining the momentum distribution have been carried out at reactor based neutron sources.[5-12] These measurements, due to the thermal spectrum of neutrons available from reactors, are limited to Q's below 10-12 Å^{-1}, although through some herculean efforts[4] there have been measurements at somewhat larger Q's. Unfortunately, at these Q's deviations from the Impulse Approximation are quite apparent and have prevented a direct observation of the condensate. Even when approximate methods are used to correct for these deviations, significant differences between the theoretical predictions for the momentum distribution and the experimental results exist.

In this review we will not attempt to comprehensively cover the great body of past work in this area. This has been very throughly and ably covered in several recent review articles.[13,14,4] Instead, we will concentrate on recent experimental studies that utilize the large flux of high energy neutrons available at spallation neutron sources. These sources, which have become available only recently, have made it possible to make detailed measurements at large enough momentum transfers that the conditions for the Impulse Approximation are approximately satisfied. While deviations from the IA are still present at these higher Q's, they are more amenable to theoretical treatment and detailed predictions are available. Thus, even though no distinct condensate peak is observed, for the first time excellent agreement with the theoretical predictions for $n(p)$ can be obtained.

REVIEW

We begin with a brief review of the current theoretical studies of the momentum distribution in liquid ^{4}He. Complete and detailed descriptions of the theoretical work are presented elsewhere in these proceedings.[15,16,17,18] Therefore, we shall limit our discussion to the most recent examples.

Liquid helium provides a very challenging system for theoretical studies. The attractive part of the helium potential is quite weak, as evidenced by the stability of the liquid even at T=0. This leads to a large overlap of the atomic wavefunctions which make quantum statistics an important factor in determining the properties of the ground state. However, while the attractive interactions are quite small the hard core repulsion between atoms is very large. This makes liquid helium a very strongly interacting system, in some sense even more strongly interacting than nuclear matter.[15] Therefore, it has been extremely difficult to develop a comprehensive microscopic theory of the dense liquid phase using perturbative techniques.[19] Some specific results can be obtained using analytic techniques. For example, a major success of these theories is the prediction of the relationship[20] between the broken Bose symmetry that characterizes the ground state when a condensate is present and the superfluid properties. In general, though, it has not been possible to develop a comprehensive microscopic theory of the ground or excited states of the liquid.

In view of the difficulty associated with analytic theories, the most detailed and comprehensive results for $n(p)$ have come from numerical calculations. Variational[16] and Greens Function Monte Carlo (GFMC)[18] techniques have been applied extensively to calculating the properties of the ground state at T=0. These first principles calculations, using realistic potentials, have obtained good agreement with several measured properties for the ground state of the liquid, such as the total energy and static structure of the liquid.

The most prominent feature in the ground state $n(p)$ is the appearance of a condensate, a δ function singularity at $p = 0$. Fig. 1 shows two recent calculations of $n(p)$ in the ground state, using GFMC[21] and variational[22] techniques. Both calculations show a condensate δ function (not visible in the variational calculation). The intensity in this δ function, which is known as the condensate fraction n_0, is $\approx 9\,\%$ of the total intensity.

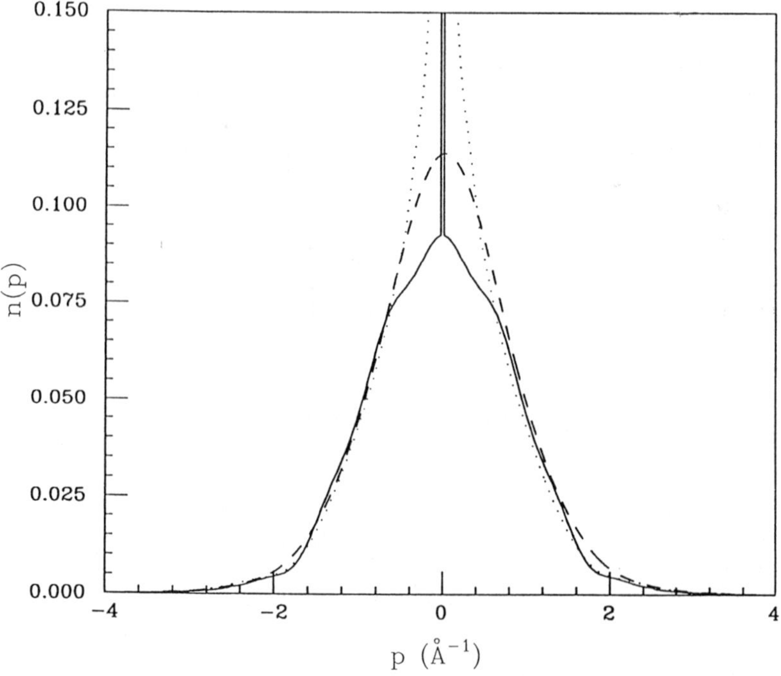

Fig. 1 Theoretical calculations of the momentum distribution. The ground state momentum distributions have been calcuated using GFMC[21] (solid) and variational[22] (dotted) techniques. The momentum distribution in the normal liquid (dashed) has been calculated using PIMC[23] techniques.

In a non-interacting gas, where the wavefunctions of the particles are free particle states and p is a 'good' quantum number, n_0 is just the fraction of atoms in the zero momentum state. This is not true in the liquid where the interactions modify the states so that they are no longer free particle states and p is no longer a 'good' quantum number. Every particle in the liquid now participates in the condensate and it is not appropriate to speak of a particular fraction of the atoms occupying the condensate, or $p = 0$ state. In the interacting liquid, the appearance of the condensate signals the development of off-diagonal long range order (ODLRO) at the one-particle level.[24,25] The condensate fraction is the magnitude of the microscopic order parameter for this phase.

The remainder of the momentum distribution, excluding the condensate δ function, is known as the uncondensed momentum distribution and exhibits several interesting features. At small p, which corresponds to long range interactions among the atoms, the effects of statistics are the dominant factor. For example, singular behavior[26,27] due to the coupling of the condensate to long wavelength collective excitations (phonons) appears. This singular behavior can be obtained exactly at small p where the phonons are well defined and goes as $1/p^2$ in the ballistic regime and $1/p$ in the hydrodynamic regime. The variational $n(p)$ in Fig. 1 explicitly shows this behavior at small p. It is not present in the GFMC results, although this is presumably due to the relatively small size of the samples used in the numerical calculations. Alternately, the large p behavior of $n(p)$ is determined primarily by the short range repulsive interaction between atoms and the statistics play little role.

The great majority of the numerical studies of the liquid have concentrated on the ground state due to the intrinsic limitations of the GFMC and variational approaches. However, Path Integral Monte Carlo (PIMC)[17] methods have recently been applied to study the liquid properties at finite temperatures. These calculations yield similar results to the ground state calculations at low temperatures i.e. a condensate fraction of approximately 9 %. However, they have the distinct advantage that they can provide results at finite temperature. For example, the condensate fraction can be obtained as a function of temperature where a rapid increase of n_0 is observed upon entering the superfluid with very little variation with temperature thereafter.

The momentum distribution may also be calculated in the normal liquid using PIMC. For example, Fig. 1 shows $n(p)$ at 3.33 K, well above the superfluid transition.[23] The momentum distribution is broad and featureless with a nearly Gaussian form, the familiar classical result. The width of the momentum distribution is determined by the quantum zero point motion of the liquid and is much larger than the width expected for classical particles. However, aside from this the shape of the momentum distribution in the normal liquid shows little effect due to quantum statistics.

DEEP INELASTIC NEUTRON SCATTERING

Inelastic neutron scattering at large momentum transfer Q provides the most direct means to obtain information on $n(p)$. In this limit, the scattering is due to single atoms and the final state of the scattering particle is assumed to be a free particle state. This is the well known Impulse Approximation (IA) and the observed scattering is proportional[15] to the Compton profile

$$J_{IA}(Y) = \frac{1}{4\rho\pi^2} \int_{|Y|}^{+\infty} pn(p)dp \qquad (3.1)$$

where ρ is the density. The scattering in the IA does not depend on the energy and momentum transfer separately, but only through the scaling variable

$$Y \equiv (M/Q)(\omega - \omega_r) \qquad (3.2)$$

where ω and Q are the energy and momentum transfer of the scattered neutron, M is the mass and $\omega_r = Q^2/2M$ is the recoil energy of the scattering atom.

In principle, the observed scattering at high Q provides direct information on $n(p)$. However, in practice the momentum and energy transfers currently attainable are not sufficient to reach the IA limit. Deviations from the IA, due to interactions of the scattering particle with its neighbors, can significantly effect the scattering. These deviations are known as final state effects (FSE) since they modify the ideal plane wave final state of the scattering particle necessary for the application of the IA.

Final state effects have been the subject of several theoretical studies[2,28-41] and considerable controversy over both the form and the importance of FSE exists. Elsewhere in these proceedings[42] we present a detailed comparison of several theories for FSE in liquid helium. The comparison is limited to large Q's where the qualitative behavior is described by the IA. We find that, while several theories provide a good description of FSE in the normal liquid, most do not accurately describe FSE in the superfluid phase. However, we do find that one of the available theories seems to provide a good description of FSE in both the normal and superfluid phases. This theory, which is due to Silver, represents FSE as a convolution of a final state broadening function with the IA result for $J(Y)$. We shall use this theory later in comparisons of the theoretical predictions with the experimental results.

Ignoring, for the moment, the complications of FSE measurements of $J(Y)$ can be used to directly determine $n(p)$. For example, if $n(p)$ is Gaussian then $J(Y)$ is also a Gaussian with the same second moment. Furthermore, the condensate, which appears as the three-dimensional delta function in $n(p)$, is a one dimensional delta function in $J(Y)$.

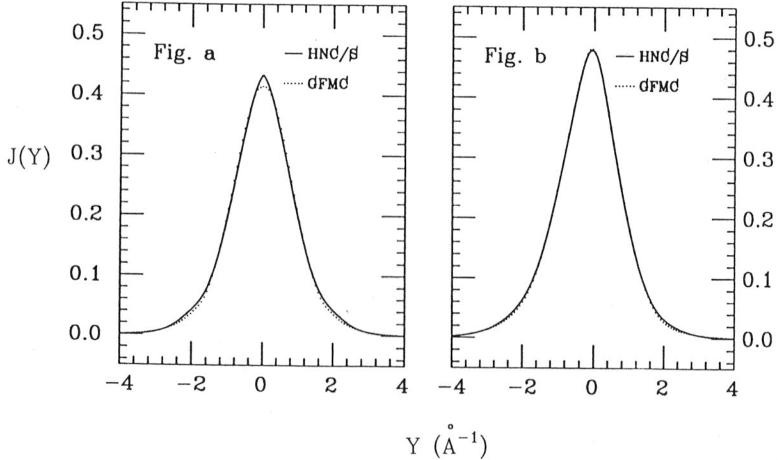

Fig. 2 Illustration of the relationship of $n(p)$ and $J(Y)$. a) shows the $J(Y)$ corresponding to the two ground state $n(p)$'s in Fig. 1. Both calculations have a condensate fraction of 9 %. However, the variational calculation (solid) shows the predicted $1/p$ singular behavior. This is lacking in the GFMC (dotted) result. b) shows the effect of instrumental resolution and FSE on the $J(Y)$ shown in a). The small differences between the predicted scattering for the two different calculations is now almost entirerly gone.

While there is a direct relationship between $J(Y)$ and $n(p)$, it is important to note that it is not a one-to-one correspondence. Features that are prominent in the momentum distribution may not be prominent in $J(Y)$. To illustrate this consider the two recent calculations of the ground state $n(p)$ discussed earlier and shown in Fig. 1. Both calculations predict a condensate fraction, which appears as a delta function with $n_0=9.2\ \%$. Both calculations also predict quite similar behavior at intermediate and large p. However, they differ markedly at small p. The variational calculation exhibits singular behavior due to coupling of long wavelength density fluctuations to the condensate which are not present in the GFMC result, presumably due to finite size effects in the calculation.

While $n(p)$ for these two calculations is quite different, the corresponding $J_{IA}(Y)$, shown in Fig. 2a, is remarkably similar. The singular behavior, which is the dominant feature in the variational $n(p)$ at small p, is quite small in $J(Y)$. When FSE and instrumental broadening are taken into account the (now small) differences between the predicted scattering for the two calculations all but disappear, as shown in Fig. 2b. The predictions of the two very different $n(p)$'s are now nearly indistinguishable! In principle, the differences between the two different $n(p)$'s we began with are still present in Fig. 2b. In practice, a measurement of the scattering would need fantastically good statistical accuracy to ever hope to observe these differences.

This insensitivity to the details at small p is a direct consequence of the IA. The Compton profile, which is proportional to the measured scattering, is the momentum distribution in the direction of the momentum transfer, with the longitudinal components averaged over. Since this is the integral of $pn(p)$ features at small p, such as singular behavior, will be suppressed due to the p in the integrand. Alternately, features at large p will be enhanced, for exactly the same reason.

To further illustrate this point, consider the problem of extracting the momentum distribution from the observed scattering. The momentum distribution may be directly obtained form the observed scattering by inverting eq (3.1), which gives

$$n(p) = -\frac{1}{2\pi p}\frac{dJ(p)}{dp} \tag{3.3}$$

again neglecting the effects of instrumental resolution and FSE broadening. However, any experimental measurement will still be affected by the statistical uncertainty of the measurements. These statistical uncertainties will translate into uncertainties in the inferred $n(p)$, although the relative magnitude of the uncertainties will not be the same throughout the spectrum.

To illustrate the effects of statistics, consider the Gaussian $J(Y)$, obtained from a Gaussian $n(p)$, shown in Fig. 3a. Statistical noise, corresponding to a statistical accuracy of 3 % at the peak center, has been added. This is characteristic of the statistical accuracy obtained in typical neutron scattering studies. The momentum distribution is obtained from $J(Y)$ using eq (3.3), which requires a differentiation followed by a division by p. Fig. 3b shows the momentum distribution obtained from the data in Fig. 3a using a simple point by point differentiation procedure.

The most striking feature of the inferred momentum distribution is the increase in the statistical noise near $p = 0$, due to the division by p. This is the same effect observed previously when considering the shape of $J(Y)$ due to different $n(p)$'s. Even large differences in $n(p)$ at small p may only cause small changes in $J(Y)$. Thus, the statistical noise present in $J(Y)$ allows a whole family of $n(p)$'s that are consistent with the observed data. This is reflected by the large errors in the inferred $n(p)$ near $p = 0$.

Admittedly, this has been an extreme example of the effects of inverting $J(Y)$ to obtain $n(p)$. Better (smoother) results can certainly be obtained by smoothing the data or using a more sophisticated differentiation procedure, but at the expense of biasing the results. This example does, however, dramatically illustrate the difficulty in extracting the momentum distribution from the observed scattering.

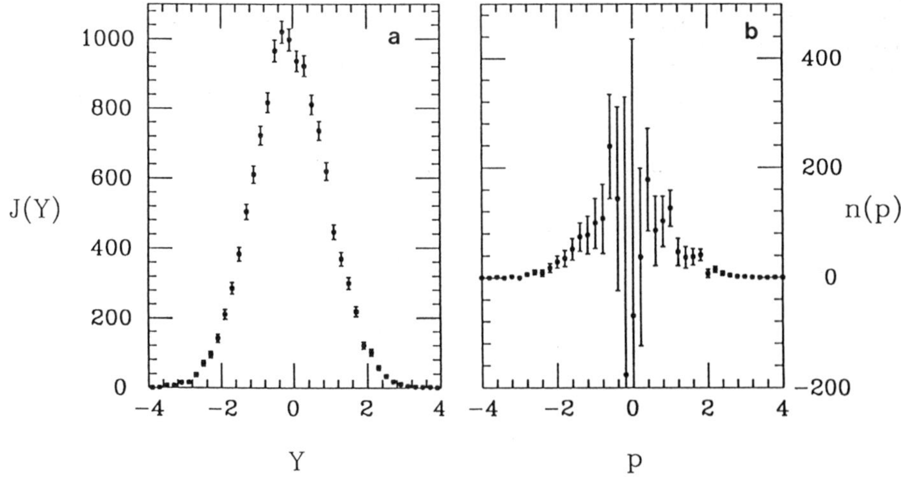

Fig. 3 a) $J(Y)$ obtained from a Gaussian $n(p)$ with statistical noise corresponding to 3 % at the peak added. b) $n(p)$ extracted from the scattering in a) using eq 3.3 using a simple point-by-pont differentiation. The errors are obtained from the statistical errors shown in a).

In view of the difficulties in extracting $n(p)$ from the experimental results, we find it more appropriate to work directly with $J(Y)$. Theoretical predictions can be compared to the experimental data using the IA, taking into account FSE and instrumental resolution. Working with $J(Y)$, as opposed to $n(p)$, has the distinct advantage that the statistical errors on the data provide a direct measure of the 'goodness-of-fit' between the theory and experiment.

COMPARISON TO THEORY

Most previous measurements of liquid helium aimed at extracting information on $n(p)$ have been carried out using reactor based instruments[5-12] where the maximum momentum transfers obtainable, with reasonable resolution, are on the order of 5-15 Å^{-1}. Unfortunately, deviations from the IA predictions, such as asymmetry in the peak and oscillations in the width as Q is varied, can still be quite significant at these Q's. In addition, accurate theoretical descriptions of FSE, particularly in the superfluid, are lacking at these low Q's.

In this review we will concentrate on recent measurements carried out using spallation neutron sources. These sources, with their high flux in the epithermal region, allow measurements at much larger Q's with relative resolutions comparable to the lower Q measurements at reactors. The higher Q's have the advantage that the scattering is consistent with the predictions of the IA. In addition, FSE are more amenable to theoretical treatment at these higher Q's.

To illustrate the results obtainable at spallation sources we will discuss recent measurements using the high resolution PHOENIX spectrometer at the Intense Pulsed Neutron Source, Argonne National Laboratory. Fig. 4 shows the measured scattering at a Q of 23 Å^{-1}, converted to $J(Y)$, at 4.2 K, in the normal liquid. The instrumental and final state broadening are also shown in Fig. 4 for reference. The instrumental resolution has been calculated using a Monte Carlo simulation of the instrument[43] and has been verified using measurements of low density helium gas. The final state broadening has been calculated by Silver for the Q used in these measurements and is discussed in detail elsewhere.[30] The instrumental and final state broadening have similar widths and are both much less than the intrinsic width of the observed scattering. An accurate knowledge of these effects is essential if we hope to extract information on the underlying momentum distribution.

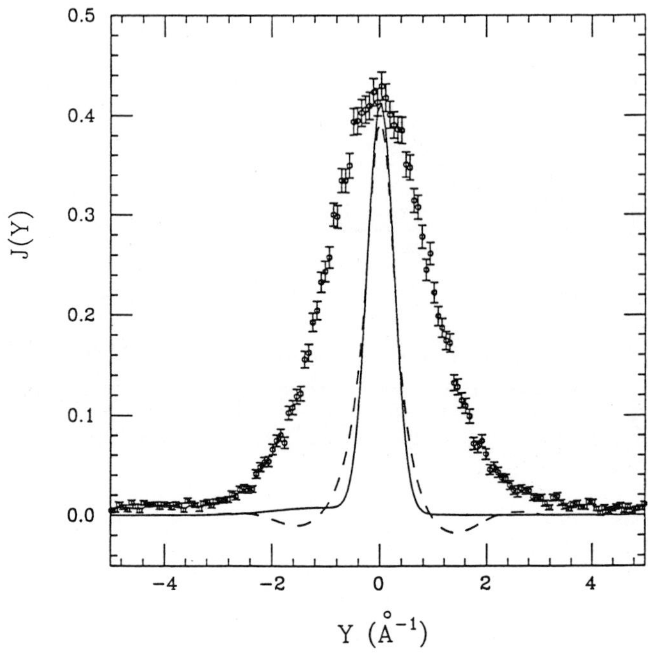

Fig. 4 The observed scattering, converted to $J(Y)$, at a Q of 23 Å^{-1} and a temperature of 4.2 K. The calculated instrumental resolution (solid line) and final state broadening (dashed line) are also shown.

The scattering shown in Fig. 4 is in qualitative agreement with the predictions of the IA. It is centered at and symmetric about $Y = 0$. In addition, measurements[44] at a variety of Q's have shown that the shape of $J(Y)$ is independent of Q above approximately 15 Å^{-1}. Thus, at least at this qualitative level the scattering is well described by the IA.

Fig. 5 shows the observed scattering at 3.5 K, in the normal liquid well above the superfluid transition, and 0.35 K, well below the superfluid transition. The scattering in both the normal liquid and the superfluid phase is broad and featureless. In the normal liquid, $J(Y)$ is nearly Gaussian, the classical result. In the superfluid, the scattering becomes visibly more peaked near $Y = 0$, but no distinct condensate peak is observed. This is consistent with the presence of a condensate peak broadened by instrumental resolution and FSE. However, due to the finite statistical accuracy of the results it is also consistent with a wide variety of momentum distributions, some which do not contain a condensate. Therefore, while the increase in scattering intensity at small Y is indicative of a condensate, it does not provide any direct evidence for a condensate fraction.

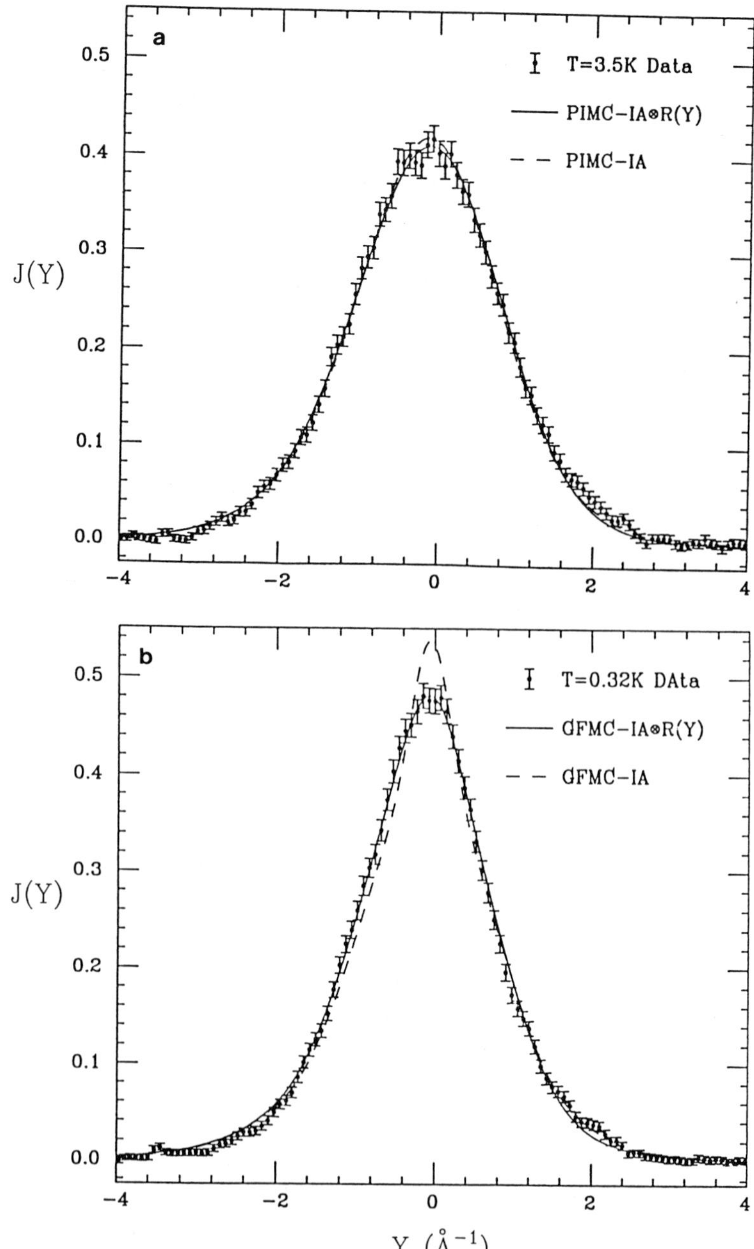

Fig. 5 The measured scattering in the normal liquid phase at 3.5 K
(a) and the superfluid phase at 0.35 K (b) of liquid ^{4}He. The
dashed curves are the theoretical predictions for $n(p)$ transformed
to $J(Y)$ and convoluted with instrumental resolution. PIMC
calculations[23] at 3.33 K have been used for comparison with the
normal liquid and are in excellent agreement with the experimental results. GFMC calculations[21] have been used for comparison
with the superfluid and large discrepancies exist near $Y = 0$. The
solid curves are again the theoretical predictions, but including
the FSE broadening[30] shown in Fig. 4. The agreement between
theory and experiment is now excellent in both the normal and
superfluid phases.

The theoretical calculations of $n(p)$ may be compared with the experimental results, providing a direct test of the calculations. The dashed line in Fig. 5a shows the theoretical prediction for $J_{IA}(Y)$ in the normal liquid using the PIMC calculations[23,45] of $n(p)$. The theoretical $n(p)$ has been converted to $J(Y)$ using the IA and broadened by the instrumental resolution. The agreement between the theoretical predication and the experiment is excellent. In this case, direct application of the IA, using the theoretical $n(p)$, provides an excellent description of the scattering in the normal liquid. FSE have little effect on the observed scattering in the normal liquid at these Q's.

The dashed line in Fig. 5b shows a similar comparison of the GFMC calculations[21] to the scattering in the superfluid. The agreement between the theoretical and experimental results is quite poor, particularly in the region of the peak center, where the condensate has the largest contribution. Based on this comparison, which has neglected FSE, we would conclude that the condensate fraction, if present at all, would have a much smaller value than the theoretical predictions.

Final state effects may be included by convoluting the theoretical predictions with the broadening function shown in Fig. 4. The solid lines in Fig. 5a and 5b are obtained when FSE are included. The predicted scattering in the normal liquid is changed very little by the inclusion of FSE. Certainly within the statistical accuracy of the measurement there is no observable change when FSE are included. Since, based on the f^2 sum rule, FSE do not change the second moment of the scattering they have little effect on the broad, nearly Gaussian $J(Y)$ in the normal liquid.

The change is much more dramatic in the superfluid phase where the momentum distribution has a sharp feature, the condensate. While FSE have little effect on the broad component of the scattering, as observed in the normal liquid, they significantly broaden the contribution from the condensate. Taking FSE into account, the agreement between theory and experiment is now excellent! For the first time *ab initio* calculations of $n(p)$ in the superfluid are in good agreement with experiment.

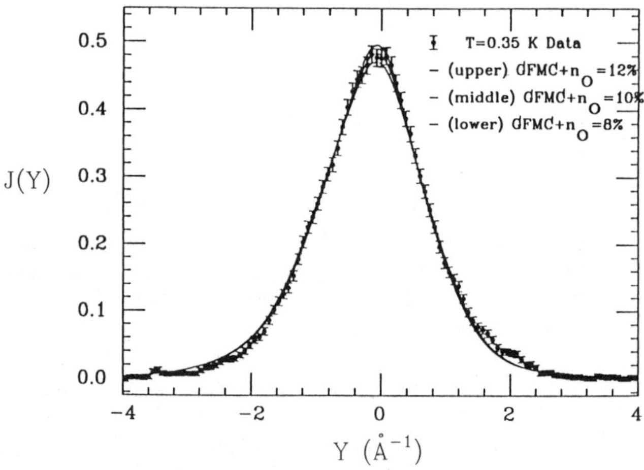

Fig. 6 Sensativity of the observed scattering to the magnitude of the condensate fraction at 0.35 K. GFMC calculations[21] have been used for the uncondensed component a narrow Gaussian to represent the condensate. The best agreement is obtained for $n_0=10$ % (central line). The two limiting values, $n_0 = 8$ and 12 % are the lower and upper lines, respectively.

An important point regarding the final state corrections is in order here. The final state broadening prediction of Silver, as shown in Fig. 4, has a narrow central peak and *negative* tails at high Y. The negative tails are essential if the broadening function is to satisfy the second moment sum rule. Thus, final state effects will not only broaden the condensate peak, they will also shift intensity around throughout the entire spectrum. For the particular FS broadening function used here, the negative tails will cause a *depletion* of the scattering at intermediate Y when a condensate is present.

In view of the discussion in the previous section regarding the relationship between $J(Y)$ and $n(p)$ it is appropriate to examine how sensitive the observed scattering is to the theoretical $n(p)$. Inevitably there is a finite statistical accuracy attached to the experimental results and a whole range of different $n(p)$'s may give equally good agreement with the data. If the statistical accuracy of the results is high then only a limited range of $n(p)$'s, all with very similar shapes, will be consistent with the data. Alternately, if the statistical accuracy is poor then the experimental results will only place very weak constraints on the underlying shape of $n(p)$.

To illustrate this, consider the scattering in the superfluid phase shown in Fig. 5b. The theoretical $n(p)$ contains a very sharp feature, the condensate δ function. By replacing the condensate δ function with a Gaussian of variable width and amplitude we can obtain some measure of the sensitivity to this particular feature. The best agreement is obtained when the width of the Gaussian is less than 0.05 Å^{-1} and n_0 is 10%, in agreement with the theoretical prediction for the condensate. Significant deviations are observed when the width is greater than 0.2 Å^{-1} and n_0 is less than 8% or greater than 12%, as shown in Fig. 6. For this particular model for the uncondensed $n(p)$ provided by the GFMC calculation we find that there is indeed a condensate with $n_0 = 10 \pm 2 \%$. However, we also point out that changing the shape of the uncondensed $n(p)$ could have an effect on the value for the condensate fraction.

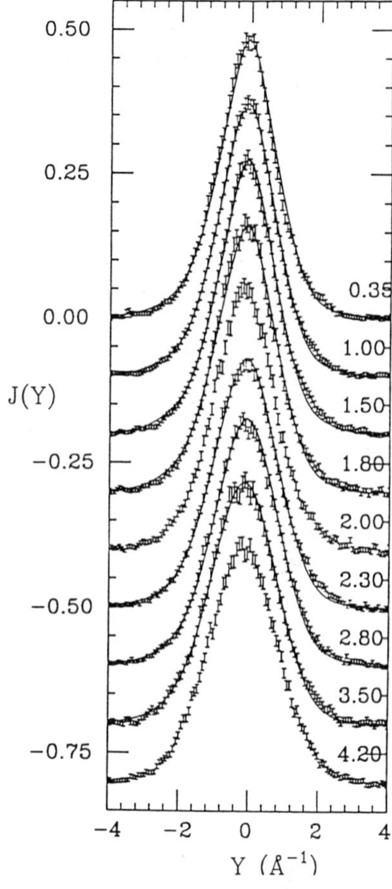

Fig. 7 Observed scattering at temperatures of 0.35, 1.0, 1.5, 1.8, 2.0, 2.3, 2.8, 3.5, and 4.2 K. The solid lines are the theoretical predictions with instrumental resolution and FSE included. GFMC calculations[21] are used for comparison with the 0.35 K results. PIMC calculations[45] are used for the remainder of the temperatures. No calculations are available for comparison with the 2.0 and 4.2 K measurements.

In a similar fashion, the sensitivity to the expected small p singular behavior in $n(p)$ can be examined. The GFMC results, which give excellent agreement with the observed scattering, do not contain the expected singular contribution. The variational $n(p)$ discussed previously explicitly includes this behavior. However, a comparison of the observed scattering with the predictions of the variational calculation yields essentially the same results as for the GFMC calculation (see Fig. 2). This is not very surprising since the weak singular behavior at small p is suppressed when $n(p)$ is transformed to $J(Y)$ as discussed earlier. Thus, the predicted small p singular behavior makes little contribution to the observed scattering and, with the experimental techniques now available, will be difficult, if not impossible, to observe.

Thus, the experimental results in the superfluid provide a clear indication of a narrow component in $n(p)$ containing approximately 9-10 % of the intensity, which is very suggestive of the condensate. Unfortunately, due to the finite statistical error inherent in any experiment, they can not definitely prove the existence of a condensate, which is formally a δ-function. Some other singular behavior, and not a condensate, could be responsible for the increase in the scattering at small p observed in the superfluid. However, as seen in the comparison with the variational $n(p)$, this would have to be a very singular behavior, much more so than the $1/p$ singularity, to obtain agreement with the experimental results. Thus, while the experimental results can not rule out a ground state $n(p)$ which does not contain a condensate, they do provide strong evidence for a very narrow feature containing $10\pm2\%$ of the total area. The excellent agreement with the theoretical predictions suggests that this very narrow feature is indeed the Bose condensate first predicted by London.[1]

To complete the comparison with the theoretical calculations, Fig. 7 shows measurements of the scattering from liquid helium at several temperatures between 0.35 K and 4.2 K. The solid lines in Fig. 7 are the theoretical predictions. The GFMC calculations[21] are used for comparison with the 0.35 K results, while the PIMC calculations[45] are used, where available, at temperatures above 1 K. The agreement is excellent over the entire temperature range! Theory and experiment appear to have converged for the momentum distribution in liquid ^{4}He at low densities.

EXTRACTION OF THE CONDENSATE FRACTION

Due to its importance in understanding the superfluid phase, considerable emphasis has been placed on determining the condensate fraction.[5-12,46-48] However, a direct determination of the condensate fraction has been frustrated by the failure to observe a distinct condensate peak. In addition, previous comparisons between the observed scattering and the theoretical predictions were not in agreement, and no value for the condensate could be inferred based on the theories. Therefore, alternate techniques to extract a value for the condensate fraction were developed. These all, in one form or another, involved modeling the momentum distribution in the superfluid to extract a value for the condensate fraction. A wide range of values for n_0 (2-17 %) were obtained, depending on the particular model used. One central feature of all these models was that they implicitly assumed the condensate exists. Therefore, taking the pessimistic point of view, all the measurements can also be viewed as consistent with some increase in $n(p)$ at small p, but with no condensate.[49]

Due to the great emphasis that has been placed on extracting a value for n_0, as opposed to the overall shape of $n(p)$, we will examine this in detail. We will not attempt to review all the previous attempts to extract a value for the condensate fraction. Instead, we will examine one of these, due to Sears et al[46], which has been used in most recent measurements aimed at extracting the condensate.[9,10,44,46] Using this technique, values for the condensate fraction of $\sim$ 10-13 % , when extrapolated to T=0, were originally obtained. These results were in resonable agreement with theoretical predictions. Recently, however, Griffin[50] pointed out that one of the central assumptions, regarding the changes induced in the uncondensed component of the momentum distribution, was incorrect. When the correct form for these changes is used, the inferred values of n_0 are 4-5 % , considerably below the theoretical estimates.

Rather than review the previous measurements, over which there is some controversy, we will illustrate the basic ideas of this procedure by applying it to the measurements discussed earlier. We begin with the model for $n(p)$ in the superfluid phase. The momentum distribution may be written in the form

$$n(p,T) = n_0(T)\delta^3(p) + (1 - n_0(T))n^*(p,T) \tag{5.1}$$

where $\delta^3(p)$ represents the condensate, $n_0(T)$ is the condensate fraction, and $n^*(p,T)$ is the uncondensed component of the momentum distribution. The observed scattering in the superfluid phase, except near $Y = 0$, is quite similar to the scattering in the normal liquid. This suggests that the momentum distribution for the uncondensed component in the superfluid may be written in the form

$$n^*(p,T) = n(p,T_\lambda) + \frac{n_0(T)}{1 - n_0(T)}\delta n^*(p,T) \,) \tag{5.2}$$

where $n(p,T_\lambda)$ is the normalized momentum distribution in the normal liquid and $\delta n^*(p,T)$ represents changes in the uncondensed $n(p)$ below T_λ.

In the absence of FSE and instrumental resolution, measurement of $J(Y)$ would provide a direct observation of the condensate. However, in the presence of these effects the condensate δ-function will be broadened. In practice, the broadening is sufficiently large the condensate peak is no longer resolved from the uncondensed component.

While, no distinct condensate peak is observed, the effects of the condensate will still be present in the observed scattering. In general, they will appear as an increase in the scattering near $p = 0$, rather than a distinct peak. The condensate fraction may then be extracted from the observed scattering in the superfluid under two assumptions. They are:

- The entire contribution of the condensate peak occurs within some small region around $Y = 0$.

- The changes in the uncondensed momentum distribution can be accounted for using theoretical calculations.

From the first assumption the scattering due to the condensate will only contribute in the region around $Y = 0$. In terms of $n(p)$, the total intensity in a region within p_c of the origin is

$$\alpha(T,p_c) = \int_0^{p_c} n(T,\vec{p}) \, d\vec{p} \,. \tag{5.3}$$

Since this region, by assumption, contains the entire contribution of the condensate we find

$$n_0(T,p_c) = \frac{\alpha(T,p_c) - \alpha(T_\lambda,p_c)}{1 - \alpha(T_\lambda,p_c) + \gamma} \tag{5.4}$$

where

$$\gamma = \int_0^{p_c} \delta n^*(p,T) \, 4\pi p^2 dp \tag{5.5}$$

Thus, knowing the form of $\delta n^*(p)$ in the region within p_c, which is where the modeling of $n(p)$ enters, we can directly obtain the condensate fraction.

One of the central assumptions in the derivation of (5.4) is that the entire contribution of the condensate is confined to the region around the origin. Instrumental resolution, which broadens the condensate peak, certainly satisfies this requirement. However, FSE are a different matter. As shown in the last section, they not only broaden the peak but they shift intensity from one part of the spectrum to another. This certainly violates the assumption above that the condensate contribution is still localized around the origin.

Final state effects must, at least in part, be taken into account if we hope to determine n_0. The most important feature of FSE to account for is the shifting of intensity caused by the final state broadening. This may be accomplished by deconvoluting the final state broadening function used earlier from the observed scattering. While, due to the statistical noise in the results, this will not entirely remove FSE it should at least account for the major effects.

Rather than attempt a numerical deconvolution, a difficult and unstable procedure with statistically noisy data, we have found it convenient to represent the scattering by a functional form and then fit this form, convoluted with instrumental resolution and FSE, to the observed scattering. A model consisting of two Gaussians, with variable amplitudes and widths and a center fixed at $Y = 0$, provides sufficient freedom to fit the data in the normal and superfluid phases. This model has the advantage that certain physical constraints, such as symmetry, positivity, etc, are implicitly included in the results of the deconvolution.

The results of deconvoluting FSE and instrumental broadening from the observed scattering are shown in Fig. 8. At high temperatures, near 4K, the scattering is nearly Gaussian. However, as the temperature is lowered toward T_λ the scattering becomes more peaked near $Y = 0$, even in the normal liquid. A large increase in the scattering intensity at small Y is observed at T_λ. The scattering in the superfluid becomes much more peaked, indicating the presence of a condensate.

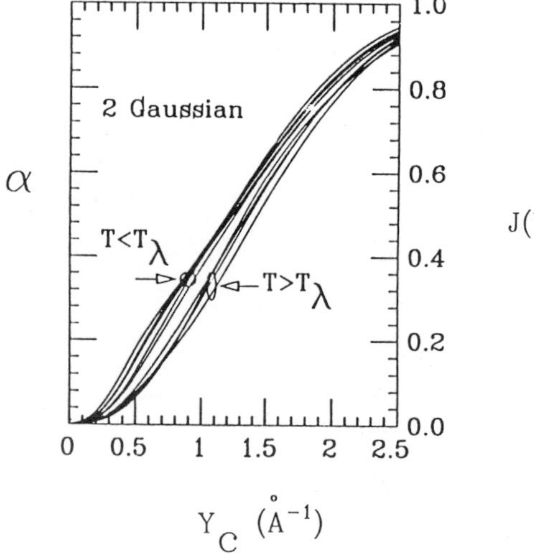

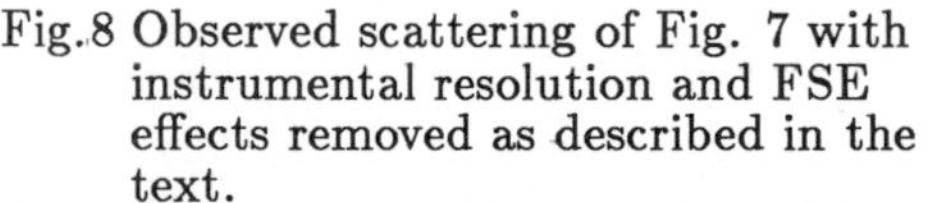

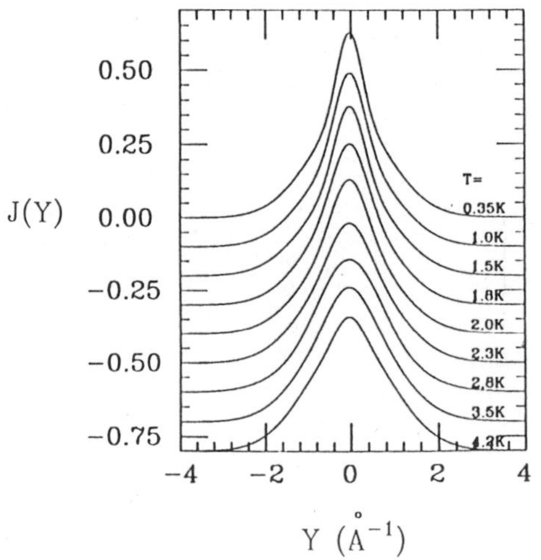

Fig.8 Observed scattering of Fig. 7 with instrumental resolution and FSE effects removed as described in the text.

Fig. 9 Integrated scattering from 0 to p_c using the scattering shown in Fig 8. The gap between the results above and below the superfluid transition is indicative of the formation of a condensate.

While the sharp increase in the scattering intensity at small p in the superfluid is consistent with the presence of a condensate, again no δ function appears in the deconvolutions in the superfluid phase. However, these same results were in excellent agreement with the theoretical predictions containing a condensate. Thus, the lack of a distinct condensate peak in the results is a direct consequence of the statistical noise present in the data. In fact, due to the statistical noise in the data, a whole series of $n(p)$'s are consistent with the results, including those with δ functions as show earlier (Fig. 5b). Thus we take the width of the narrow component in the fits, which is ~ 0.6 Å^{-1}, as indicative of the statistical smearing of the condensate peak.

In previous measurements at lower Q's, FSE are quite significant and the IA is not even qualitatively satisfied. No detailed theories for FSE are available at these lower Q's and approximate techniques such as symmetrization and averaging over different Q's have been used to remove the deviations from the IA.[9,10,46] However, as we show elsewhere in these proceedings,[42] these procedures do not properly account for the shifting of intensity by FSE. Therefore, the increase in scattering at p can not be uniquely identified with the contribution from the condensate, violating one of the central assumptions used.

The integrated intensity, α, may be obtained directly from the results in Fig. 8 using a simple integration by parts of (5.3). Fig. 9 shows $\alpha(p_c, T)$ versus p_c and T in both the normal and superfluid phases. The most prominent feature is the gap between the curves for the normal and superfluid between 0.5 and 1.5 Å^{-1}. The appearance of this gap is a dramatic signal of the formation of the Bose condensate and the size of the gap is proportional to the condensate fraction.

Changes in the uncondensed $n(p)$ due to the presence of the condensate also contribute to the scattering at small p and must be taken into account if n_0 is to be determined. These cannot be experimentally separated from the condensate contribution and we must make recourse to the theoretical predictions for the small p behavior. The main condensate induced change, at small p, is the singular behavior introduced by fluctuations in the condensate coupling to long wavelength density fluctuations (phonons). The form for the small p singular behavior[26,27,50] is

$$
\begin{aligned}
\delta n(p, T) &= \frac{mk_BT}{(2\pi)^3\hbar^2\rho n_s}\frac{1}{p^2} \qquad p < \frac{k_BT}{\hbar c} \\
&= \frac{mc}{2(2\pi)^3\hbar\rho}\frac{1}{p} \qquad p < \frac{k_BT}{\hbar c}
\end{aligned}
\tag{5.6}
$$

where ρ is the density, n_s is the superfluid fraction, and c is the velocity of sound. Griffin[50] has pointed out that the crossover between these two regimes takes place at small p ($k_BT/\hbar c$ is only .1 Å^{-1} at 2 K) so that the contribution of the $1/p^2$ term is negligible. Therefore, in the phonon region ($p <\sim 0.7$ Å^{-1}) where this result should be valid, γ is then $0.85n_0p^2$. In addition, Griffin[50] has attempted to extend this result beyond the phonon region using the measured dispersion relation, assuming the $1/p$ singular behavior is still valid.

The condensate fraction can be obtained directly from the results in Fig. 9 using the theoretical predictions for the condensate induced changes. Fig. 10 shows the condensate fraction, using Griffin's[50] calculation of γ as a function of p_c. Ideally, if instrumental resolution and FSE were absent and if $\delta n^*(p, T)$ were correct, the condensate fraction would be a constant, independent of p_c. The experimental results differ significantly from this expectation.

The inferred n_0 start at zero when $p_c = 0$, increase to a maximum value in the range of 0.6 to 0.8 Å^{-1}, and then decreases again to zero. The behavior near

$p_c = 0$ is not surprising since there is no δ function singularity in the two Gaussian fits that would give a rapid rise at $p = 0$. The width of 0.6 Å^{-1} is just the statistical broadening associated with the condensate peak in the deconvolutions. We would not expect the results to be valid for p_c less than this value. The decrease at large p_c is also not surprising since the scattering in the normal and superfluid phase is almost identical in this region.

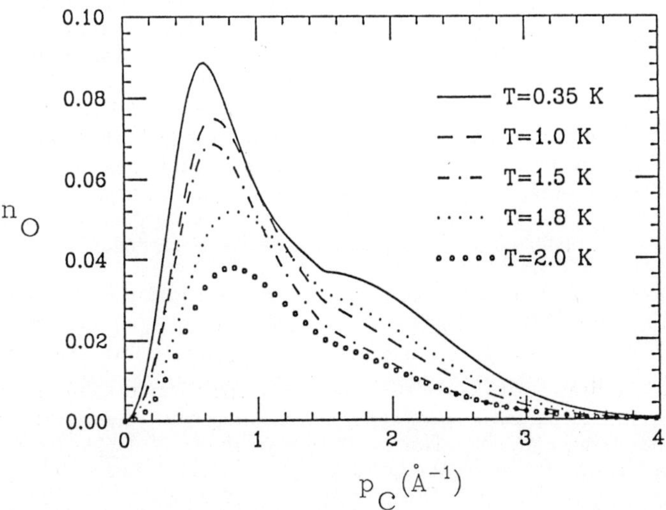

Fig. 10 Inferred values of the condensate fraction at temperatures of 0.35, 1.0, 1.5, 1.8, and 2.0 K. The curves are described in the text.

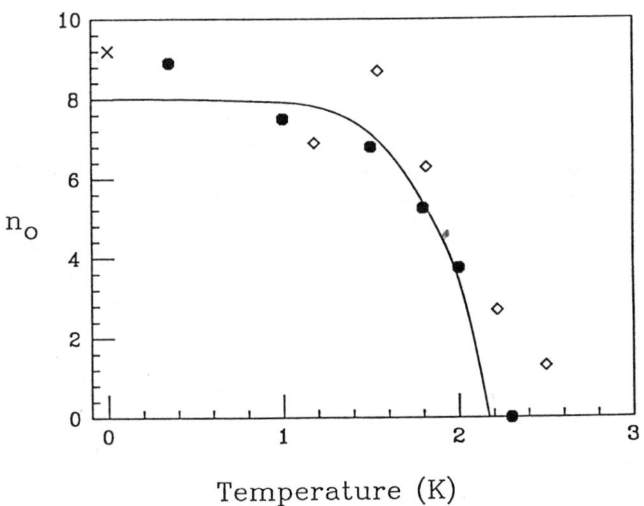

Fig. 11 Maximum of the inferred n_0 versus temperature (heavy dots). Also shown are the GFMC (cross) and PIMC (diamonds) results. The curve is a fit to the data with a critical exponent of 6 as described in the text.

There is a further limitation on the values of p_c appropriate for the extraction of n_0. The theoretical correction for the condensate induced changes γ is only valid within the phonon region, which extends to ~ 0.8 Å^{-1}. Above this region, the behavior of the condensate induced corrections have only been extrapolated. Thus, above ~ 0.8 Å^{-1} the model for the condensate induced changes is suspect.

This is precisely the region where the broad maximum in the inferred condensate occurs. Thus, we use the peak values for n_0 as a function of p_c as representative of the condensate fraction in the liquid. These values are shown in Fig. 11. At low temperatures, the inferred n_0 is ~ 10 % , in good agreement with the GFMC and variational results. In addition, the inferred n_0, shows a slow decrease with increasing temperature followed by a rapid drop at the superfluid transition, consistent with the rapid depletion near the superfluid transition predicted by PIMC studies.

Finally, we note that most previous attempts to extract the condensate[46] a different form for the small p changes in $n(p)$ has been used. This lead to values of the condensate on the order of 10-13 % at T=0, in reasonable agreement with theory. However, Griffin[50] pointed out that this form was based on an incorrect combination of the two limiting behaviors shown above. When the correct form for the small p singular behavior is used values of 4-5 % at T=0 are obtained.

The temperature dependence of the condensate fraction has often been characterized by fitting to a form such as

$$n_0(T) = n_0(0)[1 - (T/T_\lambda)^\alpha] \tag{5.7}$$

where $n_0(0)$ is the limiting value of the condensate fraction at T+) and α is a critical exponent. Past measurements have found α on the order of 3.6, indicating a fairly weak temperature dependence. However, these values were based on the incorrect form of the small p behavior of the condensate induced changes. The solid curve in fig. 12 is a fit to the results here and gives $n_0 = 8.0\%$ and $\alpha = 6$. The larger value of reflects a much sharper temperature dependence than previously reported. As can be seen in the figure, the condensate exhibits little temperature dependence until quite near T_λ and then drops quite rapidly.

In summary, it appears that resonable values for the condensate fraction in the superfluid phase can be extracted using techniques such as these. However, it is essential that both FSE broadening be properly taken into account and that a good theoretical model for the condensate induced changes be available. The values obtained for n_0 will be valid only to the extent that both these conditions are satisfied.

CONCLUSIONS

We have now reached a stage where there is excellent agreement between the theoretical results and the experimental observations for all aspects of the momentum distribution in liquid ^{4}He. In particular, the agreement between theory and experiment settles the long-standing question regarding the magnitude, and even the presence, of a Bose condensate in the superfluid. The experimental results provide convincing evidence for a Bose condensate containing 10 % of the atoms.

This recent convergence of theory and experiment has come about through several simultaneous advances. Theoretically, the availability of more powerful computational techniques and facilities for the calculation of $n(p)$ has lead to very accurate theoretical predictions. In addition, the development of accurate theoretical predictions for the FSE broadening, particularly in the superfluid, has allowed *ab initio* comparisons of the theoretical and experimental results. Experimentally, the development of spallation sources has allowed us to obtain high quality results with good statistical accuracy.

A better understanding of the strengths and weakness's of DINS as applied to determinations of $n(p)$ in quantum systems has also evolved. For example, the insensitivity of the observed scattering to some of the singular behavior in $n(p)$ is now understood. We have tried to convey an appreciation of where the measurements can provide a definite test of theories and where they are not sensitive to particular details.

Unfortunately, the original goal for much of the work in liquid helium, a direct observation of the condensate fraction, has not come to pass. In view of our current understanding of FSE in helium, it is unlikely that this goal will ever be reached in deep inelastic neutron scattering experiments. While the current experimental results do not definitively prove the existence of a condensate, they do provide such overwhelming evidence that we can now consider the problem, for the bulk liquid at least, solved.

Acknowledgements

We would like to acknowledge useful discussions with R.N. Silver, H.R. Glyde, D.L. Price, C.K. Loong, P.A. Whitlock, and R.M. Panoff. This work was supported by the National Science Foundation under grant DMR-8704288 and by OBES/DMS support of the Intense Pulsed Neutron Source at Argonne National Laboratory under DOE grant W-31-109-ENG-38.

References

[1] F. London, Nature **141**, 643 (1938).

[2] P.C. Hohenberg and P.M. Platzman, Phys. Rev. **152**, 198 (1966).

[3] for reviews see E. C. Svensson, V.F. Sears, Physics **137B**, 126 (1986.)

[4] H. Mook, contribution in these procedings.

[5] R.A. Cowley and A.D.B. Woods, Phys. Rev. Lett. **21**, 787 (1968).

[6] R.A. Cowley and A.D.B. Woods, Can. J. Phys. **49**, 177 (1971).

[7] P. Martel, E.C. Svensson, A.D.B. Woods, V.F. Sears, and R.A. Cowly, J. Low Temp. Phys. **23**, 285 (1976).

[8] H.A. Mook, Phys. Rev. Lett. **32**, 1167 (1974).

[9] H.A. Mook, Phys. Rev. Lett. **51**, 1454 (1983).

[10] H.A. Mook, Phys. Rev. B **37**, 5806 (1988).

[11] E.C. Svensson, V.F. Sears, A.D.B. Woods and P. Martel, Phys. Rev. B **21**, 3638 (1980); V.F. Sears, E.C. Svensson, A.D.B. Woods and P. Martel, Atomic Energy of Canada Limited Report No. AECL-6779 (unpublished).

[12] A.D.B. Woods and V.F. Sears, Phys. Rev. Lett. **39**, 415 (1977).

[13] H.R. Glyde and E.C. Svensson in **Methods in Experimental Physics**, Vol. 23B, D.L. Price and K. Sköld, ed. Academic Press, 1987 p.303.

[14] V.F. Sears, Can. J. Phys. **63**, 68 (1985).

[15] P.E. Sokol, R.N. Silver, and J.W. Clark, contribution in these proceedings.

[16] E. Manousakis, contribution in these procedings.

[17] D.M. Ceperley, contribution in these procedings.

[18] R. Panoff and P. Whitlock, contribution in these procedings.

[19] J.W. Clark and M.L. Ristig, contribution in these proceedings.

[20] A. Griffin, Can. J. Phys. **65**, 1368 (1987).

[21] P.A. Whitlock and R. Panoff, Can. J. Phys. **65**, 1409 (1987).

[22] E. Manousakis, V. R. Pandharipande, Q. N. Usmani, Physical Review B **31**, 7022 (1985); E. Manousakis, V. R. Pandharipande, ibid., 7029.

[23] D.M.Ceperly and E.L.Pollock, Phys. Rev. Lett. **56**, 351 (1986).

[24] O. Penrose and L. Onsager, Phys. Rev. **104**, 576 (1956).

[25] C.N. Yang, Rev. Mod. Phys. **B34**, 694 (1962).

[26] P.C. Hohenburg and P.C. Martin, Ann. Phys. (N.Y.) **34**, 291 (1965).

[27] J. Gavoret and P. Nozières, Ann. Phys. (N.Y.) **28**, 349 (1964).

[28] R.N. Silver, contribution in these proceedings.

[29] H.R. Glyde and W. Sterling, contribution in these proceedings.

[30] R. N. Silver, in Proceedings of the 11th International Workshop on Condensed Matter Theories, Oulu, Finland, 1987, Plenum Press; Rapid Communications, Physical Review B, March 1 (1988).

[31] P.M. Platzman and N. Tzoar, Phys. Rev. B **30**, 6397 (1984).

[32] H.A. Gersch, and L.J. Rodriguez, Phys. Rev. A **8**, 905 (1973); L.J. Rodriguez, H.A. Gersch and H.A. Mook, **ibid. 9**, 2085 (1974.)

[33] T.R. Kirkpatrick, Phys. Rev. B **30**, 1266 (1984).

[34] G. Reiter and T. Becher, Phys. Rev. B **32**, 4492 (1985).

[35] J.J. Weinstein and J.W. Negele, Phys. Rev. Lett. **49**, 1016 (1982).

[36] W.C. Kerr, K.N. Pathak and K.S. Singwi, Phys. Rev. A **2**, 2416 (1970).

[37] P. Martel, E.C. Svensson, A.D.B. Woods, V.F. Sears, and R.A. Cowly, J. Low Temp. Phys. **23**, 285 (1976).

[38] A.S. Rinat, Phys. Rev. B **36**, 5171 (1987).

[39] V.F. Sears, Phys. Rev. B **30**, 44 (1984).

[40] S. Stringari, Phys. Rev. B **35**, 2038 (1987).

[41] B. Tanatar, G.C. Lefever, and H.R. Glyde, J. Low Temp. **62**, 489 (1986).

[42] P.E. Sokol, T.R. Sosnick, W.M. Snow, and R.N. Silver, Contribution in these proceedings.

[43] P.E. Sokol, G.K. Kellogg, W.M. Snow, T.R. Sosnick, and J.M. Carpenter, to be published.

[44] P.E. Sokol, Can. J. Phys. **65**, 1393 (1987).

[45] D.M. Ceperly, and E.L. Pollock, Can. J. Phys. **65**, 1416 (1987).

[46] V.F. Sears, E.C. Svensson, P. Martel and A.D.B. Woods, Phys. Rev. Lett. **49**, 279 (1982).

[47] O.K. Harling, Phys. Rev. A **3**, 1073, (1971), A.G. Gibbs and O.K. Harling, *ibid.* **7**, 1748, (1973).

[48] H.A. Mook, R Scherm, and M.K. Wilkinson, Phys. Rev. A **6**, 2268 (1972).

[49] H. W. Jackson, Phys. Rev. A **10**, 278 (1974).

[50] A. Griffin, Phys. Rev. B **32**, 3289 (1985).

NEUTRON SCATTERING STUDIES OF n(p) AT REACTOR SOURCES

H.A. Mook

Oak Ridge National Laboratory
Solid State Division
Oak Ridge, TN 37831-6031

What follows is an account of neutron scattering studies on liquid helium done at Oak Ridge over the last 15 or so years. As time progressed, the measurement techniques improved considerably, but more important was our improved understanding of the measurements. In particular, a better understanding was gained of the difficulty of obtaining a condensate fraction from measurements of $S(Q,\omega)$ of liquid ^{4}He.

The idea that ^{4}He can undergo Bose Einstein condensation at low temperatures has generated great interest in much of the scientific community. It is not often one can observe the effects of quantum mechanics on a macroscopic scale. The idea behind the neutron scattering experiment to observe Bose Einstein condensation is extraordinary simple if one uses high enough neutron energies so that the impulse approximation correctly describes the scattering. The experiment is done as shown in Fig. 1 by bringing a mass 1 neutron with wave vector k_0 and energy E_0 incident on the mass 4 He and measuring the scattering at a fixed angle. If the He is at rest, one gets a peak in the scattering at an energy $\hbar\omega$ that is given by the energy and momentum conservation relations for mass 4 and mass 1 particles. However, if the He is in motion in the liquid, sometimes it moves along the incident neutron path and sometimes opposite to it so that the neutron is Doppler shifted in the scattering process, and the scattering distribution is broadened. If one has a fraction of the atoms in the He that are at rest as they would be in the case of Bose Einstein condensation, then the scattering distribution would be a sharp peak on top of the broad distribution.

In considering an experiment, the first thing to think about is how wide the scattering distribution $S(Q,\omega)$ is from the motion of the He atoms in the normal liquid. It turns out that the energy width in meV of the scattering distribution is nearly equal to the momentum transfer Q in Å^{-1} at which the experiment is performed so that the width ΔE (meV) $\simeq$ Q(Å^{-1}). Hohenberg and Platzman[1] considered such a measurement and expressed the need for high energy neutrons, so let us first consider the case for Q = 100Å^{-1}.

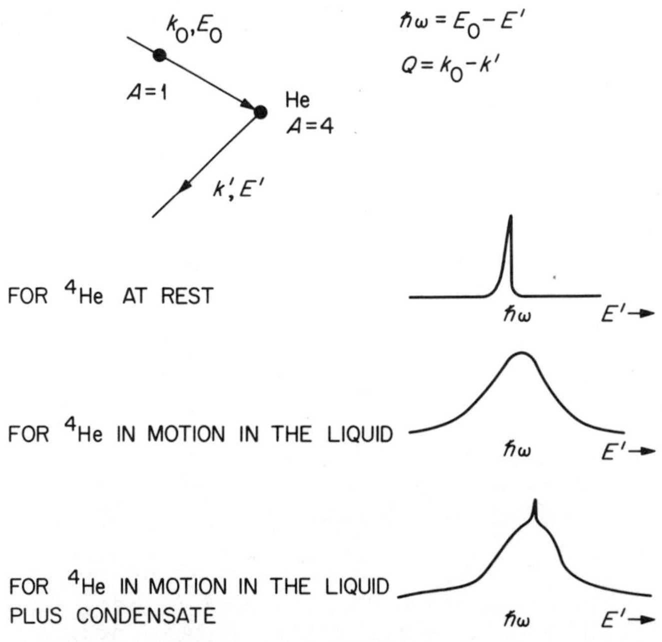

Fig. 1   Neutron scattering pattern for ^{4}He that has undergone Bose Einstein Condensation.

For a momentum transfer of 100Å^{-1}, we need incident neutrons of momentum about 100Å^{-1} or 20.75 eV in energy. The fractional energy resolution desired is roughly given by

$$\frac{0.1\,eV(width\ in\ meV)}{20.75eV} = 0.0048.$$

We really should have five resolution elements across the peak in order to see anything interesting which means we not only have to do the experiment with 20 eV neutrons, but we need resolutions on the order of 0.1%. This is an exceedingly difficult, if not impossible, prospect; so let us consider an easier experiment and try a Q of 15Å^{-1}.

For a momentum transfer of 15Å^{-1}, we need incident neutrons of momentum about 15Å^{-1} which is 460 meV in energy.

$$\frac{15(width\ in\ meV)}{460meV} = 0.032.$$

Divide this by five to get five resolution elements, and we have resolutions of somewhat under 1%. We can get 460 meV neutrons from reactor sources, and the resolution is difficult but possible to achieve. In 1971, such an experiment was undertaken[2] using a triple-axis spectrometer at the High Flux Isotope Reactor (HFIR) ; the results are shown in Fig. 2. Data were taken at 1.2K and 4.2K and the resolution is indeed such that about five resolution elements are

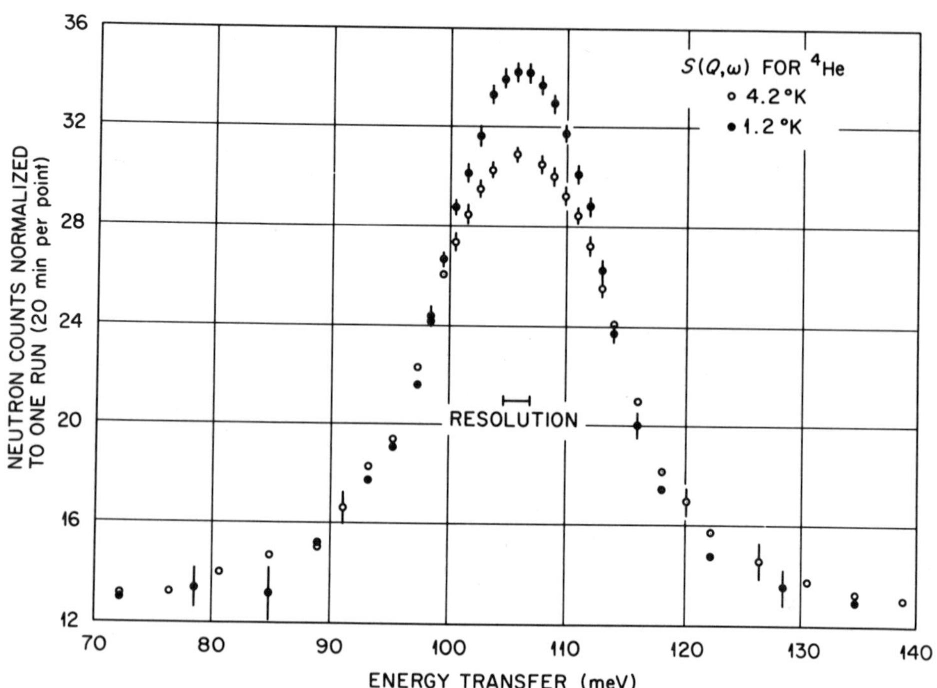

Fig. 2   Triple-axis measurement of $S(Q,\omega)$ for ⁴He at 4.2K and 1.2K.

across the peak. This is really quite an extraordinary experiment as the counting time was on the order of six months, far longer than it would be possible to count today because of the demand for beam time on the instruments. Rather good statistics were collected in the neighborhood of the peak where the condensate should be visible, while less precise data were collected in the wings of the distribution.

It is obvious that no sharp condensate peak is visible on $S(Q,\omega)$ measured at 1.2K, and indeed, as we know now, no sharp peak is expected even for momentum transfers two or three times this value. We were then left with the problem of extracting a condensate fraction. With the idea that $S(Q,\omega)$ at 1.2K should consist of two distributions, the 4.2K and the 1.2K data were least-squares fitted with two distributions. It turned out that the 4.2K data were well fitted using one distribution, but the fit only gave good results with two distributions for the 1.2K data. Assuming that the smaller of the distributions stemmed from the condensate, a condensate fraction of about 2.5% was determined. It turns out that this is not a good way to extract a condensate fraction as it ignores the change in $S(Q,\omega)$ in going from 4.2K to 1.2K, except for the very top of the peak. In the paper it was pointed out that if one considers the area between two distributions, a condensate fraction of about 10% is indicated, and it turns out that this is a better way to analyze the data.

A real breakthrough came though with the papers of Martel et al.,[3] Woods and Sears,[4] and Sears, Svensson, Martel, and Woods[5] who outlined a way to deal with data like that shown above. Their method relied on symmetrizing the data about the calculated recoil energy to minimize final

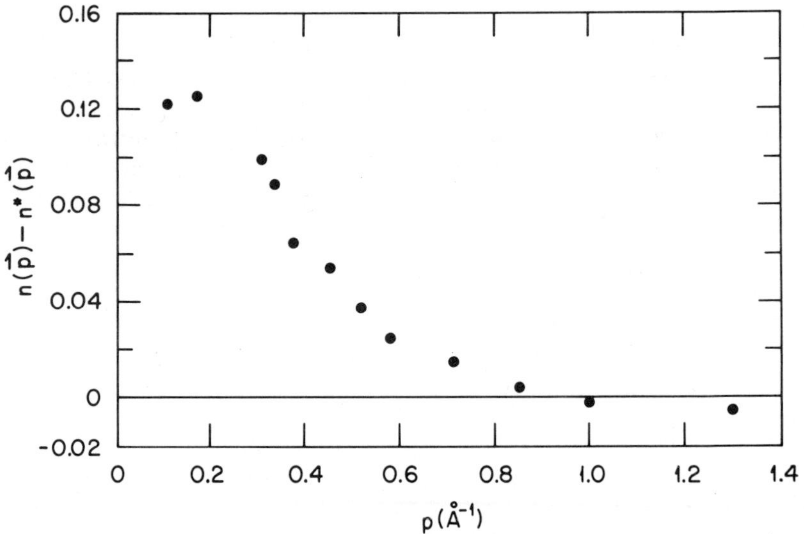

Fig. 3   n(p) at 1.2K minus n(p) at 4.2K for the triple-axis data shown in Fig. 1.

state effects, and if low Q data are used, to average data together at a number of Q values. The momentum distributions n(p) were then obtained from the data, and the condensate fraction determined from the formalism given by

$$n_0 = \frac{\epsilon}{1 - \beta + \gamma}$$

(1)

where

$$\epsilon = \int_0^{p_c} [n(p) - n^*(p)]\, p^2 dp$$

(2)

and

$$\beta = \int_0^{p_c} p^2 dp$$

(3)

where p_c is the point where n(p)-n*(p) becomes negative and γ is a correction term that becomes large near the λ temperature. Fig. 3 shows the quantity n(p)-n*(p) needed in obtaining the condensate fraction. We note that the result is fairly accurate, but that little information is available beyond $1\text{\AA}^{-1}$, as high quality data were only taken near the top of the distribution. Using the value for γ given in Ref. 5, a condensate fraction is obtained of about 10%.[6] However, it should be noted that Griffin[7] has recalculated γ, and a smaller condensate fraction would be obtained using his values.

It would clearly be of interest to obtain good measurements of n(p) over a reasonably complete p range, but to do so is difficult on triple-axis spectrometers because of the long counting times. The experiment is easier with time-of-flight since the energy resolution can be uncoupled from the Q resolution in this technique, and high Q resolution is not needed. A time-of-flight spectrometer was developed that could make fast neutron pulses by ultrasonic techniques. As shown in Fig. 4, Bragg reflection takes place from a perfect crystal according to the relation $\lambda = 2d\sin\theta$; a very narrow range of θ corresponds to a narrow range of λ, and little scattering intensity is obtained. If a high intensity ultrasonic pulse is incident on a transducer attached to the crystal, a band of λ's is obtained from each θ, and a large neutron pulse is achieved. Since the pulse is electronic, it can be repeated at any time in a precise manner. This makes it possible to use the correlation technique, as shown in Fig. 5. In a normal chopper, a pulse is made and the neutrons are counted from this pulse over some time τ long compared to the pulse. With the correlation technique, a signal function S(t) is made incident on the sample, and a convolution of the signal function and the function you wish to obtain, F(t), is counted. However, S(t) can be constructed so that F(t) can be obtained easily by the cross correlation of the counted neutrons, Z(t), with the signal function. For a continuous source, a much improved duty cycle can be obtained that can be 50% in an ideal case. For He, the gain is less, but is appreciable, especially for ^{3}He which has a very high neutron absorption cross section.

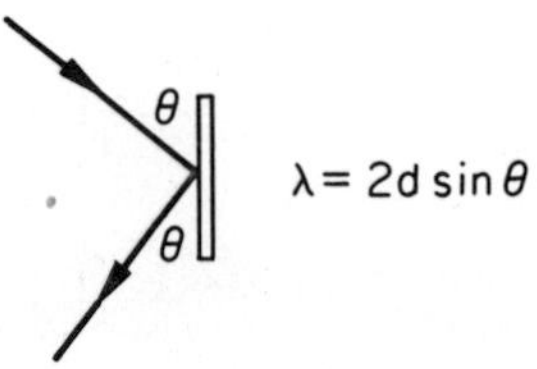

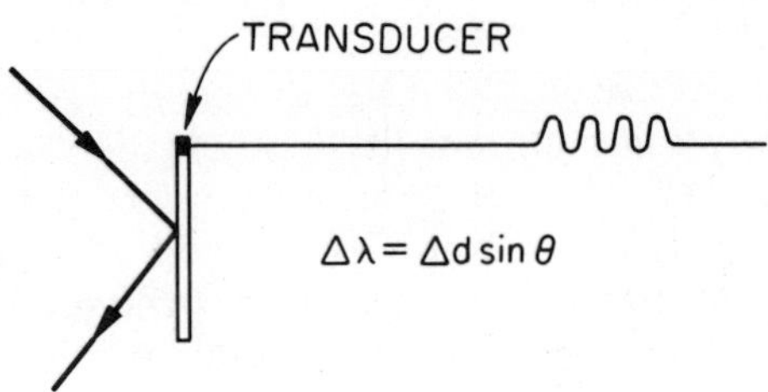

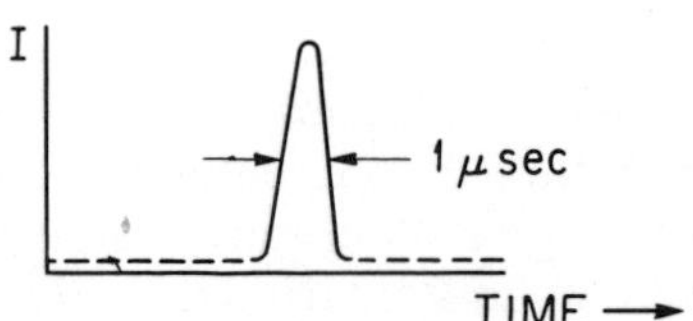

Fig. 4   Production of high intensity short time neutron pulses by ultrasonic techniques.

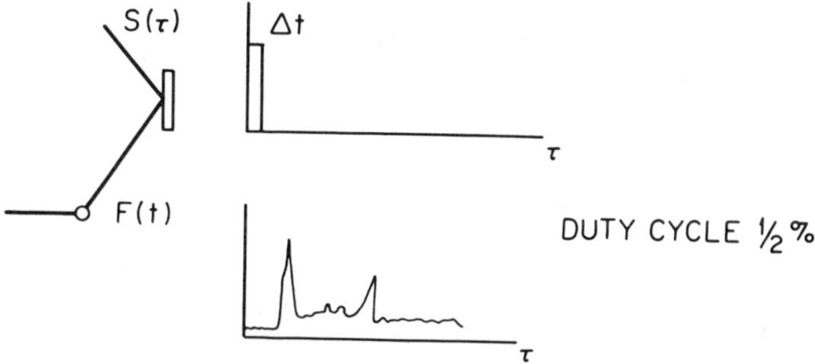

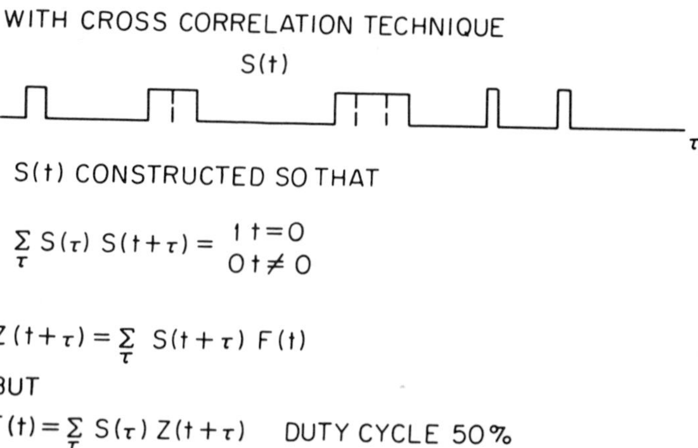

$$\sum_\tau S(\tau)\, S(t+\tau) = \begin{cases} 1 & t=0 \\ 0 & t \neq 0 \end{cases}$$

$$Z(t+\tau) = \sum_\tau S(t+\tau)\, F(t)$$

BUT

$$F(t) = \sum_\tau S(\tau)\, Z(t+\tau) \qquad \text{DUTY CYCLE } 50\%$$

Fig. 5.   The neutron cross correlation time-of-flight technique can give a high duty cycle for steady state sources.

A problem with the cross correlation technique is that the pulse must be of an exact shape or else distortions of the desired F(t) spectrum are obtained. This shape cannot be made perfectly by any technique including the ultrasonic one. However, one can correct for this by measuring the signal function that is actually generated and constructing a new function to use in the cross correlation that removes the distortions in the correlated data. The cross correlation is given by

$$F_j = \sum_i^N Z_i S_{i+j} \tag{4}$$

which can be written in matrix form as

$$F = ZS \tag{5}$$

where S is an N by N circulant matrix with elements $a_{ij} = a_{i-j}$. If the signal function as measured by a detector is given by Z, and P is the ideal shape desired in the cross correlation, one can solve for a matrix $\bar{S}$ by

$$Z\bar{S} = P .\qquad(6)$$

The matrix $\bar{S}$ is then used to cross correlate the measured data and corrects all errors for improper pulse shape. It turns out that the ideal pulse shape can be of any width so that choosing a narrow shape automatically deconvolutes the resolution from the data. Of course the statistical accuracy is decreased if a narrow width is chosen, but one is free to choose the best trade off between resolution and statistical accuracy.

The next step is to determine how to find n(p) from data taken at constant angle and not at constant Q. The cross section for scattering in the impulse approximation is given by

$$\frac{d^2\sigma}{d\Omega d\omega} = c\frac{k_f}{k_i}S(Q,\omega)$$

$$= c\frac{k_f}{k_i}\sum_p n_p \delta\!\left(\omega - \frac{Q^2}{2m} - \vec{Q}\cdot\frac{\vec{p}}{m}\right),\qquad(7)$$

where c is a constant, k_f and k_i are the final and incident neutron wave vectors, p is the momentum of the helium atom before the scattering event, m is the ^{4}He mass, and $\hbar$ has been set equal to 1.

For constant Q,

$$pn_p\big|_{p_{min}} = (4\pi Q^2 \rho / m^2)\partial S(Q,\omega)/\partial\omega,\qquad(8)$$

where ρ is the density and

$$p_{min} = [\omega - \omega(Q)](Q/m)^{-1},$$

with $\omega(Q)$ the neutron energy transfer for the momentum Q. While for constant angle,

$$pn_p\big|_{p_{min}} = \frac{\pi^2 \rho}{cm}\left(1 - \frac{\omega}{\epsilon_i}\right)^{-1/2}\left\{8\,\omega(Q)\frac{\partial}{\partial\omega} + \left[\frac{4\,\omega(Q)}{\epsilon_i - \omega} - 1 + \left(1 - \frac{\omega}{\epsilon_i}\right)^{-1/2}\cos\theta\right]\right\}\frac{d^2\sigma}{d\Omega d\omega}$$

$$\times\left\{1 + \frac{1}{8}\left(1 + \frac{\omega}{\omega(Q)}\right)\left[1 - \left(1 - \frac{\omega}{\epsilon_i}\right)^{-\frac{1}{2}}\cos\theta\right]\right\}^{-1}\qquad(9)$$

where ϵ_i is the incident neutron energy, and θ is the scattering angle.

With the cross correlation neutron time-of-flight operational , it was then possible to measure n(p) with good resolution in a reasonable amount of time. An experiment was performed with a Q of 14.79Å^{-1} and n(p) obtained from the measured $S(Q,\omega)$ using Equation 9.8 A plot of pn(p) is shown in Fig. 6 compared with a calculation by Kalos.[9] There appears to be a peak at low p in pn(p), but within the statistical error shown it probably more accurate to say that there is extra scattering at low p that is indicative of the condensate. Using similar data taken above the λ point and calculating the condensate fraction by Equation 1, a condensate fraction of about 12% is obtained. Within the measured errors and neglecting the extra scattering at low p that comes from the condensate, the measured and calculated pn(p) distributions are in good agreement.

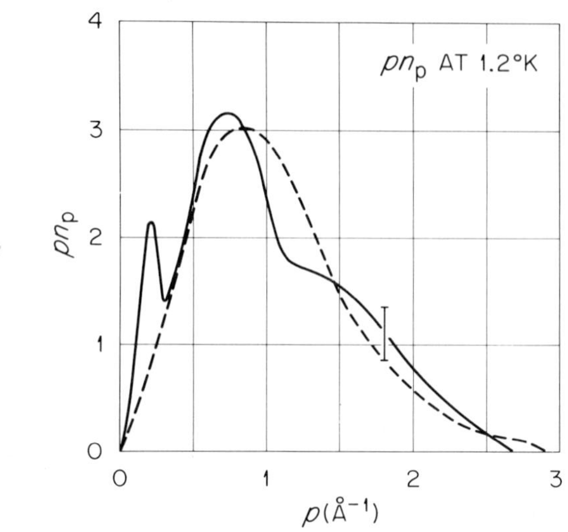

Fig. 6   pn$_p$ for ^{4}He at 1.2K. The solid line is a calculation by Kalos.

At this time, substantial improvements were made to the spectrometer. First, a large crystal in conjunction with a convergent slit was used to pulse the beam, as shown in Fig. 7. The crystal was placed in such a way that the longer wavelength, or lower energy neutrons, were reflected from a place on the crystal nearer the sample so that they arrived at the detector at the same time as the faster neutrons that started farther away from the sample. This meant that a much bigger wavelength spread could be used resulting in much higher intensity without increasing the energy resolution. Also, additional pulsing crystals were used, as shown on the bottom of Fig. 7, permitting an even bigger wavelength spread. It is only necessary to delay the pulse chain to each crystal by the proper amount so that all neutrons arrive at the detector at the same time. Also, at this time, more detectors were added to the spectrometer giving 32 separate detector banks of 5 detectors each. This makes it possible to take data at 32 separate Q values simultaneously.

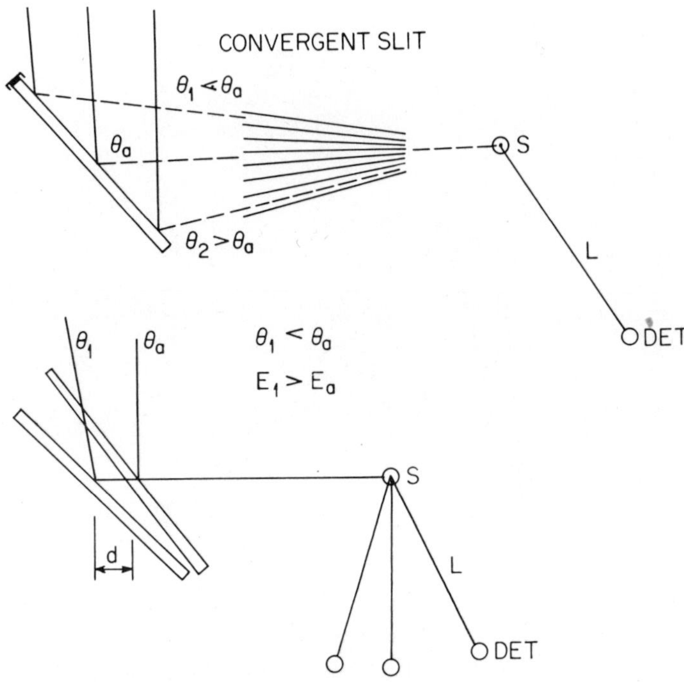

Fig. 7   A convergent slit and multiple pulsing crystals can greatly increase the neutron intensity if the pulse timing is such that all neutrons reach the detector at the same time.

With these improvements, it became possible to measure $S(Q,\omega)$ for
^{4}He with good resolution at 32 separate Q values at reasonable counting rates.
Figure 8 shows measurements of n(p)-n*(p) for three temperatures. n(p) was
obtained by symmetrizing $S(Q,\omega)$ at a number of Q's and adding the
calculated momentum distributions. A condensate fraction of about 10% was
obtained at the lowest temperature[10] in good agreement with other results.

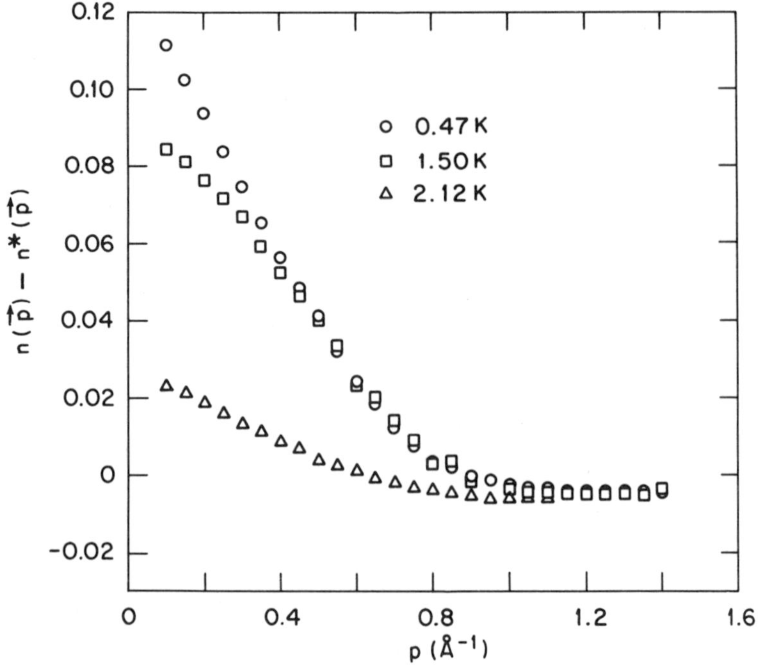

Fig. 8   n(p) for ^{4}He at low temperatures minus n*(p), the
momentum distribution measured just above the λ point.

The spectrometer improvements made it possible to perform detailed
measurements on ^{3}He for the first time.[11] ^{3}He has an absorption cross section
for neutrons of about 10,000 barns, making the measurements very difficult, so
of course the data obtained is not nearly as good as with ^{4}He. Measurements
were first made at a temperature of 1.2K which is too high to observe any
Fermi-liquid effects. In this case, $S(Q,\omega)$ should be nearly Gaussian, and Fig. 9
shows one of the measurements for a scattering angle of 81.6K. The solid line
is a least squares fit to two Gaussians, one centered at energy transfer zero
stemming from the sample container, and the other from the ^{3}He. From data
like these, the peak widths and positions were determined. The top of Fig. 10
shows the width divided by Q as a function of Q. Oscillations in this curve
have been found for ^{4}He that can be traced back to the oscillations in the ^{4}He-
^{4}He scattering cross section. The curve for ^{3}He looks like it may have a dip in it

at a Q of around 5.5Å^{-1} although the statistics are probably not inconsistent with a straight line. The minimum for ^{4}He occurs at 4.5Å^{-1} and this is certainly not observed. A more accurate experiment is needed, but the indications are that the width oscillations are different for ^{3}He than for ^{4}He, as might be expected, since the oscillations in the atomic scattering cross sections are different in the two materials.

The bottom of Fig. 10 shows the energy difference between the observed peaks in $S(Q,\omega)$ and the calculated position given by the impulse approximation. This difference is a maximum at 4.5Å^{-1} for ^{4}He but appears to be near a minimum for ^{3}He. It again would be desirable to do a more accurate experiment, but it appears that if there are oscillations in the positions of the $S(Q,\omega)$ distributions as a function of Q for ^{3}He, the oscillations have a different phase than they do for ^{4}He.

The most interesting effects in ^{3}He occur at low temperatures where the material is thought to have properties that can be described in terms of a Fermi-liquid. The sample of ^{3}He was cooled to 0.37K, which was the low temperature limit of the experiment, and $S(Q,\omega)$ determined for 32 different Q values. n(p) was then determined by symmetrizing the $S(Q,\omega)$ data and averaging the results for different Q values as was done earlier for ^{4}He. The result is shown in Fig. 11. The solid line is a least squares fit by an ideal Fermi-gas distribution for T_F = 1.8K. Again, because of the difficulty of the experiment, the error bars are larger than would be desirable. The Fermi surface is not expected to be sharp because of broadening by temperature and final state effects. Further experiments should be done at a lower temperature so that the final state effects can be isolated. Nevertheless, the experiment gives

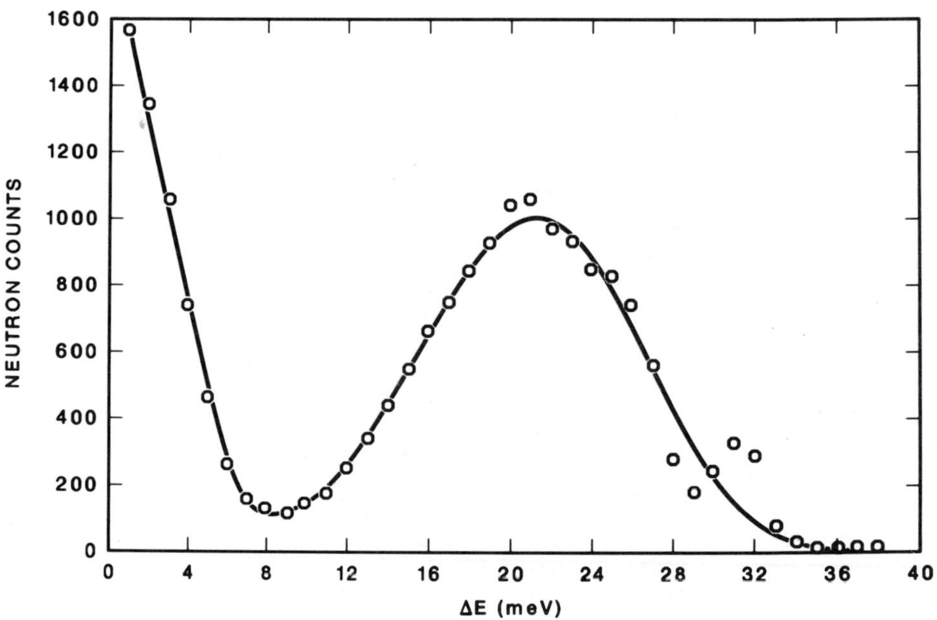

Fig. 9 $S(Q,\omega)$ for ^{3}He at 1.2K for the detector bank at 81.6 deg.

a result for the Fermi temperature that is in good agreement with generally accepted values.

We see then that we have come a long way in the last 15 years in our ability to produce reliable neutron scattering results for the quantum liquids ^{4}He and ^{3}He. Part of that gain, particularly for ^{3}He, has been the substantial improvement in instrumentation. A very important factor in the determination of the Bose condensate has been a better understanding of how to analyze the data. Probably more work needs to be done on this, but at least an analysis procedure now exists that gives consistent results for a wide variety of measurements. Measurements are only beginning on ^{3}He, but at least they appear to be possible. Certainly we can look forward to much better results in the next few years. Finally, it is very important to be able to evaluate the effects of final state effects on the measurements. It is obvious that a lot of progress has been made in this regard, but it is equally obvious that more remains to be done.

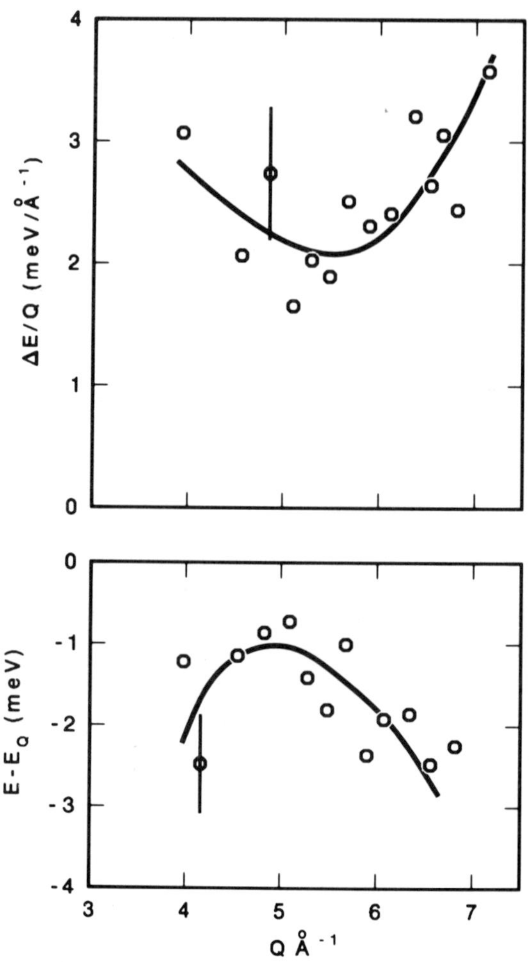

Fig. 10   Measurements of the relative width and position of $S(Q,\omega)$ for ^{3}He at 1.2K as a function of Q.

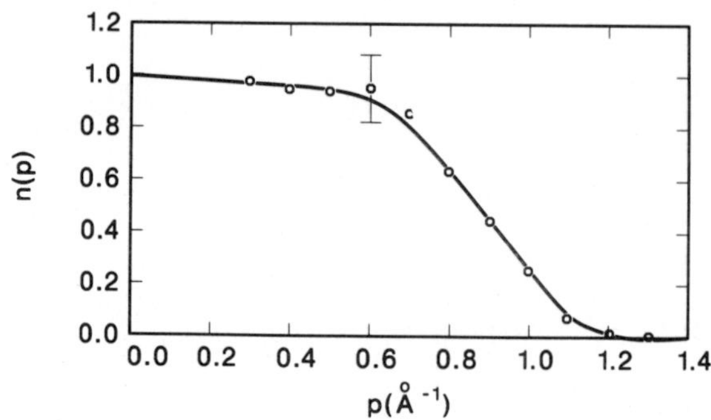

Fig. 11   n(p) for ^{3}He at 0.37K.

ACKNOWLEDGMENT

This research was supported by the Division of Materials Sciences, U.S. Department of Energy under Contract No. DE-AC05-840R21400 with Martin Marietta Energy Systems, Inc.

<u>References</u>

1.   P. C. Hohenberg and P. M. Platzman, Phys. Rev. **192**, 198 (1966).
2.   H. A. Mook, R. Scherm, and M. K. Wilkinson, Phys. Rev. A **6**, 2268 (1972).
3.   P. Martel, E. C. Svensson, A. D. B. Woods, V. F. Sears, and R. A. Cowley, J. Low Temp. Phys. **23**, 285 (1976).
4.   A. D. B. Woods and V. F. Sears, Phys. Rev. Lett. **39**, 415 (1977).
5.   V. F. Sears, E. C. Svensson, P. Martel, and A. D. B. Woods, Phys. Rev. Lett. **49**, 279 (1982).
6.   H. A. Mook, Phys. Rev. B **37**, 5806 (1988).
7.   A. Griffin, Phys. Rev. B **32**, 3289 (1985).
8.   H. A. Mook, Phys. Rev. Lett. **32**, 1167 (1974).
9.   M. H. Kalos (private communication).
10. H. A. Mook, Phys. Rev. Lett. **51**, 1454 (1983).
11. H. A. Mook, Phys. Rev. Lett. **55**, 2452 (1985).

NUCLEAR PHYSICS

MOMENTUM DISTRIBUTIONS FROM QUASIELASTIC ELECTRON-NUCLEUS SCATTERING

Ingo Sick

Dept. of Physics, University of Basel, Basel, Switzerland

1 Introduction

In electron-nucleus scattering, the main emphasis in the past has been on elastic scattering. From such data one can derive the radial distribution $\rho(r)$ of charge in the nucleus. This can be done with high accuracy, as we deal with an inclusive process with identical initial and final state. The Dirac equation for the electron in the electrostatic field of the nucleus can be solved exactly. Provided one can measure cross sections up to momentum transfers $q \simeq 4fm^{-1}$, the charge distribution can be determined in a model independent way, with error bars of order 1%, from the Fourier transform of the experimental form factors [1].

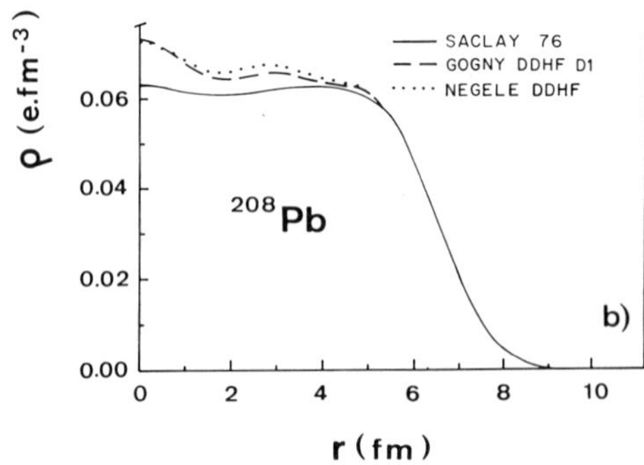

Figure 1. Experimental charge density of Pb (solid line, with width larger than error bars), compared to Hartree-Fock calculations.

As an example, we show in Fig.1 the charge density of ^{208}Pb. The experimental density [2] is compared to a number of mean-field calculations, performed in the framework of Hartree-Fock using effective, density dependent nucleon-nucleon forces. In the nuclear surface and tail-region, agreement with experiment is quite good. In the central region of the nucleus, theory predicts too much oscillatory (shell) structure of $\rho(r)$. Adding the short-range correlations between nucleons reduces their amplitude [3].

To some degree, radial distributions $\rho(r)$ and momentum distributions $\rho(K)$ contain the same information. In particular, a good reproduction of $\rho(r)$ ensures that $\rho(K)$ at momenta K less than the Fermi momentum K_F is close to reality. The charge density is, however, not as sensitive to $\rho(K)$ as on might expect. Although $\rho(r)$ for Pb reminds us of nuclear matter (constant density), the momentum distribution (Fig.2) is still quite different, due to the fact that more than 50% of the nucleons are located in the nuclear surface [4]. Nevertheless, $\rho(r)$ basically fixes $\rho(K)$ at $K < K_F$. Dedicated measurements of $\rho(K)$ are of interest mainly if they address the large-K region, or if they produce momentum distributions for individual shells.

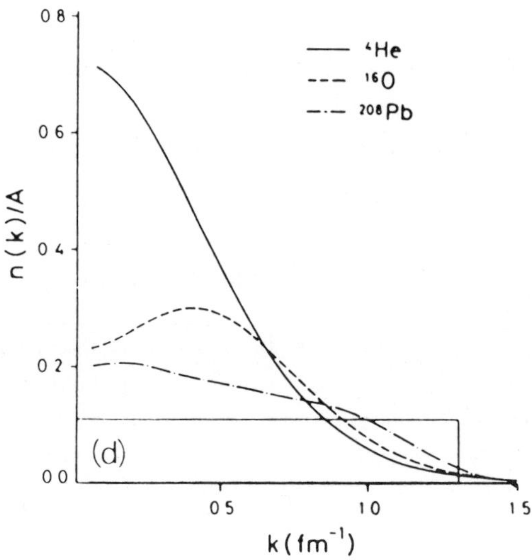

Figure 2. Momentum distribution of finite nuclei and nuclear matter. [4]

2 Coincidence reactions (e,e'p)

When measuring, in a coincidence reaction, the momenta of incident and scattered electron and recoil proton, one can deduce, in impulse approximation, energy and momentum of the initially bound nucleon. For every discrete state of the final nucleus, corresponding to well defined shell(s) of the initial nucleon, one can measure the momentum distribution Fig.3 shows an example [5] for the case of ^{40}Ca, where one can nicely distinguish between different shells; knockout from the 2s shell, for instance, yields the $\rho(K)$ with a maximum at K=0 and a node at $K = 100$ MeV/c $(0.5$ fm$^{-1})$ as expected in the independent particle model. Fig.4 shows that today these (e,e'p) experiments can be extended all the way to heavy nuclei.

The agreement between data and mean-field calculations is quite good for low to medium K. At large K, no data are available, and the calculations must break down. For low K much of the information resides in the integrated strength, the occupation number of a given state. This quantity can be compared to detailed calculations of the structure of individual states performed in the framework of the shell model which includes configuration mixing. The K-dependence can only be compared to the selfconsistent (DDHF) calculations that in general ignore mixing.

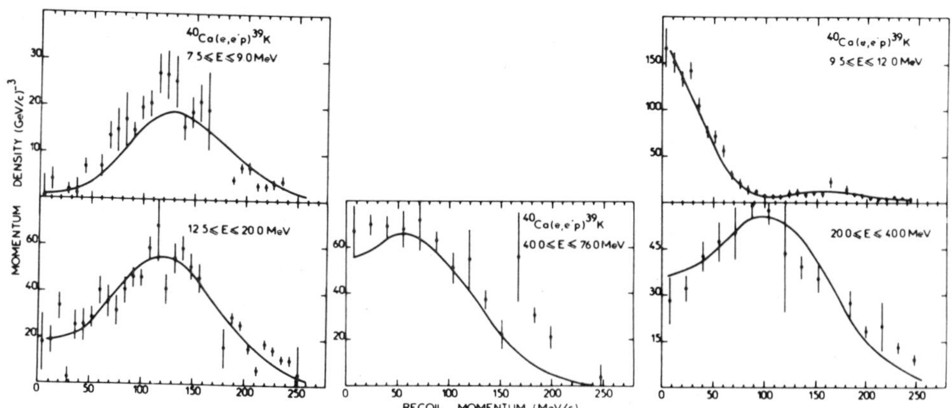

Figure 3. Momentum distributions for different
shells in ^{40}Ca. (left: $1f_{7/2}$ shell, right: $2s_{1/2}$ shell).

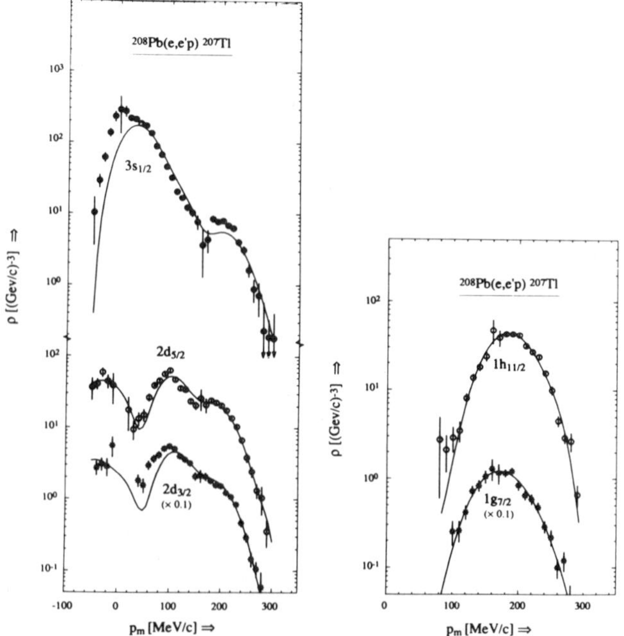

Figure 4. Momentum distributions for
different shells in ^{208}Pb ([6]).

The large-K region is, at present, not accessible experimentally, due to the low energy
and duty cycle (1%) of existing electron accelerators. With CEBAF (4 GeV, 100% duty
cycle) some of these limitations will disappear. The kinematics of experiments can be
optimized such that corrections to IA can be made smaller. In particular, experiments at
forward angles will reduce the contributions of meson exchange currents (MEC), experiments at high q will minimize final state interactions (FSI) by giving the recoil nucleon
enough energy.

Some of the basic limitations of (e,e'p) will, however, still be present. Once one detect
a recoil hadron, one has to deal with the multi-step interactions of strongly interactin
particles. Even under optimum kinematic conditions, the limitations that appear in to
day's data—effects of FSI, coherent charge exchange of outgoing nucleon, MEC— canno
be eliminated. This poses problems, particularly if one is interested in the continuur
(excitation energies of > 15 MeV). Ultimately this means, that one cannot look for sma
wave function components, the ones of highest interest if one aims at components of hig
initial momentum or separation energy. These will have to be studied via inclusive scat
tering (e,e').

3 Inclusive scattering: A=3

The general features of the inclusive spectrum are shown in Fig.5. An important differ
ence to e.g. (n,n') is immediately apparent: the large-energy loss side of the quasielasti
peak is obscured by the contribution of Δ-excitation and MEC processes. Informatio
on the momentum distribution of nucleons can only be extracted from the low-ω side, fo
energy loss $\omega < q^2/2m$.

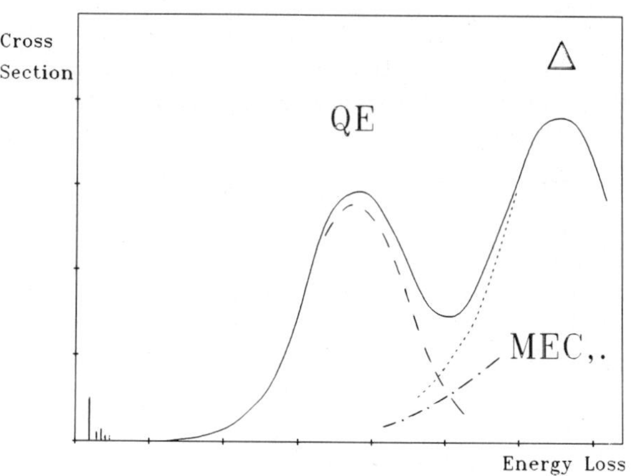

Figure 5. Schematic inclusive spectrum
for electron-nucleus scattering.

When summing over all final states, one can largely eliminate the effects of FSI, partic
ularly in the limit $q \to \infty$. This is an enormous advantage if one is looking for small wav
function components. The price one pays: one obtains only rather global informatior
which, in the main quasielastic peak for instance, is limited to average momentum (K_F
and average separation energy $\overline{SE}$. The most interesting region to study by (e,e'): ver
large q, and large (but not very large) ω. If q is large, but the energy loss $\omega \simeq (\vec{K}+\vec{q})^2/2m$
is not too large, one can study components in the wave function [7] of large initial mo
mentum $\vec{K} \approx -\vec{q}$. If q is very large, one can work at reasonably large ω where FSI hav
a small influence on the inclusive cross sections.

As an example, I want to discuss the nucleus ^{3}He. For 3 nucleons, bound by a nucleon
nucleon potential derived from NN scattering, the Schrödinger equation (Faddeev equa
tion) can be solved essentially without approximation. From such wave functions, one ca
determine the spectral function [8] S(K,SE), the probability to find in the ^{3}He nucleu

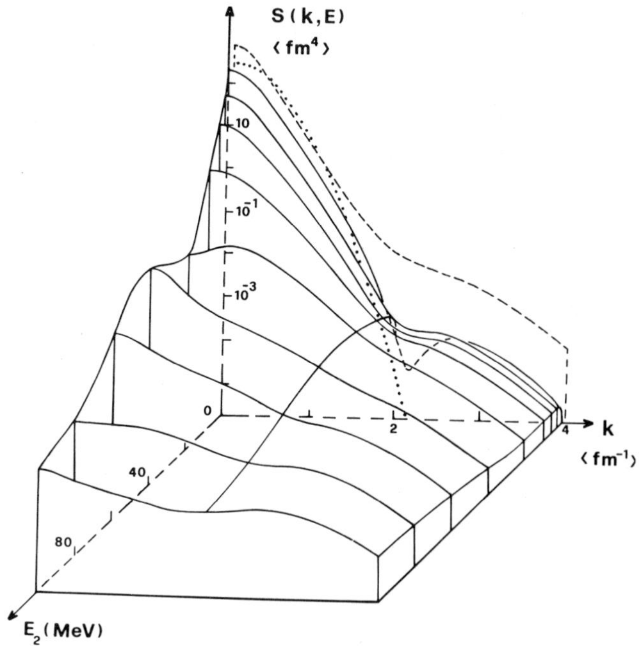

Figure 6. Spectral function S(K,SE) of ³He. The dashed curve corresponds to 2-body breakup.

a nucleon of given momentum K and separation energy SE. Fig.6 shows that nucleons of large momentum are preferentially found at large SE, in accordance with expectation. Large momenta occur due to short-range correlations between pairs of nucleons; knockout of a nucleon with $\vec{K}$ requires the correlated partner of momentum $-\vec{K}$ to get onto the mass shell, and that requires an energy of order $(-\vec{K})^2/2m$. Integration of S(K,SE) over SE yields the usual momentum distribution.

Using this spectral function to calculate cross sections, in PWIA, yields the results shown in Fig.7. The main quasielastic peak (not shown) is reproduced very well. In the tail, corresponding to large K, the calculated cross sections are much too small.

In order to understand better the results of Fig.7, we turn to y-scaling [9]. Writing the cross section

$$\sigma(q,\omega) = \int S(\vec{K}, SE)\sigma_{eN}(\vec{q}, \vec{K}) \cdot \delta(energy) \cdot d\vec{K}\, dSE$$

allows to deduce, in the limit $q \to \infty$, the scaling property where the cross section, divided by the elementary e-N cross section, becomes a function of one variable, y, only with y=K$_\parallel$.

$$\sigma(\vec{q},\omega)/\sigma_{eN}(\vec{q}, y) \cdot \partial\omega/\partial y = F(y)$$

The scaling variable is given by the argument of the energy/momentum conserving δ-function

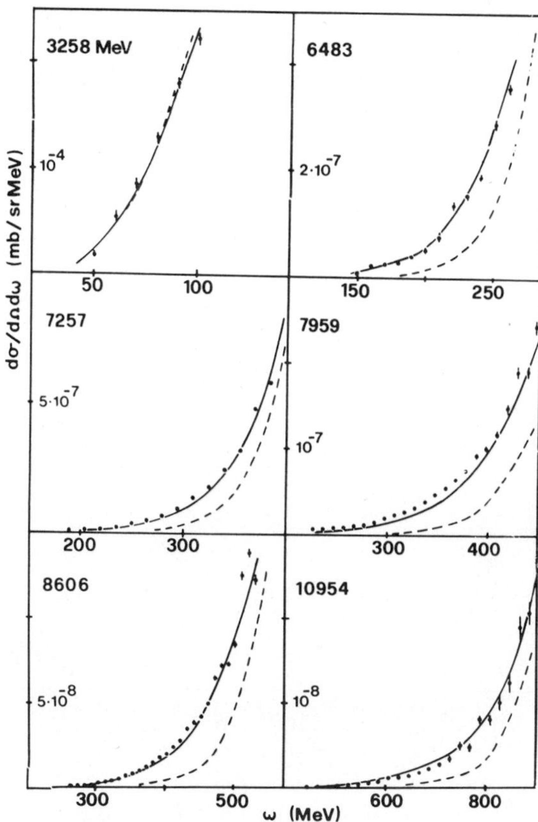

Figure 7. ^{3}He(e,e) cross sections (measured at energy (MeV) indicated, and 8°, compared to Faddeev calculations (dashed curves).

$$((q+y)^2 + m^2)^{1/2} \quad - m \quad + \quad SE \quad + \quad recoil \quad = \quad \omega$$

At this point we note that y is not the West-variable generally used. We account fo the off-shell nature of the initial nucleon (kinetic + potential energy equalling - SE), an we treat the final nucleon relativistically. Although in the limit $q = \infty$ $y \to y_{West}$, th scaling functions at values of q that are experimentally achievable differ by up to an orde of magnitude when not using the correct energy/momentum conservation.

Fig.8 shows the ^3He data in terms of F(y). Although the cross sections differ by man orders of magnitude, the data do define a unique function F(y), for $y < 0$. With thes data for ^{3}He, y-scaling in e-nucleus scattering could for the first time be demonstrate [9].

What is this observation of scaling good for? We can use it to learn something abou 3 distinct points:

- The reaction mechanism in e-nucleus scattering at large q is not clear, a priori. Be sides quasifree scattering, Δ-excitation, MEC etc. contribute. Quasifree scatterin scales, the other contributions don't, due to the occurrence of different vertex forr

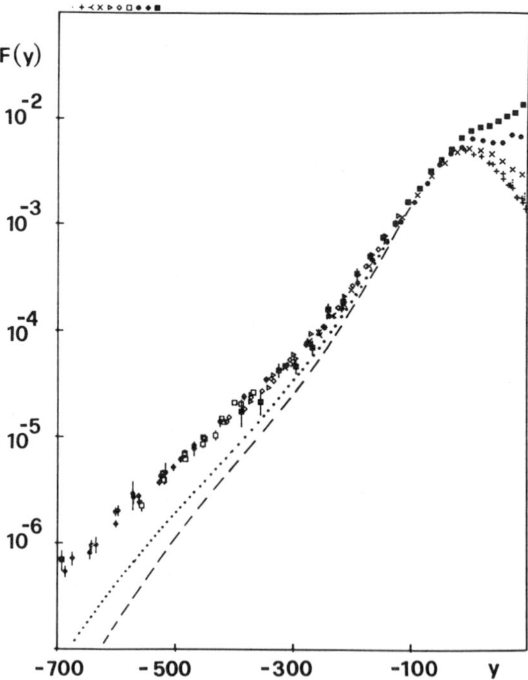

Figure 8. Scaling function $F(y)$ for ^{3}He.
Momentum transfer of data: 2-10 fm^{-1}.
Dashed curve: Faddeev prediction, dotted
curve: contribution of 3-body force added.

factors, and different energy/momentum conservation. Fig.9 demonstrates this for
the case of calculated MEC-cross sections [10]. If the data do scale, the contribution
of non-quasifree processes is less than the band width of the scaling function.

- The bound-nucleon form factor is not necessarily equal to the one known for the free
  nucleon. The scaling function $F(y)$ is independent of q only if the cross sections are
  divided by the bound-nucleon cross section of the correct q-dependence. Given a
  large q-range of the data, a sensitive measure of bound-nucleon properties is possible
  [1].

- The momentum distribution $\rho(K)$ can be determined from $F(y)$, in particular at
  large K. The comparison between experiment and Faddeev calculation of Fig.8 yields
  the same message as Fig.7, the calculated density at large $K(|\,y\,|)$ is too low.

Fig.8 shows that some of the deficiencies of the Faddeev calculation can be accounted
for [11] as a consequence of the neglect of non-nucleonic degrees of freedom; when adding a
three-nucleon force— resulting from the process of two successive two-nucleon-interactions
with an intermediary excited nucleon (Δ)— agreement with experiment is improved. The
remainder of the discrepancy of Fig.7 is not yet understood. As relativistic effects [12]
are still small in the elastic form factor at the equivalent value of q, I tend to assign the
remaining problem to additional non-nucleonic degrees of freedom.

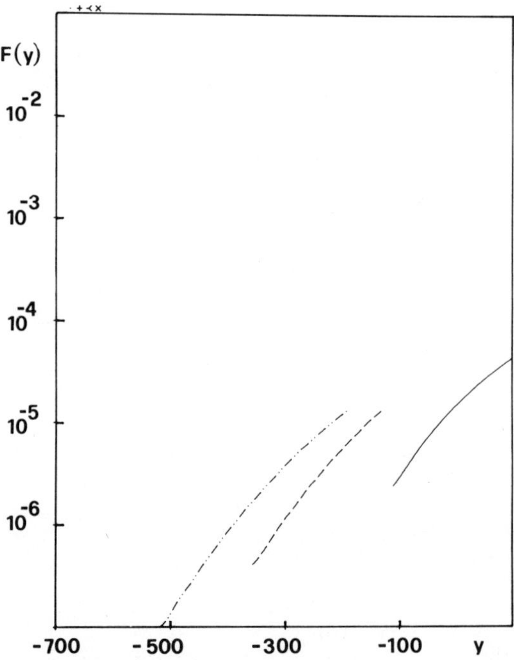

Figure 9. Cross sections for MEC (q=2-10fm^{-1}), plotted in terms of F(y).

4 Inclusive scattering: $A = \infty$

For nuclear matter, sophisticated calculations are available: Renormalized Bruckner Hartree-Fock (RBHF), Variational calculations, etc. (See talk S. Fantoni). The momentum distribution exhibits a pronounced tail at large K, reflecting the effects of (poorly known) short range NN correlations. For nuclear matter we also have the best chance to achieve a quantitative description of the final state.

Data for nuclear matter, unfortunately, are not measurable. And, as indicated by Fig.2, heavy nuclei and nuclear matter show important differences. The response function of nuclear matter can, however, be extrapolated from finite nuclei [13], as discussed in the talk of D. Day.

Two types of calculations presently are available. Fantoni and Pandharipande [14] have performed a variational calculation of the response function $\sigma(q,\omega)$. Due to the use of non relativistic kinematics, this calculation is restricted to relatively low q(low K). Butler and Koonin [15] have performed a Bruckner-Hartree-Fock calculation, again using non-relativistic kinematics. To compare to the data, they derive the scaling function F(y) using non-relativistic y, while the experimental F(y) is calculated using the relativistic y.

The convergence of F(y) to the $q \to \infty$ value is shown in Fig.10, where the experimental F(y) is compared to the calculation of Butler and Koonin, normalized to the data. The rates of convergence of experiment and theory are quite similar, and one would tend to conclude that, at a rather small q, one can measure the $q \to \infty$ value. As a matter of fact, the convergence of the data can be greatly improved if one includes in the definition of y the main effect of FSI, the presence of an (energy dependent) real potential in the final state.

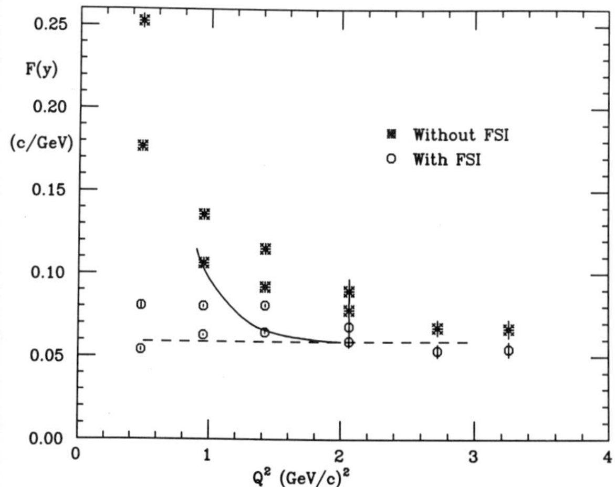

Figure 10. Convergence of F(y) as function of q. Upper points ^{56}Fe, lower points ^{197}Au. The drawn curve corresponds to the calculation of Butler and Koonin.

$$((\vec{K} + \vec{q})^2 + m^2)^{1/2} - m + SE - V(E_{rec}) = \omega$$

in which case F(y) is calculated by including the corresponding Perey factor [16]. Fig.10 shows that with this FSI correction F(y) is almost independent of q.

The main problem in the comparison between experiment and calculation: at large K the absolute values of F(y) calculated are high by factor of $\sim$3. This has been assigned to the excessive strength of tensor correlations of the NN potential employed [15]. A closer look reveals that the problem is related mainly to the use of an average separation energy. As pointed out with Fig.6, large K are preferentially linked to large SE ($\overline{SE} \simeq K^2/2M$). Large separation energies shift the strength in $\sigma(q,\omega)$ to large ω, and this has [17] a large effect upon F(y). This is demonstrated by Fig.11, which shows F(y) for ^3He for $q \sim 10fm^{-1}$ derived from cross sections calculated from the spectral function with/without SE-dependence. At large K(large y) the SE-dependence cannot be neglected!

For detailed study of such questions it always is useful to do the kind of analysis we have been doing in the past for ^3He: From S(K,SE) one can compute, in IA, $\sigma(q,\omega)$, from which F(y) can be derived. This F(y) then can be compared to $\rho(K_{\parallel})$ derived directly by integrating over S(K,SE). With this comparison one can easily check if, e.g. the definition of y is the appropriate one for the q-range considered, etc. From such studies one can draw the conclusion that the assumption of an initially on-shell nucleon, or the use of non-relativistic kinematics, leads to differences between F(y) and $\rho(K_{\parallel})$ of a factor up to 10 at large $| y |$. ($q \leq 10fm^{-1}$). A cavalier attitude in the definition of y (with the justification that it does not matter at q=∞) has grave consequences at the finite q's we experimentally can reach.

For nuclear matter, a comparison between experiment and calculation therefore will need one of two things: A calculation of $\sigma(q,\omega)$ that accounts for the relativistic energy/momentum relation of the final state. Alternatively, a calculation of the full spectral function S(K,SE) which then can be used in a relativistic IA calculation to predict $\sigma(q,\omega)$ and F(y). An S(K,SE) is presently being calculated by Fantoni et $al.$

As long as one uses PWIA, one has to worry about the effects of final state interaction of the recoil nucleon. In analogy with neutron scattering, two effects are important [18]: The scattering of the recoil nucleon leads to a shift of $\sigma(q,\omega)$ in ω, proportional to the NN scattering amplitude at 0°. It also leads to a smearing of the inelastic response, with

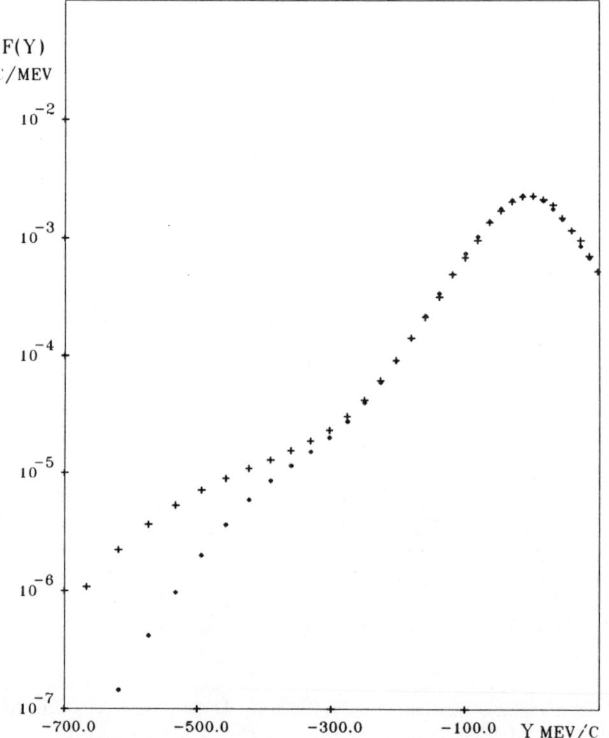

Figure 11. Scaling function of ^{3}He obtained from cross sections calculated by including (neglecting) the SE-dependence of S(K,SE).

a folding width depending on the total NN cross section. Contrary to (n,n'), the main effect in nuclear physics comes from the shift; for a steeply falling F(y) a shift of several ten MeV has major consequences. This shift is accounted for when using V(E) in the definition of y as above. The effect of the folding appears to be much smaller, as the folding widths are ten to several tens of MeV for recoil nucleon energies of 200 to 50 MeV. The question of the wings of the folding function has not been studied, however. An exploratory calculation, leading to Lorentzian tails, gives quite unreasonable results [19]. An investigation in analogy to the one done by Silver and Reiter [20] for (n,n') would be highly desirable.

5 Conclusions

For momenta $K < K_F$ the momentum distribution of nucleons in nuclei is quite well known; (e,e'p) allows to measure $\rho(K)$ with fair accuracy for individual shells. The importance of known difficulties (FSI, MEC) can be reduced in the future by going to higher electron-energy. The intrinsic limitations of processes involving the detection of strongly interacting reaction products will continue to be a problem.

For momenta $K > K_F$, little is known from experiments other than inclusive scattering (e,e'). At these large momenta, the standard mean-field calculations are inappropriate, but more quantitative calculations (BHF, variational) that do account for the short-range NN correlations, are emerging. Inclusive electron scattering at very large q is the best way to test our understanding of these short-range phenomena, and y-scaling provides a most useful way to look at the data.

Among the open problems that remain to be tackled, the one concerning the separation energy dependence of the spectral function is the most important one. An understanding of the effects of the absorptive part of the final state interaction is also desirable.

References

[1] I. Sick. *Lecture Notes in Physics*, 236:137, 1985.

[2] B. Frois, J.B. Bellicard, J.M. Cavedon, M. Huet, P. Leconte, P. Ludeau, A. Nakada, Phan Xuan Hô, and I. Sick. *Phys. Rev. Lett.*, 38:152, 1977.

[3] M.R. Strayer, W.M. Bassichis, and A.K. Kerman. *Phys. Rev.*, C8:269, 1973.

[4] M. Casas, J. Martorell, E. Moya de Guerra, and J. Treiner. *Nucl. Phys.*, A473:429, 1987.

[5] J. Mougey, M. Bernheim, A. Bussière, A. Gillebert, Phan Xuan Ho, M. Priou, D. Royer, I. Sick, and G.J. Wagner. *Nucl. Phys.*, A262:461, 1976.

[6] E.N.M. Quint. *thesis, NIKHEF-K Amsterdam*, 1988.

[7] D. Day, J.S. McCarthy, I. Sick, R.G. Arnold, B.T. Chertok, S. Rock, Z.M. Szalata, F. Martin, B.A. Mecking, and G. Tamas. *Phys. Rev. Lett.*, 43:1143, 1979.

[8] A.E.L. Dieperink, T. de Forest, I. Sick, and R.A. Brandenburg. *Phys. Lett.*, 63B:261, 1976.

[9] I. Sick, D. Day, and J.S. McCarthy. *Phys. Rev. Lett.*, 45:871, 1980.

[10] J.W. Van Orden and T.W. Donnelly. *Ann. Phys.*, 131:451, 1981.

[11] T. Sasakawa. *Lecture Notes in Physics*, 260:150, 1986.

[12] W. Gloeckle, T.S.H. Lee, and F. Coester. *Phys. Rev.*, C33:709, 1986.

[13] D. Day, J.S. McCarthy, Z.E. Meziani, R. Minehart, R. Sealock, S. Thornton, J. Jourdan, I. Sick, B. Filippone, R. McKeown, R. Milner, D. Potterveld, and Z. Szalata. *Phys. Rev. C*, to be publ.:, 1988.

[14] S. Fantoni and V.R. Pandharipande. *Nucl. Phys.*, A473:234, 1987.

[15] M.N. Butler and S.E. Koonin. *Phys. Lett.*, B205:123, 1988.

[16] I. Sick. *Comm. Nucl. Part. Phys.*, 18:109, 1988.

[17] C. Ciofi degli Atti, E. Pace, and G. Salme. *Phys. Lett.*, B127:303, 1983.

[18] Y. Horikawa, F. Lenz, and N.C. Mukhopadhyay. *Phys. Rev.*, C22:1680, 1980.

[19] S.A. Gurvitz and A.S. Rinat. *Phys. Lett.*, 197B:6, 1987.

[20] R.N. Silver and G. Reiter. *Phys. Rev.*, B35:3647, 1987.

WHAT WOULD WE LIKE TO KNOW ABOUT n(p) IN NUCLEAR PHYSICS ?

Omar Benhar

INFN, Sezione Sanitá, Physics Laboratory
Istituto Superiore di Sanitá. I-00161 Roma, Italy

Adelchi Fabrocini

Dept. of Physics, University of Pisa and
INFN, Sezione di Pisa,I-56100 Pisa, Italy

Stefano Fantoni

Dept. of Physics, University of Lecce and
INFN, Sezione di Lecce, I-73100 Lecce, Italy

Abstract.   *The momentum distribution $n(\mathbf{k})$ is analyzed in terms of the spectral function $P(\mathbf{k}, E)$. It is shown that $n(\mathbf{k})$ can be viewed as a sum of two separate terms. The first is associated with the one nucleon emission processes and give us information on the shell model picture of nuclei. The other term is due to multinucleon emission and is very much spread out in energy. This term is strongly associated with the N-N correlations in nuclei and its measurement as a function of the mass number A would be most interesting. Recent microscopic calculations of the momentum distribution and of the spectral function are reviewed.*

1. Introduction

The momentum distribution of nucleons in nuclei is a quantity of interest in the study of the nuclear structure, since it provides estimates for the cross section of inclusive $(e, e\prime)$ and exclusive $(e, e\prime N)$ reactions at large momentum transfer by the quasi-free impulse approximation. It is defined as

$$n(\mathbf{k}) = \frac{< \bar{0}|a_{\mathbf{k}}^{\dagger}a_{\mathbf{k}}|\bar{0} >}{< \bar{0}|\bar{0} >},\qquad(1.1)$$

where $a_{\mathbf{k}}^{\dagger}$ and $a_{\mathbf{k}}$ are the creation and annihilation operators of a nucleon with momentum $\mathbf{k}$ and $|\bar{0}>$ is the ground state of the nucleus.

Microscopic calculations, based on a non relativistic model of nucleons interacting via a realistic hamiltonian of the type

$$H = -\frac{\hbar^2}{2m}\sum_{i=1,A}\nabla_i^2 + \sum_{j>i=1,A}v_{ij} + \sum_{k>j>i=1,A}v_{ijk}, \tag{1.2}$$

predict large N-N correlation effects on $n(\mathbf{k})$ of nuclear matter [1], complex [2-4] and light [5,6] nuclei. As a result, the occupation probability of single particle states inside the Fermi sea results to be quenched with respect to the mean field theory estimates. Pandharipande et al. [7] have shown that such a depletion is in fair agreement with recent elastic and inelastic electron-nucleus scattering experiments [8-11].

Valuable information on the momentum distribution are obtained from the coincidence $(e, e\prime N)$ experiments [12,13], whose cross section, in Plane Wave Impulse Approximation (PWIA), is given by [14]

$$\frac{d^4\sigma}{d\epsilon_2 d\epsilon_N d\Omega_2 d\Omega_N} = \left(\frac{d\sigma}{d\Omega}\right)_{eN}(\epsilon_N + m)pP(\mathbf{k}, E), \tag{1.3}$$

where ϵ_N and $\mathbf{p}$ are the energy and the momentum of the knocked-out nucleon, ϵ_i and $\mathbf{k}_i$, $(i = 1, 2)$ those of the incident and scattered electrons and $\left(\frac{d\sigma}{d\Omega}\right)_{eN}$ is the off-shell electron-nucleon cross section. The nucleon spectral function $P(\mathbf{k}, E)$ is defined as the probability of removing a nucleon with momentum $\mathbf{k} = \mathbf{k}_1 - \mathbf{k}_2 - \mathbf{p}$ from the target nucleus leaving the final system with excitation energy $E = \epsilon_1 - \epsilon_2 - \epsilon_N - E_R$, with E_R being the recoil energy of the residual system. $P(\mathbf{k}, E)$ is given by

$$P(\mathbf{k}, E) = \frac{\sum |< \bar{0}|a_{\mathbf{k}}^{\dagger}|\bar{n}(A-1) >|^2}{< \bar{0}|\bar{0} >< \bar{n}(A-1)|\bar{n}(A-1) >}\delta(E_n(A-1) - E_0(A) - E). \tag{1.4}$$

The momentum distribution is related to $P(\mathbf{k}, E)$ via the sum rule

$$n(\mathbf{k}) = \int_{E_{min}}^{\infty} dEP(\mathbf{k}, E), \tag{1.5}$$

where $E_{min} = E_n(A-1) - E_0(A)$. Microscopic calculations of the spectral function have been performed for ^3He [15,16] and, more recently, for nuclear matter [17], which is a suitable system to study correlation and final state interaction effects in the perspective of addressing more fundamental problems, like the modifications of the nucleon form factors due to the presence of the nuclear medium. $P(\mathbf{k}, E)$ is most conveniently separated into two parts, one corresponding to the one nucleon emission processes and the other, due to multiparticle emission processes, which fournishes the *background* contribution. The nuclear matter analysis have shown that the one nucleon emission part, which contributes for $k < k_F$ only, is intimately related [18] with the hole-state strengths, whereas the background contribution is very much spread out in energy implying that high values of the removal energy components may be important to estimate $n(\mathbf{k})$ [17].

Inclusive $(e, e\prime)$ experiments performed in the region of negative values of the y-scaling variable [19], give important information on the spectral function and, consequently on the momentum distribution [20,21]. Realistic hamiltonians of the type given in eq.(1.2) must include a three-nucleon interaction. It is known that nuclear hamiltonians containing two-nucleon interactions only underbind the $A = 3, 4$ nuclei and give too large an equilibrium density for nuclear matter. Most of the calculation reviewed in this contribution have been done by using the Urbana [22] and the Argonne [23] two-body interactions and the Urbana TNI model of ref.[24] and the more realistic Urbana model VII of ref.[5] for the three-nucleon interaction.

At present, very accurate calculations of the ground state wave functions of light nuclei are possible on account of the recent progress done in the solution of the Faddeev equations [25], in the Green Function MonteCarlo method [26] and in the variational theory [5,6]. The strongly repulsive and strongly state-dependent N-N interaction induce important scalar as well spin-isospin and tensor correlations, which must be included in a variational wave function to obtain a bound nucleus. A sufficiently realistic wave function is provided by

$$|0) = \frac{G|0]}{\left[0|G^{\dagger}G|0\right]^{1/2}}, \tag{1.6}$$

where $|0]$ is the uncorrelated ground state and G is a correlation operator of the form

$$G = S \prod_{j>i=1,A} F(i,j), \tag{1.7}$$

$$F(i,j) = \sum_{n} f^{n}(r_{ij})O^{n}(i,j), \tag{1.8}$$

where S is the operator which symmetrizes $\prod_{i<j} F(i,j)$ and the operators $O^{n}(i,j)$ include the four central components $(1, \sigma_i \cdot \sigma_j, \tau_i \cdot \tau_j, \sigma_i \cdot \sigma_j \quad \tau_i \cdot \tau_j)$ for $n = 1,4$ and both the isoscalar and the isovector tensor components for $n = 5,6$. Detailed calculations with such correlated wave functions in complex nuclei are not yet possible, although they have ben carried out in some reasonable approximation [4] and new methods to treat ^{16}O are being developed [27,28]. However it is possible to perform variational calculations in nuclear matter by using hypernetted and operator chain summation techniques [29-31].

Correlated basis theories [32-35] provide for a consistent and unified treatment of the ground and the excited states of complex nuclei and nuclear matter. They are based upon the following set of correlated states

$$|n) = \frac{G|n]}{\left[n|G^{\dagger}G|n\right]^{1/2}}, \tag{1.9}$$

where $|n]$ is the generic state of an uncorrelated base; in the case of nuclear matter $|n]$ is an eigenstate of a Fermi gas hamiltonian at a given density ρ. The correlation functions $f^{n}(r_{ij})$ are determined variationally [24,36], by minimizing the energy expectation value of the hamiltonian (1.2) on $|0)$. The CB states (1.9) are not orthogonal to each other. They can be orthogonalized in such a way [35] that the diagonal matrix elements of the hamiltonian on $|n)$ (the variational estimates) are preserved. The resulting set of orthonormal states $|n>$, which are denoted as OCB states, can be used in standard perturbation theories.

In the following, nuclear matter results, obtained in the framework of correlated basis function theory, are discussed and compared with corresponding calculations performed in light and complex nuclei, the aim beeing the understanding of the N-N correlations in the hadronic matter. In section 2 the microscopic calculations of $n(k)$ are briefly reviewed, whereas section 3 is devoted to a discussion of the spectral function in nuclear matter and its connection with the hole-state strengths and with the response function.

2. Momentum distribution

A realistic estimate of $n(k)$ is obtained by using the correlated state $|0) \equiv |0>$ of eq. (1.6) to calculate the expectation value of eq. (1.1).

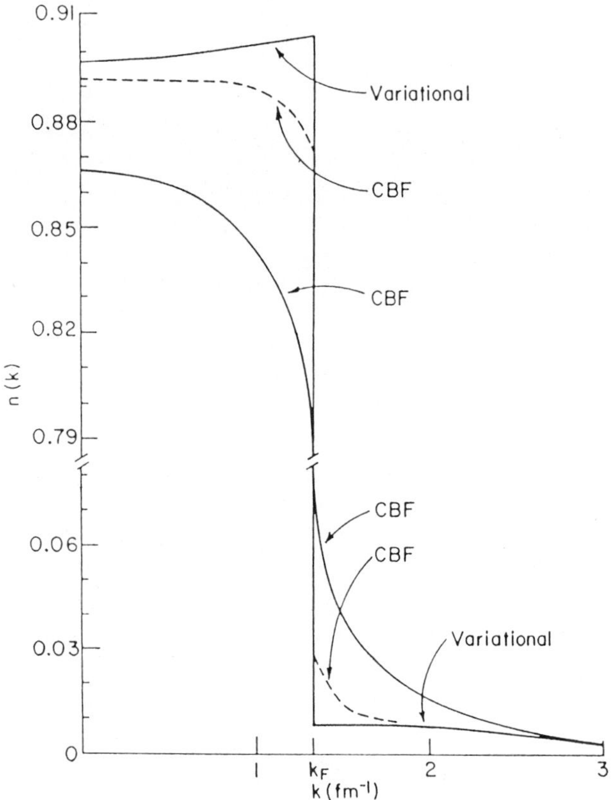

Fig. 1 *Taken from ref. [1]. The momentum distribution in nuclear matter at $k_F = 1.33 fm^{-1}$ obtained from variational and CBF perturbative calculations. The full and the broken curves display the CBF calculations with and without tensor correlations.*

In nuclear matter, calculations of $n(k)$ have been performed by using hypernetted and operator chain summation techniques. In the case of a correlation operator G

of the *Jastrow* type, namely with $F(i,j)$ depending on r_{ij} only, $n(\mathbf{k})$ is given by the sum of two terms, each of them factorizable in terms of irredicible cluster terms [37,38]:

$$n_v^{(J)}(k) = \frac{<0_J|a_\mathbf{k}^\dagger a_\mathbf{k}|0_J>}{<0_J|0_J>} = \eta^{(J)} N_c^{(J)}(k) + \Theta(k_F - k)\eta^{(J)} N_d^{(J)}(k), \qquad (2.1)$$

where both $N_c^{(J)}(k)$ and $N_d^{(J)}(k)$ are continuous functions. In the case of a non-interacting Fermi system, one has $\eta = N_d^{(J)}(k) = 1$ and $N_c^{(J)}(k) = 0$. The discontinuous piece on the r.h.s of eq. (2.1) is given by the sum of the cluster terms of the nodal type having an undressed exchange function as a nodal element.

In the more general case of a state dependent correlation operator, one can still separate $n(\mathbf{k})$ in a continuous and a discontinuous part, but one looses the factorization property, with the result

$$n_v(k) = \frac{<0|a_\mathbf{k}^\dagger a_\mathbf{k}|0>}{<0|0>} = \eta N_c(k) + \Theta(k_F - k)\eta N_d(k) + \Delta N_{comm}(k), \qquad (2.2)$$

$\Delta N_{comm}(k)$ coming from the non commutativity of the operators $F(i,j)$ [1].

Fig. 1 shows the momentum distribution calculated for nuclear matter at $k_F = 1.33 fm^{-1}$. About 70% of the quenching of $n_v(k < k_F)$ with respect to 1 is due to spin-isospin and tensor correlations. The curves labelled CBF in the Figure refer to a calculation in which the lowest order perturbative correction has been added to $n_v(k)$. Such a correction has been obtained along the correlated basis function theory [1] by including two particle-two hole OCB states admixtures into the variational ground state $|0>$.

The corresponding improved ground state wave function results to be

$$|0_2> = |0> + \frac{1}{4} \sum_{h_1,h_2,p_1,p_2} \alpha(h_1,h_2,p_1,p_2)|h_1,h_2,p_1,p_2>, \qquad (2.3)$$

$$\alpha(h_1,h_2,p_1,p_2) = \frac{<h_1,h_2,p_1,p_2|H|0>}{e_v(h_1) + e_v(h_2) - e_v(p_1) - e_v(p_2)}. \qquad (2.4)$$

Here the OCB state $|h_1,h_2,p_1,p_2>$ is obtained from uncorrelated 2p2h Fermi gas state $|h_1,h_2,p_1,p_2]$ and $e_v(h_1) + e_v(h_2) - e_v(p_1) - e_v(p_2)$ is the diagonal matrix element of the hamiltonian (1.2) on it (in the calculation of ref.[1], the orthogonalization of 2p2h states among themselves has been neglected).

The 2p2h admixtures in $|0>$ mainly correct the long range behavior of the correlation functions included in the variational ground state wave function. From Fig. 1 one can see that the perturbative corrections are quite sizeable, especially at $k \sim k_F$, and they are essential in order to reproduce the $x \log x$ ($x = \frac{|k-k_F|}{k_F}$) behavior of $n(k)$ at the Fermi surface. The dashed line is obtained by disregarding the tensor components of $F(i,j)$ in the calculation of the perturbative corrections and shows that most of the contribution to them comes from tensor correlations, which, in fact, have a longer range than the other correlations. In complex nuclei, the single particle states are also coupled to the surface vibrational states. Such a coupling

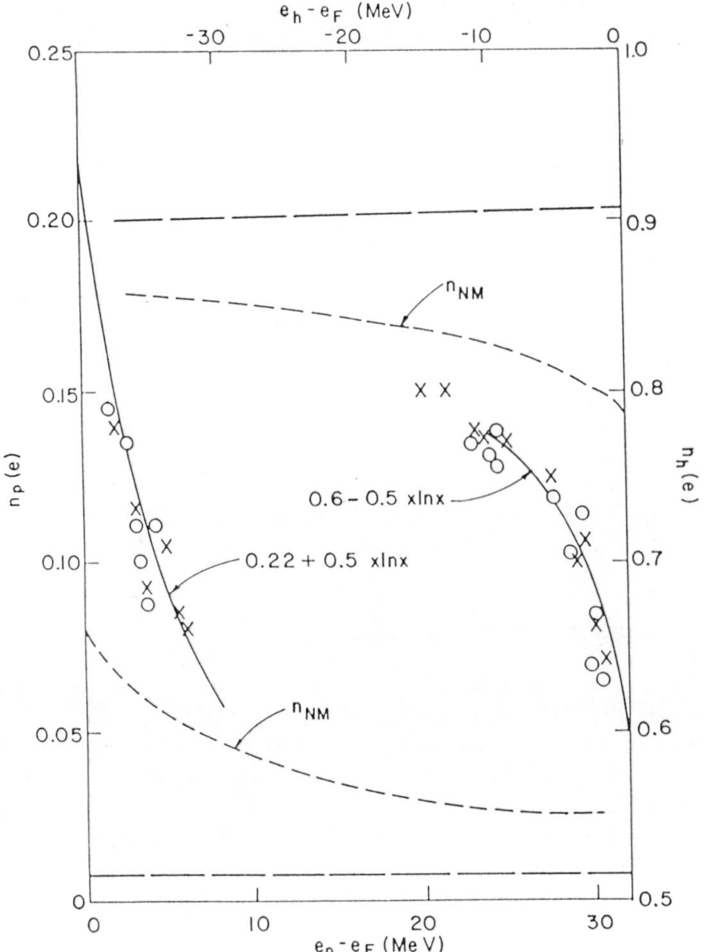

Fig. 2 *Taken from ref.[7]. Single particle occupation probabilities. The broken and dashed curves correspond to the variational and the CBF curves of Fig. 1. The occupation numbers marked with crosses (proton states) and circles (neutron states) are obtained by adding δn_{RPA} calculated in ref.[39] to the nuclear matter results to obtain an estimate of the quenching due to the coupling of the single particle states with surface vibrations. The discontinuity of the total $n(E)$ is in agreement with the experimental data.*

gives an extra quenching to the momentum distribution which has to be taken into account in order to compare it with the experimental single particle occupation probabilities. This quenching has been estimated in ref.[7] and a good agreement with the available experimental data has been found, indicating that most of the quenching of the occupation probabilities with respect to mean field estimates is due to N-N correlations rather than to strong modifications of the nucleon inside the nuclear medium.

The momentum distributions of nuclei with $A = 2, 3, 4$ are compared with that of nuclear matter in Fig. 3. The calculations for the light nuclei have been carried out by means of the Monte Carlo method [40,5].

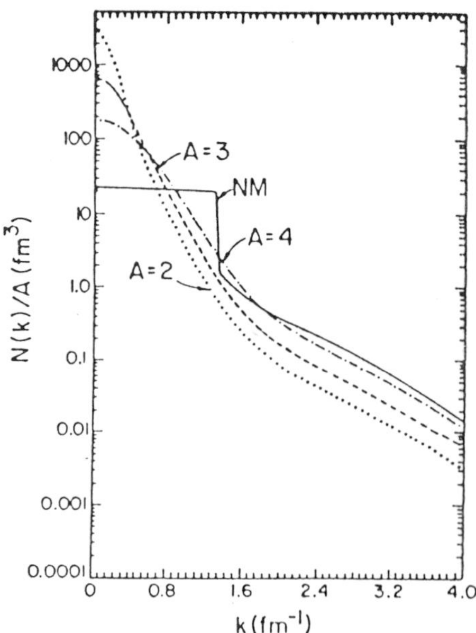

Fig. 3 *Taken from ref. [5]. The momentum distribution calculated with the Urbana v_{14} + model VII interaction in $A = 2, 3, 4$ nuclei and nuclear matter.*

The variational wave function is of the type given in eq. (1.6) in which $|0]$ are antisymmetric spin-isospin states with no spatial dependence and with the two- and three-body correlation operators having central, spin and isovectorial tensor components. The momentum distributions of the nuclei with $A = 3, 4$, for $k > k_0$, with $k_0 \sim 1.5 fm$, are assumed to be proportional to that of the deuteron, in accordance with the calculations of ref. [41] performed for ^4He, ^{16}O and ^{40}Ca by means of a phenomenological model based on harmonic-oscillator wave functions with central and tensor correlations.

A similar picture is obtained from the calculations of refs. [4,6] obtained for the Reid soft-core two-body potential, as can be seen from Fig.4. The long tail of $n(k)$ is therefore a common feature for the various nuclear systems and it is a direct manifestation of the N-N correlations in the hadronic matter. Such a feature has a non perturbative character and theories in which the strongest N-N correlations are included *ab initio* are requested to realistically describe it. The inadaquacy of Slater determinants to give accurate momentum distributions has also been pointed out in other more phenomenological approaches [41,44-46].

3. Spectral function

The nucleon spectral function (1.4) can be written in the following form

$$P(\mathbf{k}, E) = \frac{1}{\pi} \Im \frac{< \bar{0}|a_{\mathbf{k}}^{\dagger}[H - E_0 - E - i\eta]^{-1}a_{\mathbf{k}}|\bar{0} >}{< \bar{0}|\bar{0} >}, \tag{3.1}$$

A convenient perturbative scheme to calculate $P(\mathbf{k}, E)$ is obtained by splitting the hamiltonian into an unperturbed part H_0 ,diagonal in the OCB states, and the rest,

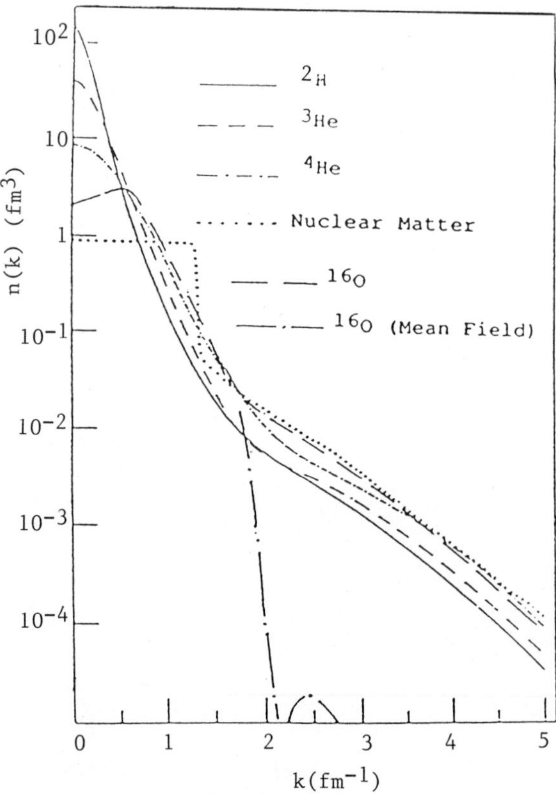

Fig. 4 *Nucleon momentum distribution in ^2H [6], ^3He [6], ^4He [42] and ^{16}O [4] obtained for RSC interaction, compared with the nuclear matter results of ref. [1] and with the mean field approach of ref. [43].*

namely

$$H = H_0 + H_I, \tag{3.2}$$

$$< i|H_0|j >= \delta_{ij} < i|H|i >= \delta_{ij} H_{ii}, \tag{3.3}$$

$$< i|H_I|j >= (1 - \delta_{ij}) < i|H|j >= \bar{H}_{ij}, \tag{3.4}$$

where $| \quad >$ are OCB states. The quantity $H - E_0$ in eq.(3.1) is splitted into $(H_0 - E_0^v) + (H_I - \Delta E_0)$, with $E_0^v = H_{00}$ being the variational estimate of the ground state energy and $\Delta E_0 = E_0 - E_0^v$ the perturbative correction to it, and then expanded in the interaction operator term $H_I - \Delta E_0$ with the result:

$$(H - E_0 - E + i\eta)^{-1} = (H_0 - E_0^v - E + i\eta)^{-1} \sum_n (-)^n [(H_I - \Delta E_0)(H_0 - E_0^v - E + i\eta)^{-1}]^n. \tag{3.5}$$

Similarly, the ground state $|\bar{0} >$ is expressed in terms of the OCB states as:

$$|\bar{0} >= \sum_n (-)^n [(H_0 - E_0^v)^{-1}(H_I - \Delta E_0)]^n |0 > . \tag{3.6}$$

The spectral function is then formally obtained by inserting $\sum_n |n >< n| = 1$ between each pair of contiguous operators.

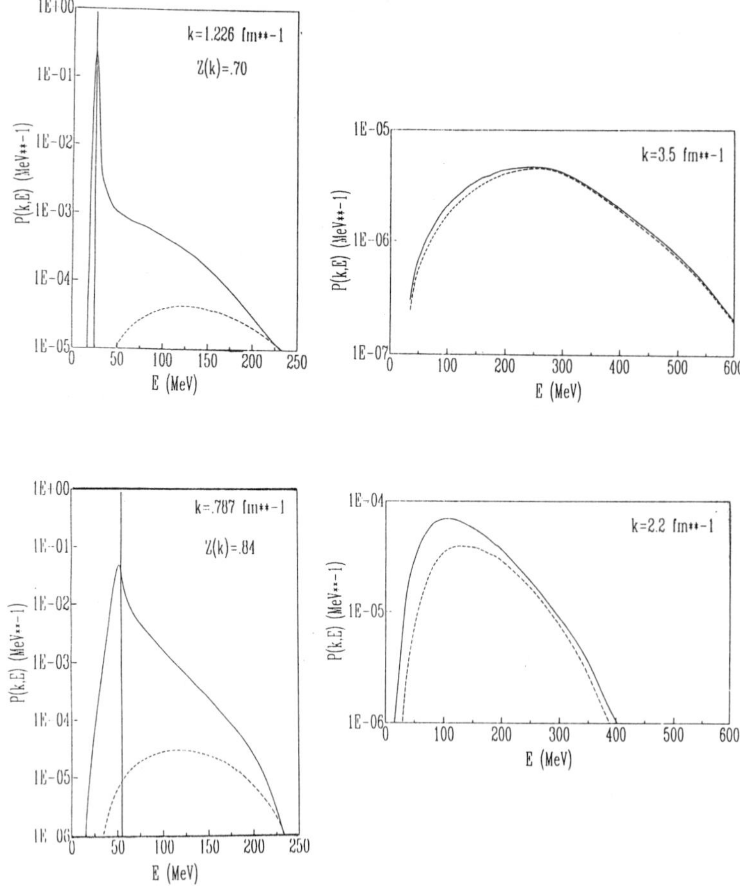

Fig. 5 *Spectral function of the CO-model of nuclear matter at $k_F = 1.33 fm^{-1}$ for $k = 0.8, 1.2, 2.2, 3.5 fm^{-1}$. The dashed lines refer to the variational estimates and the solid lines to the full calculations which include the perturbative corrections. The strenghts $Z(k \leq k_F)$ of the quasi-hole are also reported.*

Several terms appearing in both the numerator and the denominator of $P(\mathbf{k}, E)$ are higly divergent in the thermodynamic limit. The divergencies arise from the factors ΔE_0 which are of order A and from the *unlinked* parts of the non diagonal matrix elements $< 0|a_{\mathbf{k}}^{\dagger}|n >$ or $\bar{H}_{mn}$. It has been proved that all these divergent terms cancel out exactly and that the remaining connected terms can be summed up by means of hypernetted and operator chain summations [17].

A variational estimate of the spectral function is obtained at the zeroth order of the above perturbative scheme by direct use of eq. (1.4) with the OCB states. $P(\mathbf{k}, E)$ is separated into two parts given by the term in eq.(1.4) in which the OCB intermediate state $|n >$ is the $1h$ OCB state $|k >$ and the remaining term obtained with the $nh - (n-1)p$ OCB states with $n > 1$. The first one provides an estimate of the two-body break-up contribution to the spectral function and is given by

$$P_{1h}^{v}(\mathbf{k}, E) = |\Phi_{\mathbf{k}}(k)|^2 \delta(E + e_v(k)), \tag{3.7}$$

$$\Phi_{\mathbf{k}}(n) =< 0|a_{\mathbf{k}}^{\dagger}|n > . \tag{3.8}$$

This is the only contribution to the spectral function in an uncorrelated matter, where $\Phi_\mathbf{k}(\mathbf{k})$ is equal to 1. Nucleon-nucleon correlations produce a quenching of this quantity and consequently of the quasi-particle strength.

It has been proved [17] that $|\Phi_\mathbf{k}(\mathbf{k})|^2$ coincides with the discontinuous part of the momentum distribution $n_v(k)$ of eq.(2.2). It follows that $|\Phi_\mathbf{k}(\mathbf{k})|^2$ has to be regarded as the variational estimate of the quasi-particle strength $Z(k)$. In fact $Z(k_F)$ is given by the discontinuity of the momentum distribution at the Fermi surface [47].A substantial empirical evidence of a sizeable quenching of the strenght of single particle states for nuclei in the lead region has recently been provided for by both elastic and inelastic electron scattering experiments [8,13].

Moreover, in a correlated matter, one has also contributions from three- and more-break up processes, which give rise to the so called background contribution. Such a contribution has to be regarded as the most direct evidence of the N-N correlations, it is nonvanishing for $k \geq k_F$, indicating the occurrence of knock-out processes from states above of the Fermi sea and it is found to extend over a wide range of removal energy. Its energy integral fournishes the continuous part of momentum distribution $n_v(k)$. The main part of it is constituted by the three-body break-up processes given by

$$P_{2h-1p}^v(\mathbf{k}, E) = \sum |\Phi_{\mathbf{h}_i \mathbf{h}_{i'} \mathbf{p}_i}(\mathbf{k})|^2 \delta(e_v(p_i) - e_v(h_i) - e_v(h_{i'}) - E), \qquad (3.9)$$

In Fig. 5 the variational estimates $P_{1h}^v(\mathbf{k}, E)$ at $k = 0.8 fm^{-1}$ and $k = 1.2 fm^{-1}$ are represented by vertical bars whose height gives single particle strenght $Z^v(k)$. The background contribution $P_{2h-1p}^v(\mathbf{k}, E)$ is also displayed for $k = 0.8, 1.2, 2.2, 3.5 fm^{-1}$.

The perturbative correction $\delta n_{gr}(k)$ to the variational momentum distribution $n_v(k)$, due to $2h2p$ OCB admixtures in the ground state, has been found to be relevant, particularly at $k \sim k_F$ [1]. The corresponding perturbative correction $\delta P_{gr}(\mathbf{k}, E)$ to the spectral function is discussed in refs. [17,18] and it satisfies the sum rule

$$\int_{-e_v(k_F)}^{\infty} dE \delta P_{gr}(\mathbf{k}, E) = \delta n_{gr}(k). \qquad (3.10)$$

In addition, the perturbative correction $\delta P_{int}(\mathbf{k}, E)$ due to $2h1p$ OCB states admixtures in $|\mathbf{k} >$ has also to be taken into account. Its integral over the energy vanishes, therefore it contributes to the shape of the spectral function, giving a width to the single particle peaks, but not to the momentum distribution.

In Fig. 5 the results for $P(\mathbf{k}, E)$ with the perturbative corrections are shown at several values of k.

Part of the contributions to $\delta P_{int}(\mathbf{k}, E)$ corresponds to the depletion of the quasi-particle strength due to the $2h1p$ admixtures in $|\mathbf{k} >$. The integral over the energy of such part gives an extra correction $\delta Z_{int}(k)$ to $Z(k)$, which then results to be

$$Z(k \leq k_F) \approx |\Phi_\mathbf{k}(\mathbf{k})|^2 + \delta n_{gr}(k) + \delta Z_{int}(k). \qquad (3.11)$$

It turns out that $\delta n_{int}(k_F^-) = \delta n_{gr}(k_F^+)$ with the consequence that the total (variational + perturbative) background contribution is a continuous function of k. The difference between $n(k)$ and $Z(k)$ is ~ 0.1 at $k \sim k_F$ and slightly decreases for

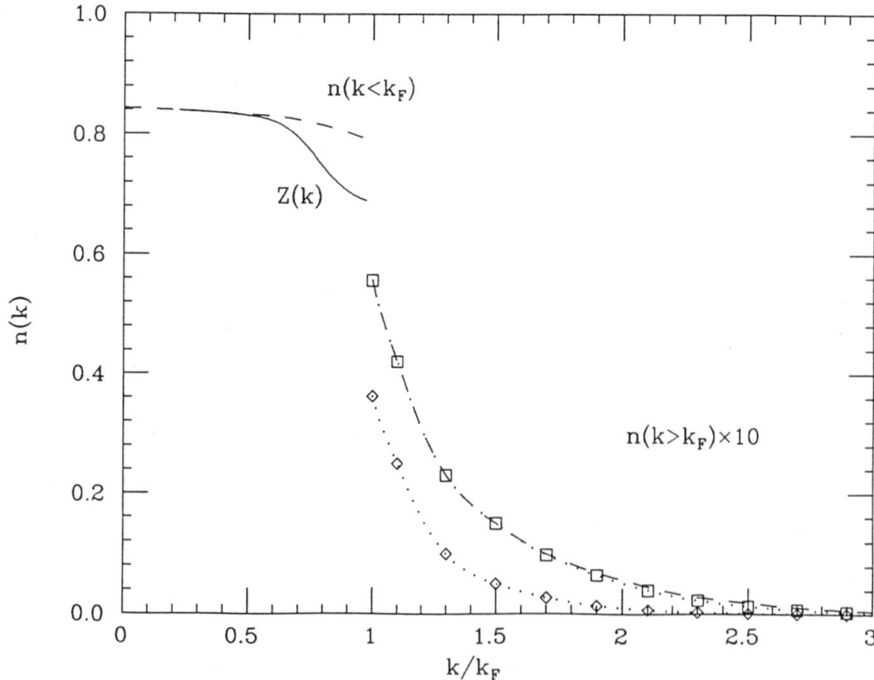

Fig. 6 *Removal energy integral of the nuclear spectral function* $n_{\overline{E}}(k) = \int_{-e_v(k_F)}^{\overline{E}} dE\, P(k, E)$ *in the CO-model of nuclear matter. The dotted lines with open diamonds and squares refer to* $\overline{E} = 100, 300 MeV$ *respectively. The solid line gives the quasi-hole strength* $Z(k)$, *whereas the dashed line is the full* $n(k)$.

$k \to 0$. If one also corrects $Z(k)$ for the coupling of the single particle states with the surface vibration modes as estimated in refs.[7,48], one gets results which are in very good agreement with the experimental data of ref.[13].

The integral over the energy of the spectral function and the quasi particle strength $Z(k)$ are displayed in Fig. 6, together with $n(k)$. The figure shows the relevance of the high removal energy tail of the spectral function, which is entirely due to N-N correlations and is a common feature of nuclear matter and light nuclei [4-6]. One can estimate the response function $S(\mathbf{q}, \omega)$ from the spectral function. If the classical kinematics is used for the outgoing nucleon the response function is given by

$$S_P(\mathbf{q},\omega) = \int_{-e_v(k_F)}^{\infty} dE \int dk P(\mathbf{k}, E)\delta(\omega - E - e_v(|\mathbf{k}+\mathbf{q}|))\Theta(|\mathbf{k}+\mathbf{q}| - k_F). \quad (3.12)$$

In the above expression the knocked-out nucleon is considered completely uncorrelated from the remaining $A-1$ nucleons. It is interesting to compare $S_P(\mathbf{q}, \omega)$ with $S(\mathbf{q}, \omega)$ at high values of of q to have an estimate of the final states effects. In Fig. 7 $S_P(\mathbf{q}, \omega)$ and $S(\mathbf{q}, \omega)$ are compared ,in the negative y-scale variable region, with a recent evaluation obtained [49] by using Brueckner theory. This last estimate appears to be much higher at $q = 1.5 GeV$.

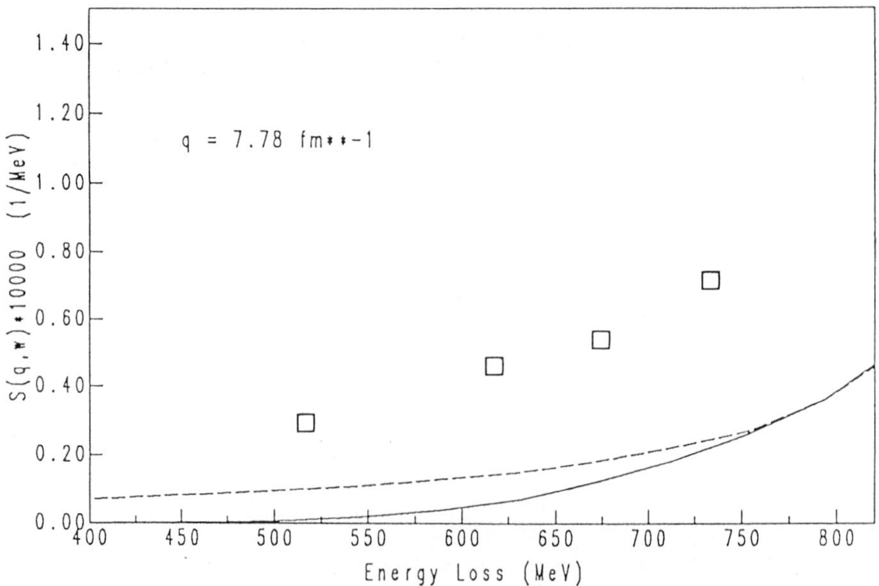

Fig. 7 $S_P(\mathbf{q},\omega)$ *(solid line) compared with* $S(\mathbf{q},\omega)$ *(dashed line) and with the results of ref. [50] (squares) for nuclear matter at the equilibrium density.*

The last equation can be readily modified to treat fully relativistically the knocked-out nucleon, which is indeed necessary to realistically estimate the response in the y-scale region.

References

[1] S.Fantoni and V.R.Pandharipande, Nucl. Phys. **A427** (1984) 473;

[2] J.G.Zabolitsky and W.Ey, Phys. Lett. **B76** (1978) 527;

[3] J.W.Van Orden, W.Truex and M.K.Banerjee, Phys. Rev. **C21** (1980) 2628;

[4] O.Benhar, C.Ciofi degli Atti, S.Liuti and G.Salme', Phys. Lett. **B177** (1986) 135;

[5] R.Schiavilla, V.R.Pandharipande and R.B.Wiringa, Nucl. Phys. **A449** (1986) 219;

[6] C.Ciofi degli Atti, E.Pace and G.Salme', Phys. Lett. **141B** (1984) 14;

[7] V.R.Pandharipande, C.N.Papanicolas, J.Wambach, Phys. Rev. Lett. **53** (1984) 1133;

[8] B.Frois and C.N.Papanicolas, Ann. Rev. Nucl. Part. Sci. **37** (1987) 4133 and references therein;

[9] J.M.Cavedon et al. Phys. Rev. Lett. **49** (1982) 978;
B.Frois et al., Nucl. Phys. **A396** (1983) 409;

[10] C.N.Papanicolas et al., Phys. Rev. Lett. **58** (1987) 2296. ;

[11] J.Lichtenstadt et al. Phys. Rev. **C20** (1979) 497;

[12] C.Marchand et al. Phys. Rev. Lett.**60** (1988) 1703;

[13] E.N.M. Quint, Thesis, University of Amsterdam (1988);
P.K.A.De Witt Huberts, *Electron-Nucleus Scattering*, A. Fabrocini *et al.* Eds. (World Scientific, Singapore,1989), 349;

[14] T.deForest, Jr., J.D.Walecka, Advances in Phys. **15** (1966) 1;

A.E.L.Dieperink and T.de Forest Jr., Ann. Rev. Nucl. Science **25** (1975) 1;

S.Frullani and J.Mougey, Adv. Nucl. Phys. **14** (1984) 1;

[15] C.Ciofi degli Atti, E.Pace and G.Salme', Phys. Rev. **C21** (1980) 805;

[16] A.E.L. Dieperink, T. de Forest Jr., I.Sick and R.A.Brandeburg, Phys. Lett. **B63** (1976) 261;

H.Maier-Hajduk, Ch.Hajduk, P.U.Sauer and W.Theis, Nucl. Phys. **A395** (1983) 332;

[17] O.Benhar, A.Fabrocini and S.Fantoni, Nucl. Phys. **A**(1988) in press; *Electron-Nucleus Scattering*, A. Fabrocini *et al.* Eds. (World Scientific, Singapore,1989), 330;

[18] O.Benhar, A.Fabrocini and S.Fantoni, preprint (1989)INFN-ISS89/2;

[19] I.Sick, contribution in this volume;

[20] I.Sick, D.Day and J.S.Mc Carthy, Phys. Rev. Lett. **45** (1980) 871;

[21] E.Pace and G.Salme', Phys. Lett. **B110** (1982) 411;

[22] I.E.Lagaris and V.R.Pandharipande, Nucl. Phys. **A359** (1981) 331;

[23] R.B.Wiringa, R.A.Smith and T.L.Ainsworth, Phys. Rev. **C19** (1984) 1207;

[24] I.E.Lagaris and V.R.Pandharipande, Nucl. Phys. **A359** (1981) 349;

[25] C.R.Chen *et al.* Phys. Rev. **C33** (1986) 1740;

[26] J.Carlson, Phys. Rev. **C36** (1987) 2026; Phys. Rev. (1988) in press;

[27] V.R.Pandharipande, private communication;

[28] E.Migli, A.Fabrocini and S.Fantoni, to be published;

[29] S.Fantoni and S.Rosati, Nuovo Cimento **A43** (1978) 413;

[30] V.R.Pandharipande and R.B.Wiringa, Rev. Mod. Phys. **51** (1979) 831;

[31] R.B.Wiringa, V.Fiks and A.Fabrocini, Phys. Rev. **C38** (1988) 1010;

[32] E.Feenberg, *Theory of Quantum Fluids*, Academic Press, N.Y., 1969;

[33] J.W.Clark and E.Krotscheck, Lect. Notes in Phys. **198** (1983) 127 and references therein;

[34] S.Fantoni, B.L.Friman and V.R.Pandharipande, Lecture Notes in Phys. **198** (1983) 289 and references therein;

S.Fantoni, Phys. Rev. **B39** (1984) 2544;

[35] S.Fantoni and V.R.Pandharipande, Phys. Rev. **C37** (1988) 1697;

[36] O.Benhar, C.Ciofi degli Atti, S.Fantoni and S.Rosati, Nucl. Phys. **A328** (1979) 127;

[37] S.Fantoni, Nuovo Cimento,**A44** (1978) 191;

[38] J.W.Clark, contribution in this volume;

[39] D.Gogny, in *Nuclear Physics with Electromagnetic interactions*, eds. H. Arhenovel and D. Drechsel, Lecture Notes in Physics **108** (1979) 88;

[40] R.Schiavilla, *Electron-Nucleus Scattering*, A. Fabrocini *et al.* Eds. (World Scientific, Singapore,1989), 236;

[41] M.Traini and G.Orlandini, Z. Phys. **A321** (1985) 479;

[42] Y.Akaishi, Nucl. Phys. **A416** (1984) 409c;

[43] J.W.Negele, Phys. Rev. **C1** (1970) 1260;

[44] M.Jaminon, C.Mahaux and H.Ngo, Phys. Lett. **B158** (1985) 103;

[45] W.H.Dickoff, contribution in this volume;

[46] O.Bohigas and S.Stringari, Phys. Lett. **B95** (1980) 9;

M.F.Flynn *et al.*, Nucl. Phys. **A427** (1984) 473;

[47] A.B.Migdal, JETP (Sov.Phys.) **119** (1960) 1153;

[48] O.Benhar,A.Fabrocini and S.Fantoni, preprint (1989) IFUP-TH 3/89;

[49] M.N.Butler and S.E.Koonin, Phys.Lett. **B205** (1988) 123;

M.N.Butler, contribution in this volume.

HYDROGENIC SYSTEMS

HYDROGEN AND HYDROGEN IMPURITIES IN RARE GASES

K. W. Herwig[†] and R. O. Simmons

Department of Physics
The University of Illinois
Urbana, IL 61801

INTRODUCTION

The hydrogen molecule in condensed systems has been extensively studied. With the exception of solid hydrogen,[1] the kinetic energy associated with translational motion of the center of mass of the molecule has not been directly measured. Deep inelastic neutron scattering (DINS) provides a direct probe of the single particle behavior of the molecule and can be used to determine the center of mass translational kinetic energy, $< KE >$.

Since hydrogen is the lightest of all molecules, the quantum mechanical nature of its rotational motion is most evident. The energy required to raise the rotational quantum number J from its ground state value of 0 to the first excited state $J = 1$ is 14.7 meV.[2] For deuterium this value is 7.4 meV and for nitrogen it is 0.49 meV.[3] The large energy separations of the rotational levels in hydrogen play an important role in the low temperature properties of both solid and liquid hydrogen. For an antisymmetric spin wave function, the rotational wave function must be symmetric, even J, and a symmetric spin wave function requires an antisymmetric rotational wave function, odd J. These two species of hydrogen are called para- and ortho-hydrogen respectively. Ortho-hydrogen is metastable at low temperatures because a conversion from ortho- to para-hydrogen requires a change in nuclear spin. The equilibrium concentration at any temperature can be easily calculated[2] and catalysts exist which allow for relatively quick achievement of the equilibrium condition. The energy levels of the excited intramolecular vibrational states are separated even more than the rotational levels, with an energy of 516 meV required to raise the vibrational quantum number n from 0 to 1.[2]

In a DINS experiment the double differential scattering cross section, $d^2\sigma/d\Omega dE$ is measured. For a system with internal degrees of freedom, such as molecular hydrogen, the scattering can be expressed as[4]

$$\frac{d^2\sigma}{d\Omega dE} = \frac{k_f}{k_0} \sum_m b_{0m}^2 S_{0m}(\vec{Q}, E) \tag{1}$$

$$S_{0m}(\vec{Q}, E) = f_{0m}(\vec{Q}) \int n(\vec{p})\delta(E - E_e - \frac{\hbar\vec{Q}\cdot\vec{p}}{M})d\vec{p} \tag{2}$$

at sufficiently high momentum transfers, Q. $n(\vec{p})$ is the normalized momentum distribution

---

[†]Present address: IPNS, Bldg. 360, Argonne National Laboratory, Argonne, IL 60439.

"

of the initial state, k_f and k_0 are respectively the final and initial momenta of the scattered neutron, and M is the mass of the molecule. b_{0m} and f_{0m} are respectively the scattering length and structure factor of the mth internal molecular level. E_e is the molecular excitation energy

$$E_e = \frac{\hbar^2 Q^2}{2M} + E_m \tag{3}$$

where E_m is the energy required to excite the internal molecular state denoted by m from the initial state. Equations 1 and 2 predict peaks in the observed scattering corresponding to transitions from the initial to excited internal states centered at E_e with a width corresponding to the center of mass translational motion of the initial state. From these widths we can extract the single particle kinetic energy of the hydrogen molecule.

We have measured $d^2\sigma/d\Omega dE$ of molecular hydrogen under a variety of conditions at the Argonne National Laboratory's Intense Pulsed Neutron Source (IPNS) using the Low Resolution Medium Energy Chopper Spectrometer (LRMECS). We measured the scattering from a cold gas of molecular hydrogen in order to determine if a model predicting the Q-dependent form factors of Equation 2 was correct and to check our treatment of the instrument resolution function. We also performed measurements to extract $< KE >$ for molecules in solid hydrogen at 4.7 K and for hydrogen molecules isolated in a matrix of solid argon.

HYDROGEN GAS

In order to analyze the data from our experiments on solid hydrogen and hydrogen in solid argon we required a model for the Q dependence of the form factors in Equation 2. We also required a testing ground for our approach to analyzing the data incorporating a numeric resolution function. By measuring $d^2\sigma/d\Omega dE$ from a cold gas of para-hydrogen molecules we expected to satisfy these two goals.

The Q-dependent form factors in Equation 2 lead to interesting behavior in the observed scattering. Young and Koppel[5] derived a formula for $d^2\sigma/d\Omega dE$ for hydrogen gas assuming no rotational-translational coupling and a system of isolated scatterers. For a ground state para-hydrogen molecule ($J = 0$) they found

$$\left(\frac{d^2\sigma}{d\Omega dE}\right)_{para} = \frac{k_f}{k_0}(\pi W)^{-\frac{1}{2}} \sum_n \sum_{J'=0,1,2,\ldots} (F_{n,J'})_{para}$$
$$\exp\{-(E - E_{J'} - nE_{vib} - \frac{\hbar^2 Q^2}{2M})^2/W\} \tag{4}$$

where

$$W = \frac{4\hbar^2 Q^2 < KE >}{3M} \tag{5}$$

and

$$(F_{n,J'})_{para} = \frac{a_{J'}^2}{n!}\left(\frac{\hbar^2 Q^2}{2M E_{vib}}\right)^n (2J'+1) \mid A_{n,J'} \mid^2 \tag{6}$$

where $a_{J'}^2 = 78.7 \; 10^{-24}$ cm^2 for J' odd and $2.05 \; 10^{-24}$ cm^2 for J' even. E_J is the energy of the Jth rotational level, M is twice the mass of a proton, and E_{vib} is the energy required to excite the first internal vibrational mode of the two protons. $A_{n,J'}$ is given by

$$A_{n,J} = \int_{-1}^{1} d\mu\, \mu^n \exp\left(-\frac{\hbar^2 Q^2 \mu^2}{4M E_{vib}} + \frac{iQ\hbar a\mu}{2}\right) P_J(\mu) \tag{7}$$

where a is 0.7411 Å, the equilibrium distance of the two protons, and P_J is the Legendre polynomial of order J. A similar set of equations can be given for a ground state ortho-hydrogen molecule ($J = 1$) where $a_{J'}^2 = 78.7\ 10^{-24}$ for all J'. We note that as in Equations 1 and 2, the above equations predict peaks in $d^2\sigma/d\Omega dE$ centered at an excitation energy which is a sum of the recoil energy $\hbar^2 Q^2/2M$ and the energy of an internal molecular excitation. The widths of the peaks are predicted to be proportional to the center of mass kinetic energy of the molecule. Samples of predominantly para-hydrogen ground state molecules are desired because only peaks corresponding to final states of odd J' will be apparent and the data may be more easily analyzed.

Although the above formulas have been in the literature for quite some time, to the best of our knowledge no experiments have been performed before ours to check their validity. Neutron scattering experiments on hydrogen gas have focused on measurements of the total cross section[6,7] or on low incident energy, quasi-elastic experiments.[8] We have measured $d^2\sigma/d\Omega dE$ of gaseous hydrogen at 20 K.

Our sample consisted of gaseous hydrogen contained in four vertical aluminum tubes connected at one end to a manifold filled with a nickel silicate catalyst. By slowly flowing the room temperature supply of gas over the cold catalyst, the equilibrium concentration of para-hydrogen molecules could be achieved. The temperature of the catalyst was maintained at less than 20 K while the gas was added. At this temperature the catalyst should prepare a sample which is better than 99.8% para-hydrogen.[2] Throughout the course of collecting data the temperature of the sample cell was maintained at 18.0 ±0.5 K. The gas then was in equilibrium with a liquid contained in the catalyst container located outside the incident neutron beam. The gas had a molar volume of 3.3 10^{-4} cc/mole.[9]

Data were collected from the predominantly para-hydrogen sample at scattering angles, ϕ, from 18.6 to 72.9 deg using LRMECS. At IPNS a pulse of 500 MeV protons was directed onto a depleted uranium target resulting in a large flux of neutrons by a process called spallation. These neutrons were moderated in a liquid methane moderator. LRMECS has a sample position 6.84 m from the moderator face and a bank of ^{3}He filled detectors located 2.5 m from the sample position. The incident neutron energy is determined by phasing a Fermi chopper located in front of the sample position with the proton pulse. In this experiment, the chopper was phased such that neutrons with a nominal incident energy of 504 meV were selected.

Data from groups of three detectors were added together to increase the counting rates with a range of scattering angle of ±0.9 deg from the mean scattering angle. Background data were taken from the same sample cell filled with enough ^{3}He gas to simulate the attenuation of the incident neutron beam by the hydrogen sample. This allowed a very clean subtraction of the aluminum sample cell from the data. The range of scattering angles covered in this experiment corresponded to a range in Q of 5.0 to 15.5 Å^{-1}.

Figure 1 shows the data (with sample cell subtracted) taken at a scattering angle of 25.8 deg. Clearly visible are the scattering peaks due to transitions from the ground state ($J = 0$) para-hydrogen molecule to excited rotational states of $J = 1$ and 3. Also shown is a fit to the data for a model consisting of a Gaussian momentum distribution and Q-dependent form factors such as are expressed in Equations 1 and 2. The fit includes the instrumental resolution function.[10,11] Figure 2 compares the predictions of the Young and Koppel[5] model with the results of our experiment. In general excellent agreement is observed over all scattering angles. The slight discrepancies could be due to some small sample dependent backgrounds or to a small fraction of ortho-hydrogen molecules in the sample. The kinetic energy extracted from the widths of the peaks was 26 ± 3 K, in reasonable agreement with an ideal gas value of $\frac{3}{2}$T = 27 K. The good agreement of the kinetic energy values indicates that our treatment of the instrumental resolution broadening works quite well.

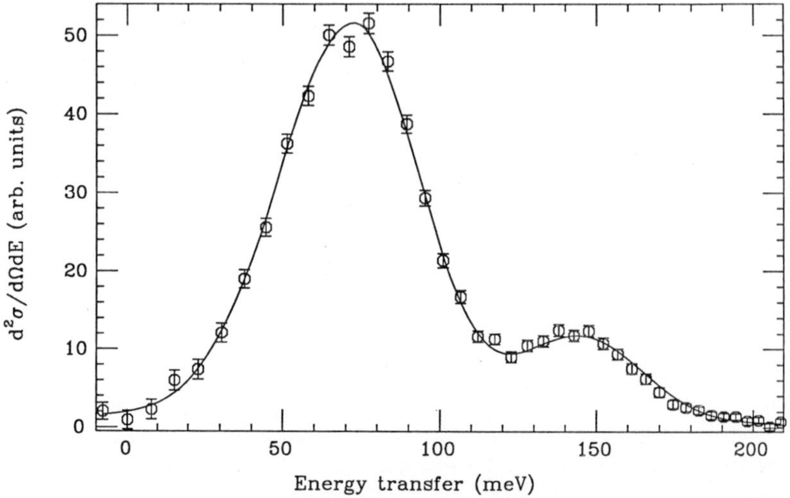

Figure 1. Scattering data from gaseous hydrogen at 25.8 deg. The solid curve is a fit to the data including instrumental resolution. The scattering peaks are due to internal molecular rotational excitations.

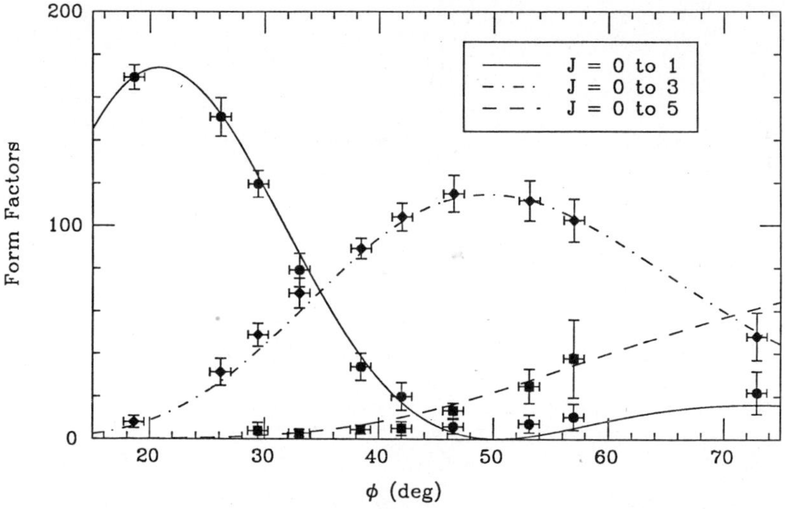

Figure 2. The dependence of the rotational form factors on scattering angle. • are the data for the $J = 0$ to 1 transition. ◆ and ■ are the data for the $J = 0$ to 3 and 5 transitions respectively. The smooth curves are from the Young and Koppel model.[5]

SOLID HYDROGEN

Hydrogen, like other quantum solids, is characterized by a single particle zero point kinetic energy, $< KE >_0$, which is a relatively large fraction of the total energy per particle in the solid. This property plays a large role in the low temperature thermodynamic properties of the solid. DINS is the only technique which can directly measure $< KE >_0$.

The nearest neighbor distance in a quantum solid lies well beyond the minimum of the pair potential in order to minimize the free energy per particle resulting in a depression of the melting temperature.[12] The potential at a lattice site due to all the other molecules in the crystal located at their appropriate lattice positions is actually a slight local maximum rather than a minimum. In a standard harmonic theory describing lattice vibrations, the square of the phonon frequency is related to the Laplacian of the potential.[13] For a quantum solid, this Laplacian is actually negative for some positions, resulting in imaginary phonon frequencies for at least part of the Brillouin zone. However, phonons do exist in hydrogen[14] and self-consistent phonon theories generate real phonon frequencies.[15]

Quantum solids are also characterized by large RMS zero point excursions of the particles from their mean lattice positions due to their light mass and relatively weak binding energies. The orientational properties of molecular hydrogen in a low temperature solid are affected by these large excursions. The anisotropic coupling in ortho-hydrogen is effectively renormalized by averaging over the large zero point motion.[16] Additionally, theories attempting to describe the low temperature properties of all quantum solids must include short range correlations between at least nearest neighbors because the large zero point excursions would otherwise allow large overlap of the particles' wave functions.

These effects make a determination of the ground state properties of solid hydrogen of interest. A previous experiment by Langel and co-workers[1] measured the single particle kinetic energy for solid hydrogen using DINS. They determined a value for $< KE >_0$ of 76 $\pm$ 9 K for a sample at a temperature of 10 K. We performed a similar experiment with a solid at 4.7 K. From this measurement we have determined the single particle kinetic energy of solid hydrogen at 4.7 K.

The hydrogen sample was condensed into 20 vertical aluminum tubes of 1.59 mm outer diameter with 0.13 mm walls attached at the bottom to a manifold. The sample cell was attached at the top to a cold finger connected to a liquid ^{4}He cryostat. Before entering the manifold the hydrogen passed through a container filled with a nickel silicate catalyst which was kept at a temperature of approximately 20 K. Gas was added until the sample cell was filled with liquid hydrogen. The cell temperature was kept at 17.7 K for 19 hours, so that further contact of the hydrogen with the catalyst would increase the para-hydrogen concentration. The solid sample was prepared by lowering the temperature of the sample cell and adding hydrogen keeping the pressure at approximately 0.28 MPa. The solid sample was maintained at a temperature of 4.7 K for three days before data were taken. While data was taken, the sample cell temperature was maintained at 4.73 $\pm$0.03 K. The molar volume of the sample was 23.2 cc/mole.

Data were taken using LRMECS with the chopper phasing such that 509 meV incident energy neutrons were selected. Groups of three detectors were added together in order to increase the counting rates. As in the previous experiment, background data was taken on the sample cell filled with enough ^{3}He to simulate the attenuation of the incident beam by the solid hydrogen. Usable data were obtained from scattering angles ranging from 21.6 to 63.6 deg.

Figure 3 is a plot of the data (with the sample cell subtracted) collected at a mean scattering angle of 42.0 deg ($Q = 10.5$ Å^{-1}). The data is comprised of three peaks corresponding to transitions from the ground state ($J = 0$) para-hydrogen molecule to excited rotational states of $J = 1, 3,$ and 5. The peaks are quite broad due to the large $< KE >_0$ and some broadening due to the instrumental resolution. The solid curves are fits to the data using the Young and Koppel[5] model to predict the Q-dependent form factors for each transition. Each curve represents a different $< KE >_0$ with the center curve being the best fit and the upper and lower curves representing the errors in determining $< KE >_0$ from this particular data. The fits also include a slight background term which we attribute to multiple scattering of neutrons from the hydrogen sample and its surroundings (radiation shields, ^{4}He dewar, etc.).

Figure 4 gives the values of $< KE >_0$ for all the scattering angles analyzed and the errors in determining $< KE >_0$ from the data at each angle. Our resulting value is 77 K with an associated error of 4 K. This is in excellent agreement with the 76 $\pm$9 K value determined by Langel and co-workers[1] for their sample at 10 K indicating that there is no detectable

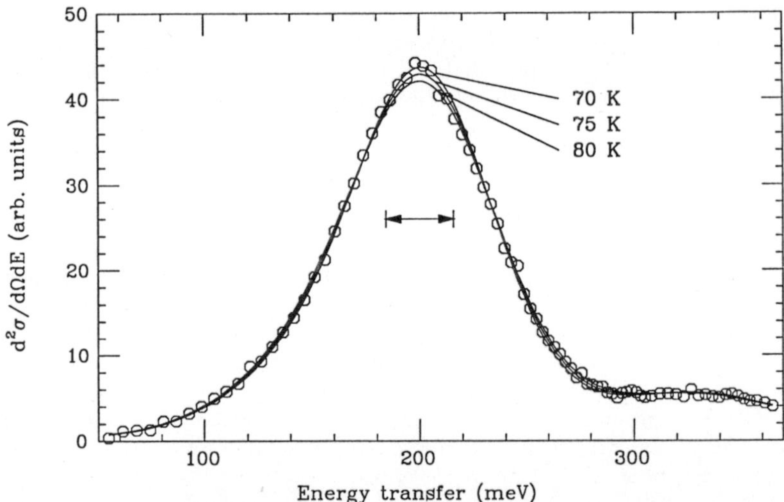

Figure 3. Fits to the solid hydrogen data for a scattering angle of 42.0 deg. The errors associated with the data are approximately the same size as the symbols. The arrow between the two horizontal lines represents the full width at half the maximum intensity of the instrumental resolution.

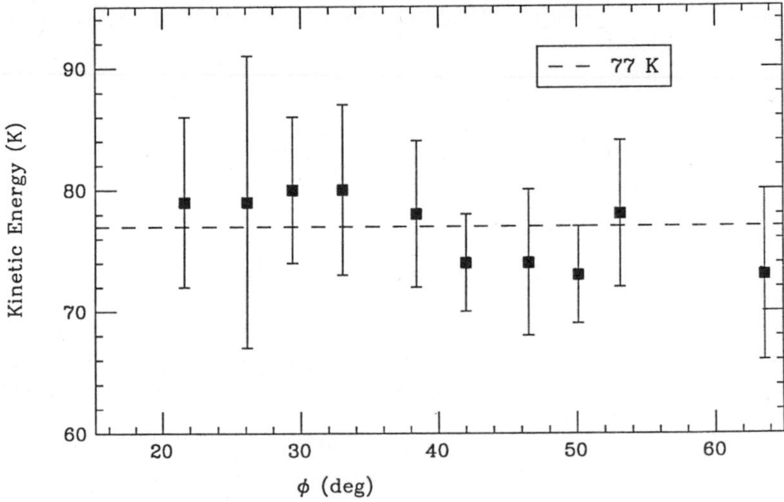

Figure 4. Kinetic energy values from fits to the solid hydrogen data as a function of scattering angle. The dashed line shows the resulting average kinetic energy.

temperature dependence between 4.7 and 10 K for $< KE >$ and that our measurement was indeed of the ground state of the system.

There has not been a great deal of theoretical effort devoted to calculating $< KE >_0$ for solid hydrogen but two comparisons can be readily made. In the simplest, a Debye model of the solid, the ground state energy per particle is[17]

$$E_0 = \frac{\int_0^{\omega_D} (\frac{1}{2}\hbar\omega)\omega^2 d\omega}{\int_0^{\omega_D} \omega^2 d\omega} \tag{8}$$

$$= \frac{9}{8}\Theta_D \tag{9}$$

where Θ_D is the Debye temperature, ω_D is the Debye frequency, and ω is a phonon frequency. $<KE>_0 = \frac{1}{2}E_0$ for a harmonic system, and the Debye ground state single particle kinetic energy is just $\frac{9}{16}\Theta_D$. Θ_D is 121.4 K for solid hydrogen at a molar volume of 23.2 cc/mole[18] giving a value of 68.3 K for the zero point single particle kinetic energy. This value lies well outside the errors in our experimentally determined value. This is not too surprising as de Boer and Blaisse[19] noted that this model gave total zero point energies for the quantum solids which were too small.

Bruce[20] calculated the ground state properties of solid hydrogen as a function of molar volume using a variational technique. His trial wave function was a product of Gaussians localizing the particles at lattice positions and a Jastrow factor to incorporate pair correlations. He used a Lennard-Jones 6-12 pair potential. Monte Carlo methods were employed to calculate the many body integrals with a system of 256 particles. A second order polynomial fit to his theoretical values gives 75 K for $<KE>_0$ at a molar volume of 23.2 cc/mole in excellent agreement with our experimentally determined value. This model also gives reasonable agreement for the isothermal bulk modulus at molar volumes greater than 19 cc/mole. Below this molar volume, the hard core of the Lennard-Jones pair potential makes the model too stiff and the calculated bulk modulus rises well above the experimentally determined values.

HYDROGEN IN SOLID ARGON

The properties of hydrogen as a dilute impurity in noble gas solids and liquids have been studied extensively. Hydrogen as a light, energetic impurity in solid argon has been examined by NMR.[21,22] The diffusive motion of molecular hydrogen in liquid argon has been probed by neutron scattering.[23,24,25] A great deal of study has been performed on the infrared absorption spectrum of dilute hydrogen molecules in solid argon.

The infrared experiments demonstrated the existence of a localized lattice excitation for the hydrogen molecule at an energy of 161 K for a sample at 82.15 K.[26,27] A latter low temperature experiment found that the energy of this excitation was 170 and 157 K for ortho- and para-hydrogen molecules respectively.[28] Kriegler and Welsh[26] solved the Schroedinger equation for the hydrogen molecule in the spherically averaged potential due to the surrounding argon atoms fixed at their lattice sites and calculated an energy of 157 K for this excitation. This value was the difference between the ground and first excited states. Chandrasekharan and co-workers[29,30] used a variational technique to calculate a ground state value for the kinetic energy of the hydrogen molecule of 43.2 K. This calculation also demonstrated that the presence of the hydrogen molecule as a substitutional impurity does not greatly change the mean positions of neighboring argon atoms. The smaller size of the hydrogen molecule is compensated somewhat by large RMS excursions of the molecule from its lattice position. We have used DINS to measure the kinetic energy of the initial state of the hydrogen molecule embedded in a matrix of solid argon. This is in contrast to the infrared absorption experiments which measure excitations from this initial state.

Our sample was a mixture of hydrogen and argon condensed into the same sample cell as used in the hydrogen gas experiment described earlier. The infrared experiments had previously demonstrated that solid samples with a 1 % molar concentration of hydrogen molecules could be readily prepared. The large incoherent cross section of the hydrogen molecule made a 1 % molar concentration of hydrogen in argon a reasonable choice for the sample. Gas entering the sample cell had to first pass through a copper container filled with a nickel silicate catalyst cooled to 20 K. Hydrogen gas which condensed as a liquid in this catalyst container. The sample cell and catalyst container were cooled to 16 K and the liquid hydrogen was kept in contact with the catalyst for eight hours in order to prepare a sample of almost pure para-hydrogen molecules. The cell was then warmed up gradually and the para-hydrogen gas allowed to enter the four aluminum sample tubes. Argon was condensed into the cell as a liquid at 85 K until the cell was full. While maintaining a supply of argon at a pressure of 1.7 MPa the temperature of the sample cell was lowered and a solid sample prepared. The sample cell was cooled to 25.0 ± 0.5 K and data collected.

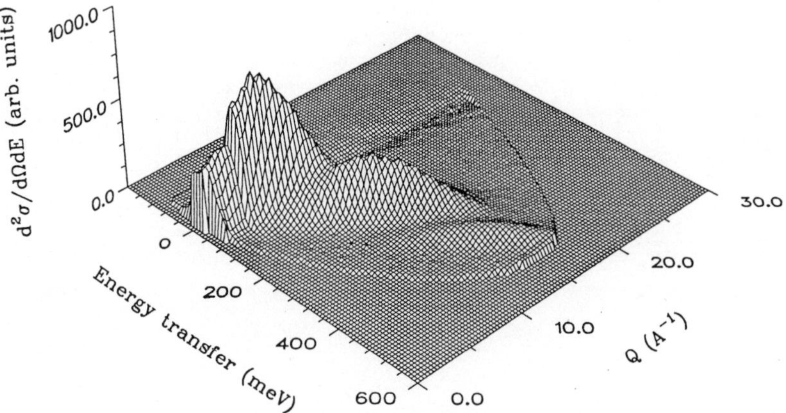

Figure 5. The differential scattering cross section for hydrogen molecules in solid argon. The incident energy was 500.7 meV. The sample and argon signals have been subtracted. The transitions to excited rotational states of $J = 1$, 3, and 5 are clearly visible.

We measured $d^2\sigma/d\Omega dE$ for the sample and a background run of pure argon using LRMECS with the chopper phased to select neutrons with a mean incident energy of 500.7 meV. Data were collected at scattering angles from approximately 6.0 to 116.4 deg. Data from groups of three detectors were added together to increase counting rates. Figure 5 shows the dynamic structure factor $d^2\sigma/d\Omega dE$ for the sample minus the argon background run. Clearly visible are peaks in $d^2\sigma/d\Omega dE$ corresponding to transitions from the $J = 0$ ground state of the molecule to excited states of $J = 1$, 3, and 5. Fits to the data using the Young and Koppel[5] model to predict the Q-dependent form factors for the rotational transitions were performed for data taken at scattering angles from 21.9 to 53.7 deg. The kinetic energy we extract from the widths of these peaks is 62 ± 5 K. This is considerably larger than the 43.2 K zero point kinetic energy calculated by Chandrasekharan and co-workers.[29,30] Our data, however, were taken from a sample at a temperature of 25.0 K. As the temperature is increased, lattice modes are excited which include the para-hydrogen molecule. Our measured value includes the contribution from these thermally excited lattice vibrations and is not unreasonable when compared to the 43.2 K zero point value. A low temperature (4.2 K) measurement is clearly required to verify the variational calculation.

210

CONCLUSION

In conclusion we would like to emphasize that DINS does provide a means to determine the single particle kinetic energy of various systems. In particular the hydrogen molecule in a series of different environments is an interesting case to study. From our measurements on gaseous hydrogen we conclude that the Q-dependent form factors governing the intensity of the rotational lines are modeled quite well by the calculation of Young and Koppel.[5] These calculated form factors allow us to extract the kinetic energy from data where the rotational peaks are not as well resolved as those of the gas (due to a higher $< KE >$).

Using these form factors we have measured a value for the zero point kinetic energy of solid hydrogen of 77 $\pm$4 K in excellent agreement with the 75 K result of a variational calculation.[20] We have extended the measurements of $< KE >_0$ for solid para-hydrogen described in this article to cover a range of molar volumes down to 17.0 cc/mole.[4,31] Our measurement of $< KE >$ for hydrogen molecules embedded in solid argon of 62 $\pm$5 K for a sample temperature of 25.0 K is greater than the result of a 0 K variational calculation.[29,30] We have also measured $< KE >$ for para-hydrogen embedded in solid argon for temperatures greater than 25 K,[32] but a low temperature measurement is required in order to confirm the variational calculation.

Acknowledgements

This research was supported in part by the U.S. Department of Energy, BES-Materials Sciences, under contracts DE-AC02-76ER01198 and W-31-109-ENG-38 and by the U.S. National Science Foundation under NSF DMR86-12860.

REFERENCES

1. W. Langel, D. L. Price, R. O. Simmons, and P. E. Sokol, accepted for publication in Phys. Rev. B.
2. I. F. Silvera, Rev. Mod. Phys. **52**, 393 (1980).
3. M. Bloom and J. A. Morrison, in *Surface and Defect Properties of Solids*, Vol. 2. (Billing and Sons, Ltd., London 1973).
4. K. W. Herwig, M. C. Schmidt, J. L. Gavilano, and R. O. Simmons, *LT-18 Proceedings*, Japanese J. Appl. Phys. **26**, 445 (1987).
5. J. A. Young and J. U. Koppel, Phys. Rev. **135A**, 603 (1964).
6. G. L. Squires and A. T. Stewart, Proc. Roy. Soc. A **230**, 19 (1955).
7. T. L. Houk, D. Shambroom, and R. Wilson, Phys. Rev. Lett. **26**, 1581 (1971).
8. S. H. Chen, T. A. Postol, and K. Sköld, Phys. Rev. A **16**, 2112 (1977).
9. P. C. Souers, *Cryogenic Hydrogen Data Pertinent to Magnetic Fusion Energy*, Lawrence Livermore Laboratory report UCRL-52628, (1970).
10. C. -K. Loong, S. Ikeda, and J. M. Carpenter, Nuc. Instr. and Meth. **A260**, 381 (1987).
11. K. W. Herwig, Ph.D. thesis, The University of Illinois at Champaign-Urbana (1988).
12. J. Van Kranendonk, *Solid Hydrogen*, (Plenum Press, New York, 1983).
13. M. Born and K. Huang, *Dynamical Theory of Crystal Lattices*, (Oxford University Press, Oxford, 1985).
14. M. Nielsen, Phys. Rev. B **7**, 1626 (1973).
15. V. V. Goldman, J. Low Temp. Phys. **38**, 149 (1980).
16. V. V. Goldman, Phys. Rev. B **20**, 4478 (1979).
17. V. Deitz, J. Chem. Phys. **2**, 296 (1934).
18. J. K. Krause and C. A. Swenson, Sol. State Comm. **31**, 833 (1979).
19. J. deBoer and B. S. Blaisse, Physica **14**, 149 (1948).
20. T. A. Bruce, Phys. Rev. B **5**, 4170 (1972).
21. M. S. Conradi, K. Luszczynski, and R. E. Norberg, Phys. Rev. B **19**, 20 (1979).

22. M. S. Conradi, K. Luszczynski, and R. E. Norberg, Phys. Rev. B **20**, 2594 (1979).

23. V. F. Sears, Proc. Phys. Soc.**86**, 953 (1965).

24. V. F. Sears, Proc. Phys. Soc.**86**, 965 (1965).

25. O. J. Eder, S. H. Chen, and P. A. Egelstaff, Proc. Phys. Soc. **89**, 833 (1966).

26. R. J. Kriegler and H. L. Welsh, Can. J. Phys. **46**, 1181 (1968).

27. J. De Remigis and H. L.Welsh, Can. J. Phys. **48**, 1622 (1970).

28. J. A. Warren, G. R. Smith, and W. A.Guillory, J. Chem. Phys. **72**, 4901 (1980).

29. V. Chandrasekharan, M. Chergui, B. Silvi, and R. D. Etters, Physica **131B**, 267 (1985).

30. B. Silvi, V. Chandrasekharan, M. Chergui, and R. D. Etters, Phys. Rev. B **33**, 2749 (1986).

31. K. W. Herwig, J. L. Gavilano, M. C. Schmidt, and R. O. Simmons, to be published.

32. K. W. Herwig, R. C. Blasdell, M. C. Schmidt, and R. O. Simmons, to be published.

HYDROGEN POTENTIAL IN β-V$_2$H STUDIED BY DEEP INELASTIC NEUTRON SCATTERING

R. Hempelmann,[1] D.L. Price,[2] G. Reiter,[3] and D. Richter [4]

[1]Institut für Festkörperforschung
KFA Jülich, West Germany

[2]Materials Science Division
Argonne National Laboratory
Argonne, IL 60429, U.S.A.

[3]Physics Department
University of Houston
Houston, TX 77204, U.S.A.

[4]Institut Laue Langevin
Grenoble, France

Two complementary techniques of deep inelastic neutron scattering were
used to study hydrogen in β-V$_2$H: (i) By means of neutron vibrational
spectroscopy we measured hydrogen vibrations up to the fourteenth order;
from these data we derived the effective single-particle potential, the
shape of which is a parabola with a flattened bottom, and the hydrogen
wave functions. (ii) By means of neutron Compton scattering we determined
the kinetic of energy of the hydrogen; the value agrees with that calculated from the vibrational ground-state wave function.

INTRODUCTION

Hydrogen dissolves in nearly all metals in large quantities, occupying interstitial sites which are energetic minima of the hydrogen potential. The form of this potential, usually expressed as an effective
single-particle potential, is on of the important questions in fundamental
research on metal-hydrogen systems. Its knowledge is essential for a
microscopic understanding of a large variety of phenomena such as transport properties, superconductivity and thermodynamics, and correspondingly
there is considerable theoretical interest in it. Experimentally, this
potential can be determined by neutron vibrational spectroscopy (NVS) (1)
and, recently, by neutron Compton scattering (NCS), i.e. either by a
measurement of the excitation energies of the localized hydrogen vibrations or by a measurement of the hydrogen momentum distribution. In this
brief communication we report an experiment combining both methods on a
single system, which on the one hand leads to a rather detailed understanding of the potential and on the other hand allows a critical
comparison of the capabilities of these two methods of deep inelastic

neutron scattering with respect to hydrogen in metals. For the system to
be investigated we chose β-V$_2$H, an ordered hydride phase with hydrogen on
pseudotetragonal octahedral sites, because from the literature (2) and
from feasibility studies of our own we expected large a harmonic effects.

The experiments were performed on the chopper spectrometers HRMECS
and LRMECS at the Intense Pulsed Neutron Source of Argonne National
Laboratory. The vibrational spectroscopy part has already been published
(3) and will be only briefly summarized here.

Neutron Vibrational Spectroscopy

Figure 1 gives a synopsis of our spectroscopy results. The spectrum
with E_i = 494 meV was taken with multi-domain single crystals, whereas the

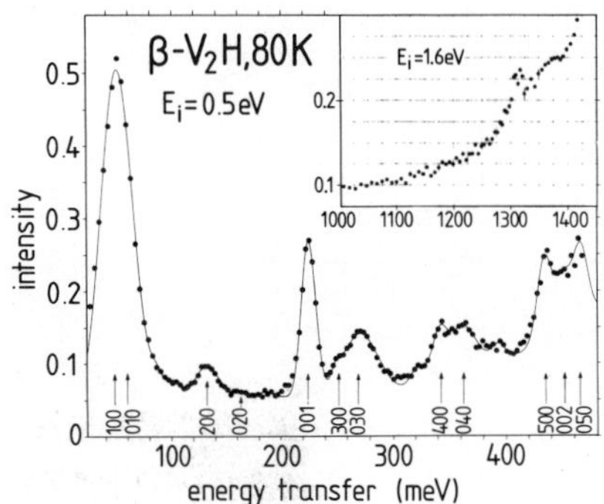

Fig. 1. Neutron spectra from β-V$_2$H at different
neutron energies E_i. The line is the
result of fitting Gaussians to the experimental spectrum in order to define the peak
positions. The indices lmn mark our assign-
ments for the different transitions.

inset spectrum with E_i – 1600 meV is from a single-domain single crystal.
The striking features are: (i) the regular repetition of double peaks (at
lower energy transfer the splitting is only resolved with lower incident
energy), and (ii) the sharp excitation at 223 meV. The peak positions in
these spectra as well as in those taken with 110 meV, 317 meV, 743 meV and
975 meV incident neutron energy were determined by a fitting procedure and
are plotted in the left-hand side of Figure 2 versus quantum number. At
high energies the excitation energies of the soft x and y directions (only
x is plotted) are practically equidistant and indicate that in this energy
range the potential in these directions is rather harmonic. At the lowest
energies, however, the spacing closely resembles that of a square well.
Therefore, in the soft directions the hydrogen potential has to be a
parabola with a flattened bottom. In the stiff z direction, however, the
sequence of the observed vibration is consistent with a harmonic oscillator.
From a quantitative data evaluation in terms of the empirical ansatz

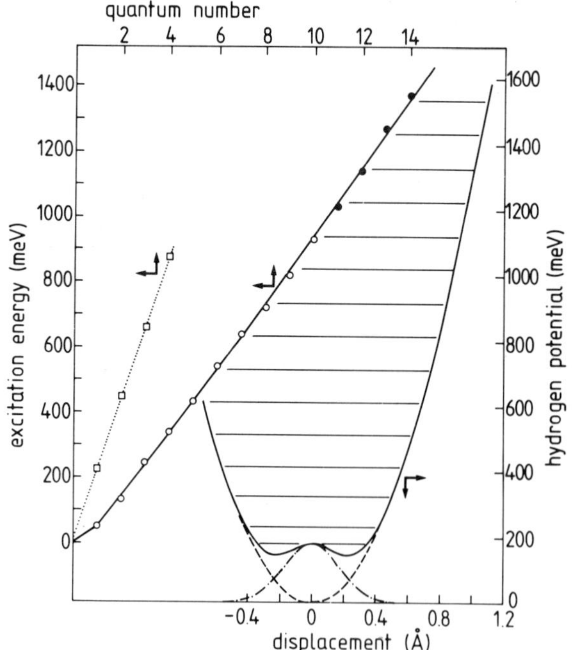

Fig. 2. Left-hand side: Hydrogen excitation
energies in the stiff (001) direction
(squares) and in the soft (100) direction
(multidomain crystals, open circles;
single-domain crystal, filled circles).
Right-hand side: resulting H potential
in the soft (100) direction and the
corresponding energy levels.

$$V(x) = \frac{1}{2} M \omega_o^2 x^2 + \left(b/(2\pi\sigma^2)^{1/2}\right) \exp(-x^2/2\sigma^2), \tag{1}$$

we derived an effective single-particle hydrogen potential (its x component is shown in the right-hand side of Fig. 2) characterized by the
following parameters:

x direction: $\hbar\omega_o$ = 104 meV, b = 78.4 Å meV, σ = 0.174 Å

y direction: $\hbar\omega_o$ = 107 meV, b = 66.6 Å meV, σ = 0.170 Å

z direction: $\hbar l_o$ = 223 meV (harmonic oscillator).

The wave functions are linear combinations of the wave functions χ_n of the
harmonic oscillator; for the ground state we obtain:

$$\psi_o(x) = 0.964\ \chi_o(x) + 0.267\ \chi_2(x) - 0.009\ \chi_4(x) \pm \ldots$$

$$\psi_o(y) = 0.976\ \chi_o(y) + 0.216\ \chi_2(y) + 0.013\ \chi_4(y) \pm \ldots$$

Via the formalism of creation and annihilation operators the ground state
kinetic energy can easily be calculated, yielding

$$E_{kin}^x = 45.0\ \text{meV}, \quad E_{kin}^y = 16.2\ \text{meV}, \quad E_{kin}^x = 55.8\ \text{meV}.$$

Neutron Compton Scattering

General Remarks. At very high energy and momentum transfer the interaction of the neutron with the proton in the collision process is so short
in space and time that only the dynamic short-time properties of the
proton are probed. This is, classically, the instantaneous velocity/
momentum distribution. Neither the periodicity of a vibration originating
from the potential nor any correlations to other protons are relevant on
this short time scale. The chemical binding (i.e. in our case the
hydrogen potential) plays a role only to the extent that it governs the
momentum distribution in the ground state. For an ensemble of protons
with the momentum distribution n_p the scattering function is given by

$$S(Q,\omega) = \int_{-\infty}^{\infty} n_p \; \delta \left(\hbar\omega - \frac{\hbar^2 Q^2}{2M} - \frac{\hbar}{M} \, \mathbf{p} \cdot \mathbf{Q} \right) d^3 p. \tag{2}$$

If an H Atom in a metal is vibrating in a harmonic potential, the ground
state wave function $\psi_o(x)$ is known to be a Gaussian; $\psi_o(p)$, the Fourier
transform of $\psi_o(x)$, is again a Gaussian, and the same is true for the
momentum distribution $n_p = |\psi(p)|^2$:

$$n_p \propto \exp\left(-p^2 / 2\langle p^2 \rangle \right). \tag{3}$$

The protons contribute to the scattering function only if their momentum
takes values such that the argument of the δ function in Eq. (2) vanishes.
Hence

$$S(Q,\omega) \propto \exp \frac{-\left(\hbar\omega - E_r\right)^2}{2\hbar^2 \langle (\mathbf{Q} \cdot \mathbf{p})^2 \rangle / M^2}, \tag{4}$$

and the scattering function is a Gaussian centered at the recoil energy

$$E_r = \hbar^2 Q^2 \,/\, 2M \tag{5}$$

with variance

$$\sigma^2 = \frac{\hbar^2}{M^2} \langle (\mathbf{p} \cdot \mathbf{Q})^2 \rangle = \frac{\hbar^2 Q^2}{3M^2} \langle p^2 \rangle = \frac{4}{3} E_r \, E_{kin}. \tag{6}$$

The second equality of Eq. (6) is obtained by spatial averaging.

The statements of Eqs. (4-6) – here derived for a harmonic oscillator
– are generally valid: the lineshape of the Compton profile corresponds to
the momentum distribution $\psi(p)^2$, the center of the Compton profile allows
a mass determination, and the second moment – which in the case of a
Gaussian is simply the Gaussian parameter – is proportional to the kinetic
energy of the oscillator.

Results. Compton scattering data of V_2H were taken with the incident
neutron energy of 1600 meV for two different orientations of the single-
domain single crystal: (i) x and y directions in the scattering plane;
(ii) x and z directions in the scattering plane. In the former case the
stiff z vibration does not contribute, so H behaves almost like an isotropic oscillator; the observed Compton profile is displayed in Figure 3.

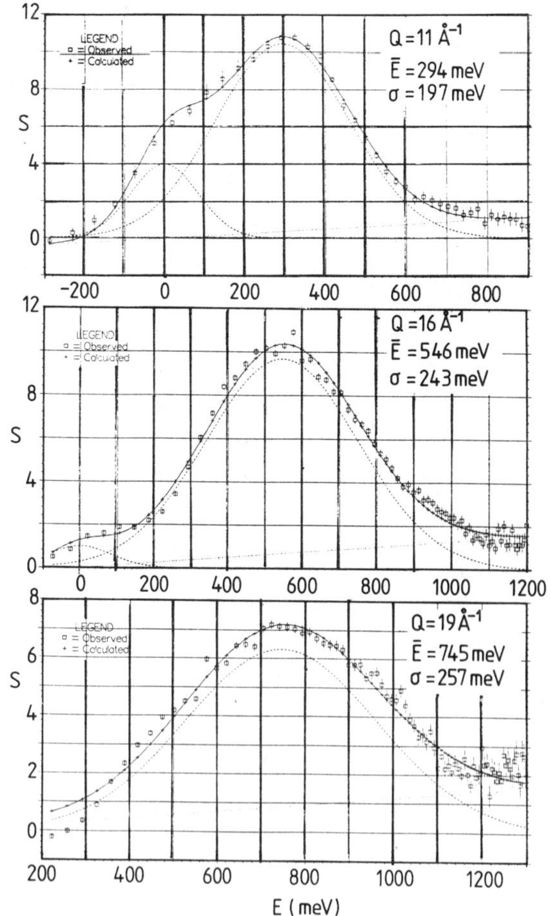

Fig. 3. Neutron Compton profile of H
in a V_2H single-domain single
crystal with the xy vibration
s in the scattering plane.
Solid line: Gaussian fit.

The peak positions and linewidths for both orientations, resulting from
Gaussian fits, are plotted in Figures 4 and 5 versus Q^2 and Q, respectively. From the respective slopes we obtain M_H^{xy} = 1.03 ± 0.03 amu;
M_H^{xz} = 1.12 ± 0.03 amu; E_{kin}^{xy} = 16.0 ± 0.4 meV; E_{kin}^{xz} = 25.5 ± 0.8 meV.

Discussion. The apparent directional dependence of the hydrogen mass
can in our opinion be considered as a criterion for the validity of the
impulse approximation: the impulse approximation is valid if the apparent
mass equals the physical mass. With the xy vibrations of the crystal in
the scattering plane, where we obtain a hydrogen mass of 1.03 ± 0.03 amu,
we are just in the regime of the impulse approximation, whereas with the
xz vibrations in the scattering plane, where the hydrogen mass appears to
be 1.12 ± 0.03 amu, we are clearly outside that regime. Equivalently,
these statements can be expressed in terms of the Reiter criterion (4)
which requires that Q should be larger than, say, $5Q_c$, where the critical
momentum transfer $Q_c = \sqrt{2m\omega_o}/\hbar$. For $\hbar\omega_o$ = 5- meV (x vibration) we get
$5Q_c$ = 25 Å⁻¹, which corresponds to the largest Q value of our experiment

(see Fig. 5), whereas for $\hbar\omega_o$ = 223 meV (z vibration) we get $5Q_c$ = 55 $\mathring{A}^{-1}$ which is clearly beyond our Q range.

The model-independent determination of the <u>kinetic energy</u> in β-V_2H at 80 K allows a microscopic check of the spectroscopic results described in the previous paragraph. The xy average amounts to 15.6 meV. In addition, the contribution of the acoustic phonons has to be taken into account which we estimate in the following way: V_2H has a Debye temperature of 429 K (5); we consider the vanadium atoms approximately as Einstein oscillators with excitation energy $\hbar\omega_v$ = $k_B\theta_D$ = 37 meV and obtain for their kinetic energy E_{kin}^v = $\hbar\omega_v/4 - 9$ meV; the hydrogen atoms follow the vanadium atoms and more or less perform the same acoustical vibration; because of the smaller hydrogen mass, by a factor of 50, the acoustical kinetic hydrogen energy then approximately equals 0.2 meV. Hence, spectroscopically, the total average kinetic energy in the xy plane amounts to 15.8 meV, comparing well with the independently determined Compton result of 16.0 ± 0.4 meV. The total spectroscopic average kinetic energy in the xz plane equals to 35.6 meV, i.e. it is considerably larger than for the xy plane. Also with the Compton results the xz value is much larger than the xy value, but does not agree with the spectroscopic result. This might originate from the invalidity of the impulse approximation in the z vibrational direction.

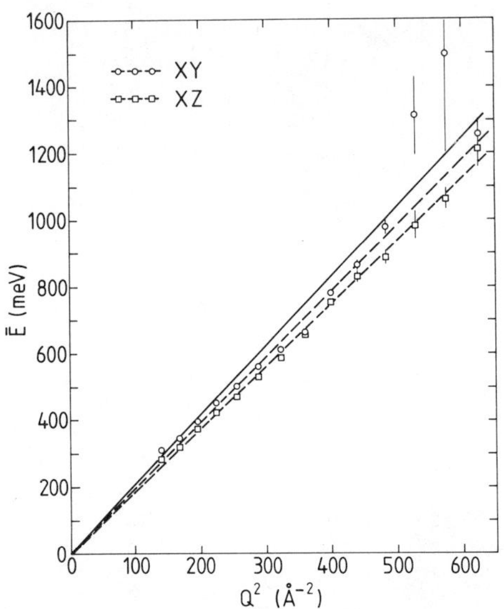

Fig. 4. Q^2 dependence of the center of the Compton profiles (xy vibrations in the scattering plane, circles; xz vibrations in the scattering plane, squares). The solid line indicates M = 1 amu.

<u>Conclusion.</u> The combination of two methods of deep inelastic neutron scattering provides on the one hand side rather detailed information about the effective single-particle hydrogen potential in V_2H, and on the other

218

hand it allows a critical evaluation of the respective capabilities of the
two techniques. Neutron vibrational spectroscopy provides the excitation
energies, which are important for the partition function in statistical

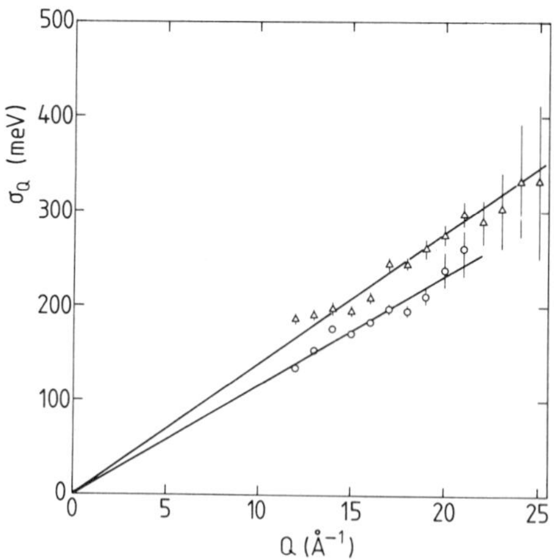

Fig. 5. Q dependence of the second moments
of the Compton profiles (xy vibrations in the scattering plane,
circles; vz vibrations in the
scattering plane, triangles).

thermodynamics, and the hydrogen potential in the energy range of the
measurement; statements about the ground state are an extrapolation.
Neutron Compton scattering at low sample temperature, on the other hand,
provides direct information about the ground state; as an integral quantity we obtain the kinetic energy, and the agreement with the spectroscopic result is very satisfactory. In principle, the wave function can
be obtained from a line-shape analysis; this would be most desirable but
does not seem to be feasible with the present statistical accuracy.

Finally, we should like to emphasize that, with the neutron scattering experiment described in this contribution, the characteristics of a
neutron spallation source have been fully utilized:

availability of high neutron incident energies;
small Q values   → epithermal spectroscopy (3);
large Q values   → Compton scattering;
variation of the magnitude of Q   → phonon form factor or transition
                                        matrix elements (6);
variation of the direction of Q   → variation direction (3).

Acknowledgement

We thank C.-K Loong, R. Kleb and the IPNS Operations Staff for kind
and skilled support for the experiments. The work at Argonne was
supported by the U.S. Department of Energy, BES-Materials Sciences, under
contract #W-31-109-ENG-38.

References

(1) R. Hempelmann and J. J. Rush, in G. Bambakidis and R. C. Bowman (eds.). "Hydrogen in Disordered and Amorphous Solids," NATO ASI Series B: Physics Vol. 136, Plenum Press, New York, p. 283
(2) D. Klauder, V. Lottner, and H. Scheuer, Solid State Commun. 32: 617 (1979)
(3) R. Hempelmann, D.Richter, and D.L. Price, Phys. Rev. Lett. 58:1016 (1987)
(4) G. Reiter and R. Silver, Phys. Rev. Lett. 54:1047 (1985)
(5) D. Ohlendorf and E. Wicke, J. Phys. Chem. Solids 40:721 (1979)
(6) R. Hemplemann, to be published.

THEORY OF MOMENTUM DISTRIBUTIONS AND THEIR MEASUREMENT

FOR HYDROGEN IN SOLIDS

George Reiter

Physics Department
University of Houston

The availability of sources of high energy neutrons has made possible inelastic scattering measurements from solids and liquids at transferred energies on the order of 10 eV. At these energies, $S(q,\omega)$, as we will show, is well described by the impulse approximation

$$S(q,\omega) = \int \delta(\omega - hq^2/2m - \vec{q}\cdot\vec{p}/m)\,n(\vec{p})\,d\vec{p}$$

(1)

where $n(\vec{p})$ is the momentum distribution function. The result (1) shows that the measured neutron cross section is the Radon transform[1] of the momentum distribution, and such a transform may be inverted to give the momentum distribution. For light ions, such as hydrogen, in a heavy matrix, where one can think of the light ion as subject to a fixed external potential provided by the matrix, this distribution is determined by the potential. If the temperature is sufficiently low that the ion can be regarded as being in the ground state of the potential well, one can invert the Schroedinger equation and obtain the potential from the measured momentum distribution. Pulsed neutron sources provide therefore, a means of direct measurement of Born Oppenheimer potentials. We believe this is in fact, a practical measurement with existing machines, and will discuss the theoretical problems that arise when one imagines doing such a measurement. In particular, we will consider first the question of how high a momentum transfer is needed for equation (1) to be valid. Given that it is valid, we will discuss how large a mass ratio between the scattered particle and the matrix is needed so that one can neglect the motion of the matrix. Given that that is the case, we will consider how best to extract the momentum distribution from the measurements, and invert the Schroedinger equation to obtain the potential. Because of its high incoherent cross section, low mass, and importance in many interesting systems, hydrogen will be the first candidate for such measurements, but the theory applies equally well to any light ion.

The limits of validity of the impulse approximation can be made clear using a quasiclassical approximation.[2] The scattering function, $S(q,\omega)$ can be expressed approximately, for a single particle in a potential $V(x)$, as the fourier transform of $S(\vec{q},t)$

$$S(q,t) = \int e^{-i\vec{q}\cdot\left[\left(\frac{h\vec{q}}{2m} + \frac{\vec{p}}{m}\right)t - \int_o^t \int_o^{t'} \frac{\vec{F}(\vec{x}(t''))}{m}dt''dt'\right]} f(\vec{x},\vec{p})\,d\vec{x}\,d\vec{p}$$

(2)

where $\vec{x}(t)$ is a trajectory that begins at $\vec{x}$ with momentum $-\vec{p} + h\vec{q}/2$, and $f(\vec{x},\vec{p})$ is the average of the Wigner density operator, and may be thought of loosely as the probability of finding the particle at the point $\vec{x}$ with momentum $\vec{p}$. $\vec{F}(\vec{x}) = -\vec{\nabla}V(\vec{x})$. Rigorously $\int dx f(\vec{x},\vec{p}) = n(\vec{p})$.

If we neglect the forces altogether, $S(\vec{q},t)$ will decay to zero in a time $t_c = m/q\Delta p$, where Δp is the width of the momentum distribution in the $\vec{q}$ direction. We assume for simplicity that this is essentially the same in all directions, so that there is only one characteristic time. As q approaches infinity, t_c approaches zero. The force integral involves two integrations over time, and hence is proportional to t^2. The contribution to the phase coming from the free particle motion is proportional to t. Therefore, as t_c approaches zero, i.e. q approaches infinity, one can always neglect the force term. The remainder, when fourier transformed, gives the impulse approximation, Eq. (1).

A problem arises for a potential for which there are no unbound states, such as an harmonic oscillator, and for which all the classical orbits are periodic. The particle eventually returns to its original position, and the phase mixing is undone. In a strict sense, the impulse approximation in the form (1) cannot be correct. This can also be seen from the fact that the spectrum is discreet, and hence $S(q,\omega)$ must consist of a series of delta functions, whereas (1) gives a continuous function. The resolution of this difficulty is that the impulse approximation gives the amplitude of the delta functions, that is, the envelope of the spectrum. To see this , consider a harmonic oscillator of frequency ω_0 . The quasiclassical approximation is exact in this case, and one can show from (2) that

$$S(q,t) = e^{-hq^2/2m\omega_0[(1 - \cos \omega_0 t) + i \sin \omega_0 t]} \tag{3}$$

$S(q,t)$ is clearly periodic, for any q, and can be written as

$$S(q,t) = \sum_{n=-\infty}^{\infty} A_n(q)e^{-in\omega_0 t} \tag{4}$$

The coefficients An(q) are given by

$$An(q) = \frac{\omega_0}{2\pi} \int_{-\pi/\omega_0}^{\pi/\omega_0} e^{in\omega_0 t}S(q,t)dt \tag{5}$$

In order for the impulse approximation to hold, that is, that the particle behaves like a free particle for the characteristic time t_c, we clearly require that $\omega_0 t_c < 1$. Since $\Delta p^2 = hm\omega_0/2$ in this case, the value of q that must be exceeded to fulfill this requirement, is

$$q_c = (2m\omega_0/h)^{1/2} \tag{6}$$

where $\omega_0 t_c = q_c/q$.

When $q/q_c \gg 1$, we find

$$An(q) = \omega_0 \int_{-\infty}^{\infty} \delta\left(n\omega_0 - \frac{hq^2}{2m} - \frac{qp}{m}\right)n(p)dp$$

where $n(p) = (2\pi\Delta p)^{1/2} e^{-p^2/2(\Delta p)^2}$.

The factor of ω_0 gives the spacing between the levels. For a more general potential , I would expect that

$$\int_{\omega_n-\varepsilon}^{\omega_n+\varepsilon} S(\vec{q},\omega)d\omega \equiv A_n(\vec{q}) = \Delta\omega_n \int_{-\infty}^{\infty} \delta\left(\omega_n - \frac{hq^2}{2m} - \frac{\vec{q}\cdot\vec{p}}{m}\right) n(\vec{p})dp \tag{7}$$

where $\Delta\omega_n$ is the spacing between the levels in the vicinity of the nth level. The criterion for being in the impulse approximation for a general potential, where the period of oscillation varies with the amplitude, would be $\omega(q)t_c < 1$ where $\omega(q)$ is the period when the momentum transfer is q. When applied to the extreme anharmonic potential, the infinite square well, this criterion gives an ambiguous result $\omega(q)t_c \simeq 1$ for all large q. However, direct calculation shows that (7) holds for the square well. Since this represents the limit of extreme anharmonicity, I expect that (7) holds for any potential.

Realistic potentials will be closer to an harmonic oscillator than a square well. To get some idea of how fast the impulse approximation is approached, I have compared the envelope of the intensities for an harmonic oscillator for several values of q/q_c, with the exact result, in Fig. (1). Clearly, $q/q_c > 5$ would suffice to give a sufficiently accurate picture of the underlying momentum distribution, for most purposes.

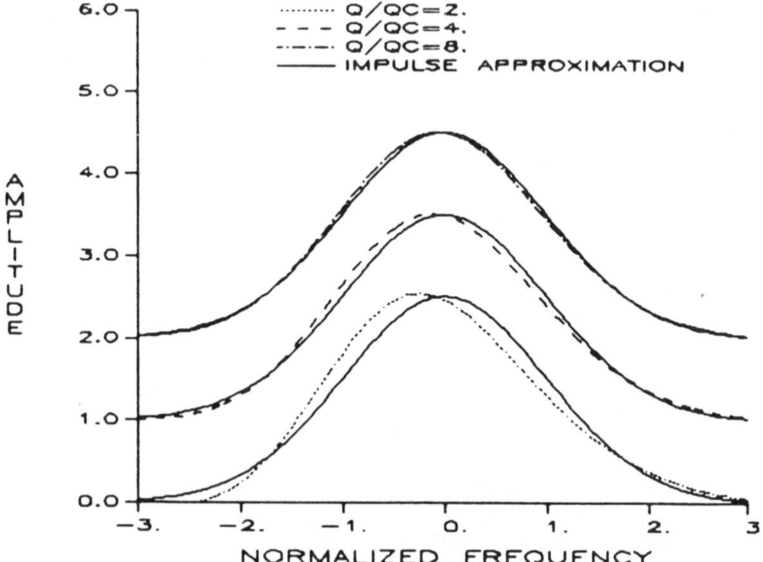

Fig. 1. A comparison of the envelope function of $S(q,\omega)$ for an harmonic oscillator compared with the impulse approximation result for several values of q/q_c.

If one considers particles in an harmonic lattice, rather than a single particle in a well, then Eq. (3) should be replaced by

$$S(\vec{q},t) = e^{-\sum_\lambda h(q\cdot\vec{\varepsilon}(\lambda))^2/2\omega_\lambda M[(1 - \cos(\omega_\lambda t)) \coth h\beta\omega_\lambda/2 + i \sin \omega_\lambda t]} \tag{8}$$

where $S(\vec{q},t)$ is the incoherent scattering function. $\vec{\varepsilon}(\lambda)$ are the eigenvectors where λ labels the eigenvalues ω_λ. I have included the effects of finite temperature in (8).

In the present case, phase mixing between modes of different frequencies will cause $S(\vec{q},t)$ to vanish as t approaches infinity, and the original statement of the impulse approximation will hold. A conservative estimate of q_c in this case would replace ω by the largest normal mode frequency, but a more reasonable estimate would use the center of mass of the density of states. In any case, for $q/q_c \gg 1$,

$$S(\vec{q},t) = e^{\frac{-ihq^2}{2M}t}\, e^{\frac{-q^\alpha q^\beta}{4M}\sum_\lambda (\varepsilon_\lambda^\alpha \varepsilon_\lambda^\beta \omega_\lambda)t^2} \qquad (9)$$

which is the impulse approximation result in the time domain. The momentum distribution is

$$n(p) = \frac{1}{(2\pi)^3}\, |\Delta p|^{1/2} e^{-1/2 p^\alpha p^\beta (\Delta p^{-1})_{\alpha\beta}} \qquad (10)$$

where

$$\Delta p_{\alpha\beta} = \frac{hM}{2} \sum_\lambda \varepsilon^\alpha(\lambda)\varepsilon^\beta(\lambda)\omega_\lambda \qquad (11)$$

For an anharmonic lattice, the result would be more difficult to calculate, but we believe the impulse approximation would be correct, and become valid at roughly the same value of q/q_c as for the harmonic lattice. Indeed, the quasiclassical approximation makes clear that the impulse approximation will hold unless there is a "zero time scale" in the problem. That is, something that can produce significant deviations from free particle behavior on any time scale, no matter how short. For helium, for instance, the hard core in the potential makes it possible that this will occur. If the scattered atom is a very small distance from a background atom, the scattered atom can collide with the background and experience a very large deflection in an arbitrarily short time. A hard core gas, therefore, does not have the impulse approximation as the limit of the scattering function for large q.[3,4] This is graphically illustrated by a gas of hard core bosons in one dimension, where both the momentum distribution, and $S(\vec{q},\omega)$ can be exactly calculated.[4] This is shown in Fig. (2) where, we have included also the quasiclassical approximation.

Another example is that of a heavy particle in a light background. As the mass ratio becomes infinite, this becomes a model of an harmonic oscillator subject to Brownian noise. The Brownian trajectories are never free, and as a consequence, one does not have an impulse approximation limit. This was shown explicitly by Dattagupta[5] and the author, for the model of a damped harmonic oscillator. I show in Fig. (3) a comparison of the exact $S(\vec{q},\omega)$ with the impulse approximation result when the damping is small, for the same values of q/q_c as in Fig. (1). The deviations do not approach zero for large q.

It might be argued that even in a solid, that the surrounding atoms have hard cores, and the scattered ion can collide with them, and this is the case. However, as in the liquid,[7] the equilibrium are such that the scattered ion is unlikely to be near such a core, initially, and must travel for some time before it strikes one. The final state corrections that persist for large q, are therefore much smaller than would be the case in the gas, where there are no equilibrium correlations that have the effect of keeping the particles separated. We expect the final state corrections to be essentially negligible, in solid helium, for instance although, in principle, present.

An interesting case that has been considered in some detail by Warner, Lovesey and Smith[8] is that of a light particle in a heavy lattice. They considered only the case that the light particle was harmonically coupled to the harmonic lattice, but the results will be valid for a light particle in an anharmonic well, coupled to an harmonic lattice as well.

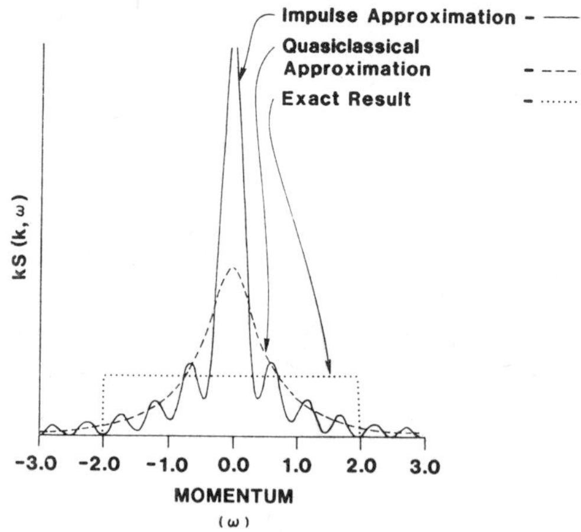

Fig. 2. A comparison of the impulse approximation prediction of $S(q,\omega)$ with the exact result and the quasiclassical approximation for a hard core bore gas in 1-dimension. From Reiter and Becher, Ref. 4.

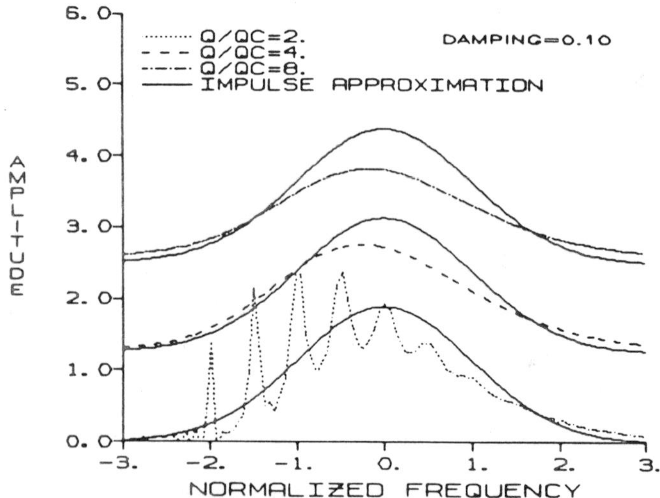

Fig. 3. The $S(q,\omega)$ for a damped harmonic oscillator using the model of Dattagupta and Reiter, showing the failure of the exact result to approach the impulse approximation limit.

They show that $S(\vec{q},t)$ is the product of a factor of the form (3), with the light particle mass m increased by a factor of 1/h(M/m), and a factor of the form (8), with a modification of the density of states due to the presence of the light atom.

The factor h(M/m) is essentially one for mass ratios in excess of 10, and the primary effect of the lattice is to broaden the delta function peaks in $S(\vec{q},\omega)$ for the light particle into gaussians. In the impulse approximation limit, the corrections to the width of the momentum distribution of the light particle due to its coupling to the heavy lattice is of order (1-h(M/m)) and hence negligible for large mass ratios. For materials such as hydrogen in Vanadium or Niobium, with mass ratios of 50 or 100, one may, to a very good approximation, regard the hydrogen as individual particles in fixed potential wells, although the broadening due to the lattice permits one to use the impulse approximation in its original form, (1) for sufficiently high momentum transfer. We shall therefore assume that $S(\vec{q},\omega)$ is well represented by (1), and ask what we can learn from it.

We turn now to the second main point of the paper, the extraction of $n(\vec{p})$ from data described by Eq. (1). In the case of an harmonic crystal, both the momentum distribution and the scattering function in the impulse approximation limit are gaussians, and are simply related to one another. For anharmonic potentials, one is faced with the problem of extracting the momentum distribution from the scattering function. This is the problem of inverting the Radon transform. It appears in tomography, and in precisely the same mathematical form, in X-ray Compton scattering, positron annihilation and NMR zeugmatrography. The problem is, given limited data, with some noise, to find an algorithm that gives an approximate inversion that is useful for the purposes intended, and does not require an unacceptable number of data points to work. These criterion are clearly not precise mathematical requirements. They can, nevertheless, be met in our case, by a procedure that was originally developed by Davison,[8] Grunbaum,[9] and Louis[10] for its mathematical interest.

If we express the frequency in terms of the dimensionless $s = (\omega - hq^2/2m)\tau_c$, $\tau_c = m/q\Delta p$, the momentum in terms of $\vec{p}' = \vec{p}/\Delta p$, and the direction of $\vec{q}$ by the unit vector $\hat{q}$, then our problem is, given $S(\hat{q},s)$ to find $n(\vec{p})$. Davidson's solution may be expressed as follows. Expand the observed function in Hermite polynomials, $H_n(s)$, and spherical harmonics, $Y_{lm}(q)$. That is, determine the coefficients A_{nlm} such that

$$S(\hat{q},s) = \sum_{n,l,m} A_{nlm} \frac{e^{-s^2}}{\pi^{1/2}} H_{l+2n}(s)\, Y_{lm}(\hat{q})$$

(12)

Then the function that gives the observed Radon transform is

$$n(p\hat{q}) = \sum_{nlm} B_{nlm} \frac{e^{-p^2}}{\pi^{3/2}} p^l L_n^{l+1/2}(p^2) Y_{lm}(\hat{q}) \ ,$$

$$B_{nlm} = 2^{2n+l} n!(-1)^n A_{nlm}$$

(13)

where $L_n^{l+1/2}$ are Laguerre polynomials. The normalizations and definitions are as in Magnus and Oberhettinger.[11] The implication is, that in order to invert the transform, one need only find the coefficients A_{nlm} in Eq. (23), use the calculated B_{nlm}, and one has the original function in the representation in Eq. (24).

Davison's proof is indirect, but the result may be shown by direct calculation. Using a fourier series for the delta function in the definition of the Radon transform, and the standard expansion of a plane wave in spherical harmonics, we find for any particular term in the series in Eq. (23),

$$\int \delta(s-\hat{q}\cdot p) \frac{e^{-p^2}}{\pi^{3/2}} p^l L_n^{l+1/2}(p^2) Y_{lm}(\hat{q}) d\vec{p} =$$

$$\int e^{-ist_l l} \frac{2^{3/2}}{t^{1/2}} J_{l+1/2}(tp) e^{-p^2} p^l L_n^{l+1/2}(p^2) dp\, Y_{lm}(\hat{q})$$

(14)

The integral over p can be done explicitly,[12] as can the fourier transform that results, and we find that the integral in Eq. (14) is

$$\frac{-1^n}{2^{2n+l} n!} \frac{e^{-s^2}}{\pi^{1/2}} H_{2n+l}(s)$$

(15)

demonstrating the theorem. $J_n(x)$ is the Bessel function of order n.

The implication is , that in order to invert the transform, one need only find the coefficients Anlm in Eq. (12) from the data, use the calculated Bnlm, and one has the original function in the representation in Eq. (13). In our case, $S(\vec{q},s)$ is $S(\vec{q},\omega)$, properly rescaled, and the theorem says that to obtain the momentum distribution, also rescaled, one finds the series of the form in Eq. (12) that best represents the data, and the momentum distribution is then given in the form in Eq. (13).

What makes this a method of inversion particularly well suited to the problem of measuring momentum distributions in solids is that the series expansion can be expected to have a small number of terms. In fact, if the potential is harmonic, there will be only one term, and for weakly anharmonic crystals, the coefficients Anlm can be related directly to the anharmonic terms in a Taylor series expansion of the potential. It is the apparently fortuitous fact that the eigenfunctions of an isotropic harmonic oscillator, in any dimension, are also the product of gaussians, Laguerre polynomials and spherical harmonics, that makes the inversion scheme described here so well suited to the problem at hand. As a consequence, if the anharmonic and anisotropic terms in the potential are represented as

$$\Delta H(r\hat{q}) = \sum_{nlm} h_{nlm}\, r^l L_n^{l+1/2}(r^2) Y_{lm}(\hat{q})$$

(16)

then the coefficients in the expansion in Eq. (23) of $S(q,\omega)$ are, to first order,

$$A_{nlm} = \frac{-h_{nlm}}{2n+l}\; 1/2\; i^l\; \frac{1 + (-1)^{m+l}}{2^{2n+l} n!}$$

$$A_{000} = (4\pi)^{1/2}\; .$$

(17)

One can measure small anharmonicity directly by an accurate measure of $S(\vec{q},\omega)$. Even for strongly anharmonic double well potentials, however, $S(\vec{q},\omega)$ will be a well localized function with a few wiggles, and can be reasonably well represented by a series such as in Eq. (12) with a few terms.

Of course, if the anharmonicity is small, one may have to have very good statistics to extract the coefficients Anlm accurately. One can in fact, give a definite answer to the question of just how long one must count to measure anharmonicity of a certain size. We will assume that the anharmonicity is sufficiently small that Eq. (17) may be used, so the problem is simply to extract an accurate value of Anlm from the data.

First, observe that in a crystal, the Anlm for different values of m are not all independent, and in particular, we need only consider the combinations of $Y_{lm}(\hat{q})$ that have the symmetry of the crystal.[13] Then denoting these lattice harmonics by $U_i(\hat{q})$ we would have

$$S(s,\hat{q}) = \sum_i g_i(s) U_i(\hat{q})$$

,

$$g_i(s) = \sum_n A_{ni} H_{2n+l(i)}(s)\; \frac{e^{-s^2}}{\sqrt{\pi}}$$

(18)

where

$$\overline{H}_n = H_n/(2^n n!)^{1/2}$$

If the series is terminated to have n such harmonics, then the $g_i(s)$ can be extracted by measuring along n directions $\hat{q}_K$, and using[14]

$$g_i(s) = \sum_k [U]_{i,k}^{-1} S(s,\hat{q}_k)$$

(19)

where $[U^{-1}]_{i,k}$ is the inverse of the matrix $U_{ij} = U_i(\hat{q}_j)$. The n directions must be chosen so that this matrix is non-singular, of course, and could be optimized in various ways. If the data were known precisely, at every point q, then one could calculate A_{ni} from the formula

$$A_{ni} = \int_{-\infty}^{\infty} \overline{H}_{2n+i(l)}(s)g_i(s)ds$$

(20)

Of course, the data is not known precisely, nor is it known at every point. We will assume it is known at a sufficient number of points that the integral can be approximated by

$$\hat{A}_{ni} = \sum_k \overline{H}_{2n+i(l)}(s_K)\, \hat{g}_i(s_K)\Delta s$$

(21)

where Δs is the spacing between frequency points, assumed uniform. The points are assumed to cover a large enough interval so that the contribution to the integral in (20) from the region outside the interval is negligible. Since the standard deviation of the gaussian leading term in (21) is $1/\sqrt{2}$, we can take the interval to be $[-3/\sqrt{2}, 3/\sqrt{2}]$

The $\hat{A}_{ni}$, $\hat{g}_i(s_K)$ are the measured values for a particular experiment. Then, expressing the measured values $\hat{g}_i(s_K)$ in terms of their average value, $g_i(s_K)$ and a fluctuation $\delta g_i(s_K)$, using the fact that

$$<\delta g_i(s_m)\delta g_i(s_n)> = \delta_{mn}g_i(s_n)$$

(22)

and approximating $g_i(s_K)$ by the dominant term in the series $A_{oo}\, e^{-s^2}/\pi$, we obtain

$$<(\delta \hat{A}_{ni})^2> = A_{oo}\, \Delta s$$

(23)

This is a very conservative estimate for $i \neq 0$. The extraction of a particular coefficient A_{ni} will be possible when

$$A_{ni}/A_{oo} > <\delta \hat{A}_{ni})^2>^{1/2}/A_{oo} = (\Delta s/A_{oo})^{1/2}$$

(24)

Now $A_{oo}/\sqrt{\pi}$ is the maximum count rate at the peak, which we will denote by N_0. If there are M channels, evenly spaced then we require

$$A_{ni}/A_{oo} > (6/\sqrt{2\pi}N_0 M)^{1/2}$$

(25)

to be able to make a measurement. Thus with 50 channels, and a maximum count rate of 1000, one could extract coefficients that are as small as .07% of the dominant gaussian intensity.

Eq. (22) assumes Poisson statistics, which is not correct for filter difference spectrometers. The analysis is not as simple for that case, but as we show in the next example, one can obtain satisfactory results with readily obtainable count rates for these spectrometers as well.

An interesting problem, in which the anharmonicities are not small, is the determination of the potential well for hydrogen in hydrogen bonds. There has been a great deal of theoretical and experimental work devoted to this question.[15] Deep inelastic neutron scattering (DINS) has significantly different characteristics from neutron spectroscopy or crystallography, techniques that have been used in these studies, and should provide complementary information. For instance, one of the issues in the hydrogen bond problem that should be resolvable by DINS is the question of whether observed bimodal distributions of hydrogen in the bond are the result of a ground state wavefunction in which the hydrogen is in both sites, or a statistical mixture of bonds in which the hydrogen is in one site or the other. The DINS scattering functions are very different for the two cases, containing oscillations at a wave vector proportional to the inverse of the distance between the centers for the two sites in the bond in the first case (see Fig. 4), and no oscillations in the second case. The maps determined from crystallography are identical. Furthermore, if a system can be found in which the hydrogen self traps in one well or the other at low temperatures, as demonstrated in various theoretical calculations[16] one could perhaps observe a cross over from one type of behavior to the other as the time scale for tunneling increased.

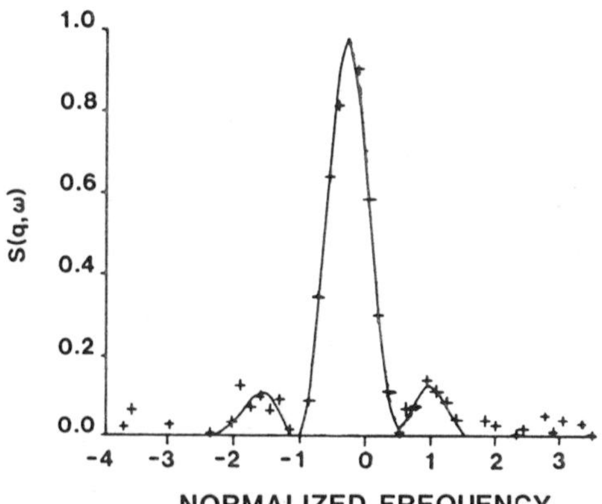

Fig. 4.   Synthetic data for the $S(q,\omega)$ appropriate to the double well potential shown in Fig. 5, obtained by the filter difference method, with a counting rate such that the rms noise level is 4% of the peak count. The data has been fit to the form in Eq. (12) using eight terms in the expansion.

As an illustration of the practicality of the inversion method, and its stability to noise, we have calculated the potential for a simple model of a hydrogen bond from simulated data. If we make the assumption, that transverse to the direction of the bond the potential is harmonic, with a force constant independent of the distance along the bond, the problem separates into three one-dimensional problems. The scattering function for momentum transfer along the bond is just the momentum distribution itself, rescaled by q. There is then no need to do an inversion, but to test the method, we have fit the data to a series of the form in Eq. (12) and used the series to calculate the potential from the Schroedinger equation

$$\frac{\displaystyle\int \frac{p^2}{2m}\, e^{i\vec{p}\cdot\vec{x}/t}\,\phi(\vec{p})d\vec{p}}{\displaystyle\int e^{i\vec{p}\cdot\vec{x}/t}\,\phi(\vec{p})d\vec{p}} = E - V(\vec{x}) \tag{26}$$

There is a sign ambiguity in going from $|\phi(\vec{p})|^2$, which is what is measured, to $\phi(\vec{p})$, but since the ground state wave functions can be chosen to be real, and are relatively simple functions, this can be easily resolved. An incorrect choice will generally produce grossly incorrect results for V(x). The wave function was assumed to be two unit width gaussians, displaced from the origin by a distance twice

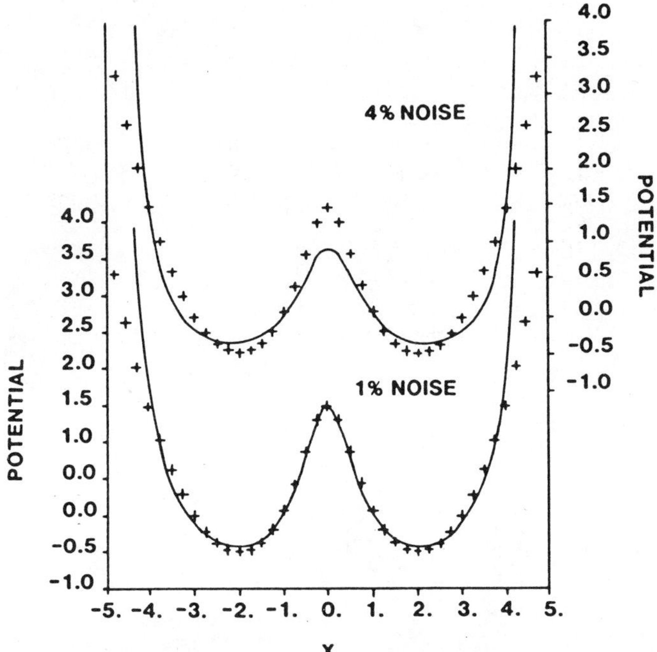

Fig. 5. The potential reconstructed from the synthetic data shown in Fig. 4, using Eq. (26) and the fit to the data of the form in Eq. (12), for two values of the noise level. The solid lines are the reconstruction, the + signs the original potential.

their variance. The potential that would produce this wave function is essentially two coupled harmonic wells. The momentum distribution was calculated, and "experimental" $S(q,\omega)$'s were generated by adding gaussian noise of varying intensities. This is the sort of noise that would be appropriate to the filter difference method, since the signal is obtained as the difference of two counting rates. The intrinsic Poisson statistics of an ordinary spectrometer would be considerably more favorable, for the same peak counting rates. The fitted "data" is shown in Fig. 4. Fifty data points were used throughout. The results for the reconstructed potential are shown in Fig. 5. Note that even for the largest noise level, a clear picture of the double well potential emerges. The smallest noise level corresponds to a maximum total counts of 10^4, which is quite practical, on existing machines, for scattering from hydrogen. It is

230

interesting to note that the inversion breaks down first at large distances. This is what one would expect from the quasiclassical approximation, since the proton must have a high momentum to penetrate a large distance into the barrier, and the noise occludes the high momentum information first.

We conclude that DINS is a practical means of directly measuring Born Oppenheimer potentials for hydrogen in solids, at least in systems in which there is only one inequivalent hydrogen site per unit cell.

REFERENCES

1.   D. Ludwig, Comm. Pure and Appl. Math 19, 49 (1966).
2.   G. Reiter and R. Silver, Phys. Rev. Letts. 54, 1047 (1985).
3.   J. Weinstein and J. W. Negele, Phys. Rev. Letts. 49, 1016 (1982).
4.   G. Reiter and T. Becher, Phys. Rev. B 32, 4492 (1985).
5.   S. Dattagupta and G. Reiter, Phys. Rev. A 31, 1074 (1985).
6.   R. Silver and G. Reiter, Phys. Rev. Rapid Comm. 35 (1987).
7.   M. Warner, S. Lovesey and J. Smith, Z. Phys. B 51, 109 (1983).
8.   M. E. Davison, Numer. Funct. Anal. Optimiz. 3, 321 (1981).
9.   M. E. Davison and F. A. Grunbaum, Comm. Pure and Appl. Math 34, 77 (1981).
10. A. K. Louis, SUNY Buffalo Report # MIPG42, 1981.
11. W. Magnus and F. Oberhettinger, Formulas and Theorums for the Function of Mathematical Physics (Chelsea, New York, 1954).
12. M. Abromowitz and S. Stegun, Handbook of Mathematical Functions, p. 785, (Dover, New york, 1972).
13. W. V. Houston, Revs. Mod Phys. 20, 161 (1948).
14. G. P. Das, K. V. Bhagwat and V. C. Sahni, Phys. Rev. 36, 2984 (1987).
15. P. Schuster, The Hydrogen Bond (North Holland, Amsterdam, 1976).
16. S. Chakravarty and A. J. Leggett, Phys. Rev. Lett. 52, 5 (1984).

ELECTRONIC SYSTEMS

MOMENTUM DISTRIBUTIONS IN STRONGLY CORRELATED ELECTRON SYSTEMS: HEAVY FERMIONS, MOTT INSULATORS AND HIGH TEMPERATURE SUPERCONDUCTORS

S. Doniach

Dept. of Applied Physics
Stanford University
Stanford, CA 94305

Abstract

We discuss effects of strong Coulomb interactions in narrow band materials on the electron momentum distribution. Correlations reduce the height of the Fermi step by a renormalization factor z which may be very small in heavy fermion compounds. In a Mott insulator, correlations open up a band gap for a 1/2-full band. Nevertheless at practical resolution the resulting momentum distribution may still look like that of a free Fermi surface. In particular, for the high temperature superconductors it may be difficult to distinguish the Fermi surface from the antiferromagnetic Brillouin zone boundary.

Introduction

The physics of many particle interactions in electron systems - principally solids - shows substantial differences from that for many particle interactions in helium (both 3 and 4) and nuclear matter. These major qualitative differences arise from two features:

1) The long range Coulomb interaction between electrons, to be contrasted with the hard core and short range attractive potential relevant for helium and nuclei, and

2) The existence of an ionic lattice in which the electrons move. The latter of course leads to the atomic shell structure for electrons, which in turn gives rise to the huge variety of phenomena found in condensed matter systems.

An important feature of the physics of electrons in solids is that in many cases the effective electron density in a given sub-band can be quite low owing to low overlap between orbitals on different atoms. This effective low electron density (of course, electron densities are high on individual atoms) leads to a situation where the effective kinetic energy and potential energy become comparable:

$$< \frac{\hbar^2 k_{eff}^2}{2m} > \simeq < \frac{e^2}{a} > \tag{1}$$

or in other words large r_s.

Under these conditions many instabilities occur such as magnetism, superconductivity, charge density waves and the like. These are situations where straightforward density functional theory for the inhomogeneous electron gas using the local density approximation breaks down.

In this paper we'll be interested in three classes of strongly correlated electron systems:

transition metals and compounds (3d-shell)

rare earth metals and compounds (4-f shell)

actinide metals and compounds (5-f shell)

2. Density functional theory of the inhomogeneous electron gas - limits of the local density approximation

The idea of computing electronic states in solids using a self consistent local potential can be traced back to Hartree, Fock and Slater. The density functional approach formulated by Hohenberg and Kohn has become a very powerful tool through the use of the local density approximation.

It was pointed out some time ago by Lam and Platzman [1] that, while the electron momentum distribution

$$n(p) = \sum_\sigma < c_{p\sigma}^\dagger c_{p\sigma} >$$
(2)

is a property of the ground state wave function and hence should in principle be exactly calculable within the general density functional approach, the usual formulation of LDA needs to be generalized to deal with the momentum distribution problem since the density functional $E[\rho]$ does not explicitly depend on $n(p)$. This led to the formula of Lam and Platzman for computing $n(p)$ within the LDA (see Bauer [2] for an alternative derivation):

$$n(p) \simeq \sum_{k_{occ}} |\int d^3r \phi_k(r)|^2 + \Delta n^{LDA}(p)$$
(3)

where $\phi_k(r)$ is a one electron band wave function computed using LDA, and

$$\Delta n^{LDA}(p) = \int d^3r \{n_{int}(\rho(r), p) - n_0(\rho(r), p)\}$$
(4)

with $n_{int} = $ local $n(p)$ for the interacting homogeneous electron gas and $n_0 = $ same for noninteracting gas.

This formula has been applied by Bauer and Schneider [3] to compute the Compton profile for single crystal Cu metal (see Fig. 1). What is plotted here is the quantity measured in the Compton cross section, namely the projection of $n(p)$ on the scattering vector direction $\hat{n}$:

$$J_{\hat{n}}(q) = \int d^3p \, n(p)\delta(q - \hat{n} \cdot \vec{p})$$
(5)

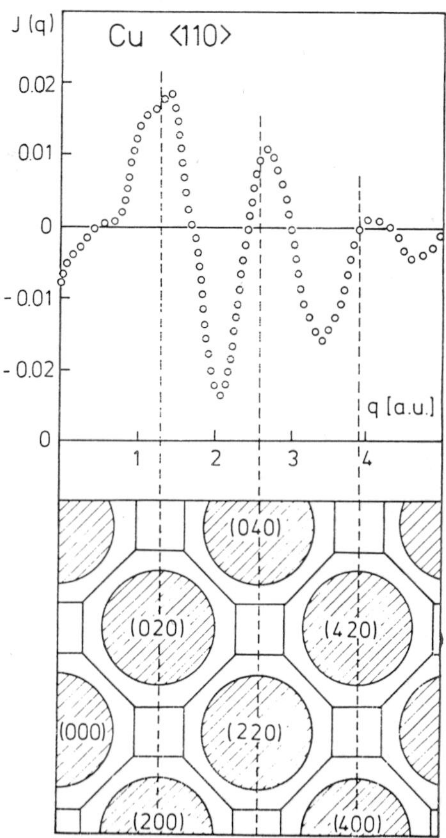

Fig. 1. Projection of $n(\vec{p})$ on a vector for Cu. Upper part: difference between the local density-functional theory and the experiment. Lower part: XY plane of the Brillouin zone and the Fermi surface of copper in the repeated zone scheme. (From Ref. 3.)

As may be seen in the figure for $J_{theory} - J_{expt}$ there are significant deviations (though in absolute value only on the scale of a couple of percent) between the observations and the theory, particularly in the unoccupied region of the Brioullin zone.

3. Breakdown of the local density approximation: mixed valence and heavy fermions

In solids in which there are very narrow bands present, such as $CeAl_3$, $CeCu_2Si_2$, UPt_3, etc., extremely strong correlation effects occur which manifest themselves in effective masses for conduction electrons on the scale of 100 times the masses predicted from self consistent one - electron theory.

Thus the LDA gives a very poor description of the physics of these systems even though, quite surprisingly, band structure calculations of the Fermi surfaces in materials like $CeSn_3$ and UPt_3 work amazingly well when compared to de Haas-van Alphen data [4].

Nevertheless it was pointed out some time ago by Gutzwiller [5] that the effects of strong electron-electron Coulomb repulsion leading to very strong correlation will completely change the character of the momentum distribution relative to what would be predicted for an effective one - electron model.

In this section I discuss a more recent approach to highly correlated electron systems based on a slave boson operator representation.

Let me start by considering an Anderson lattice model for a set of N-fold degenerate f state atoms in which a conduction band represented by $c_{k\sigma}$ operators hybridizes strongly with a local f-band. Very strong on-site Coulomb repulsion is represented by $\sum_{mm'} U n_m^f n_{m'}^f$. It is convenient to represent this model in terms of a functional integral approach in which the action is written as

$$A = \int_0^\beta d\tau \sum_{im} f_{im}^\dagger (\frac{\partial}{\partial \tau} - \varepsilon_f^0) f_{im} + \sum_{k\sigma} c_{k\sigma}^\dagger (\frac{\partial}{\partial \tau} - \varepsilon_k^0) c_{k\sigma}$$

$$+ \frac{v}{\sqrt{N}} \sum_{im} (f_{im}^\dagger c_{im} + hc) + U \sum_{imm'} n_{im}^f n_{im'}^f.$$

(6)

Here $n^f = f_{im}^\dagger f_{im}$ and c_{im} is an appropriate spherical harmonic projection of the conduction band onto a local Wannier function. The advantage of using the functional integral approach is that, in the limit of very large orbital degeneracy for the f state, it can be shown that a "quasi-symmetry breaking" occurs in which the phase of the correlation $< c_k^\dagger f_k >$ becomes well defined. It has been shown however this is only a quasi-symmetry breaking since infrared fluctuations will in general average over these phases [6].

These effects may be understood in the strong correlation $U \to \infty$ limit by replacing the f electron annihilation and creation operators in terms of a product $\tilde{f}^{\dagger}_{im} b_i$ where the b_i are a set of slave boson operators whose job is to keep track of the inhibition of double occupany for the f level in the large U limit. This is guaranteed by enforcing the constraint:

$$\sum_m n^{\tilde{f}}_{im} + b^{\dagger}_i b_i = 1 \tag{7}$$

on each site of the lattice, so that the occupation of the f-sites can either be zero or one but never greater than one. In the limit of large degeneracy one finds that the saddle point of the functional integral

$$\mathcal{Z} = \int \mathcal{D}\tilde{f}\mathcal{D}b\mathcal{D}\lambda \exp\{A + \int_0^{\beta} d\tau \sum_{im} \lambda_i(n^{\tilde{f}}_{im} + b^{\dagger}_i b_i - 1)\} \tag{8}$$

is represented by a symmetry breaking of the phase of $< b >$. Physically this is equivalent to a time dependent Hartree-Fock approximation in which

$$b_i \to < b > e^{-\Delta\varepsilon_f \tau}. \tag{9}$$

Here $\Delta\varepsilon_f$ represents an energy shift of the f band from its unrenormalized energy ε^0_f. If the functional integral is rewritten as a Feynman integral in which $\tau \to it$, then Equation (3) would represent a time dependent hybridization between the f band and the conduction band. The electrons can propagate from conduction to f state by a synchronous oscillation on all sites of the lattice. By absorbing the time dependent phase factor into the definition of the pseudo particle creation operator

$$\tilde{f}_{im} \to \tilde{f}_{im} e^{-\Delta\varepsilon_f \tau} \tag{10}$$

as shown by Read and Newns [7], the saddle point then appears in the "renormalized band picture" in terms of an f level shifted to lie above the Fermi level and considerably narrowed by a factor $< b^2 >$ (see Fig. 2).

It can then be shown that the renormalized band width is given by

$$bandwidth \propto < b^2 > \frac{v}{N} N(\varepsilon_F) \tag{11}$$

corresponding to a mass renormalization $\frac{m^*}{m} = 1/ < b^2 > \equiv 1/z$. Self consistency yields the value

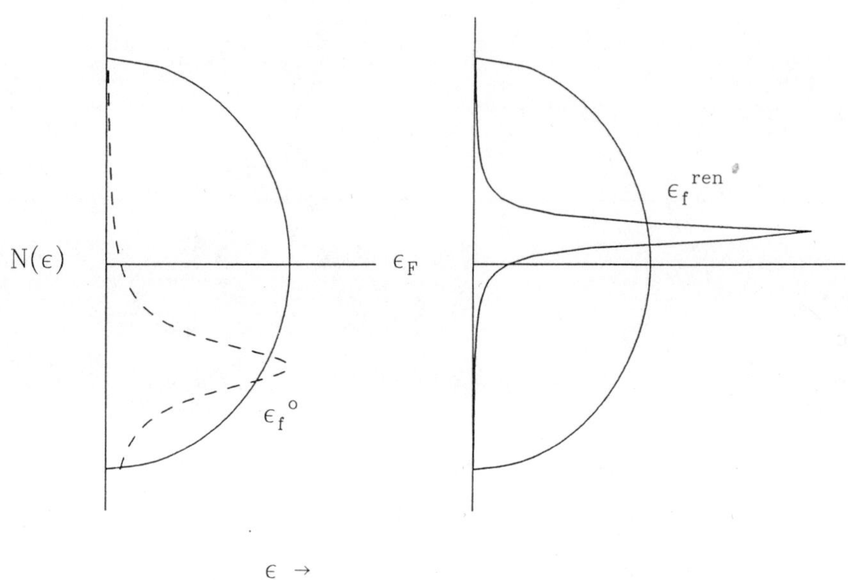

Fig. 2. Renormalization of a narrow band hybridized with a broad band due to correlation: left panel, noninteracting; right panel, renormalized.

$$< b^2 > \simeq e^{-\varepsilon_f^0/v^2 N(\varepsilon_F)} \equiv T_K/\varepsilon_F \tag{12}$$

where T_K is the Kondo temperature for the strongly coupled f-spins. So the renormalized mass is just the inverse of the Kondo temperature, measured relative to the bare f-band width, which can become very large.

4. Effect of strong correlation on the momentum distribution of Fermi liquids

The momentum distribution is represented by the equal time limit of the f-electron Green's function

$$\lim_{\tau \to 0^-} G_p^f(\tau) = \int_{-\infty}^{\varepsilon_F} d\varepsilon \mathcal{I}m G_p^f(\varepsilon). \tag{13}$$

In terms of Fermi liquid theory, the Fermi surface is defined by the moment $\vec{p}$ at which

$$\mathcal{I}m\Sigma_p(\varepsilon - \varepsilon_F) \propto (\varepsilon - \varepsilon_F)^2,$$
$$\left.\frac{\partial \Sigma_p(\varepsilon)}{\partial \varepsilon}\right|_{\varepsilon_F} = \frac{1}{z} - 1 \tag{14}$$

In other words $\mathcal{I}m G_p(\varepsilon) \to G_p^{incoh}(\varepsilon) + Sgn(\varepsilon - \varepsilon_p)\delta(z(\varepsilon - \varepsilon_F))$ as $p \to p_F$, where $G_p^{incoh}(\varepsilon)$ is smoothly varying in the vicinity of ε_F. Thus, quite generally, Fermi liquid theory leads to a step in $n(p)$ at the Fermi surface whose height is renormalized by the factor z.

For the correlated band, G is expressed in terms of quasiparticle operators as

$$G_p^f(\tau) = < T\{\tilde{f}_p(\tau)b^\dagger(\tau)b(0)\tilde{f}_p^\dagger(0) > . \tag{15}$$

However, one cannot simply substitute the average value $< b >$ for the boson operator since this violates the sum rule for the one particle Green's function [8]. So to evaluate G_p^f one has to include both the average value of $b = < b > + \delta b$ and the fluctuation contribution δb:

$$G_p^f(\tau) = < T\{\tilde{f}(\tau)\tilde{f}^\dagger(0)\} > < b^2 > + < T\{\tilde{f}(\tau)\delta b^\dagger(\tau)\delta b(0)\tilde{f}^\dagger(0) > . \tag{16}$$

In energy space this leads to a spectral density $\mathcal{I}m G_p^f(\varepsilon)$ which contains a mean field part with considerably reduced spectral weight $< b^2 >$, representing the quasiparticle pole, together with a broad incoherent part arising from the boson fluctuation which tends to center around the unrenormalized f-level position [9] (Fig. 3).

On integrating over energy, as in (13), the resulting form for n(p) then contains a considerably reduced step at the Fermi level whose height is measured by the weight of the quasiparticle peak which is given by the renormalization factor z. Thus we see that the momentum distribution is changed drastically from the usual Fermi step to a continuous function in which the Fermi surface is represented by a step with considerably reduced amplitude. It seems clear that considerable improvements in resolution for measurement of n(p) will be needed in order to see these effects in heavy Fermion compounds where z could be as small as 1 percent.

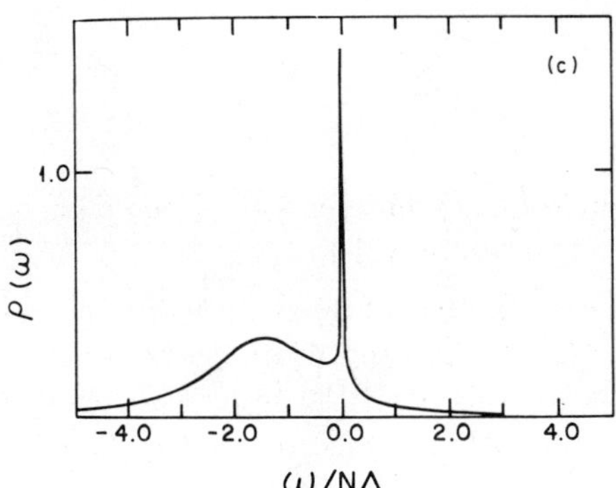

Fig. 3. f-spectral weight for an f-level interacting with a conduction band. The calculation is for a single f-impurity; one would expect a similar form for an f-band (from Coleman [9]).

5. n(p) for Mott insulators

The simplest model for a Mott insulator is that proposed by Hubbard in which a single band of d-like electrons represented in a tight binding model contains a strong on-site Coulomb repulsion with energy U:

$$H = - \sum_{\sigma,<ij>} t\, c_{i\sigma}^\dagger c_{j\sigma} + U n_{i\uparrow} n_{i\downarrow} \tag{17}$$

In the very strong correlation limit, $U >> t$ it has been shown by a number of authors [10] that a simple canonical transformation of the Schrieffer-Wolff type leads to a replacement of the Hamiltonian by the form

$$H \simeq -t \sum_{\sigma<ij>} \tilde{c}_{i\sigma}^\dagger b_i b_j^\dagger \tilde{c}_{j\sigma} + J \sum_{ij} \vec{S}_i \vec{S}_j \tag{18}$$

where again the slave bosons are used to rigorously exclude double occupany of the d-states:

$$\sum_\sigma n_{i\sigma} + b_i^\dagger b_i = 1. \tag{19}$$

The second term in Equation (17) represents the Heisenberg superexchange between the d-electrons on different sites, in which $J = t^2/U$.

Exactly at $n_{i\sigma} = 1$ i.e. half-filling of the single band, the boson constraint requires the condition that $b^\dagger b = 0$, i.e. the band width of the conduction band is rigorously zero in the half full case. At this point the system becomes an insulator and at the same time an antiferromagnet through the Heisenberg superexchange term.

What is the effect of these correlations on $n(p)$? In order to treat the strong correlation limit for finite U, it would be necessary to introduce more than one slave boson[11]. Since this is fairly complicated, we resort to a simpler way to get a qualitative idea of what happens by looking at the weak to intermediate correlation regime where U is less than or roughly equal to t. In this regime, a reasonable approximation is the simple Hartree-Fock approximation in which the Coulomb repulsion term is replaced by an average one-particle potential term

$$\sum_i U n_{i\uparrow} n_{i\downarrow} \rightarrow \sum_{i\sigma} U < n_{-\sigma} > n_{i\sigma}. \tag{20}$$

For the antiferromagnetic state, this then leads to a doubling of the original unit cell for the single Hubbard band with resulting tight binding Hamiltonian for spin up and spin down electrons on sublattices A and B respectively

$$\begin{aligned}
H_{HartreeFock} = \varepsilon_0 \sum_i (n_i^A - n_i^B) &- t \sum_i c_{Bi}^\dagger (c_i^A + c_{i+x'}^A + c_{i+y'}^A + c_{i+x'+y'}^A) \\
&- t \sum_i c_{Ai}^\dagger (c_i^B + c_{i-x'}^B + c_{i-y'}^B + c_{i-x'-y'}^B)
\end{aligned} \tag{21}$$

where $\varepsilon_0 = U < n_\uparrow >$. The resulting tight binding wave function has the form

$$\psi_{k\uparrow,\downarrow}(kr) = \sum_i e^{ik\cdot R_i}[u_k\phi^0(r - R_i^A) \pm v_k\phi^0(r - R_i^B)]. \tag{22}$$

This Hamiltonian can be diagonalized straightforwardly leading to energy bands satisfying

$$\begin{vmatrix} \varepsilon^0 - \varepsilon_k & \tilde{t} \\ \tilde{t}^* & -\varepsilon^0 - \varepsilon_k \end{vmatrix} = 0 \tag{23}$$

where

$$\tilde{t} = e^{i(k_x+k_y)/2}[2\cos\frac{(k_x - k_y)}{2} + 2\cos\frac{(k_x + k_y)}{2}] \tag{24}$$

leading to

$$n(p) = |\phi_p^0|^2 \sum_{k_{occ},G} \delta(\vec{k} - \vec{p} - \vec{G})(1 + \sin 2\theta_k e^{i\frac{(k_x+k_y)}{2}} e^{-i\frac{(p_x+p_y)}{2}}). \tag{25}$$

Here k_x and k_y are defined in the antiferromagnetic Brillouin zone, $\vec{k}$ and $\vec{p}$ are measured in units of the antiferromagnetic unit cell spacing $\tilde{a} = a\sqrt{2}$, and

$$\sin 2\theta_k = \frac{2D\,t}{\tilde{t}^2 + D^2}, where\ D = \sqrt{\varepsilon_0^2 + \tilde{t}^2} - \varepsilon_0. \tag{26}$$

$\phi^0(p)$ is the Fourier transform of the atomic form factor in Equation (22).

This formula represents the form factor for the antiferromagnetic electrons and as such it goes continuously over from the Fermi distribution in the original noninteracting tight-binding band to an insulating type form factor for finite U/t. Form factors of this tight-binding type were discussed by Berko and Plaskett in 1958 [12]. What seems clear is that as the correlation switches on, the Mott insulator $n(p)$ form will still have the same symmetry and quite possibly a rather similar form whether the system is an insulator or a conductor. Therefore at finite resolution it may be quite easy to confuse an apparent Fermi surface in a Mott insulator with what is in reality a form factor for a full antiferromagnetic sub- band. (See Fig. 4).

6. The Cu perovskite superconductors

In the case of the high temperature superconductors it seems necessary to include a second band in order to represent the effect of copper-oxygen hybridization in the CuO_2 conducting planes of these materials. This is represented by a simplified two-band Hubbard model as

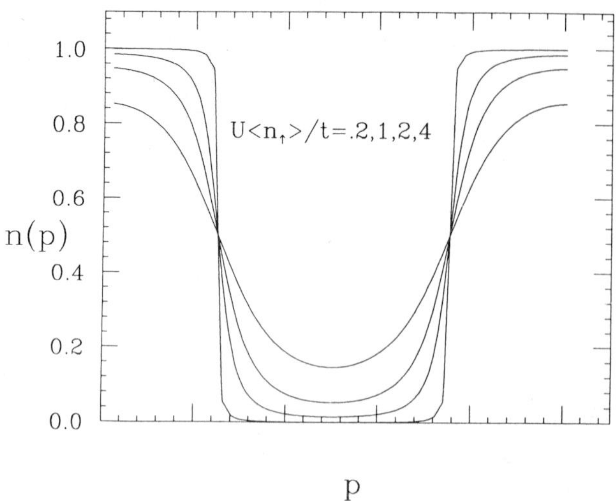

Fig.4. n(p) for a Mott insulator calculated using Hartree Fock.[A factor $|\phi^\circ(p)|^2$ has been omitted - see equation (25)]. As U/t is increased an antiferromagnetic gap opens up. However, the antiferromagnetic form factor still shows a sharp decrease at the Brillouin zone boundary.

$$H = -t \sum_{<ij>\sigma} c_{i\sigma}^\dagger d_{j\sigma} + U \sum_i n_{i\uparrow}^d n_{i\downarrow}^d + \varepsilon_{Cu}^0 \sum_{i\sigma} d_{i\sigma}^\dagger d_{i\sigma} + \varepsilon_{Ox}^0 \sum_{i\sigma} c_{i\sigma}^\dagger c_{i\sigma} \qquad (27)$$

As in Section 3, the effects of strong correlation will be to lead to a set of renormalized bands [13] in which the copper band floats up toward the bottom of the oxygen band and acquires a considerably narrowed band width $< b^2 > = 1 - n_d$ which is proportional to the hole concentration.

In the case of $n_d = 1$, or half-full limit, the system will be an antiferromagnet with $n(p)$ form factors as discussed in Section 4.

The interesting question is now what happens when the system is doped with holes as in the case of $La_{2-x}Sr_xCuO_4$ or $YBa_2Cu_3O_{6+y}$. In this case one wants to consider a tight-binding antiferromagnetic band, renormalized as discussed above with finite width, and the question is: where do the first holes go in the Brillouin zone?

A "seat of the pants" argument is based on the idea that the conducting orbitals will tend to go along the oxygen diagonal which corresponds to points $(\frac{\pi}{2}, \frac{\pi}{2})$.

Thus we expect to see "bites" taken out of the four sides of the antiferromagnetic Fermi surace leading to a small region of the antiferromagnetic form factor which will have a vertical slope of height z where z is proportional to the hole concentration. In fact, one would expect the effects of strong correlations to further reduce this height since a step at height z would correspond to a band of non-interacting holes. In practice, interaction with a magnetic background, which may be disordered but which we still expect to show strong antiferromagnetic short range correlations [14], will be expected to increase the effective mass of the holes and hence to further reduce the height of the Fermi step in the $n(p)$ distribution.

From these arguments it seems rather unlikely that, with practical resolution[15], it will be easy to distinguish a true Fermi surface from the form factor of the strongly correlated Mott insulator electrons which form the majority of electrons in the hybridized copper-oxygen bands which lead to superconductivity.

Acknowledgment:I am grateful to the Office of Naval Research and the National Science Foundation for support for this work.

References

1) L. Lam and P. M. Platzman, Phys. Rev. B **9**, 5122 (74).

2) G. E. W. Bauer, Phys Rev. B **27**,5912(83).

3) G. E. W. Bauer and J. R. Schneider, Phys. Rev. Lett. **52**, 2061 (83), Z. Phys. B **54**17 (83).

4) D. Koelling, Solid State Comms., **42**, 247 (82). T. Oguchi, A. J. Freeman, and G. W. Crabtree, J. Magn. Mag. Matls. **63-64**, 645 (87); C. S. Wang, M. R. Norman, R. C. Albers, A. M. Boring, W. E. Pickett, H. Krakauer, and N. E. Christensen, Phys. Rev. B **35**, 7260 (87).

5) M. C. Gutzwiller, Phys. Rev. **137**A, 1726(65); see also C. M. Varma, W. Weber and L. J. Randall, Phys. Rev. **33**, 1015(86).

6) N.Read, J. Phys. C,**18**,2651(85); P. Coleman, Phys. Rev. B **35**, 5072(87); A. J. Millis and P. A. Lee, Phys. Rev. B **35**, 3394(87).

7) N. Read and D. M. Newns, J. Phys. C. **16**, 3273(83).

8) A. E. Ruckenstein and S. Schmitt-Rink, Phys. Rev. B, Rapid Communications **38**, 7189 (1988).

9) P. Coleman, Phys. Rev. B **29**, 3035(84).

10) See J. Hirsch, Phys. Rev. Lett. **54**, 1317.

11) G. Kotliar and A. E. Ruckenstein, Phys. Rev. Lett. **57**, 1362(86).

12) S. Berko and J.S. Plaskett, Phys. Rev. **112**, 1877(58).

13) G. Kotliar, P. A. Lee, and N. Read, Proc. Interlaken Conf., Physica C **153 - 155**, 538, (88).

14) M.Inui, S. Doniach, and M. Gabay, Phys. Rev. B, **38**p. 6631 (88).

15) M. Peter, L.Hoffman, and A. A. Manuel, Proc. Interlaken Conf., Physica C **153-155**, 1724(88).

X-RAY COMPTON SCATTERING FROM CONDENSED MATTER SYSTEMS

P. M. Platzman

AT&T Bell Laboratories
Murray Hill, New Jersey

Inelastic x-ray scattering at large momentum transfers is in principle an almost ideal probe of the ground state properties of electrons in solids.[1] In this talk we will present the basic theory of the scattering. We will then discuss the results of experiments on some simple systems which illustrate the kinds of physics one can extract from such systems.

X-ray's of reasonable energies 10 keV $< \hbar\omega_1 <$ 100 keV interact weakly with bulk samples so that their inelastic scattering may be accurately treated in a weak coupling Born approximation. In this limit the scattering cross section is completely specified by a momentum transfer $\mathbf{k}$, and an energy transfer ω, ($\hbar=1$) the polarization of the incoming (ε_1) and outgoing photon (ε_2) and in some cases the energy of the initial photon ω_1 if it happens to be near an absorption edge in the material. As we shall see the typical energy transfer at high momentum transfers $k\lambda_c> >1$ (where λ_c is a characteristic length for the scatterer for example the interparticle spacing is ($\hbar=1$),

$$\omega \cong k^2/2m \cong (2\sin^2\theta/2)\,\omega_1\,(\omega_1/mc^2). \tag{1}$$

Since we are in a non-relativistic regime where $\hbar\omega_1/mc^2 \leq .1$ the energy transfer is much smaller than the energy itself and it is a good approximation to assume, as in Eq. (1), that the momentum transfer is determined by the scattering angle. The dependence on polarization is crucial for x-ray Compton scattering from ferromagnetic solids.[2] The dependence on ω_1 is more subtle and has only been discussed recently.[3] We will discuss it briefly in this talk. As we shall see it gives us a handle on the phase of the scattering amplitude.

The scattering cross section in Born approximation from a system of N electrons is given by the golden-rule[4] as,

$$\frac{d\sigma}{d\Omega d\omega} = \left(\frac{e^2}{mc^2}\right)^2 \overline{\sum_{f\,i}} \left| <f| \sum_{j=1}^{N} e^{i\mathbf{k}\cdot\mathbf{r}_j} M_j |i> \right|^2 x \tag{2}$$

$$\delta(E_f-E_i-\omega).$$

The states $<f|$, $<i|$ are the final and initial many body states of our N particle system. The bar over the sum is the usual average over initial states while M_j is the scattering amplitude from the jth electron which can have a momentum $\mathbf{p}$. In the weakly relativistic case ($\hbar\omega_1/mc^2 \cong p/mc \leq .1$) this amplitude is,

$$M_j = A + i\mathbf{B}\cdot\sigma_j + \mathbf{C}\cdot\mathbf{p}_j/m \qquad (3)$$

The amplitude A is the normal charge scattering i.e. electric dipole absorption followed by electric dipole emission. It gives parallel polarization. The **B** term which is pure imaginary relative to A gives us spin information and consists of cross polarization pieces. It may be identified with other multipoles such as magnetic dipole and electric quadrupole.[5] The **C** term is an orbital piece which allows us to separate the spin and orbital contributions to the static moment of an antiferromagnetic solid.[6] In inelastic scattering it plays little role.

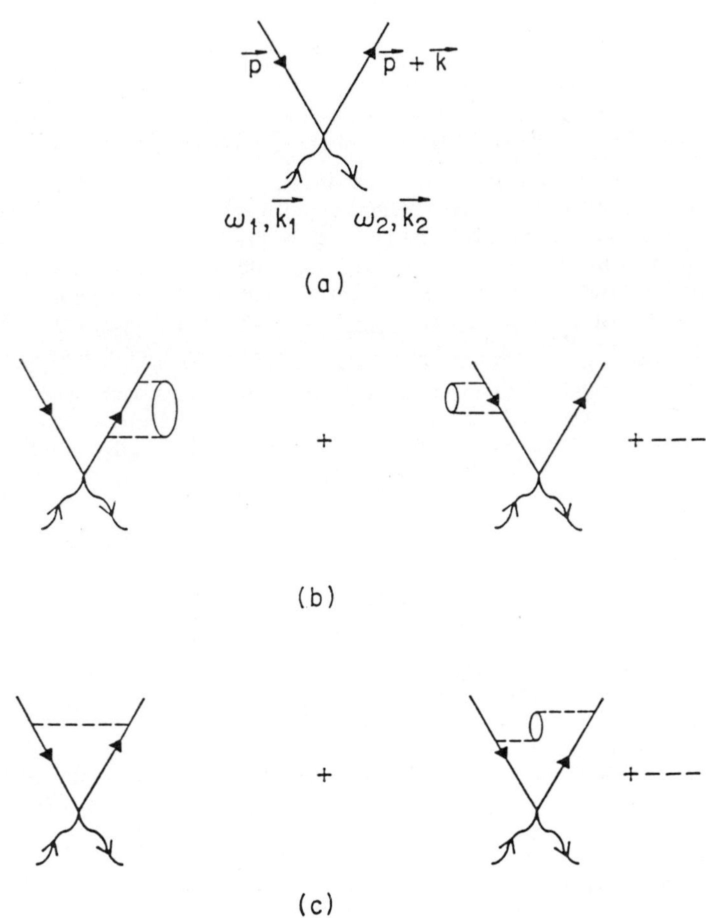

Fig. 1 Feynman diagrams showing schematically the various matrix elements involved in Compton scattering a) the basic impulse approximation (IA) b) typical recoiling electron and hole final state effects c) electron-hole (excitonic) final state effects.

If we think of the initial state in Eq. (2) as the "full vacuum" then the final state consists of some number of electron hole pairs. No pairs present corresponds to the elastic amplitude. If the hole is one of the filled core levels we call the process Raman scattering. If all holes are in the outer valence bands and the momentum transfer is large $k\lambda_c >> 1$ then we call the process Compton scattering. In this limit it is a very good approximation to consider the scattering from single electron with momentum $\mathbf{p}$ spin up ($\uparrow$) or down ($\downarrow$), and weight that scattering with the *probability* of finding it. In this limit we obtain[7]

$$\frac{d\sigma}{d\omega d\Omega} = \left[\frac{e^2}{mc^2}\right]^2 \times \int \left[(\boldsymbol{\varepsilon}_1 \cdot \boldsymbol{\varepsilon}_2)^2 \left[n_{\mathbf{p}\uparrow} + n_{\mathbf{p}\downarrow}\right] + \right. \tag{4}$$

$$\left. 2 \, \text{Im}(A^*B_z) \left[n_{\mathbf{p}\uparrow} - n_{\mathbf{p}\downarrow}\right] \times \delta(\omega - k^2/2m - \mathbf{k}\cdot\mathbf{p}/m)\right] d^3\mathbf{p}$$

where $n_{\mathbf{p}\uparrow} = <a^+_{\mathbf{p}\uparrow} a_{\mathbf{p}\uparrow}>, etc.$

Eq. (4) is the conventional impulse approximation formula with a small but important spin dependent piece proportional to $\text{Im}(A^*B_z)$ where A and $\mathbf{B}$ are defined by Eq. (3) and $\hat{z}$ is the direction of the macroscopic magnetization in a ferromagnetic sample. This piece exists only if the polarization vectors are complex i.e. have some degree of circular polarization.

Since there has been a great deal of discussion at this meeting regarding the validity of the impulse approximation in hard core Helium systems and in nuclei we will briefly discuss the validity of Eq. (4) in these very "soft" Columbic systems. It is somewhat easier to talk about the corrections by resorting to a Feynman diagram description of the scattering amplitude (see Fig. 1). Fig. 1(a) shows the basic diagram which gives the impulse approximation. Simply put it says that the scattering by x-rays creates with matrix element M (solid circle) a *free* particle hole pair. There are only three physically distinct types of final state effects.

1) The outgoing hot particle momentum $\mathbf{k}+\mathbf{p}$ ($k \gg p$) may interact with the remaining soup of particles in the ground state Fig. 1b. Crudely speaking such scattering produces a smearing of the energy of this particle i.e.,

$$\text{Im}E_{\mathbf{p}+k} \cong \left|\frac{(\mathbf{p}+\mathbf{k})}{m}\right| n\sigma \cong \frac{k}{m} n\sigma \tag{5}$$

Since σ, the two particle scattering cross section is a constant for hard core systems the broadening increases linearly with k. For Columbic systems, for example the electron gas with plasma frequency ω_p and Fermi momentum p_F this is not so and[8]

$$\text{Im } E_\mathbf{k} \cong \alpha(\hbar\omega_p) (p_F/k)^3. \tag{6}$$

The const α is a function of the dimensionless density $r_s \sim n^{-1/3} \, me^2$ which is roughly unity for metallic systems. In this case α is also of order unity. This final state effect, since it falls off as k^{-3} may be made completely negligible.

The change in the real part of the self energy (Fig. 1b) for a Columb system is[8]

$$\text{Re } E_k \cong \gamma (\hbar\omega_p)^2/(k^2/2m), \tag{7}$$

with γ of order one. Since it falls off as k^{-2} it is also negligibly important at high momentum transfers.

2. The interaction between particle and hole (Fig. 1c) may also be shown to fall off as k^{-3} for Columb scattering.

3. The interaction of the hole with its surroundings (Fig. 1b), may be included exactly. A discussion may be found in ref. 7. The modified IA formula which includes such effects is (no spin dependence),

$$\frac{d\sigma}{d\omega d\Omega} \sim \int d^3p d\omega' \, \text{Im} G_p(\omega') \, \delta \, (\omega-(p+k)^2/2m + \omega') \tag{8}$$

where

$$G_p(\omega') = \int dt e^{i\omega' t} < a_p^+(t) a_p(0) > \tag{9}$$

Insofar as we can neglect the ω' in the argument of the δ function i.e. ω' is negligible compared to $(p+k)^2/2m$ we arrive at the standard IA formula. These effects have been called separation ratio effects in the talk by Prof. Sick. They express the fact that for a well defined momentum the hole may have a variety of energies. Since the energy scale of the core hole is the Fermi energy E_F for the outer electrons, these corrections in the context of the other corrections are of order one and are the most important. In the tails of the momentum distribution where the spectrum is changing rapidly they will become more significant and interesting.

If the initial frequency is near an edge for the excitation of a core electron then there is a further complication which we only briefly mention.[3] Suppose for the purpose of simplicity that the many electron state can be characterized as a Slater determinant of one electron orbitals $\Phi_i(r)$ then

$$n(p) \sim \sum_i \left| \int e^{ip\cdot r}\Phi_i(r)d^3r \right|^2 \equiv \sum_i | \, g_i(p) \, |^2 \tag{10}$$

The usual impulse cross section is independent of the phase of the one particle orbital. In a tight binding picture this phase contains information on the position of the wave function. If we consider only one tight binding orbital, i.e. one electron localized at a position $\mathbf{a}$ relative to the nucleus then including the resonance term but neglecting corrections to impulse gives

$$\frac{d\sigma}{d\Omega d\omega} \sim \int d^3p \left| \varepsilon_1 \cdot \varepsilon_2 e^{ip\cdot a}\tilde{g}(p) + \frac{2(k_2\cdot\varepsilon_2)(k_1\cdot\varepsilon_1)h(0)}{m(\omega_1-E_{Th}-\varepsilon_{p+k}+i\Gamma_S/2)} \right|^2 \tag{11}$$

$$\delta\left[\omega-k^2/2m-\frac{k\cdot p}{m}\right]$$

where,

$$h(0) = \left[\frac{1}{3}\left(\frac{a_S}{2}\right)^{3/2}(2\pi)^{1/4}\right]\left[\int d^3r(\mathbf{r}\cdot\nabla\psi_S)\right]\Phi_0(0) \tag{12}$$

$g(\mathbf{p}) \equiv e^{ip\cdot a}\tilde{g}(\mathbf{p})$ and $\tilde{g}(\mathbf{p})$ is a real number.

Let's now switch to what one would actually see in a typical experiment and then display and discuss a number of experiments. Fig. 2 shows a hypothetical spectra. In 2(a) we show what one might see if one had only "outer electrons" for example solid hydrogen or solid helium.

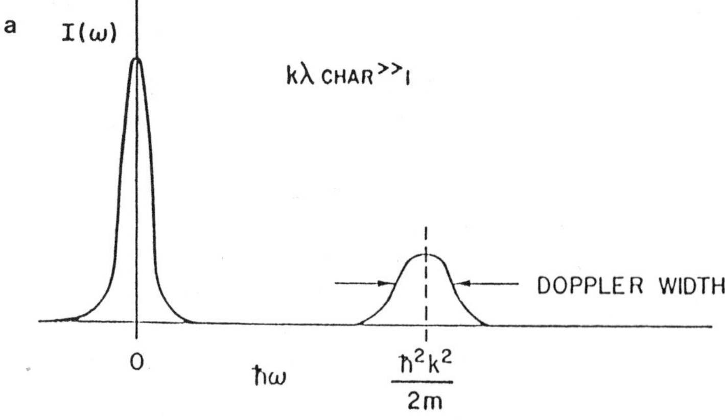

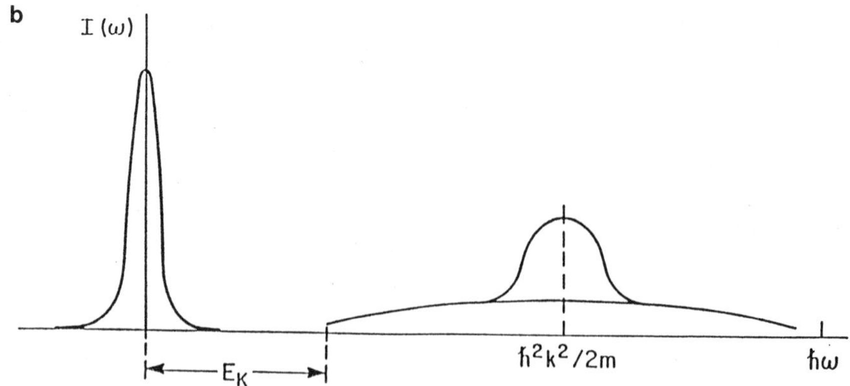

Fig. 2   A schematic Compton profile (CP for a) Hydrogen b) A
simple light metal like Be, the broad piece with an edge
comes from the 1S core.

Around $\omega=0$ there would be an unresolved blob of stuff which would correspond to the excitations of vibrations, rotations, etc. of the nuclei. This is the domain discussed in earlier talks using hot neutrons. It has an energy spread of tens to hundreds of millivolts. At much higher energies centered around $k^2/2m_e$ appears the impulse part of the electronic spectrum, the Compton profile (CP). For solid hydrogen this would be closely related to the Fourier transform of the hydrogen atom 1S wave function, i.e. a Lorentzian. In 2(b) we show what might be expected for a simple metal like Be. The relatively narrow free-electron like piece characterizing the Fermi-distribution sits on top of a broader background characterizing the core electrons. For modest recoil energies the cores show a threshold as indicated and are well described by atomic theory. The outer electrons are more complicated in that they will be anisotropic relative to the crystal axis, have magnetic properties for some materials etc. While the contribution from the valence electrons may in principle be separated from the core, for example by looking at the anisotropic part of the spectrum or at the temperature and magnetic field dependence, in practice it is non-trivial since there are in general many more core electrons. Thus very good statistics are required to do the subtraction. We will hear a discussion of positron annihilation in solids. This technique as Professor Berko will explain to us measures a related quantity, i.e. a correlated momentum profile between positron and electron. Since the positron is excluded from the ion cores by the large Columb repulsion it is far simpler to subtract core contributions.

The energy resolution of the outgoing x-rays is simply related to the momentum resolution (see Eq. (4))

$$\frac{\Delta\omega_2}{\omega_2} = \Delta p_z \left(\frac{\hbar}{mc} \right)$$

(13)

or in atomic units,

$$\Delta p_z \cong 1.25 \times 10^2 (\Delta\omega_2 / \omega_2).$$

(14)

Solid state detectors have proved to be the "spectrometers" which to date have been most useful.[1] Since Compton scattering is spread in angle and energy conventional crystal spectrometers give very low count rates even at existing synchrotron sources. We will, of course, hear about this from M. Cooper. The problem with the solid state detectors however is that they typically, at 30 keV, have resolutions of 300 eV i.e. a momentum resolution of one atomic unit. New bolometers[9] and tunnel junction detectors[10] with energy and momentum resolutions ten to a few hundred times better are under development. Such objects along with crystal spectrometers[11] coupled to area detectors could lead to a complete revolution in this field.

Now let's examine some typical CP and briefly consider their content. These profiles are generally defined by a,

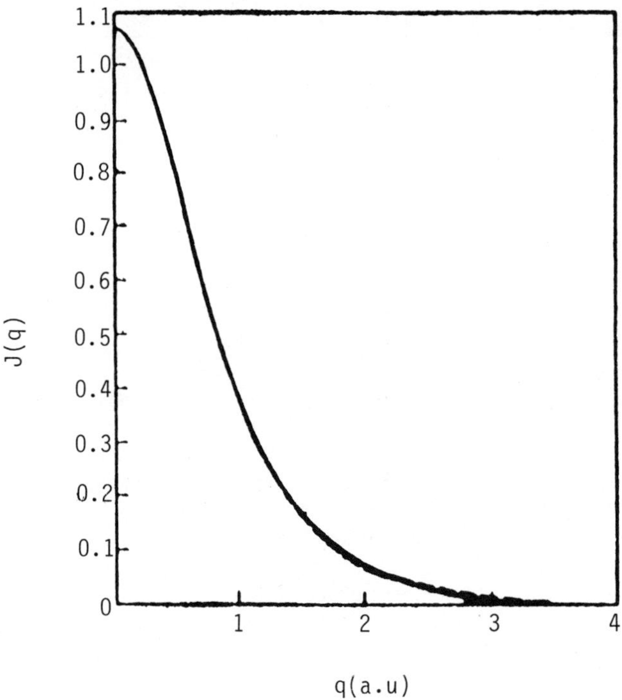

Fig 3   Comparison of the experimental CP and that calculated by the IA. The areas under the two were made equal to 2, the number of electrons per atom in helium. The solid line is the experimental profile, and the calculated profile is the dotted line.

$$J(\mathbf{q}) = \int n(p_x, p_y, \mathbf{q})dp_x dp_y \tag{15}$$

which is independent of the direction of $\mathbf{q}$ for non oriented molecules, liquids etc. In Fig. 3 we show half (the spectrum is essentially symmetric) of a profile for atomic He. The solid line is the experimental profile and the calculated profile is the dotted line. As expected (nothing interesting) the agreement is essentially perfect.

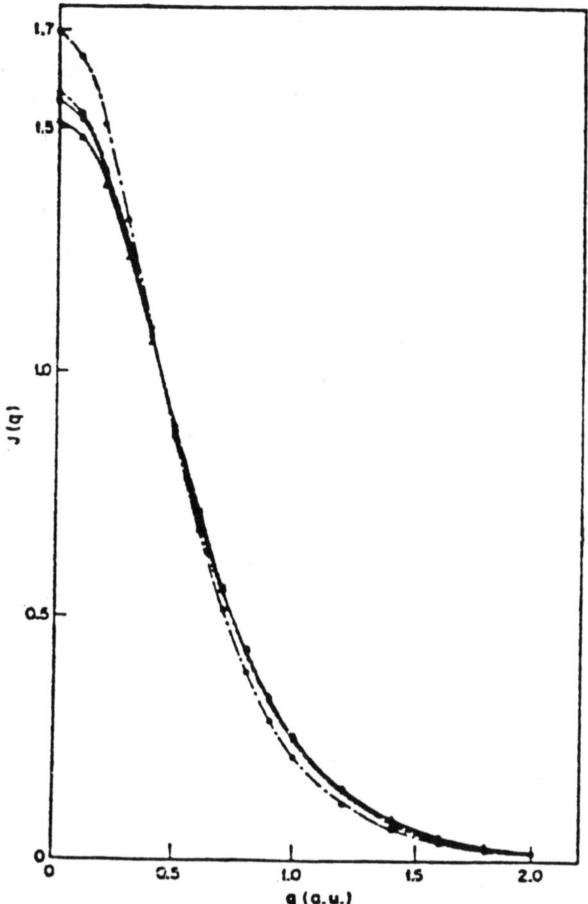

Fig. 4 Comparison of theory and experiment for H_2. Experimental ————; theory for two hydrogen atoms (2H) — · —; theory based on Hartree-Fock self consistent-field (HF-SCF) wave functions — · · —; theory based on multiconfiguration self-consistent-field wave functions · · · · · ·

In Fig. 4 we compare experiment (solid curve) with three increasingly sophisticated theories of the Hydrogen molecule.[12] At $q=0$ the largest disagreement occurs for theoretical wave functions for two hydrogen atoms. Somewhat better agreement is obtained for a theory based on a Hartree Fock wave functions. Only slightly better agreement obtains for the best multi-configurational wave function. A study of the differences between the spectra at the one percent level clearly give important information on electron correlation.

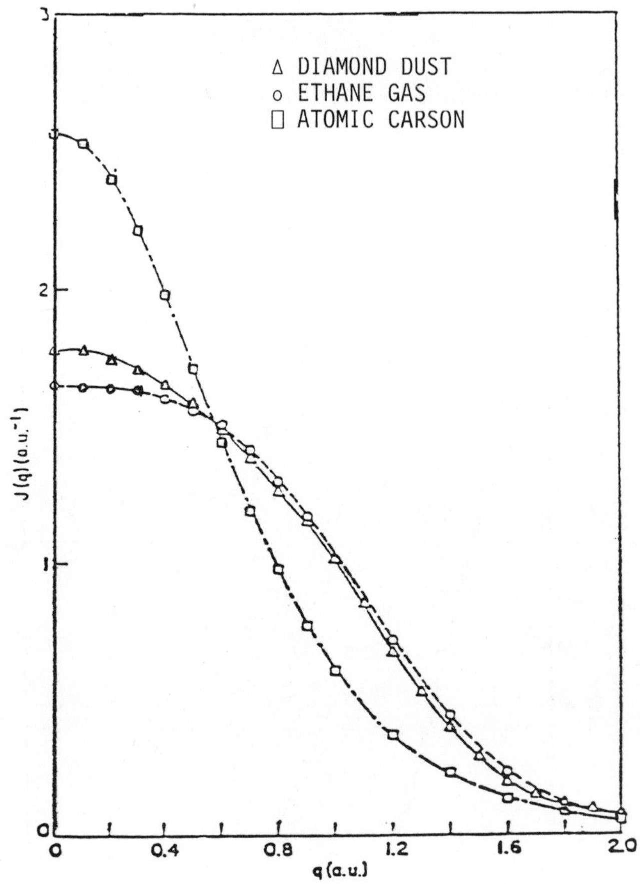

Fig. 5 CP for three types of carbon electrons.

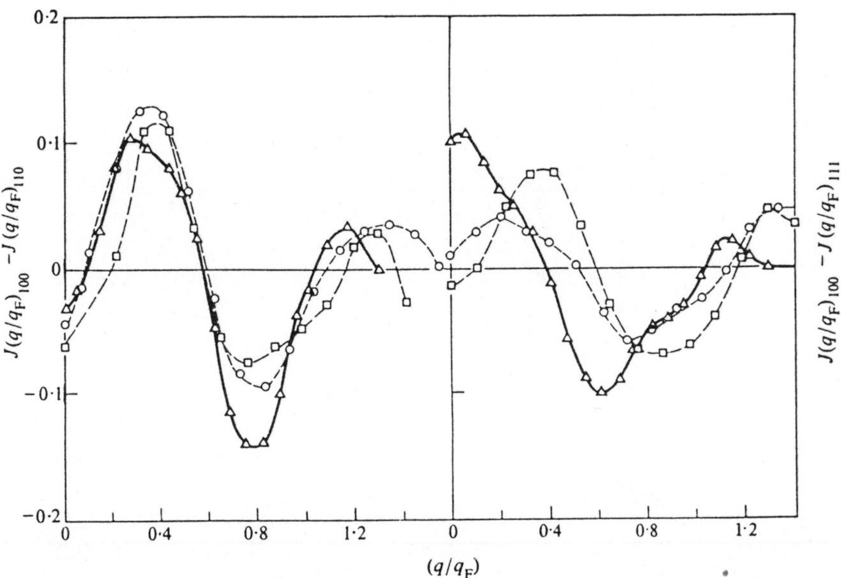

Fig. 6 The anisotropy of the CP for Si ($\bigcirc$), Ge ($\square$), and Diamond ($\triangle$).

In Fig. 5 we show Compton profiles for three "types" of carbon electrons.[13] The atomic carbon curve is theoretical.   The profile for the electrons in the carbon carbon single bond (diamond dust) and for the carbon carbon double bond come from a beautiful set of experiments performed by P. Eisenberger and W. Marra.[14]

In Fig. 6 we show some examples of anisotropic data for our two favorite covalently bonded elemental semiconductors Si and Ge along with one for diamond. There has by now been a reasonable amount of theoretical work on such profiles using local density approximation for the wave functions.[15]

For simple metals there has been several interesting experimental and theoretical investigations.   The three dimensional non-interacting Fermi distribution when converted to a J($\mathbf{q}$) becomes a parabola cutoff at $q = P_F$. (See Fig. 7).

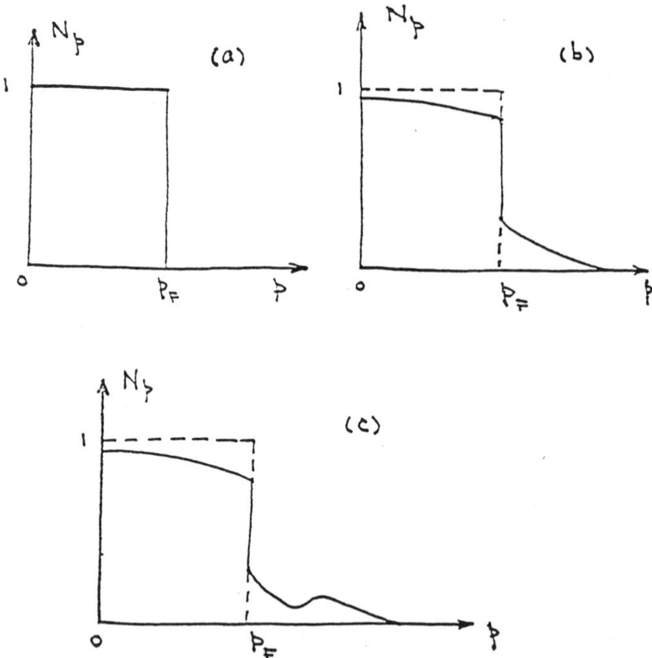

Fig. 7   The 3-D momentum density for a) a non
          interacting electron gas, b) an interacting electron
          gas in the absence of a periodic potential and
          c) an interacting gas in a real solid.

The break in slope simply reflects the size of the break in the 3-D n($\mathbf{p}$) at $p = P_F$. In the interacting case the cutoff parabola develops tails (see Fig. 6b) and the break is decreased. The size of the break called Z_{p_F} and the shape of the tail are all of interest in theories of the interacting electron gas immersed in a uniform positive background of charge. In a real crystal the periodic potential introduces additional complications. Simply stated such periodicity introduces images of the first Brillouin zone profile at all other reciprocal lattice points Q_R. This effect is illustrated in a 3D momentum sketch for a monovalent solid (Fig. 7). Here the Fermi surface does not touch the first

Brillouin zone. The kink for $q>Q_R$ reflects the higher zone contributions. For weak periodic potentials it is a small effect. In Fig. 8 we show a Compton profile of polycrystalline Na taken by Eisenberger et al.[16] with a resolution of $\Delta p_z \cong .1$ a.u. The dashed curve is the best fit (not very good) to a sophisticated theory of the interacting electron gas. Band structure effects are neglected.

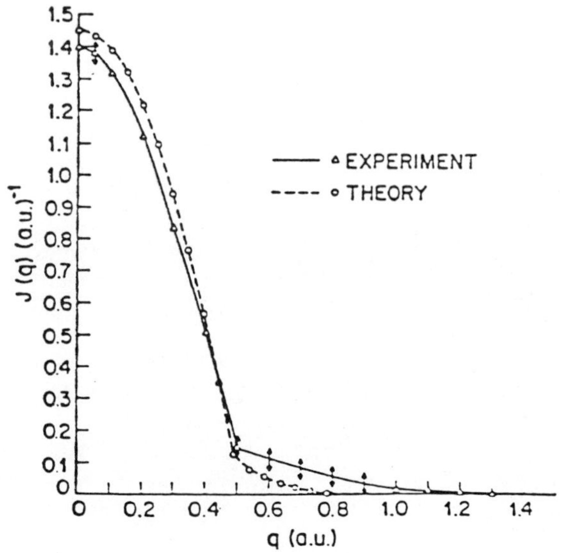

Fig. 8   CP of polycrystalline Na: solid line, experiment; dashed line, theory.

In Fig. 9 we show the first usable magnetic Compton spectrum taken by Sakai and Ono[17] using a radioactive source and a solid state detector. Denny Mills, in this volume, will tell us much more about such spectra. The characteristic dip at the origin tells us something about how the minority spins (s-p electrons) in such systems are confined in space i.e. they are spread out relative to the majority spins (d-electrons). The dashed curve shows the best available (at the time) local density spin polarized calculation.[18] The data is too poor to make any real comments.

This brief summary of data is illustrative of most of the spectra which are currently available. The big question is what is next. Well my crystal ball says that very soon we will be able to make ordinary Compton measurements in reasonable periods of time i.e. count rates of 10^2–10^4/sec with resolutions of 10^{-2} a.u. This will enable us to look at the Fermi-surfaces of ordinary metals and to look at correlation effects for example momentum tails. It will enable us to investigate in detail the nature of the covalent bonds in semiconductors. Resonant Compton will surely give us some new and interesting information on such systems. Of course exotic materials like transition metal oxides will be among the first to be investigated. Such experiments will give us a new insight on the metal insulator transition and possibly partially answer the question of the existence of a Fermi surface in the high T_c materials.

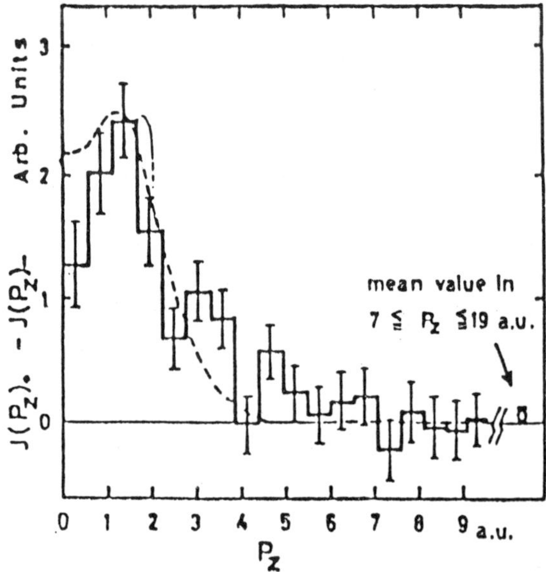

Fig. 9 The spin polarized Compton profile of polycrystalline Fe. The dashed curve is the local density theoretical results of Ref. (18).

There is no question that some of the most interesting experiments will involve magnetic Compton scattering. Such experiments naturally (flipping ratios) get rid of background from core electrons. They will enable the experimenter to study the true ground state wave functions in for example Ni and Fe. With 10^{-2} a.u. resolution one will be able to see temperature effects as one passes through the Curie temperature. In the so called heavy Fermion materials where, magnetism, strong electron correlations and superconductivity[19] are all intertwined there are a variety of possibilities. In particular a dramatic change of n(p) due to temperature and magnetic field will give us a new handle on the ground state properties of such systems. Finally my crystal ball says that the electronic structure of bonds particularly in oriented molecules i.e. molecular crystals will become an important field of research. In summary, I believe that thanks to synchrotron sources and new spectrometers we are on the verge of a real revolution in Compton Scattering experiments.

REFERENCES

1. *Compton Scattering* - Ed. Brian Williams, McGraw Hill, N.Y. 1977.
2. P. M. Platzman and N. Tzoar Phys. Rev. B 2 3556 (1970). F. DeBergevin, M. Brunel, Acta. Cryst. A 37 314, 325 (1981), D. Gibbs, D. E. Moncton, K. L. D'Amico, J. Bohr and B. Grier, Phys. Rev. Lett. 55, 234 (1985).
3. P. M. Platzman, Phys. Rev. Brief Reports, to be published.
4. *Collision Theory*, M. L. Goldberger and K. M. Watson, John Wiley and Sons, N.Y. (1964).
5. M. Gell-Mann and M. L. Goldberger, Phys. Rev. 96 1433 ((1954).
6. M. Blume, J. Appl. Phys. 57, 3615 (1985).
7. D. Gibbs, D. R. Harshman, E. I. Isaacs, D. B. McWhan, D. Mills and C. Vettier, Phys. Rev. Lett. 61, 1241 (1988).

8. *Interacting Fermi System*, P. Nozieres. W. A. Benjamin, Inc., N.Y. (1964).

9. P. M. Platzman and N. Tzoar, Phys. Rev. B **2**, 3556 (1970).

10. S. H. Mosley, R. L. Kelley, R. J. Schoelkopf, A. E. Szymkowiak, and D. McCammon. *IEE Transactions on Nuclear Science* **35**, (1) p. 59.

11. D. Twerenbold and A. Zehnder. J. Appl. Phys. **61**, (1) 1 (1987).

12. N. Sakai, N. Shiotani, M. Ito, H. Kawata, Y. Amemiya, M. Ando, S. Yamamoto and H. Kitumura, (to be published).

13. P. Eisenberger, Phys. Rev. A2 1678 (1970).

14. P. Eisenberger (private communication).

15. P. Eisenberger and W. Marra, Phys. Rev. Lett., **27**, 1413 (1971).

16. M. Y. Chou, M. L. Cohen and S. G. Louie, Phys. Rev. B **33**, 6619 (1986).

17. P. Eisenberger, L. Lam, P. M. Platzman and P. Schmidt, Phys. Rev. B**6**, 3671 (1972).

18. N. Sakai and K. Ono. J. Phys. Soc. Jpn. **42** 770 (1977).

19. S. Wakoh, and Y. Kubo, J. Magn. Magn. Mater. **5**, 202 (1977).

20. S. Doniach, these proceedings.

MEASURING THE MOMENTUM DISTRIBUTION OF THE UNPAIRED SPIN ELECTRONS IN FERROMAGNETS USING SYNCHROTRON RADIATION

Dennis M. Mills

Advanced Photon Source, Argonne National Laboratory
Argonne, IL 60439

ABSTRACT

The dominant term in the x-ray Compton cross-section of an electron is the interaction of the photon and the electron's charge. Platzman and Tsoar many years ago pointed out that there is also an interaction between an x-ray and the electron's spin and in principle this interaction can give information on the momentum distribution of the unpaired spin electrons in the solid. Unfortunately the spin sensitive term is not only small compared to the charge term, but in addition couples to the photons in first order only with that components of the x-ray beam that is circularly polarized. A lack of intense sources of circularly polarized x-rays combined with the relative small size of the spin sensitive term makes measurements of the momentum distributions of unpaired spin electrons difficult resulting in little experimental progress initially made in spin or magnetic Compton scattering.

In the past several years interest in spin sensitive Compton scattering has been revived due in large part to the availability of intense beams of high energy photons from synchrotron radiation sources. The radiation from storage ring sources has well defined polarization states; highly linearly polarized in the orbital plane and elliptically polarized above and below the plane of the orbit of the circulating particles. The high flux and unique polarization properties of synchrotron radiation sources have greatly facilitated measurements of the momentum distributions of the unpaired spin electrons in ferromagnetic solids. Recent results of the work of several groups will be presented along with some thoughts on the impact that the next generation of storage rings, such as the Advanced Photon Source, and insertion devices specifically designed to produce circularly polarized x-ray beams will have on the field of magnetic Compton scattering.

---

*This work supported by the U.S. Department of Energy, BES-Materials Science, under contract no. W-31-109-ENG-38.

261

INTRODUCTION

For over fifty years Compton scattering has been used as a tool in
the study of electronic momentum distributions in condensed systems.
After the initial work by Dumond and Kirkpatrick in the 1930's, however,
there were few significant advances in the experimental technique until
the advent of the high resolution solid state detector (SSD) in the
1960's. SSD's permitted the use of high energy (although low flux)
radioactive gamma sources which markedly improved the quality of the
data and extended the range of sample materials that could be studied.
In the last decade work has proceeded to develop the necessary apparatus
to extend Compton scattering experiments from investigating the momentum
distribution of all the electrons to selectively looking at only the
unpaired spin electrons. To progress further researchers need the magnetic Compton scattering equivalent of the SSD. It is my opinion that
the new generation of high energy synchrotron radiation sources to be
built will provide that impetus. The copious quantities of highly
(linearly and circularly) polarized, collimated, short wavelength x-rays
these storage ring sources can provide are precisely what is needed for
progress in this area.

THEORETICAL BACKGROUND

When an incident beam of monochromatic x-rays strikes an atom, some
of the photon's energy can be transferred to an electron. If this electron were at rest, the energy spread of these inelastically scattered
photons would depend only on the energy width of the incident beam. In
any real material, however, the electron will not be a rest and the energy spread of the Compton scattered photons will be Doppler broadened
by the electron's velocity. Hence the line shape of the inelastically
scattered photons, the so-called Compton profile, contains information
on the velocity or momentum distribution of the electrons in the sample.
(A more detailed analysis of the relationship between the shape of the
Compton profile and the momentum distribution of the electrons and the
approximations under which this relationship holds can be found in several recent articles.[1,2]) Utilizing the Born approximation, the
differential cross-section for scattering of x-rays from a single electron can be written as[3]:

$$\frac{d^2\sigma}{d\Omega d\omega} = r_o{}^2 \left(\frac{\omega_2}{\omega_1}\right) S(\underline{k},\omega) \tag{1}$$

where r_o is the classical radius of the electron, ω_1 (ω_2), the incident
(scattered) x-ray frequency, $\omega=\omega_1-\omega_2$, and $S(\underline{k},\omega)$ is the generalized scattering factor. $S(\underline{k},\omega)$ can be expressed as:

$$S(\underline{k},\omega) = \sum_{m,n} |M_{mn}|^2 \, \delta(E_n - E_m - \omega) \tag{2}$$

where M_{mn} are the matrix elements $\langle m|e^{i\underline{k}\cdot\underline{r}}|n\rangle$. In 1970 Platzman and
Tsoar[4] showed that when relativistic effects are taken into account, M_{mn}
can be written (to order $\hbar\omega/mc^2$) as

$$M_{mn} = A\delta_{mn} + i\left(\frac{\hbar\omega_1}{mc^2}\right)\underline{B}\cdot\underline{\sigma}_{mn} \tag{3a}$$

where

$$A = (\underline{\varepsilon}_1 \cdot \underline{\varepsilon}_2),$$ (3b)

$$\underline{B} = -\left(\underline{\varepsilon}_1 \cdot \underline{\varepsilon}_2(\hat{\underline{k}}_1 \times \hat{\underline{k}}_2) - \frac{1}{2}(\hat{g} \cdot \hat{g})(\underline{\varepsilon}_1 \times \underline{\varepsilon}_2) - \hat{g} \times (\hat{g} \times \underline{\varepsilon}_1 \times \underline{\varepsilon}_2)\right); \quad \hat{g} = \hat{\underline{k}}_1 - \hat{\underline{k}}_2)$$ (3c)

and σ is the electron spin. The first term yields the standard Thompson cross-section and the second leads to the spin-dependent effects. Two things are immediately apparent: (1) the magnetic or spin term is scaled relative to the charge term by the factor $(\hbar\omega_1/mc^2)$ and (2) the terms are out of phase by 90°. The physical origin of these factors comes about due to the different interactions that are involved in the electronic and the spin scattering. When eqn. 3a is squared to produce the differential cross-section; the cross term, i.e. the term linear in the spin, will vanish unless A has an imaginary component. In this single electron model, this occurs only if the polarizations are complex; namely circularly polarized x-rays are required to pick up the leading magnetic scattering term. It should be noted that the pure magnetic term, the term proportional to $(\underline{B} \cdot \underline{\sigma})^2$, does not require circularly polarized x-rays to be observed. However, as shall be seen later this term is very small compared to the electronic scattering intensity and is not amenable to the technique that will be employed to isolate the spin from the charge scattering since it is an even function of the spin.

The explicit form for the differential cross-section, to first order in the spin, is given below following the notation of Lipps and Tolhoek.[5]

$$\frac{d^2\sigma}{d\Omega dp_z} = \frac{r_o^2}{2}\left(\frac{k_2}{k_1}\right)\left[(\Phi_o + P_\ell\Phi_\ell)J^+(P_z) + P_c\Phi_c(\underline{\sigma})J^-(P_z)\right]$$ (4a)

$$\Phi_o = (1+\cos^2\phi)+(k_1-k_2)\frac{\hbar c}{m_o c^2}(1-\cos\phi)$$ (4b)

$$\Phi_\ell = \sin^2\phi$$ (4c)

$$\Phi_c(\underline{\sigma}) = -(1-\cos\phi)\underline{\sigma}\cdot(\underline{k}_1\cos\phi + \underline{k}_2)\frac{\hbar c}{m_o c^2}$$ (4d)

where $k_1(k_2)$ is the incident (scattered) wavevector, $P_\ell(P_c)$ the linear (circular) degree of polarization and ϕ the scattering angle. The information on the momentum distribution are contained in the $J^\pm(P_z)$:

$$J^\pm(P_z) = \iint \sum_{bands} [n\uparrow(\underline{p})\pm n\downarrow(\underline{p})]dp_x dp_y$$ (5)

where $n\uparrow(\underline{p})$ $(n\downarrow(\underline{p}))$ is the spin-up (spin-down) momentum distribution. As can be seen from the above equations, with a typical incident x-ray energy of tens of kilovolts, the ratio of the spin-to-charge Compton intensities is only a few percent. This problem is further compounded in an atom where essentially all the electrons contribute to the charge scattering but only those unpaired spin electrons participate in the magnetic scattering. Hence, even in Fe, which is a relatively low z material with a high moment per atom (2.2 μ_B), the spin component of the

Compton profile is less than one percent of the entire profile. From an
experimental viewpoint then, one of the major challenges in measuring
spin dependent Compton profiles is the problem of extracting it from the
much larger electronic component. The most common technique in use is
that of a spin flip method. Data is taken first with the spins aligned
in one direction (spin-up) and then reversed (spin-down). From equations
4 it can be seen that the spin flip will have no effect on the Φ_0 and Φ_ℓ
but will change the sign of the spin dependent term. Subtraction of
spin-up and spin-down data will then eliminate the charge contribution
leaving a remainder proportional to the differences in the spin-up and
spin-down momentum distributions. This spin flipping is achieved by an
application of an external magnetic field on the sample.

Note that a sign reversal can also be produced in the spin dependent
term if the handedness of circularly polarized radiation is changed.
This technique would also serve as a means to isolate the spin dependent
Compton profile and might be particularly useful for those materials
where it is difficult to reverse the direction of the moment with external magnetic fields. One can also get the individual Compton profiles
from subtracting data taken in a spin-up (or spin-down) configuration
from a pure electronic Compton profile of the same material.[6] (A pure
electronic Compton profile can be obtained by orienting the sample so
that the moment is perpendicular to the scattering plane for example.)
Regardless of the technique employed, the key to success of these methods
is large quantitites of circularly polarized x-rays.

SOURCES OF CIRCULARLY POLARIZED X-RAYS

The pioneering experimental measurements of spin dependent momentum
distributions were performed using nuclear oriented radioactive sources.
When sufficiently cooled and placed in an external magnetic field, it is
possible to enhance the probability of those nuclear transitions that
emit circularly polarized gamma rays. Starting initially with ^{57}Co
sources (122kev) and later switching to ^{191}Os (129 kev) sources, Sakai
and coworkers[7,8] have achieved a degree of circular polarization of 0.80
and fluxes of about 10^6 photons/cm^2 onto the sample; the count rate being
limited by self heating effects of the radioactive source. With these
count rates data collection times of several hundred hours are necessary
in order to collect one magnetic Compton profile.

Needless to say, these studies were extraordinarily difficult and
progress in the area of experimental determined spin dependent Compton
profiles was slow. Clearly if any major inroads were to be made, new
sources of high intensity, circularly polarized x-rays were required.
Fortunately x-rays generated from high energy storage rings have been
able to supply that increased flux. The possibility of exploiting the
unique properties of synchrotron radiation for use in magnetic Compton
scattering studies is still in its infancy. Interestingly enough two
different methods of obtaining circularly polarized radiation has already
been successfully employed to obtain magnetic Compton profiles with
statistical accuracies comparable to (or better) that the original data
taken with gamma ray sources but with measurement times an order of
magnitude less! Figure 1 shows the polarization properties of the radiation emitted from a dipole magnet. On axis, that is in the plane of the
orbiting particle, the radiation intensity is a maximum and the polarization is nearly completely linear. As one goes above (below) the orbital
plane, the radiation becomes increasingly right hand (left hand) circular polarized; but with a loss of flux as compared to the on-axis
intensity. Cooper and his colleagues[9,10,13] working at the Synchrotron

Radiation Source (SRS) at Daresbury, UK have exploited this property of
the radiation by arranging their apparatus to see only radiation above
(or below) the orbital plane. Using this "inclined view" geometry they
have collected data on several transitions metal ferromagnets and
recently collected the first directional spin dependent Compton profiles. (See section on EXPERIMENTAL DATA for further details). With the
SRS running at 2.5 Gev and 100 ma they obtained, from a high field
superconducting wavelength shifter, an intensity at the sample of 10^9
photons/cm^2 with the degree of circular polarization about 0.80 at 60
kev.

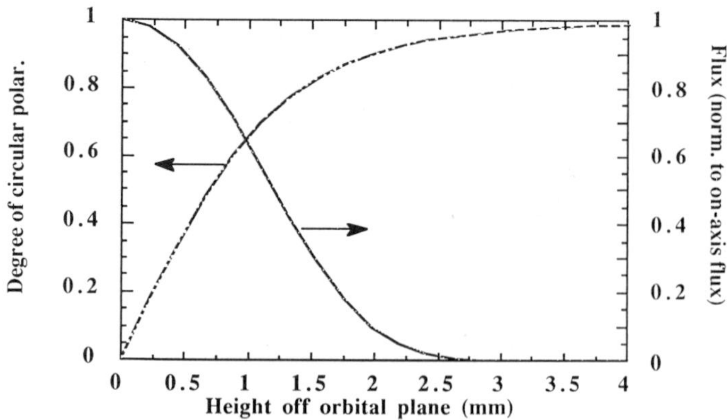

Fig. 1. Degree of circular polarization is plotted as a function of
height above (or below) the orbital plane 30 meters from an APS
band magnet. Also plotted is the height dependence of the flux
normalized to the on-axis flux valve. This has been calculated
at 39 keV, twice the critical energy of the bend magnet radiation. The effects of finite emittance and slit size have been
neglected in these calculations.

The standard technique for increasing photon flux at storage ring
sources is through the introduction of insertion devices; wigglers or
undulators. Insertion devices are magnetic structures with alternating
magnetic field directions perpendicular to the orbital plane of the particle. As the particle travels through the insertion device it is forced
to oscillate back and forth (The magnetic fields are carefully trimmed so
that the particles net deflection after going thought the device is
zero.) In the case of a wiggler, one can think of each pole as a source
point of radiation, so that if a device has N poles, the photon flux is N
times that of a single dipole source point. Unfortunately these alternating fields, the very origin of the increased flux, destroy the circular polarization above and below the orbital plane. Because they produce
alternately righthand and lefthand polarization above the orbital plane,
the resulting off-axis radiation from wigglers is almost totally unpolarized. However, one can make use of the ample quantity of linear polarized x-rays by conversion of the linearly polarized radiation to circular
polarized radiation via x-ray phase plates. Mills[11] has successfully
produced circularly polarized x-rays at 40 kev with a degree of circular

polarization between 0.40 and 0.70 with an x-ray phase plate and used
these x-rays to collect spin dependent Compton profiles of the transition
ferromagnets (Fe, Co, and Ni) and Gd.[12] This data was collected at the
Cornell High Energy Synchrotron Source (CHESS) on a dipole source (5.5
Gev and 25 ma) with an incident beam intensity of greater than 10^9
photons/cm^2.

EXPERIMENTAL SET-UP

   The general experimental set-up is shown in Figure 2. As can be
seen from equations 4 for a given incident x-ray energy, the spin
contribution to the cross-section is maximized in a back scattering geometry with the spin direction parallel to the incident beam. (See Fig.
3) An electro-magnet is used to align and reverse the spins in the
sample. Depending on the details of the experiment, either a reflection
or transmission geometry can be used.

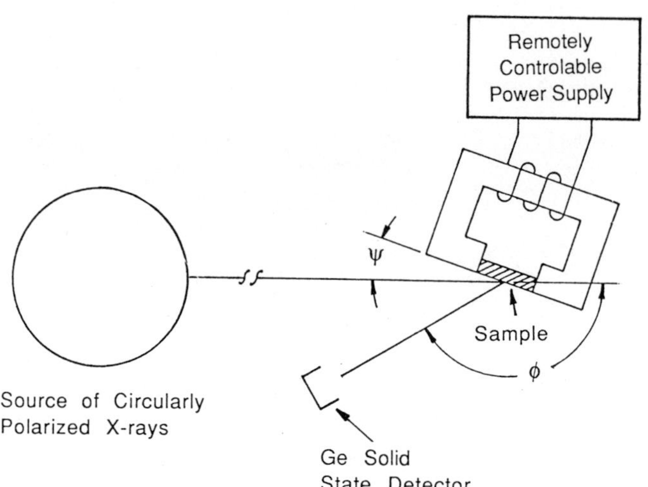

Fig. 2   Schematic of the experimental layout for magnetic Compton
          scattering. Shown here is a reflectance geometry (entrance and
          exit beam on same side of sample); however, in some instances a
          transmission geometry (entrance and exit beams on different
          sides of the sample) is more advantageous.

   To date all magnetic Compton scattering profiles have been collected
with energy dispersive solid state detectors due to the dearth of magnetically scattered photons. This has necessarily resulted in experimental
data being of 'modest' momentum resolution; typically 0.7 to 0.9 a.u.
(Recall that Compton profiles are generally only several au's wide!).
With current detector technology, improvements in momentum resolution
will be difficult to achieve unless large increases in the flux of circularly polarized x-rays can be made available so that SSD's can be replaced with energy dispersive crystal analyzers.

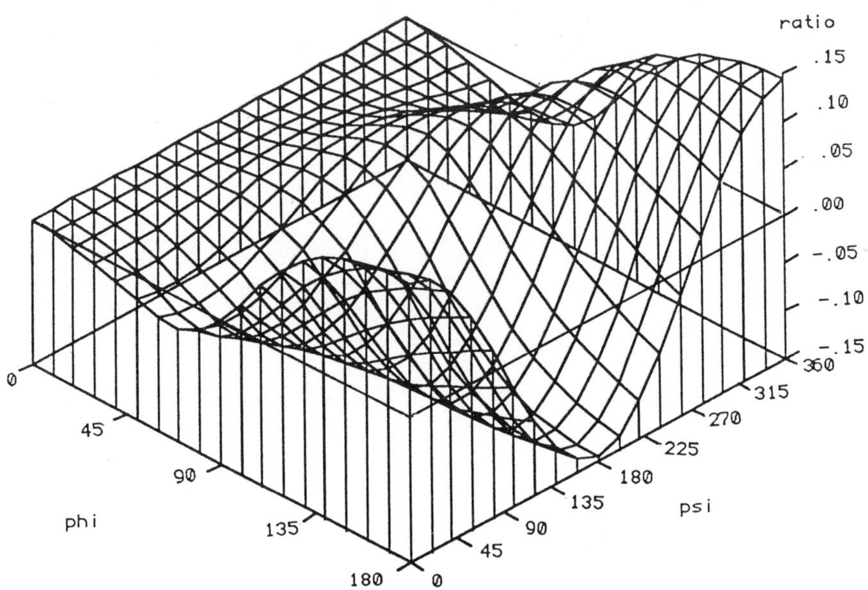

Fig. 3. The magnetic-to-electronic cross section ratio for a single
electron plotted as a function of scattering angle ϕ for several
different spin orientations ψ. The spin is assumed to lie in
the scattering plane, $\psi=0$ is along the incident beam direction. These were calculated for an incident x-ray energy of 40
keV. The ratio scales directly with the photon energy.

EXPERIMENTAL DATA

Because of the experimental difficulties involved, to date magnetic Compton scattering experiments have been limited systems of
relatively high magnetic moments per atom and low atomic numbers, i.e.
Fe [7,8,9,12,13], Co [12,13], Ni [12] and $MnFe_2O_4$ [8], with the single exception
being gadolinium. A representative sampling of that data is shown in
Figs. 4, 5 and 6. All data from the pure ferromagnetic transition metals
show a dip in the momentum distribution about $p_z=0$; a manifestation of
the relative abundance of the negative spin density far from the atomic
core location. This dip has been also been predicted from theoretical
studies [14,21] and the source of this reverse magnetization has been associated with the sp electrons. In the calculations for Fe, where a majority of the theoretical work has been carried out, the negative spin components is consistently underestimated. Some recent work by Cooper and
colleagues [10] have indicated where the discrepancy between the measurements and the calculations may lie. They have recorded, for the first
time, directional Compton profiles and have been able to compare them
directly with calculations, identifying the crystal orientations where
the major differences appear. (See Fig. 6).
In an attempt to contrast the difference in the profiles between
intinerate and localized spin electrons, Sakai and Sekizawa [8] have measured the spin dependent Compton profiles of Fe and ferrimagnetic Mn
ferrite. Contrary to what might be expected in a ferrimagnetic insulator, they found that Mn ferrite also displayed the characteristic dip
about $p_z=0$ normally associated the the itinerate nature of the unpaired
spin electrons. They have conjectured that this may be due to different
momentum distributions of the 3p and 4s electrons between Mn^{+2} and Fe^{+3}.

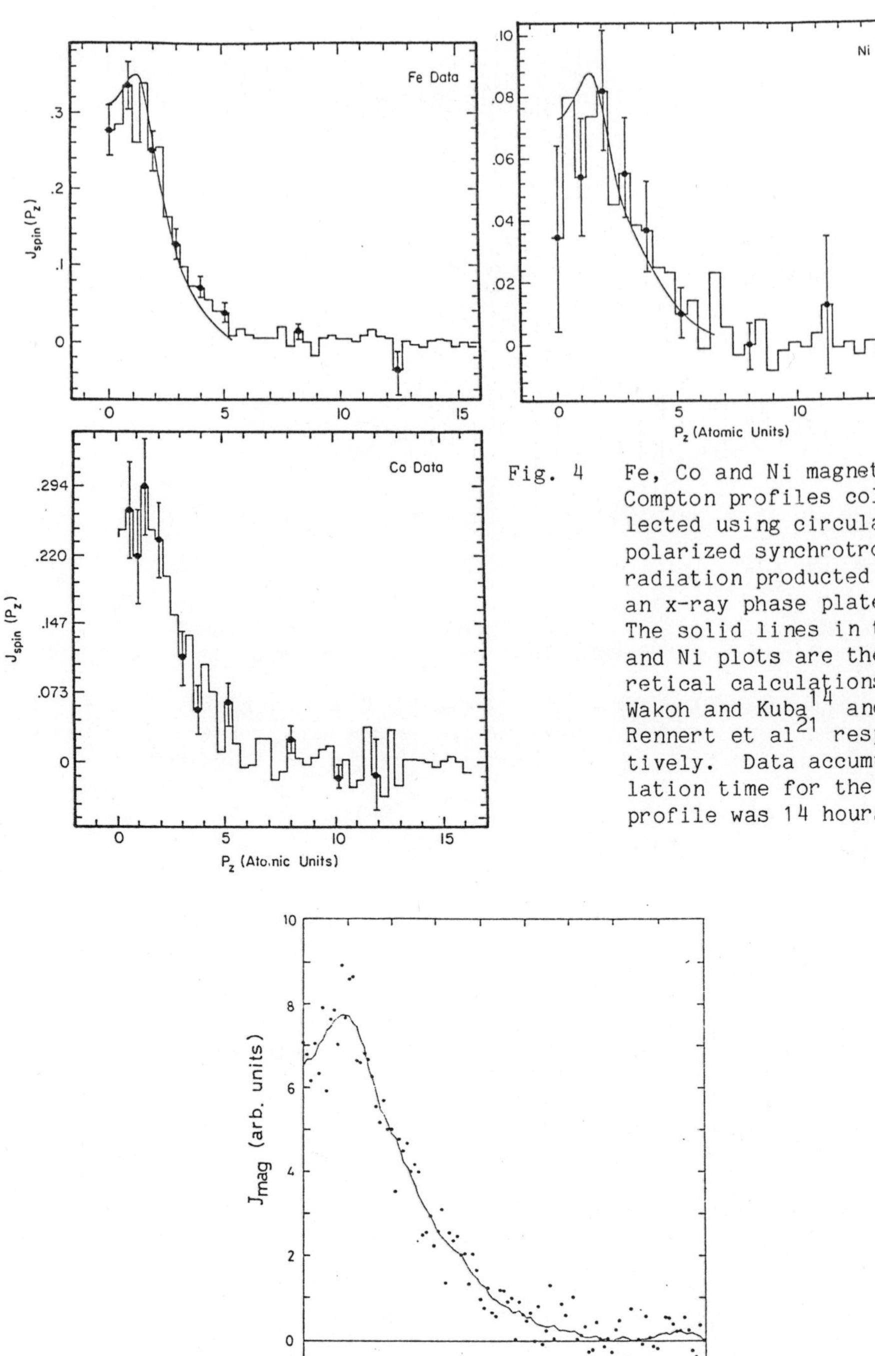

Fig. 4   Fe, Co and Ni magnetic Compton profiles collected using circular polarized synchrotron radiation producted with an x-ray phase plate.[12] The solid lines in the Fe and Ni plots are theoretical calculations by Wakoh and Kuba[14] and Rennert et al[21] respectively. Data accumulation time for the Fe profile was 14 hours.

Fig. 5   Magnetic Compton profile of $MnFe_2O_4$ collected using a radioactive [191]Os source.[8] The solid line is the experimental data after a smoothing routine had been applied. Data accumulation time was 333 hours.

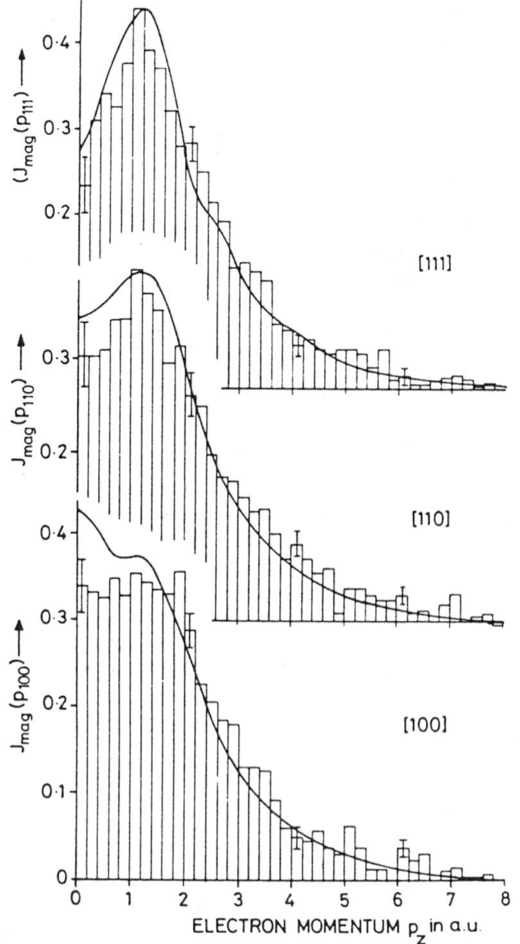

Fig. 6   Directional magnetic Compton profiles
for Fe taken with off-axis synchrotron
radiation.   The solid line is an APW
calculation[10] convoluted with the experimental resolution function.   Data
collection time for each profile as
about 20 hours.

Preliminary investigations into the spin Compton profile of Gd, a
rare earth ferromagnet with atomic-like magnetism, has indicated that[6,12]
the dip about the origin is absent in this material; showing infact that
all spin-dependent Compton profiles don't look alike.

FUTURE DIRECTIONS

Unless more potent sources of circularly polarized x-rays become
available so that detailed and systematic studies of materials can be
undertaken, magnetic Compton scattering experiments will soon be re-
legated to that class of experiments which I call proof of principle;
i.e. interesting from a experimental viewpoint but of little use to the
better understanding of the physical properties of the sample.   I believe

that storage ring sources can provide that increase in flux required to take this technique from a proof of principle status to a useful and incisive tool for the study of the momentum distributions of the unpaired spin electrons in condensed matter systems. The next generation of synchrotron radiation sources, the Advanced Photon Source (APS) to be built at Argonne and the European Synchrotron Radiation Facility (ESRF) being built at Grenoble, are high energy storage rings (6-7 Gev) that have been specifically designed for use with insertion devices. Both these facts are important to Compton scattering researchers; the high beam energy meaning increased flux at shorter x-ray wavelengths and the compatibility with insertion devices meaning the possibility of tailoring the emitted radiation to the specific needs of the experimenter. Many advances are being made in the area of exotic insertion devices designed explicitly for the production of circularly polarized x-rays. Helical[15] undulators, crossed planar undulators,[16] crossed-retarded field undula-tors and wigglers[17] and asymmetric wigglers[18] have all been proposed for the production of intense beams of circularly polarized x-radiation. In the very near future a distorted helix wiggler is to go on the 6 Gev accumulator ring, TRISTAN, at KEK in Japan for the production of high fluxes of elliptically polarized x-rays at energies between 40 and 60[19] kev. We excitedly await the the results from this new insertion device design.

The construction of these new storage ring sources has precipitated a renewed interest in x-ray optics; in particular short wavelength x-ray optical components. Many of these novel components can be installed at currently operating synchrotron radiation sources. For instance, only recently has work been done on x-ray phase plates that operate at ener-gies above 40 kev and this study was one a dipole magnet source which had a critical energy (half power point) of ~10 kev. The transfer of the same device to the wiggler beamline at CHESS should afford a 50 fold increase in intensity due to the higher critical energy (29 kev) of the wiggler. (One of the experimental obstacles of working on these high field insertion devices is the handling of the power they put out, often many kilowatts!) Focusing optics can produce considerable increases in flux on the sample over that of the flat double crystal monochrometers available on most x-ray beamlines. Here the problem is one of producing efficient focusing optical components capable of working at 40 kev and above. An optimized match of the band-pass of the monochromator to the resolution of the detector can significantly improve the quality of mag-netic Compton scattering profiles. Given the current energy resolution of solid state detectors (several hundred ev in the energy region of interest) experimental momentum resolution is not compromised if the bandpass of the monochromated radiation is increased from the current value of several ev's (typical of perfect single crystal monochromators) to that approaching the detector resolution. In this fashion an order of magnitude increase in flux on the sample realizeable with virtually no effect on the resultant momentum resolution. To increase the momentum resolution over what one can obtain with a SSD, dispersive analyzing crystals are required. Analyzing crystals used in a transmission mode in conjunction with a 1D detector (film, image plates, wire detectors, CCD's) can provide momentum resolutions to better that 0.1 au but at an exorbitant cost in counting rates. The effective operation of x-ray phase plates, focusing optics, wide band-pass monochromators and analyzing crystals (or perhaps even combinations of these devices) at short x-ray wavelengths presents a stimulating challenge to designers of modern x-ray optical components. Concurrent with improvements in the area of x-ray production and optical components, advances in semiconductor micro-fabrication techniques are allowing detectors to be produced with resolutions 10 times better that currently available SSD's.[20] The converge of all these technologies, insertion device development, x-ray optical components, and detector development in the next several years will most certainly produce new and thought providing studies in the field of spin-dependent momentum distribution measurements.

ACKNOWLEDGEMENTS

The author would like to thank both Profs. Sakai and Cooper for their discussions and insights into the field of spin dependent magnetic Compton scattering. The author would also like to thank the organizers of the Workshop on Momentum Distributions for inviting him to participate in their meeting. Support for this work was provided by the Department of Energy under contract number No. W-31-109-ENG-38.

REFERENCES

1. "Compton Scattering," B. W. William, ed., McGraw Hill, New York 1977.
2. Cooper, M., Rep. Prog. Phys., 48 p. 415, 1985.
3. Eisenberger, P. and Platzman, P., Phys. Rev. A, 2, p. 415, 1970.
4. Platzman, P. and Tsoar, N., Phys. Rev. B, 2, p. 3556, 1970.
5. Lipps, F. W. and Tolhoek, H. A., Physica xx, p. 395, 1954.
6. M. Cooper, private communication.
7. Sakai, N. and Ono, K., Phys. Rev. Letts. 37, p. 351, 1976.
8. Sakai, N. and Sekizawa, H., Phys. Rev. B 36, p. 2164, 1987.
9. Cooper, M. J., Laundy, D., Cardwell, D. A., Timms, D. N., Holt, R. S. and Clark, G., Phys. Rev. B 34, p. 5984, 1986.
10. Cooper, M. J., Collins, S. P., Timms, D. N., Brahmia, A., Kane, P. P., Holt, R. S. and Laundy, D., Nature, 333, p. 151, 1988.
11. Mills, D. M., Nuc. Inst. and Meths. A226, p.531, 1988.
12. Mills, D. M., Phys. Rev. B. 36, p. 6178, 1987.
13. Timms, D. N., Brahmia, A., Collins, P., Collins, S. P., Cooper, M. J., Holt, R. S., Kane, P. P., Clark, G. and Laundy, D., J. Phys. F, 18, p. L57, 1988.
14. Wakoh, S. and Kubo, Y., Jour. of Mag. and Mag. Matls. 5, p. 202, 1977.
15. Kincaid, B. M., J. Appl. Phys. 48 p. 2684, 1977.
16. Kim, K. J., Nuc. Inst. and Methods, A219, p. 425, 1986.
17. Onuki, H., Nuc. Inst. and Methods. A246, p. 94, 1986.
18. Goulon, J. Elleaume, P. and Raoux, D., Nuc. Inst. and Methods, A254, p. 192, 1987.
19. Yamamoto, S. Shiga, T., Sasaki, S., and Kitamura, H., to be published in Rev. Sci. Inst.
20. McCammon, D., Juda, M. Zhang, J., Holt, S. S., Kelly, R. L., Mosley, S. H. and Szymkowiak, A. E., Jap. Journ. of Applied Phys., 26, Suppl. 26-3 (1987).
21. Rennert, P., Carl, G. and Hergert, W., Phys. Stat. Sol. (b) 120, p. 273, 1983.

POSITRON AND POSITRONIUM MOMENTUM SPECTROSCOPY IN SOLIDS: AN OVERVIEW

Stephan Berko

Department of Physics
Brandeis University
Waltham, MA 02254

Abstract

After a brief introduction to the experimental techniques used in positron annihilation spectroscopy and an outline of the underlying theory, we describe the applications of the angular correlation of annihilation radiation method to the study of the
momentum density of electrons in solids, and in particular of the Fermi surfaces of
some pure metals and alloys. The use of low energy positron beams for surface studies
will also be briefly discussed, with emphasis on the newly developed angle-resolved
positronium emission spectroscopy.

Introduction

The behavior of positrons in condensed matter has been the subject of intense
experimental and theoretical investigation during the last decades, and the use of
positrons as a probe of electronic structure has been well documented and reviewed.[1,2]
By 1988, positron (e^+) annihilation spectroscopy (PAS) in solids could be divided into
two main categories: a) bulk studies, using mainly fast positrons from radioactive β^+
sources, and b) surface and near-surface studies, using the more recently developed
monochromatic, variable energy e^+ beams.[3,4] Information about the electrons sampled
by the positron is gained by studying the various properties of the annihilation quanta
or, as sometimes is the case with slow e^+ beams, by detecting directly the scattered
positrons.

In bulk measurements, the fast positrons injected into the sample from a long-
lived β^+ source such as ^{22}Na, ^{58}Co or ^{64}Cu lose their energy by ionizing collisions and
phonon scattering, and reach thermal equilibrium with the solid prior to annihilation,
penetrating[5] the sample well into the bulk to mean depths of 10-100 μm. We note that
there is no exclusion principle between the electron and its anti-particle, the positron.
In well annealed metals the thermalized positron takes the form of a delocalized Bloch
wave, with maxima at interstitial regions due to core repulsion, while, if defects are
present, the positron can localize around low density defects such as vacancies, dislocations and microvoids. The same core repulsion leads to the possibility of slow positrons
being ejected from a surface corresponding to a negative work function[3,4]–the effect
used for the production of slow e^+ beams.

The annihilation process, well understood theoretically (QED), results in two quanta (2γ) from e^+–e^- spin singlet overlap, or in three quanta (3γ) from spin triplet overlap. In metals 2γ decays predominate, since the 3γ decay is $\sim 10^3$ times less probable. In some insulators and in liquids[6] the positron can capture an electron prior to annihilation, forming the hydrogenic e^+–e^- positronium (Ps) atom.[7] Ps annihilates subsequently, with a characteristic signature, and is thus detectable by studying the annihilation γ-s.

2. Experimental Techniques

Four typical experiments can be performed with the annihilation γ-s in order to gain information about the electrons the e^+ annihilates with: a) 2γ angular correlation experiments; b) the measurement of the energy spectrum of one of the γ-s from 2γ decay; c) e^+ lifetime experiments; and d) $3\gamma/2\gamma$ yield measurements.

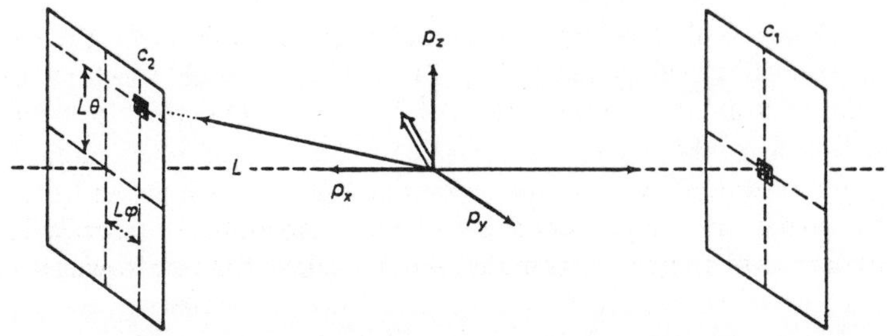

Fig. 1. Geometry of the 2γ angular correlation experiment. The double arrow represents the momentum $\boldsymbol{p}$ to be conserved by the two annihilation γ's.

a) The 2γ angular correlation of annihilation radiation (ACAR) depends, by momentum and energy conservation, on the momentum distribution $\rho^{2\gamma}(\boldsymbol{p})$ of the annihilating e^+–e^- pair. Given a typical electronic momentum $\boldsymbol{p}$ ($|\,\boldsymbol{p}\,|\sim 10^{-2}mc$) (Fig. 1), the angle between the two γ-s deviates only by a few milliradians from anti-collinearity, since each γ has momentum $\sim mc$ (energy $\sim mc^2$). Thus θ and ϕ are directly related to the $p_z (= \theta mc)$ and $p_y (= \phi mc)$ components of $\boldsymbol{p}$. The two-dimensional (2D) ACAR spectrum is measured by γ detectors at a large distance L from the sample (typically 5-15 meters). Using multiple γ detectors or position sensitive γ cameras,[1,2] one obtains the projection of $\rho^{2\gamma}(\boldsymbol{p})$ (i.e. p_x is not measured):

$$N(p_y, p_z) = \left[\int \rho^{2\gamma}(\boldsymbol{p})dp_x\right] \otimes R(p_y, p_z) \tag{1}$$

where $\otimes$ R corresponds to the convolution by the 2D angular resolution R of the setup. In the examples shown p_y and p_z are plotted in mrad $\equiv$ mc $\times\ 10^{-3} \simeq 0.137$ a.u. Typical resolutions obtainable are $\sim (0.2\text{-}0.7)$ mrad. Many of the earlier ACAR experiments were performed with "long-slit" detectors, thus integrating over an additional component of $\rho^{2\gamma}(\boldsymbol{p})$ (i.e. $N(p_z) = \iint \rho^{2\gamma}(\boldsymbol{p})dp_x dp_y$).

b) The momentum component along the direction of the annihilation γ-s is reflected in a Doppler shift of a few keV from their mc^2 (511 keV) energy; thus $N(p_x)$ can be measured by observing the energy spectrum of one of the γ-s, usually with a high efficiency Ge spectrometer. Since the effective momentum resolution of such spectrometers is an order of magnitude poorer than that of an ACAR setup, the technique is not used in Fermi surface measurements. Because of simplicity and fast data acquisition, Doppler profile spectroscopy is however often used in defect studies, where the finer details of $\rho^{2\gamma}(\boldsymbol{p})$ are not of paramount importance.[1]

c) The positron lifetime depends sensitively on the electronic density at the site of the positron. In solids one obtains e^+ lifetimes of $\tau \sim 10^{-10} - 10^{-9}$ s, compared to the typical thermalization times of $10^{-12} - 10^{-11}$ s. Lifetimes are obtained with fast nuclear counting techniques using as a timing signal a nuclear γ (1.28 Mev) that follows promptly the emission of a positron from the ^{22}Na β^+ source; the precision of a few percent can be obtained for a $\sim 10^{-10}$ s lifetime.

d) The measurement of the $3\gamma/2\gamma$ yield is important when Ps is present or, if the e^+ and e^- spins are relevant, as in the studies of magnetically ordered samples using spin-polarized positrons.[1]

In the following paragraphs we focus on 2D ACAR theory and experiments in solids[8]; because of limited space we only discuss a few examples – for thorough, up-to-date reviews see Refs. 1, 2, and references therein.

3. Various Applications of the 2D ACAR Technique

In Fig. 2 we assembled four representative 2D ACAR spectra corresponding to four very different physical systems. Fig. 2a represents one of the first full 2D ACAR spectra[9] obtained from an oriented crystal of Al; the nearly perfect inverted hemisphere (in the extended zone) reflects the "nearly-free-electron" nature of the conduction band in Al. High precision ACAR experiments in metals such as Al, Cu and the alkali metals are used to check the importance of $e^+ - e^-$ many-body correlations and to study the dynamics of positron thermalization as well as the positron's effective mass.[10]

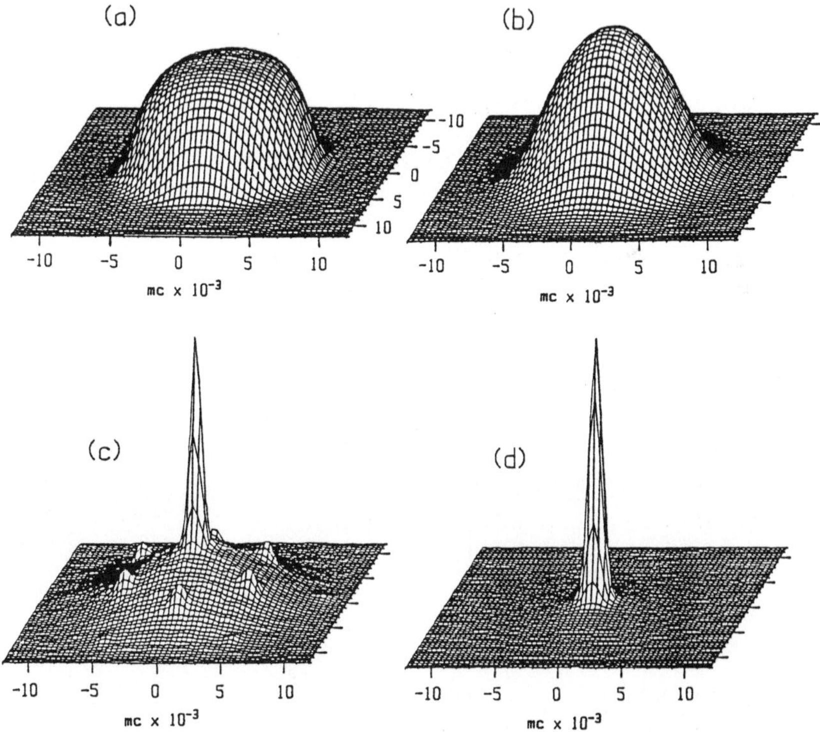

Fig. 2. 2D ACAR spectra in a) Al at 70 K, delocalized positron; b) Al at 900 K, positron localized in thermally activated vacancy; c) oriented crystal of SiO_2, delocalized positronium and d) positronium localized in radiation produced microvoid in vanadium.

Discontinuities in the ACAR spectra reflect the position and topology of the Fermi surface. Since the e^+ annihilation technique does not require long coherence lengths as do standard solid state Fermiology techniques (de Haas-van Alphen, magneto-resistance, etc.), 2D ACAR experiments are uniquely well suited for Fermi surface studies of substitutionally disordered alloys[1,2] and of metals where high sample purity is impossible to achieve, or when there are low temperature phase transitions.

In Fig. 2b we plot the 2D ACAR spectrum[11] from an oriented single crystal of Al raised to a temperature of ~ 900 K; at this elevated temperature sufficient number of vacancies are thermally created to trap almost all the positrons. A detailed study of the 2D ACAR shapes as a function of temperature was used to study the equilibrium vacancy ensemble in Al, and to test theoretical models of electronic structure around vacancy sites.[11] Positron annihilation spectroscopy is extensively used in metallurgy, particularly with the lifetime and Doppler profile techniques applied to defect studies[1,2] such as vacancy formation energies, dislocations, annealing cycles, radiation damage, etc. The detailed use of high precision 2D ACAR experiments such as in the above example[11] awaits further exploration.

Figure 2c represents the 2D ACAR spectrum[9] exhibited by an oriented crystal of SiO_2. The narrow $p=0$ momentum component is characteristic of thermalized positronium annihilation. The satellite peaks appear exactly at the reciprocal lattice vectors of the SiO_2 crystal, and indicate that positronium is formed in a delocalized Bloch state[12], as in some other insulators and in some ionic crystals; the satelite peaks correspond to the higher momentum components of the excitonic-like positronium Bloch wave. High precision 1D ACAR work has been recently reported[13] on the temperature dependence of the Ps momentum distribution in alkali halides, that even permits the measurement of the effective mass of the Ps Bloch state.

In Fig. 2d we plot the 2D ACAR spectrum from positrons annihilating in vanadium that has a high concentration ($\sim 10^{16} cm^{-3}$) of microvoids produced by neutron bombardment.[14] It is found that nearly 50% of all positrons injected into the metal end up annihilating via the narrow component, indicating high efficiency Ps formation in the void. The 2D resolution is narrower than the width of this component, permitting the momentum distribution measurement of the Ps due to its confinement in the microvoid (~ 60Å dia.). It was pointed out[14] that "the observation of Ps formation in voids of V is not only of interest to void physics and thus to radiation damage problems in metals, but can lead to the possibility of voids becoming a "micro-laboratory" for the study of the unusual many-Ps state of matter." Should the future intensity of slow e^+ beams be increased substantially,[4] it will indeed become possible to produce more than one Ps atom at any time within a void, and thus study Ps–Ps interactions. With a high density Ps gas even Bose condensation could be achieved, as discussed by P. M. Platzman.[15] A very high resolution 2D ACAR measurement would then allow one to measure the Bose-condensate fraction from the amplitude of the δ function peak expected, in analogy to the neutron scattering experiments[16] from liquid He that form the subject of several papers in this volume.

Although many interesting and important ACAR experiments have been performed in amorphous systems,[17] the bulk of 2D ACAR work to date is in oriented single crystals. We outline briefly the theory of such experiments.

4. Theoretical Considerations

Since the earliest theoretical formulation of positron annihilation in metals[18] it was clear that positrons are not "ideal" probes to study electrons in solids – particularly in metals – because of the strong e^+–e^- attraction and many-body correlation problem. As pointed out by Brandt,[19] instead of the "classical" exchange and correlation hole of the interacting electron gas, in the case of a positron immersed in an electron gas the Coulomb attraction produces a "Coloumb cusp" in the electron density at the positron site. It turns out that, in spite of these strong correlations, the momentum density is reasonably well described by the independent particle model (IPM); of greatest importance is the fact that if a Fermi surface is present, its position is not shifted by the e^+–e^- correlations (at least within perturbation theory).

4.1. <u>The interacting one positron–many electron gas</u>. The theory of annihilation probability of a positron in a homogeneous electron gas has had a long history in many-body physics; the various approaches have been reviewed thoroughly by Carbotte,[20] Arponen and Pajanne,[21] and Stachowiak.[22]

The fundamental formalism of the partial annihilation rate $\Gamma(\boldsymbol{p})$ was given by Ferrell[23] in terms of the two-particle electron-positron Green's function G_{ep}. $\Gamma(\boldsymbol{p})$ is directly related to the e^+–e^- momentum density by $\Gamma(\boldsymbol{p})d\boldsymbol{p} = \pi r_o^2 c/(2\pi)^3 \rho^{2\gamma}(\boldsymbol{p})d\boldsymbol{p}$, where $r_o = e^2/mc^2$. The total e^+ annihilation rate (inverse lifetime λ) is then given by $\Gamma = \lambda^{-1} = \int d\boldsymbol{p}\Gamma(\boldsymbol{p})$. The first study of $\Gamma(\boldsymbol{p})$ vs the average electron gas density (i.e. vs r_s) was obtained by Kahana, and Kahana and Carbotte[20] using Bethe-Goldstone type ladder diagram summation. Their result indicated a substantial piling up of electrons at the site of the positron, leading to large annihilation rate increases over the IPM prediction, and a characteristic $\Gamma(\boldsymbol{p})$ shape with increased enhancement as $\boldsymbol{p} \to \boldsymbol{p}_{Fermi}$ (Fig. 3). Problems encountered in this theory with decreasing gas densities have been studied in some detail by Stachowiak et al.[22] Arponen and Pajanne[21] use a boson formalism to analyse the e^+–e^- problem; their result agrees with Kahana and Carbotte's for the high density gas ($r_s \sim 2$), but differs substantially in the shape of $\Gamma(\boldsymbol{p})$ for lower densities. Positronium formation is predicted in most theories only for low densities ($r_s > 6$). We should also mention the completely different approaches based on variational theories[21,23]; Maldague[24] for example starts by constructing trial wavefunctions using effectively a correlated Ps interacting with the electron gas.

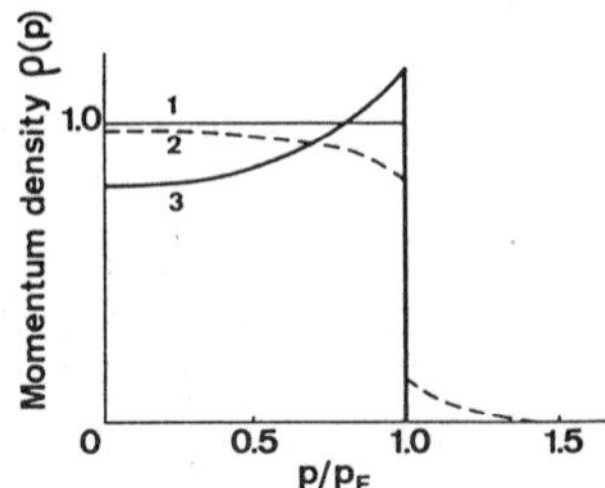

Fig. 3. Theoretical momentum densities $\rho(\boldsymbol{p})$ for free electrons (1) and interacting electron gas (2), and $\rho^{2\gamma}(\boldsymbol{p})$ for the one positron–many electron gas (3) normalized to equal volume, $r_s = 4$.

These e^+–e^- gas theories were tested by studying the lifetime in simple (nearly-free-electron) metals (i.e. Γ vs r_s) as well as the ACAR spectra in such metals to check the $\Gamma(\boldsymbol{p})d\boldsymbol{p}$ predictions.[1,10] It is customary in real metal systems to compute $\Gamma(\boldsymbol{p})$ and Γ from independent particle model (IPM) theories and compare them with the many-body results; one usually speaks of the many-body "enhancement" of Γ and the "momentum dependent enhancement" in $\Gamma(\boldsymbol{p})$.

4.2 <u>The IPM equations for periodic potentials</u>. In the independent particle model the e^+–e^- momentum density is given by

$$\rho^{2\gamma}(\boldsymbol{p}) = \text{const.} \sum_j \left| \int d\boldsymbol{r} \exp[-i\boldsymbol{p} \cdot \boldsymbol{r}] \, \psi_+(\boldsymbol{r}) \, \psi_{-j}(\boldsymbol{r}) \right|^2 \tag{2}$$

where $\psi_+(r)$ is the positron ground-state and $\psi_{-j}(r)$ stands for the jth electron. One notes that with a constant positron wavefunction $\rho^{2\gamma}(p)$ reduces to the true electronic momentum density $\rho(p)$ (in IPM), as measured in a Compton scattering experiment.[25]

In a periodic potential (i.e. a crystaline sample), one obtains

$$\rho^{2\gamma}(p) = \sum_{k,\,j} n_j(k) \sum_g \mid A_j(k,g) \mid^2 \delta(p - k - g) \qquad (3)$$

Here the A-s are the Fourier coefficients of the positron $(k^+=0)$ and electron $(k, j$ band index) Bloch wave product; $n_j(k)$ is the occupation number of the jth band, i.e. the stepfunction $\theta(E_f - E_{k,j})$ at T=0, reflecting the Fermi surface, if any. One thus measures the Fourier transform of the electrons sampled by the positron in the extended zone scheme. An analysis of $\rho^{2\gamma}(p)$ (as well as $\rho(p)$) for various simple bands (NFE and tight binding) have been given a long time ago.[26] The contribution to $\rho^{2\gamma}(p)$ due to non-zero reciprocal lattice vectors g are called "Umklapp annihilations". To emphasize the Fermi surface one can form[27] the periodic "reduced momentum density"[8] $\tilde{\rho}^{2\gamma}(p) = \sum_g \rho^{2\gamma}(p - g)$; in the reduced zone one obtains:

$$\tilde{\rho}^{2\gamma}(p) \;\propto\; \sum_j n_j(k)\delta(p - k) \left\{ \sum_{g'} \mid A_j(k,g') \mid^2 \right\} \qquad (4)$$

The g' sum in the {} brackets reduces[28] to $\int |\psi_{k,j}(r)|^2 |\psi_+(r)|^2 \, dr$ i.e. it is proportional to the IPM annihilation rate with the $\psi_{k,j}$ electron in real space representation. In the approximation that this term varies only slowly with k, $\tilde{\rho}^{2\gamma}(p)$ reflects mainly the Fermi surface–the so called LCW theorem.[27] Since in many high precision 2D ACAR experiments one forms $\tilde{\rho}^{2\gamma}(p)$ or its equivalent 2D projections, in order to study Fermi surfaces, the approximate validity of the LCW theorem has been recently re-studied in detail in several papers.[29,30,31] The ground state $(k^+=0)$ positron Bloch state $\psi_+(r)$ is usually computed with various band-theoretic techniques, using the periodic Coulomb potential obtained from a self-consistent electron charge density computation.[1,2]

4.3 e^+-e^- many-body effects and band theory. In order to understand in detail the high precision 2D ACAR spectra, various theories have been developed to apply the e^+-e^- many-body techniques from the jellium model to the more realistic case of periodic crystals. In nearly free electron metals it was predicted some time ago[32] that the e^+-e^- effects will <u>not</u> alter appreciably the intensity ratio of the "Umklapp annihilation" components $(g\neq0)$ of the IPM predictions (Eq. 3). This has been born out experimentally in the study of the high momentum components (HMC) of aluminum[33] and the alkali metals.[34] In Fig. 4 and Fig. 5 we plot the enlarged 2D ACAR spectra for Li[34] and Al,[8] showing the good agreement in the shape and size of the HMC between experiment and band theoretic computations that include a $k^+=0$ positron state.

To incorporate e^+-e^- correlations directly into band computations, a local density functional theory has been recently developed by Chakraborty[35] and Boronski and Nieminen,[36] based on a two-component charged fermion formalism. Except for the original applications to Al and to vacancy trapped positrons, these theories need to be tested in more detail on more complex metallic systems. Several more empirical theories of the momentum dependent e^+-e^- correlation effects have been introduced by various authors, used mainly to study transition metals.[37,38,39] In the careful analysis of the IPM band theoretical fits to various 2D ACAR spectra it is found that the e^+-e^- many-body effect is different between "s-p" and "d" character bands. In particular, one notes a many-body "dehancement" of the annihilation rate for d bands in transition

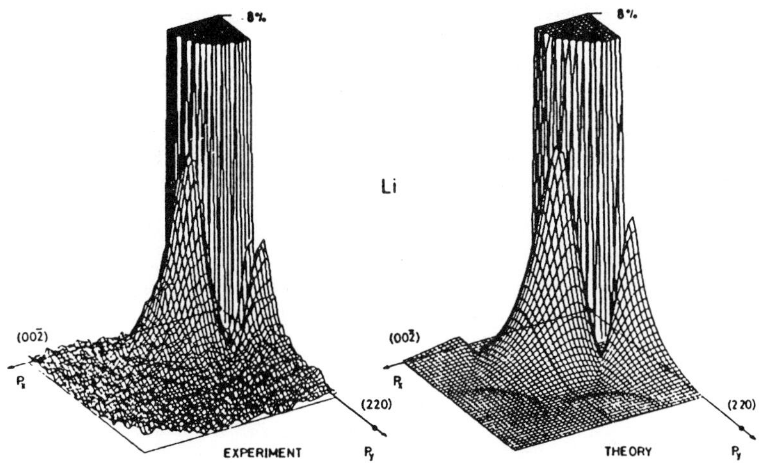

Fig. 4 High momentum component in the experimental and (KKR) theoretical 2D ACAR spectra in lithium (Ref. 34).

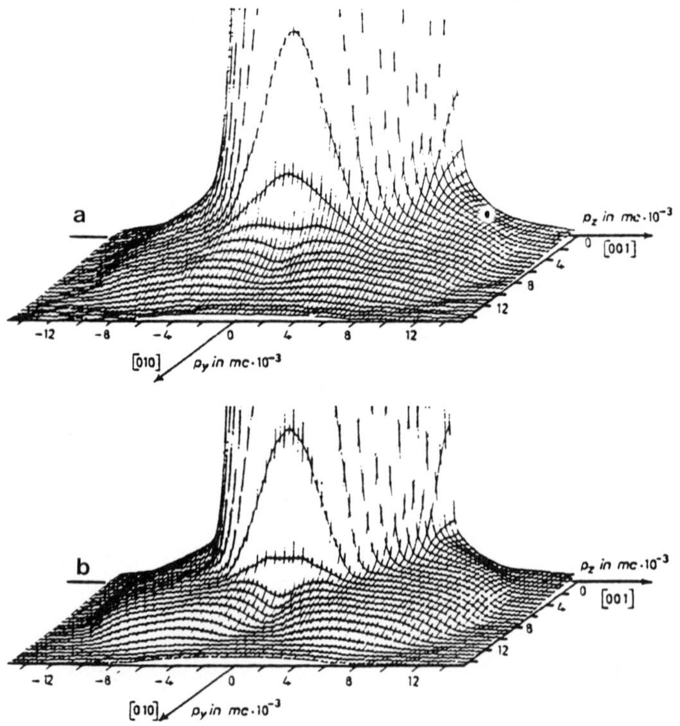

Fig. 5 High momentum region of a) experimental and b) (OPW) theoretical 2D ACAR Al spectra. The experimental spectrum corresponds to a vertical enlargement of the scale of Fig. 2a (from Ref. 8).

metals,[37,38] as opposed to the "enhancement" found in "s-p" bands. It is clear that more fundamental theoretical work will be required to clarify the details of the e^+-e^- correlation effects. Such work is particularly necessary if one is to use the 2D ACAR technique to study in detail the highly correlated electronic systems such as heavy Fermions and the high T_c superconductors.

5. 2D ACAR and Fermi Surface Studies

5.1 <u>Pure metals and compounds</u>. In order to search for discontinuities due to a Fermi surface (FS), one either studies the 2D ACAR spectra and compares them directly to theoretical predictions (when available), or one folds the spectra by the LCW technique, obtaining projections of the reduced density–Eq. 4. A more complete procedure is to <u>reconstruct</u> $\rho^{2\gamma}(p)$ from various crystallographic projections, using tomographic methods.[1,2,40] In Fig. 6 we plot contours of such a reconstruction for pure Cu in two planes, used as a test case for the technique.[8] The observed discontinuities fit the known Cu FS to within 3%. A cross section of this data along different directions reveals directly the e^+-e^- many-body enhancement as $p \rightarrow p_f$, as in Fig. 3. Such measurements allow a quantitative test of the e^+-e^- correlation effects–for a review see Ref. 22.

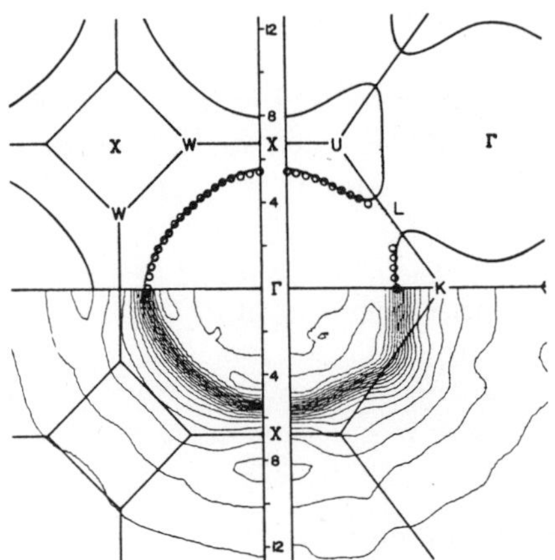

Fig. 6. Contour map of the reconstructed $\rho^{2\gamma}(p)$ in Cu (bottom half) for the (100) (left side) and the (110) (right side) planes (Ref. 8). Top half: the Fermi surface of Cu (solid line) and the measured loci of discontinuities from $\rho^{2\gamma}(p)$ (open circles). The energy gap at the neck (L) produces a smearing of $\rho^{2\gamma}(p)$ into the next zone, as predicted theoretically (Ref. 26) in the general case (Ref. 8).

The success of the $\rho^{2\gamma}(p)$ reconstruction technique can be best seen in a recent high precision study of vanadium by the University of Geneva positron group.[40]

Fig. 7. The reconstructed experimental $\rho^{2\gamma}(p)$ of vanadium (top), on a plane as shown in the insert, compared to the theoretical momentum density (bottom). Other cuts have also been reconstructed and compared with theory. The complex $\rho^{2\gamma}(p)$ was reconstructed using only four crystallographic projections[40] (i.e. 2D ACAR spectra). A good agreement is obtained between theory and experiment without the use of many-body corrections.

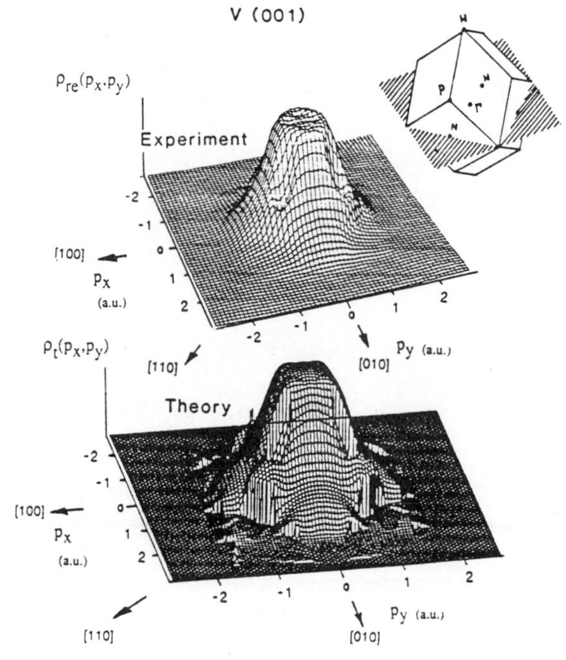

Fermi surfaces have been obtained by these PAS technique in many metals where the standard FS techniques are not feasible. The most detailed FS data are obtained[8,41] by reconstructing the 3D $\rho^{2\gamma}(p)$, and then LCW folding for the full $\tilde{\rho}^{2\gamma}(p)$. Notable among these high precision studies are the measurements of the A15 superconducting compounds[42,43] such as V_3Si. The complex Fermi surfaces obtained in these compounds are in good agreement with the theoretical LMTO band predictions.[43] Other recent high precision FS 2D ACAR experiments have been performed[44] on the hexagonal transition metals Ti, Ru and Re and on rare earth hexaborides.

A recent study of 2D ACAR spectra in paramagnetic vs. antiferromagnetic chromium shows, for the first time, a difference of the FS topology between the two phases.[45] Another interesting and important PAS application has been the study of the intermetallic compound[46] Ni_3Ga, a strong paramagnet of itinerant electrons. Theoretical APW computations show that in such compounds large variations in $\tilde{\rho}^{2\gamma}(p)$ can appear (Eq. 4) due to contributions from the filled bonding-type bands; these "wave function terms" can dominate over the effect of the Fermi surface term, and only a very high precision, high angular resolution 2D ACAR study can separate these effects. This problem can become particularly important in the 2D ACAR studies of high T_c superconducting compounds. The existing high T_c experiments[43,47] are discussed in this volume by L. C. Smedskjaer. The important question: "is the ground state of high T_c superconductors a Fermi liquid?" can be answered by examining the existance and position of Fermi surfaces and to check Luttinger's theorem. In our opinion the present 2D ACAR experiments[47] do not have yet high enough precision to provide a unique answer to this fundamental question. Also relevant to such studies are the ACAR experiments in other metal oxides, in particular in the incomplete perovskite structured tungsten bronze[48] Na_2WO_3.

5.2 <u>Ferromagnetic studies with spin polarized positrons</u>. Due to parity nonconservation, positrons emitted from β^+ sources are partially spin polarized along their direction of motion, and retain much of their polarization upon slowing down in a ferromagnet. This led to the early development of their use to study spin-aligned electrons in solids, and of the underlying theory.[8,49,50] In a typical experiment one measures ACAR spectra with a magnetic field aligning the majority electrons parallel ($\uparrow$) or antiparallel ($\downarrow$) to the positron spin; because only spin singlet overlap can lead to 2γ decay, a change in the spectra upon field reversal is observed. Forming $\Delta\rho^{2\gamma}(p) = \rho_\uparrow^{2\gamma}(p) - \rho_\downarrow^{2\gamma}(p)$ or the equivalent differences in the projections $N_{\uparrow,\downarrow}$, one obtains expressions for the majority vs. minority momentum densities. Thus spin labeled Fermi surfaces can be derived from 2D ACAR data, such as in gadolinium,[51] nickel[52] and iron.[53] Another interesting recent application of this technique has been[54] on the "half-metallic ferromagnet" NiMnSb, possessing a metallic majority spin band, while the minority spin band shows a gap. One notes that these spin-labeled ACAR experiments have their counterpart in the spin-dependent Compton profile measurements using circularly polarized x-rays.

5.3 <u>Substitutionally disordered alloys</u>. Perhaps the most important applications of the ACAR technique have been to the study of the electronic structure of random alloys.[55] The many PAS experiments in this field have been thoroughly reviewed.[56] The theory of random alloys has undergone a major change in the last decade, with the development of the coherent-potential-approximation (CPA),[55] in particular in its most advanced "KKR-CPA" form.[57] Since k is not a good quantum number in disordered alloys, even the existence of a Fermi surface has been questioned in the past. One essential conclusion of CPA is that in most alloys an effective Fermi surface is still definable; the smearing due to disorder, although appreciably larger than the effect of the temperature, remains a reasonably small fraction of the FS dimension.

Central to CPA is the configuration averaged Green's function and the "Bloch spectral density function" $A(\boldsymbol{k}, E)$. For a pure metal $A(\boldsymbol{k}, E) = \sum_j \delta(E - E_{\boldsymbol{k},j})$; the effect of disorder is to broaden the δ functions at each $\boldsymbol{k}$. The electronic momentum densities $\rho(\boldsymbol{p})$ can also be obtained within the various forms of CPA; the inclusion of the positron leads to some difficulty ("vortex corrections"), but can be incorporated within a certain approximation to obtain $\rho^{2\gamma}(\boldsymbol{p})$. These studies, with an emphasis on the physics of the electron and positron states in disordered alloys have been recently reviewed.[58]

Most ACAR experiments have been performed on the various phases of Cu-based alloys.[56] The changes in the FS dimensions have been traced out with increasing alloy concentration. In several α-phase systems these changes follow rigid band behavior (such as in Cu-Zn); on the other hand in alloys with transition metals (Cu-Ni for example), large deviations from rigid band behavior are found, as predicted by CPA theory.[56,58] The sharpness of the FS also follows the trend computed by CPA. A set of very recent 2D ACAR studies on Cu-Ge[59] and Cu-Pd[60] are indicative of the success of the technique to test detailed predictions of the theory; small deviations in the 2D ACAR spectra reveal the influence of the impurity band below the Fermi level, and disorder induced FS "smearing" is substantially larger in the Cu-Ge alloy. Notable also are the momentum reconstructed experiments[61] on the hexagonal Hume-Rothery phases of Cu-Ge and Ag-Al, phases that need more theoretical study.

A high precision 2D ACAR study of the Nb-Mo system[62] shows the ability of the technique to obtain details of the FS in a more complex alloy system. The folded momentum densities are in good agreement with the theoretical densities obtained by a KKR "virtual crystal approximation" method. In Fig. 8 we plot the experimental FS discontinuities obtained from a LCW folded 2D ACAR spectrum, compared to the 2D projections of the edges of the theoretical 3D FS sheets onto the [110] plane.

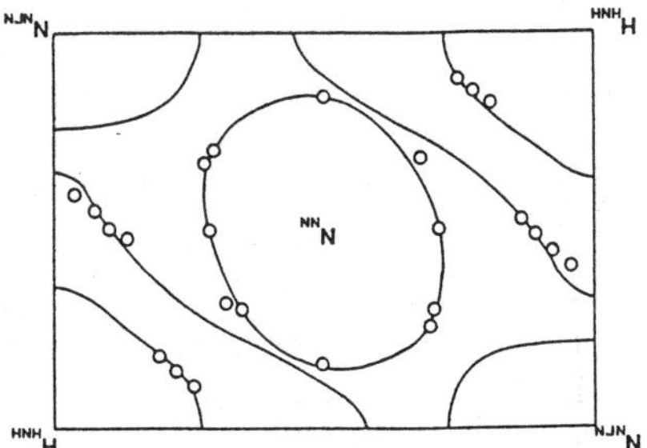

Fig. 8. A section of the 2D $\boldsymbol{k}$ space density for Nb$_{50}$Mo$_{50}$. The curves correspond to the loci of the FS discontinuities predicted by theory, the circles show the positions of the corresponding extrema in the experimental 2D folded momentum density (Ref. 62).

5.4 <u>Other ACAR studies</u>. Except for a few early ACAR studies in semiconductors,[63] most recent PAS applications have been focused on semiconductor defect physics.[64] In graphite, a semimetal, 1D and 2D ACAR experiments exhibit an anisotropic momentum spectrum with respect to the c axis, showing that the positrons sample preferentially the π band electrons in this layered system.[65] Several PAS experiments have also been performed in alkali metal graphite intercalates, exhibiting the appearance of a high-intensity, highly anisotropic narrow momentum component. Recent 2D ACAR results indicate that this unusual momentum spectrum is mainly due to the formation of quasi 2D positronium,[65] leading to some potentially interesting two-dimensional many-body phenomena.

We finish this review by briefly describing the recent application of the 2D ACAR technique to surface studies using high intensity, variable energy positron beams.

6. Surface Studies with Slow e$^+$ Beams: 2D ACAR Experiments

During the last decade the development of slow positron beams led to a wealth of interesting surface physics phenomena,[3,4,66] due mainly to the work of K. F. Canter at Brandeis University, K. G. Lynn at the Brookhaven National Laboratory, (BNL), A. P. Mills, Jr. at AT&T Bell Laboratories and their collaborators. The production of slow, monoenergetic positrons relies on the use of "fast-slow moderators": when fast positrons are implanted into a well annealed sample, those within ≈ 1000 Å of the surface diffuse back and are ejected nearly monoenergetically by their negative work function. Typical moderator efficiencies are $\sim 10^{-4} - 10^{-3}$. The e$^+$ beam energy is varied electrostatically and the beam is guided by magnetic or electrostatic optical systems onto the surface to be studied. The following picture emerges at a metal surface[3]: positrons are either a) ejected, or b) trapped into an "image-potential induced" surface state, or c) pick up an electron to form a Ps of a few eV energy leaving the surface. These three channels can have roughly equal probability. In addition, upon heating the sample, the surface-trapped positrons can thermally "desorb" by electron capture, leading to a thermalized Ps emission. The study of these phenomena has been mainly performed with "table-top" e$^+$ beams yielding $\sim 10^5 - 10^6$ e$^+$/s. By now slow e$^+$-s have been successfully used to study the e$^+$ surface states, near surface and interface defects, ferromagnetic surfaces, etc. – see the review by Schultz and Lynn.[4] Slow positrons are also elastically scattered, leading to low energy e$^+$ diffraction (LEPD) experiments.[67] The experimental implementation of the "brightness enhancement" proposed by Mills[3] has resulted in a high brightness positron microbeam at Brandeis that is being used to develop a brightness-enhanced "positron reemission microscope".[68]

The extension of the ACAR technique to the study of surfaces had to wait for the development of ultra-high intensity slow e$^+$ beams. By 1985 two such facilities were functioning: one uses the LINAC beam at Livermore[69] and the other uses high intensity ^{64}Cu sources at BNL.[70] The beam intensities reached at BNL were up to 10^8 e$^+$/s. In Fig. 9 we show the 2D ACAR spectrum obtained from 200-eV e$^+$-s incident on a clean Al(100) surface.[70] At this energy $\sim 98\%$ of the positrons diffuse back to the surface.[3] The incident e$^+$ beam is from the left in Fig. 9. The momentum spectrum shows a clear asymmetry between the $+$ and $- p_\perp$ axis, due to the 2γ annihilation of the emitted spin-singlet Ps, with a $p_\perp > 0$ component. The total spectrum can be decomposed into a symmetrical component about $p_\perp = 0$, due to trapped positrons annihilating from the surface state, and a component with only $p_\perp \geq 0$, due to Ps leaving the surface. Thus one can study independently the nature of the surface state and that of the Ps emission mechanism.

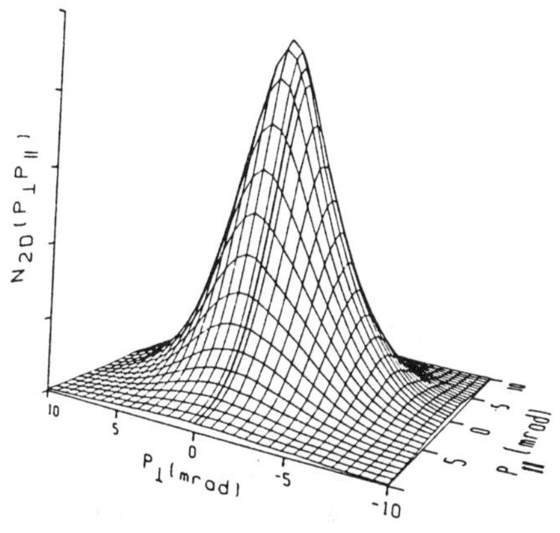

Fig. 9. 2D ACAR spectrum for 200-eV positrons incident upon a clean Al(100) surface. $p_{\perp,\parallel}$ are momenta $\parallel$ to the surface. Note that the positive $p_\perp$ axis is on the left of $p_\perp = 0$.

Experiments indicate that the e^+ surface state produces a nearly isotropic momentum spectrum, even though most theories predict an anisotropy due to the extended 2D nature of the positron on a clean surface. The many existing theories of the surface state range from an "image-correlation-potential" well model,[4,71] to the picture of a surface physisorbed[72] Ps. It is clear that the subject calls for further experimental and theoretical investigation.

The recent detailed study[73] of the emitted Ps momentum spectrum from several crystallographic orientations of Al indicated that Ps emission appears to occur primarily via the sudden capture of an electron near the surface by the thermalized positron, as was also shown by Ps time of flight experiments.[4] The momentum spectrum exhibits anisotropic lobes that were shown[73] to be produced by the energy gaps in the surface projected band structure of Al. The metal is left in a one-hole excited state; inelastic many-body effects seem weak and the Ps spectra reflect reasonably well the electron density of states near the surface.[74] Direct Ps emission from a graphite surface is permitted energetically, but is forbidden by $p_\parallel$ momentum conservation due to the nature of the electronic bands. Ps emission was nevertheless observed,[75] albeit with a large temperature dependent intensity. A simple theory involving phonon-assisted Ps emission describes well the T dependence and the observed momentum spectrum. These 2D ACAR experiments indicate that angle-resolved positronium emission can become a useful electronic surface spectroscopy, perhaps paralleling the success of the 2D ACAR experiments in bulk studies. A 10-100 fold increase in e^+ beam intensity is technically feasible; such future beams, incorporating brightness enhanced microbeam technology will provide, no doubt, more detailed information about the rich physics of e^+ and Ps surface interactions, and lead to further surface studies with positrons.

Acknowledgements. I am indebted to E. C. von Stetten and X. S. Li for reading the manuscript and to Jane Jordan for the hard work of typing it in its several incarnations. This review could not have been written without the many illuminating discussions I have had with colleagues from the positron physics community. The work was supported by the U. S. National Science Foundation through Grant No. DMR-8814185.

References

1. See the reviews in "Positron Solid-State Physics – Proc. Int. School E. Fermi," edited by W. Brandt and A. Dupasquier (North-Holland, Amsterdam, 1983).
2. See also the reviews in the Proceedings of the International Conferences on Positron Annihilation (ICPA) published every three years – "Positron Annihilation – ICPA-7," edited by P. C. Jain, R. M. Singru and K. P. Gopinathan (World Scientific, Singapore, 1985); "Positron Annihilation – ICPA-8," edited by M. Dorikens, L. Dorikens-van Praet and D. Segers (World Scientific, Singapore), to be published 1989.
3. Review by A. P. Mills, Jr., in Ref. 1, p. 432.
4. Review by P. J. Schultz and K. G. Lynn, Rev. Mod. Phys. **60**, 701 (1988).
5. See the review by R. Paulin in Ref. 1, p. 565.
6. For positron annihilation studies in liquids and condensed gases, see the reviews in "Positron Annihilation Studies of Fluids," edited by S. C. Sharma (World Scientific, Singapore, 1988).
7. Positronium reviews by S. Berko and H. N. Pendleton, Ann. Rev. Nucl. Part. Sc. **30**, 543 (1980); A. Rich, Rev. Mod. Phys. **53**, 127 (1981).
8. See the review by S. Berko in Ref. 1, p. 64.
9. S. Berko, M. Haghgooie and J. J. Mader, Phys. Lett. A **63**, 335 (1977).
10. Review by A. Stewart in "Positron Studies of Solids, Surfaces, and Atoms," edited by A. P. Mills, Jr., W. S. Crane and K. F. Canter (World Scientific, Singapore, 1986), p. 28.

11. M. J. Fluss, S. Berko, B. Chakraborty, K. R. Hoffmann, P. Lippel and R. W. Siegel, J. Phys. F: Met. Phys. **14**, 2831 (1984).

12. Review by A. Dupasquier in Ref. 1, p. 510; T. Hyodo in "ICPA-7" (Ref. 2), p. 643.

13. J. Kasai, T. Hyodo and K. Fujiwara, J. Phys. Soc. Japan **57**, 329 (1988).

14. M. Hasegawa, Y. J. He, K. R. Hoffmann, R. R. Lee, S. Berko and T. Takeyama, "ICPA-7" (Ref. 2), p. 260, and references therein. See also E. Kuramoto, K. Kitajima and M. Hasegawa, Phys. Lett. **86A**, 311 (1981).

15. P. M. Platzman, in "Positron Studies of Solids, Surfaces, and Atoms," edited by A. P. Mills, Jr., Wm. S. Crane and K. F. Canter (World Scientific, Singapore, 1986), p. 84.

16. As originally suggested by P.C. Hohenberg and P.M. Platzman, Phys. Rev. **152**, 198 (1966).

17. See the review by N. Shiotani, "Positron Annihilation – ICPA-6," edited by P. G. Coleman, S. C. Sharma, L. M. Diana (North Holland, Amsterdam, 1982), p. 561.

18. S. DeBenedetti, C.E. Cowan, W.R. Konneker and H. Primakoff, Phys. Rev. **77**, 205 (1950).

19. See W. Brandt in Ref. 1, p. 1.

20. J. Carbotte, Ref. 1, p. 32.

21. J. Arponen and E. Pajanne in "ICPA-7" (Ref. 2), p. 21.

22. H. Stachowiak, Helvetica Phys. Acta **61**, 415 (1988).

23. See the early review by R. A. Ferrell: Rev. Mod. Phys., **28**, 308 (1956).

24. P. F. Maldague, Phys. Rev. **20**, 21 (1979).

25. "Compton Scattering" – edited by B. Williams (McGraw Hill, N.Y. 1977)

26. S. Berko and J. S. Plaskett, Phys. Rev. **112**, 1877 (1958).

27. D. G. Lock, V. H. C. Crisp and R. N. West, J. Phys. **F3**, 561 (1973).

28. N. Shiotani, Solid State Phys. (Japan) **19**, 526 (1981).

29. L. P. L. M. Rabou and P. E. Mijnarends, Solid State Commun. **52**, 933 (1984).

30. A. K. Singh and R. M. Singru, J. Phys. F: Met. Phys. **14**, 1751 (1984).

31. J. H. Kaiser, R. N. West and N. Shiotani, J. Phys. F: Met. Phys. **16**, 1307 (1986).

32. B. B. J. Hede and J. P. Carbotte, J. Phys. Chem. Solids **33**, 727 (1972).

33. J. Mader, S. Berko, H. Krakauer and A. Bansil, Phys. Rev. Lett. **37**, 1232 (1976).

34. L. Oberli, A. A. Manuel, R. Sachot, P. Descouts, M. Peter, L. P. L. M. Rabou, P. E. Mijnarends, T. Hyodo and A. T. Stewart, Phys. Rev. **B31**, 1147 (1985).

35. B. Chakraborty, Phys. Rev. **B24**, 7423 (1981).

36. E. Boronski and R. M. Nieminen, Phys. Rev. **B34**, 3820 (1986).

37. See the review by G. Kontrym-Sznajd and M. Šob, J. Phys. F: Met. Phys. **18**, 1317 (1988).

38. T. Jarlborg and A. K. Singh, Phys. Rev. **B36**, 4660 (1987) and references therein.

39. M. Matsumoto and S. Wakoh, J. Phys. Soc. Japan **56**, 3566 (1987).

40. A. K. Singh, A. A. Manuel, R. M. Singru, R. Sachot, E. Walker, P. Descouts and M. Peter, J. Phys. **F15**, 2375 (1985); L. M. Pecora and A. C. Ehrlich, A. A. Manuel, A. K. Singh, M. Peter and R. M. Singru, Phys. Rev. **B37**, 6772 (1988).

41. A. A. Manuel, Phys. Rev. Lett. **49**, 1525 (1982).

42. W. S. Farmer, F. Sinclair, S. Berko, and G. M. Beardsley, Sol. St. Comm. **31**, 481 (1979); S. Berko, W. S. Farmer, and F. Sinclair, in "Superconductivity in d- and f-band Metals," edited by H.Auhl and M. B. Maple (Academic Press, NY, 1980).

43. T. Jarlborg, A. A. Manuel, and M. Peter, Phys. Rev. **B27**, 4210 (1983); see also the review by A. A. Manuel, Helvetica Phys. Acta, **61**, 451 (1988).

44. S. Tanigawa, S. Terakado, K. Ito, A. Morisue and N. Shiotani, "ICPA-7" (Ref. 2), p. 282.

45. A. K. Singh, A. A. Manuel and E. Walker, Europhysics Lett. **6**, 67 (1988).

46. N. Shiotani, C. R. Bull, R. N. West, N. Kawamiya, Y. Kubo and S. Wakoh, J. Phys. Soc. Japan **55**, 1961 (1986).

47. For a general review see M. Peter in "ICPA-8" (Ref. 2), to be published 1989.

48. T. Akahane, K. R. Hoffmann, T. Chiba, and S. Berko, Sol. St. Comm. **54**, 823 (1985) and references therein.

49. S. Berko and A. P. Mills, J. de Phys. **32** C1, 287 (1971), and references therein.

50. P. E. Mijnarends, Physica **63**, 248 (1973).

51. K. R. Hoffmann, S. Berko and B. J. Beaudry in "ICPA-6" (Ref. 17), p. 325, 1982.

52. T. Jarlborg, A. A. Manuel, Y. Mathys, M. Peter, A. K. Singh and E. Walker, J. Magn. Magnetic Materials **54**, 1023 (1986).

53. P. Genoud, A. K. Singh, A. A. Manuel, T. Jarlborg, E. Walker, M. Peter and M. Weller, J. Phys. F: Met. Phys. **18**, 1933 (1988).

54. K. E. H. M. Hanssen and P. E. Mijnarends, Phys. Rev. B**34**, 5009 (1986); see also review by Mijnarends in "ICPA-8" (Ref. 2), to be published 1989.

55. "Electrons in Disordered Metals and at Metallic Surfaces," edited by P. Phariseau, B. L. Györffy and L. Scheire, (Plenum Press, N.Y., 1979).

56. S. Berko in "Noble Metal Alloys," edited by T. B. Massalski, W. P. Pearson, L. H. Bennett and Y. A. Chang, (Met. Society AIME, 1986) p.109; R. N. West in "ICPA-7" (Ref. 2), p. 11.

57. See the review by G. M. Stocks and H. Winter in "Electronic Structure of Complex Systems," edited by P. Phariseau and W. M. Temmerman, (Plenum Press, N.Y., 1985).

58. A. Bansil in "ICPA-6" (Ref. 17) p. 291; P. E. Mijnarends, Phys. Stat. Sol. **102**, 31 (1987); B. L. Györffy in "ICPA-8" (Ref. 2), to be published 1989.

59. P. E. Mijnarends, L. P. Rabou, K. E. Hanssen and A. Bansil, Phys. Rev. Lett. **59**, 720 (1987).

60. L. C.Smedskjaer, R.Benedek, R. W. Siegel, D. G. Legnini, M. D. Stahulak and A. Bansil, Phys. Rev. Lett. **59**, 2479 (1987).

61. S. Koike, M. Hirabayashi, To. Suzuki and M. Hasegawa, Phil. Magazine B**45**, 261 (1982).

62. J. H. Kaiser, P. A. Walters, C. R. Bull, A. Alam, R. N. West and N. Shiotani, J. Phys. F: Met. Phys. **17**, 1243 (1987).

63. M. A. Shulman, G. M. Beardsley and S. Berko, Appl. Phys. **5**, 367 (1975) and references therein.

64. Review by G. Dlubek and R. Krause, Phys. Stat. Sol. **102**, 443 (1987).

65. R. R. Lee, E. C. von Stetten, M. Hasegawa and S. Berko, Phys. Rev. Lett. **58**, 2363 (1987) and references therein.

66. K. F. Canter and K. G. Lynn, J. Vac. Sci. Technol. **A2**, 916 (1984).

67. A. H. Weiss, I. J. Rosenberg, K. F. Canter, C. B. Duke and A. Paton, Phys. Rev. B**27**, 867 (1983).

68. G. R. Brandes, K. F. Canter and A. P. Mills, Jr., Phys. Rev. Lett. **61**, 492 (1988); see also J. Van House and A. Rich, Phys. Rev. Lett. **61**, 488 (1988).

69. R. H. Howell, P. Meyer, I.J. Rosenberg and M. J. Fluss, Phys. Rev. Lett. **54**, 1698 (1985).

70. K. G. Lynn, A. P. Mills, Jr., R. N. West, S. Berko, K. F. Canter and L. O. Roellig, Phys. Rev. Lett. **54**, 1702 (1985).

71. A. P. Brown, A. B. Walker and R. N. West, J. Phys. F**17**, 2491 (1987); Lou Yongming in "ICPA-8" (Ref. 2), to be published 1989, however, predicts near isotropy using IPM.

72. P. M. Platzman and N. Tzoar, Phys. Rev. B**33**, 5900 (1986).

73. D. M. Chen, S. Berko, K. F. Canter, K. G. Lynn, A. P. Mills, Jr., L. O. Roellig, P. Sferlazzo, M. Weinert and R. N. West, Phys. Rev. Lett. **58** 921 (1987).

74. For theoretical references see Ref. 4; also A. Ishii, J. B. Pendry and D. K. Saldin in "ICPA-8" (Ref. 2), to be published 1989.

75. P. Sferlazzo, S. Berko, K. G. Lynn, A. P. Mills, Jr., L. O. Roellig, A. J. Viescas and R. N. West, Phys. Rev. Lett. **60**, 538 (1988).

FERMI SURFACES BY POSITRON ANNIHILATION IN YBa$_2$Cu$_3$O$_x$

L.C. Smedskjaer

Materials Science Division, Argonne National Lab, Argonne IL 60439

Abstract

We discuss how Fermi-surface signatures may obtained from 2-D ACAR positron annihilation data. We report 2-D ACAR measurements in both c- and a-axis projections on single crystal YBa$_2$Cu$_3$O$_x$ with oxygen concentration x$\approx$7 (T$_c$ = 93.6 K) and a set of Fermi surfaces are proposed.

Introduction

In spite of the intensive scrutiny of the past two years, the mechanism of superconductivity in the cuprates remains elusive. An important step in the determination of this mechanism is the characterization of the electronic structure of these materials. A set of Fermi surface sheets for YBa$_2$Cu$_3$O$_7$ have been derived from 2-D ACAR [1,2] measurements that are in substantial agreement with band calculations [3,4]. Furthermore, calculations of the two-gamma momentum density, including the positron wave function, show good overall agreement with the experimental results [5].

Results

2-D ACAR spectra have been measured at room temperature in both c-axis and a-axis projections for a single-crystal YBa$_2$Cu$_3$O$_7$ (2×0.7×0.3 mm^3)[1]. Since previous experiments on polycrystalline material revealed that the sample properties changed slightly under the experimental conditions, the exposure of the sample to the vacuum and the positron irradiation was kept to a minimum[2]. For this reason only 4×10^6 counts were measured for each projection. The experimental resolution consists of two contributions, one from the instrument of 0.3 mrad, the other (0.52 mrad) from the thermal motion of the positron at room temperature. The effective resolution is therefore 0.6 mrad and thus being dominated by the thermal motion of the positron, rather than the instrument. For presentational purposes the data shown in the present work have been further smoothed, raising the effective resolution to about 0.8 mrad.

---

[1]Note that the single crystals are twinned and that the measurements cannot distinguish between the a and b directions.

[2] The value of T$_c$ = 93.6 K was determined <u>after</u> the completion of the positron experiments.

Following the correction for the measured camera efficiency, the center for the data and their symmetries were determined. For the c-axis projection a C_{4v} symmetry was found, while the a-axis projections showed C_{2v} symmetry. The normalized Chi-square sum for the symmetry hypothesis were 1.02 with an expected standard deviation of 0.02 for the Chi-square sum. Thus, the spectra had the required symmetries to within uncertainty; specifically the Chi-square sum indicates that an upper limit for the overall instrumental artefacts is14% <u>of the standard deviation</u> on the data. However, the possibility exists that the artefacts might be large in some areas and not in other, thereby resulting in this very low overall value for the artefact. This possibility was addressed by measuring a glue sample which should result in a perfectly radial symmetric distribution. It was concluded that instrumental artefacts do not exceed 20% of the standard deviation of the data, even after symmetrization. The spectra were therefore symmetrized around their centers, augmenting the statistical precision considerably (a factor of 8 on the total counts for the c-axis projection, and a factor of 4 for the a-axis projection). Thus the statistical precision in an octant (quadrant) for the symmetrized data is equivalent to that of 32×10^6 counts (16×10^6 counts) for the total spectrum.

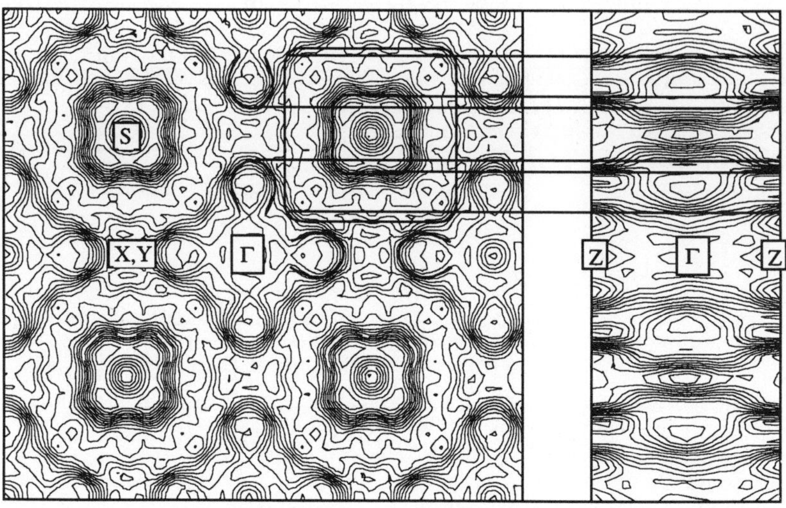

Figure 1. Left: Total 2-D ACAR spectrum for $YBa_2Cu_3O_7$ in c-axis projection. Right: Anisotropic part of 2-D ACAR spectrum for $YBa_2Cu_3O_7$ in c-axis projection.

With the purpose of exposing the structure in the data more clearly a radially isotropic and smooth distribution was subtracted. Since the subtracted function is smooth this procedure does not contribute to, nor create, structures in the data remaining after subtraction. The result, often called the anisotropic distribution, is shown in fig 1 for a c-axis projection. The amplitude of the anisotropic distribution is 3% of the original data amplitude.

The data are then "folded" back into the first Brillouin zone according to

$$\eta(k_x,k_y) = \sum_{G_x,G_y} n(k_x + G_x, k_y + G_y)$$

where n(p_x,p_y) is the anisotropic distribution, while G is a reciprocal lattice vector in the projection plane and η is the resulting distribution. Contour plots of the results for c- and a-axis projections are shown in a periodic zone-scheme representation in fig 2.

<u>Discussion</u>

The striking feature in fig 2 are the regions were the contours run closely together. This means that the electron momentum density varies rapidly at these locations. The common interpretation,in case of a simple metal, would be that the variations were caused by one or more underlying Fermi surfaces. However, $YBa_2Cu_3O_7$ is substantially more complicated than a simple metal, and a further investigation is needed.

First, an examination of the a-axis projection shows that the structures are cylindrical in the c-axis direction, since all contour lines are almost parallel to this axis. Thus, the structures are seen "on end" in the c-axis projection, and from the side in the a-axis projection. It is noted that the "diameter" of the cylinders, when seen from the side in the a-axis projection, corresponds closely to the "diameter" when seen "on end". Since the a- and c-axis data are independent of one another and have different symmetries, but yet are in agreement, the possibility of artefacts due to the data treatment seems remote. Thus, the structures are "bona fide" structures. It is noted that in the a-axis projection the small "raindrop" feature and the "pill-box" feature are seen on top of one another and cannot be distinguished in this projection due to the finite resolution of the instrument. A small dispersion along the c-axis, possibly associated with one or both of these features is also seen. It is, however, not clear from the data how to associate this dispersion, especially in view of the finite instrument resolution.

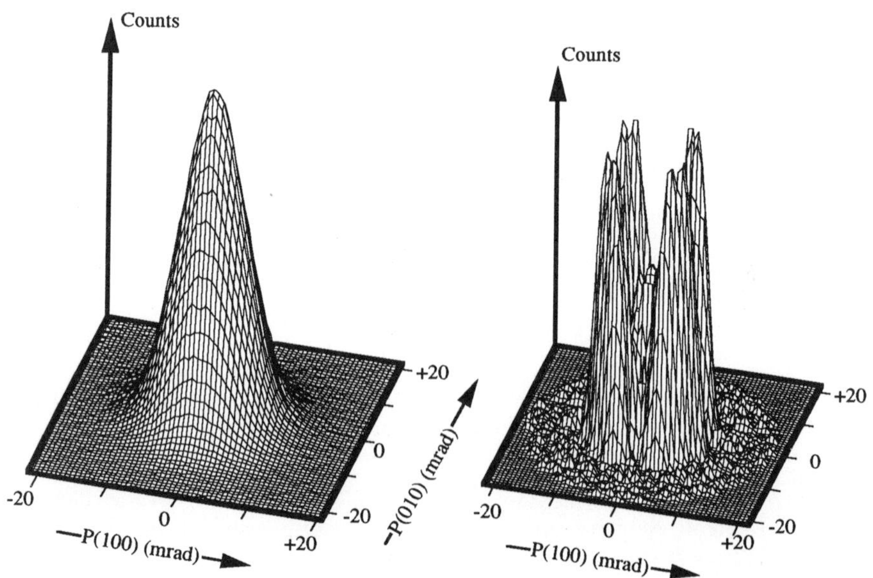

Figure 2. Contour plot of LCW folded anisotropic data. The approximate positions of the Fermi surface are indicated. Left: c-axis projection, right: a-axis projection. Note that the a-axis projection plot has been been expanded in the c-direction for clarity. The horizontal lines are drawn in order to compare the positions of features in the a- and c-axis projections.

Secondly, a more detailed examination of sections through fig 2 shows that the slopes of the data in the regions of dense contours are consistent with the hypothesis of an underlying discontinuity in momentum space (e.g. Fermi surface) when accounting for the instrumental resolution.

Thirdly, the structures seen in fig.2 can also be found in sections through the anisotropic data (fig 1). The structures are less pronounced in this representation, due to the reduced statistical sensitivity. However, the slopes are also here consistent with an underlying discontinuity in momentum space. Most importantly, the umklapp's of the structures can be observed as well. This demonstrates that the structures are not just local, but that they are consistently found throughout momentum space as for example a Fermi surface would be. The "backside" of the peaks seen in fig.1 deserves a further comment: the slope is steep, but not steep enough in relationship to the height of the peak to qualify as a <u>single</u> Fermi surface. This feature does, however, not contribute to the plot in fig 2, since it follows a higher zone boundary very closely. In this regard the $YBa_2Cu_3O_7$ differs strongly from a simple metal, where structures in the zone boundaries are unusual.

It has been demonstrated that the structures seen in the 2-D ACAR data have, as far as the data goes, the necessary properties to be interpreted as Fermi surfaces. This interpretation is somewhat limited by the instrumental resolution, since the transition width of the underlying structures need not be zero (Fermi surface) in order to produce the observed transition widths. It is estimated that an upper limit for the transition width of the underlying structures would be the instrument resolution. The finite resolution might also give rise to systematic errors in the position of the Fermi surfaces shown in fig 2. This will especially occur at points were two or more Fermi surfaces are close together. In this case a "mixing" of features will take place resulting in ambiguities regarding the shapes of the Fermi surfaces[1]. For this reason the Fermi surfaces shown in fig 2 cannot be considered unique, but has been chosen over the "diamond" pattern[2] because they give a more detailed agreement with the data. Also, the count of electrons in the partially filled bands would be 2.5 (diamond non degenerate) or 3.2 (doubly degenerate) for the diamond pattern. This contrasts the model proposed in fig 2, which gives 4 electrons/unit cell while assuming the large square hole to be doubly degenerate. This is in agreement with the predictions of bandtheory.

We consider now the Fermi surface topology deduced from the present experiments (fig. 2). Since the a-axis projections show approximate cylindrical symmetry along the c-axis, the Fermi surface sheets will be assumed cylindrically symmetric in our further discussions. We first draw attention to a small square-like hole surface centered at S (the "pill box"). This sheet corresponds to a theoretically [3,4,5] predicted Fermi surface associated with a relatively heavy-mass Cu-O chain band. A larger hole surface, the "belly", is also centered at S, and is associated with the (nearly degenerate) Cu-O plane bands that cannot be distinguished within experimental resolution. Finally, a raindrop-shaped electron surface is observed along the

Y-Γ axis. This feature may result from an (unresolved) superposition of the Fermi surface of the light and the heavy mass Cu-O chain bands. In conclusion, we note the substantial overall agreement between the presently measured Fermi surface and those predicted by band theory[3,4] for $YBa_2Cu_3O_7$. In addition, detailed comparisons between the experimental results and two-gamma momentum-density calculations [5] also show good agreement. The latter calculations also confirm the experimental observation (fig. 2) that the "belly" is less visible near the Y-S (X-S) axis, whereas the "pillbox" is clearly visible.

In conclusion the experimental data indicates the presence of structures consistent with an underlying Fermi surface in $YBa_2Cu_3O_7$; in addition, the structures are close to the locations of Fermi surfaces predicted by bandtheory. It therefore seems reasonable to propose that the features in the data are Fermi surfaces[3].

This work was supported by the U.S. Department of Energy, Basic Energy Sciences - Materials Sciences, under Contract No. W-31-109-Eng-38.

---

[1] The problem may be mitigated by an improved instrumental resolution.

[2] For the diamond pattern, the Fermi surface is visualized as a diamond-like Γ-centered electron sheet and a smaller square-like hole sheet at the S-point.

[3] Note added in proof: Recent results by photoemission (A.J. Arko et al) confirms the notion of Fermi surfaces in the $YBa_2Cu_3O_7$. Even more recent angle resolved photoemission results (J.C. Campuzzano et al.) show a Fermi surface in rather close agreement with the present results.

<u>References</u>

1) Hoffmann L., Manuel A.A., Peter M., Walker E., and Damento M.A., Europhys. Lett. <u>6</u>, 61 (1988)
2) Smedskjaer L.C., Liu J.Z., Benedek R., Legnini D.G., Lam D.J., Stahulak M.D., Claus H., and Bansil A., Physica C <u>156</u> 269 (1988)
3) Yu J., Massida S., Freeman A.J., and Koelling D.D., Phys. Lett. A <u>122</u>, 203 (1987)
4) Krakauer H., Pickett W.E., and Cohen R.E., J. Supercond. <u>1</u>, 111 (1988)
5) Bansil A., Pankaluoto R., Rao R.S., Mijnarends P.E., Dlugosz W., Prasad R., and Smedskjaer L.C., Phys Rev Lett. <u>61</u> 2480 (1988)

EXPERIMENTAL TECHNIQUES FOR MOMENTUM DISTRIBUTIONS

NEUTRON EXPERIMENTAL TECHNIQUES FOR n(p)

Ron Holt, Jerry Mayers and Andrew Taylor

Neutron Science Division
Rutherford Appleton Laboratory
Chilton, Didcot, Oxon. OX11 0QX, UK

INTRODUCTION

The determination of electron momentum density distributions in condensed matter is now a well established technique and the experimental methods employed have been adequately summarised in this volume by M J Cooper [1] (recent x-ray and electron techniques) and by S Berko [2] (positron annihilation studies). These well established experimental techniques contrast sharply with the more recent inelastic scattering studies involving high energy neutrons. The requirement for intense high energy neutron beams emerged from the theoretical work of Hohenberg and Platzman [3] who showed that such energetic beams were necessary in order to approach the limits imposed by the Impulse Approximation. Moreover, the Impulse Approximation was a necessary fulfillment of the experimental conditions to facilitate the interpretation of the results without recourse to complex final state corrections [4] so evident in the earlier momentum density work on reactor sources [5]. With the advent of pulsed neutron sources, and their abundant flux of epithermal neutrons, there have naturally been significant developments in the experimental methods used to obtain atomic momentum distributions, n(p). Such techniques are now more commonly referred to as Deep Inelastic Neutron Scattering (DINS) or Neutron Compton Scattering. It is therefore the purpose of this chapter to highlight the basic neutron experimental methods with particular emphasis on pulsed source techniques.

EXPERIMENTAL TECHNIQUES

Reactor Source Methods

Steady state reactors, such as those at ILL and Chalk River, produce a neutron spectrum which is Maxwellian with a peak which is determined by the temperature of the moderator near the reactor core. A reactor therefore provides a continuous source of thermal neutrons with little intensity in the epithermal region (see figure 1). During the last decade the determination of atomic momentum densities by the inelastic scattering of thermal neutrons have been undertaken using a variety of crystal spectrometers. The majority of these reactor based experiments have been concerned with determining the Bose condensate fraction in liquid helium-II [5-8].

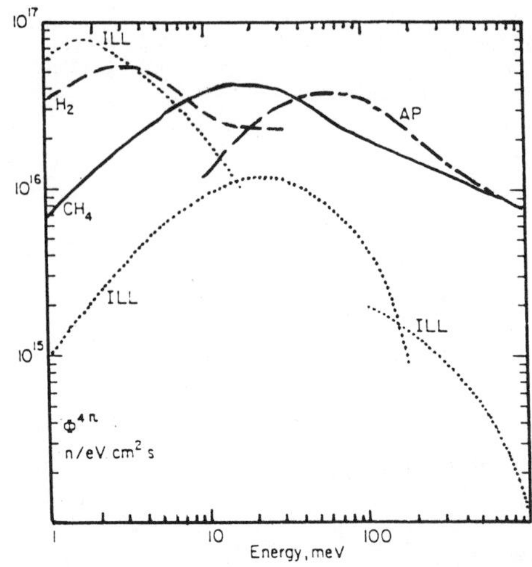

Figure 1. Comparison of the total neutron flux emitted after moderation from a reactor source (ILL) and a pulsed neutron source (ISIS). AP represents the ambient water moderator, H_2 the liquid hydrogen moderator and CH_4 the liquid methane moderator.

The early measurements carried out by Cowley and Woods [9] and Martel et al [5] used the triple-axis crystal spectrometer. Here the required incident neutron energy is obtained by diffraction from a monochromating crystal. The resulting beam is subsequently scattered from a sample and the scattered beam analysed using a second crystal in step with a neutron detector (figure 2). Examples of monochromator/analyser crystals combinations include Be(201)/Cu(220) and Be(110)/Ge(220), the latter having superior resolution characteristics. Incident energies are however restricted to below ~250 meV resulting in q values typically of order 12Å^{-1}. This is below the Impulse limit and sophisticated final state corrections are necessary before the momentum density can be extracted with confidence. Generally only one point in (q,ω) space can be collected at one time and hence long experiments are necessary in order to collect sufficient statistics.

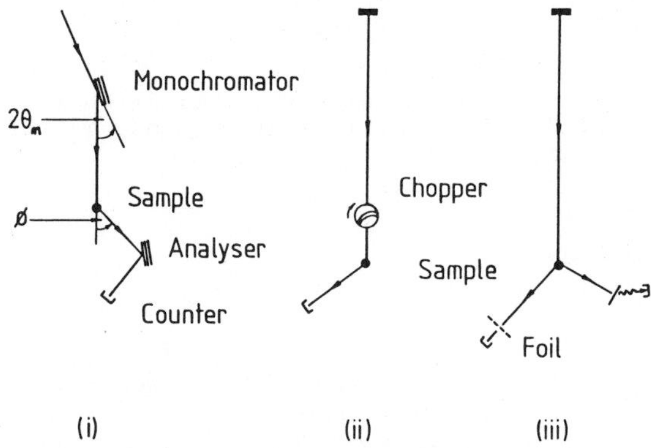

Figure 2. Inelastic scattering spectrometers used on reactor and pulsed neutron sources for the determination of atomic momentum distributions in condensed matter.

296

An advantages of a triple axis instrument is, however, the ability to collect data in constant q using a fixed scattered neutron energy, E_1, and variable incident energy, E_0, and scattering angle, θ. Another advantage is that the instrument resolution function is simply a convolution of the geometry of the path followed by the neutrons with the mosaic spread of the analysing crystals used; to first order the resolution can be described simply in terms of a Gaussian function. This can be determined theoretically or from measurements of the elastic scattering carried out on vanadium ; typical FWHM values of between 5–7 meV are readily obtained, equivalent to ~0.7 Å^{-1}.

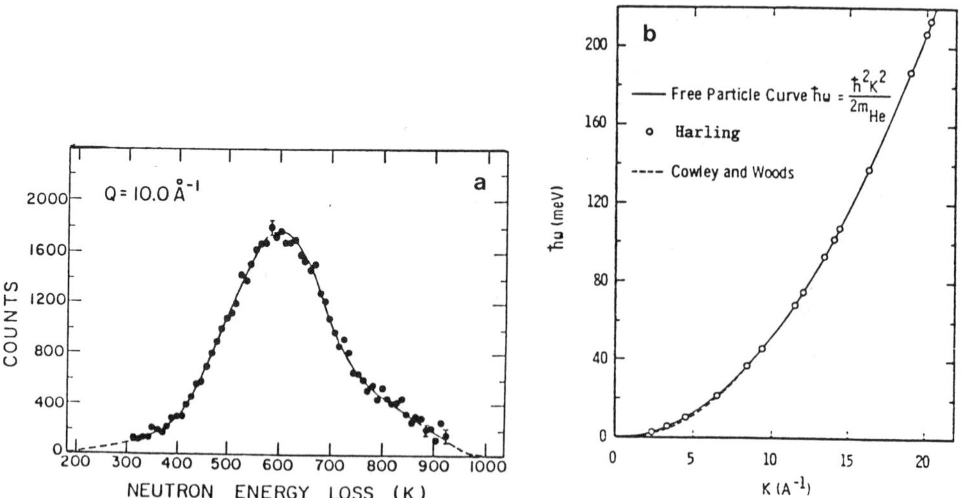

Figure 3. (a) The momentum distribution in liquid ^{4}He obtained on a triple-axis crystal spectrometer and (b) the recoil peak position as a function of momentum transfer using the crystal-chopper technique of Harling.

An alternative approach on reactor sources has been developed by Mook using a cross-correlation time-of-flight spectrometer that utilizes a magnetically pulsed beam. This technique, which can be programmed for any type of correlation pulse chain, has been developed extensively to incorporate both multi-crystal focussing devices and multidetector arrays (see Mook, this volume [10]). Despite the significant improvements in both count rate and resolution (~0.1–0.3 Å^{-1}) the range of available q is still restricted to between 6–15 Å^{-1}.

A third crystal spectrometer which has been used to study the Bose condensation in helium II is based on a rotating aluminium crystal and chopper time-of-flight spectrometer [6,7]. Although initial energies were again restricted (up to ~350 meV) the range of q was extended to beyond 20 Å^{-1}. Examples of the results obtained with these spectrometers on liquid helium are shown in figure 3.

297

On a pulsed neutron source a similar Maxwellian spectrum (as observed on a reactor source) is also produced in addition to an intense pulse of epithermal neutrons (see figure 4) resulting from the collision of high energy protons (550–800 MeV) with a heavy metal target (usually uranium or tantalum). The pulse rate is determined by the synchrotron machine and the extraction characteristics. The resulting high energy neutrons are slowed down within a moderator to give a neutron pulse which has the characteristics of diffusion. It is this rich and intense source of epithermal neutrons which has allowed the development of spectrometers that can determine the momentum distribution in atomic nuclei.

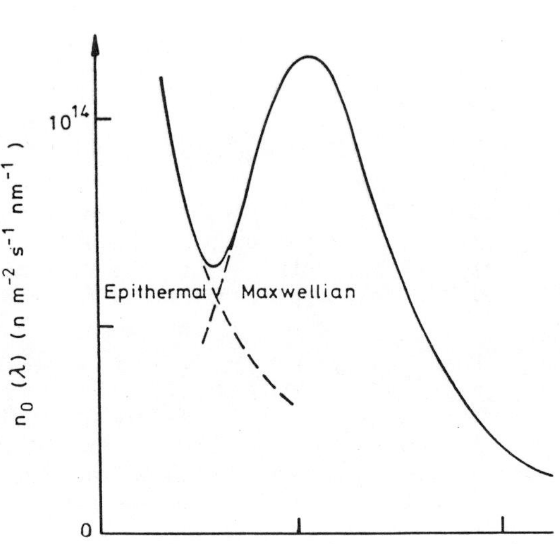

Figure 4. The pulse structure of the neutron spectrum showing the intense epithermal neutron flux necessary to carry out deep inelastic neutron scattering experiments within the Impulse Approximation.

Two basic types of inelastic spectrometers on pulsed neutron sources have been developed (see figure 2). The Direct Geometry (DG) spectrometer uses a monochromating device, such as a fast rotating chopper, to define the incident neutron energy, E_0. The Inverse Geometry (IG) spectrometer uses an analysing device, such as a crystal or absorption resonance, to define the final energy, E_1 (the remaining energy in both systems being defined by time-of-flight). Whatever system is used the most important contributions to the resolution arise from uncertainties associated with the energy ΔE, the time Δt and the flight path lengths ΔL (uncertainties associated with the angular contribution are considered later).

(a) Direct Geometry Spectrometers

A DG chopper spectrometer on a pulsed neutron source is one which selects neutrons within a specific time window from the incident epithermal beam. Several such rotating devices are available i.e. disk choppers and Fermi choppers (see figure 5(a)). However, due to mechanical limitations the rotational speeds are not rapid and the time for a chopper window to sweep across the face of the moderator gives rise to a lighthouse effect to the incident beam.

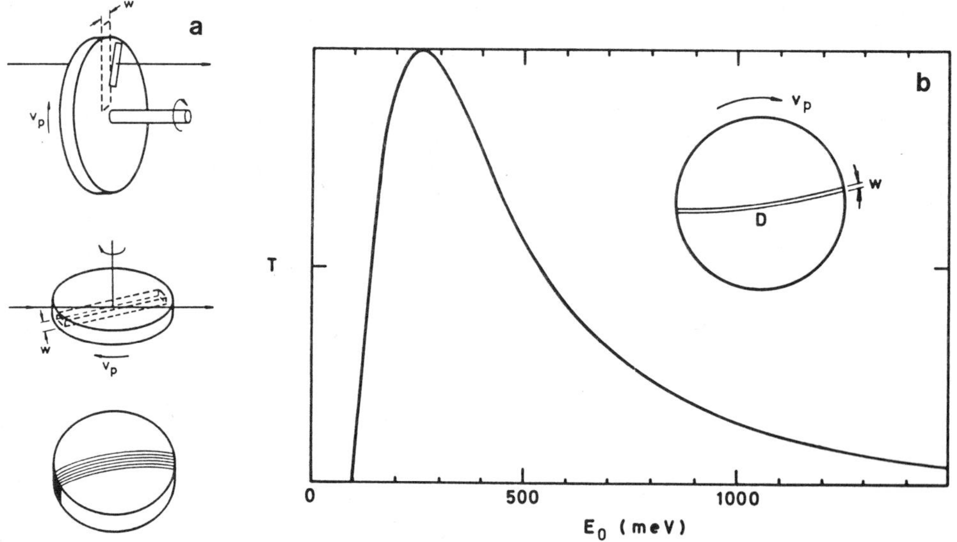

Figure 5. (a) The main types of neutron chopper. In practice the curved Fermi choppers gives the best performance for epithermal neutrons. (b) The transmission function for a curved slit Fermi chopper optimised for 250 meV incident neutrons.

Disk choppers have several disadvantages when used with epithermal neutron beams (see Windsor [11]). On the other hand the fast Fermi chopper provides for a more precise definition to the pulse and the transmitted beam intensity is a product of the initial flux, the moderator area, the slit package area, the angular divergence of the slit package and the ratio of slit width to slit plus absorber width. The energy resolution for a chopper spectrometer is therefore a complex convolution of the energy spread in the incident beam (time width of the chopper transmission) with the geometry of the spectrometer. The asymmetry in the incident pulse is reflected in both energy and time distribution of the neutron after passing through the chopper and resulting in a complicated instrumental resolution function. No simple analytical forms can be used to describe the resolution and only calculations using sophisticated Monte Carlo routines can determine the exact shape of the instrument function. These simulation studies have found that the resolution function varies across the recoil scattering peak.

The present range of curved fast Fermi chopper devices can only effectively operate with incident neutron energies up to a maximum of ~2 eV. Higher energies produce unacceptable transmissions which tend to broaden the incident energy and severly degrade the resolution. Slit packages used in such devices are usually made from boron fibre–aluminium components optimised to give peak transmission at well defined energies (see figure 5(b)). The chopper can also be spun at reduced multiplets of the pulsed source frequency producing monochromatic beams with energies between 100 meV and 2 eV.

Chopper spectrometers offer high intensity with good resolution characteristics at q values up to ~40 Å^{-1}. In most spectrometers a large angular range is covered and this is important in recoil scattering measurements where it is necessary to determine the form of the sample

induced background and extent of multiple scattering over an extended range
in $S(q,\omega)$. Figure 6 shows some recent results obtained on the PHOENIX
chopper spectrometer at IPNS facility on liquid ^{4}He at $q\sim24$ Å^{-1}. The
resolution FWHM was 0.5 Å^{-1} and its functional form was determined from
Monte Carlo calculations. Also shown in figure 6 is the width of the
atomic momentum density obtained on a single crystal of Be using the ISIS
chopper spectrometer, HET. Both figures illustrate the quality of data now
attainable and the extended range of q values now available on pulsed
source instruments.

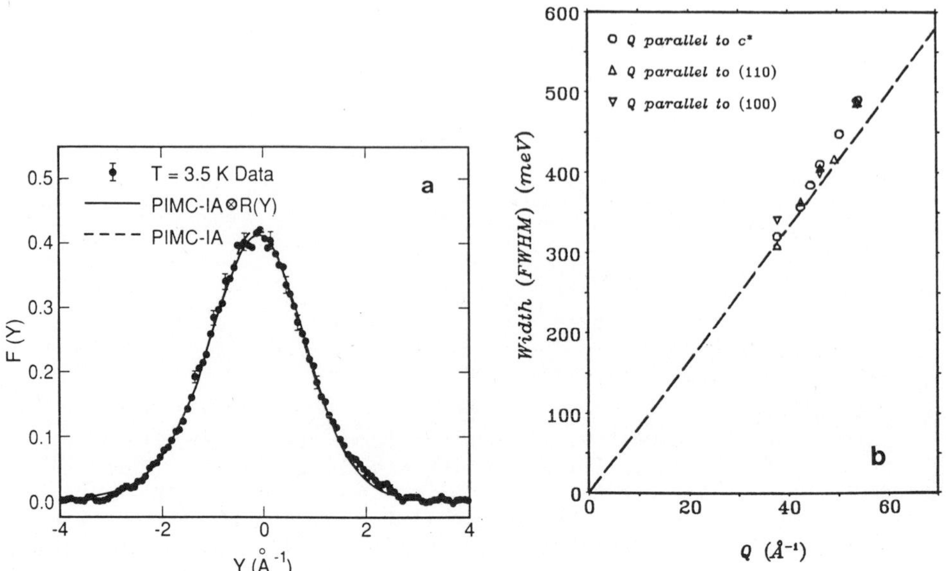

Figure 6. (a) The experimental momentum distribution in He^4 at
3.5K obtained on the PHOENIX chopper spectrometer at IPNS. (b)
The observed recoil width in a single crystal of Be as a function
of q obtained in a single run using the HET chopper system at ISIS.

(b) Inverse Geometry Spectrometers

Inverse geometry instruments are specially suited for pulsed neutron
sources since all of the incident neutron pulse is used, and the secondary
flight path is relatively short. However, crystal spectrometers have not
been used in recoil scattering because of the limitation in energy transfer
of ~500 meV and the low crystal reflectivity due to the decrease in Bragg
scattering at higher energies. Effects associated with multiple Bragg
scattering are also important. All of these imposes severe restrictions on
the achievable count rate and the q range accessible.

An alternative device, which is well suited for recoil scattering, is
based upon strong absorption resonances in the neutron–nuclear cross
section. Many of these resonances appear at high energies, typically
between 0.1 eV and 100 eV and as such can be used to define the scattered
neutron energy, E_1. The most useful resonances are those which have
absorption cross sections in excess of 10000 barns; the ideal resonance for
recoil scattering would comprise of a very strong resonance (for high
sensitivity) with a very narrow width (for high resolution). In practice
the experimental technique is a compromise between count rate and
resolution requirements.

Two analyser instruments have been developed based on this technique and these are shown schematically in figure 7(a) : a Resonance Detector Spectrometer (RDS), using the high energy gamma-rays emitted in the neutron capture process and detected in scintillation counters, and a Resonance Filter-Difference Spectrometer (RFS), in which all the scattered neutrons are detected within the solid angle defined by the width of the helium tube or collimator. In the former the recoil spectrum is obtained directly and is especially useful for line shape analysis. As three or four gamma-rays are emitted per capture event it is possible to increase the solid angle for detection by increasing the foil area and number of photon detectors. In the latter method the recoil spectrum is obtained indirectly from the difference between measurements with and without the resonance material (figure 7(b)). Typical resonances are 0.87 eV (Sm), 4.28 eV (Ta), 4.096 eV (Au) and 6.67 eV (U).

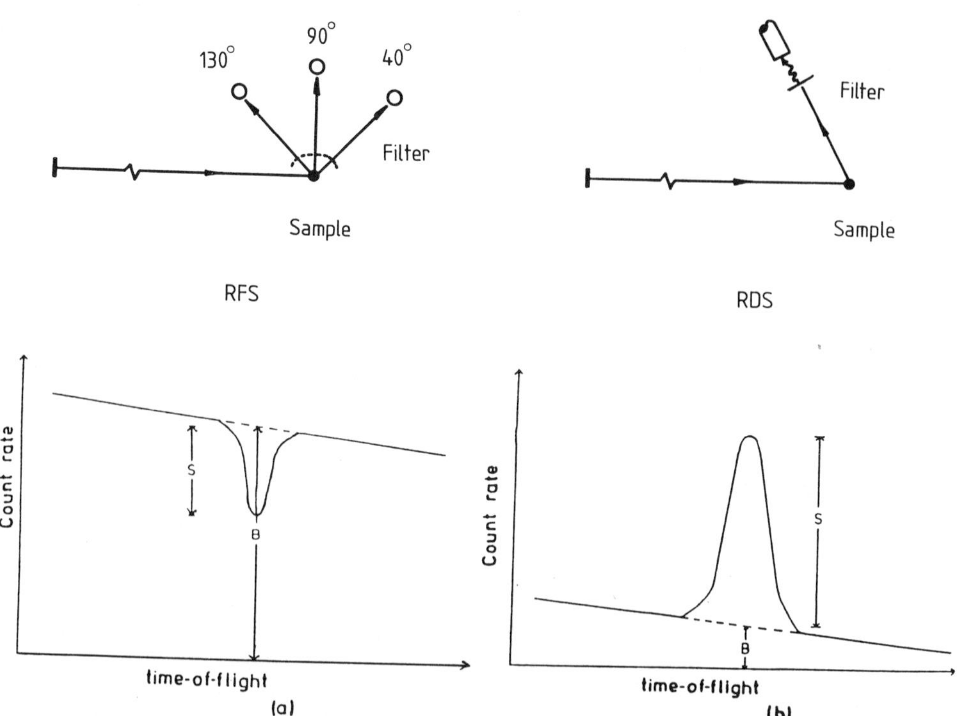

Figure 7. Resonance absorption spectrometers used on pulsed neutron sources which determine the recoil scattering spectrum either directly or by taking a difference between two spectra.

In recoil scattering it is the total cross section of the atom which contributes to the signal and hence the inherent limitation of the RFS method with a single detector (at a single well defined scattering angle) is not a problem owing to the large signals involved. Shielding of the detectors from the fast neutron and gamma-ray background associated with the main beam and the halo of neutrons produced by scattering from the primary beam collimators can be achieved using appropriate B_4C and boron loaded lead but this is only effective for resonances up to ~10-15 eV.

Because of the large resonance energies involved any region of (q,ω) space can, in principle, be reached. However, it is the combination of resonance material and spectrometer geometry which dictates the range of energy and momentum transfer. Recoil energy and momentum transfer curves for the 4.28 eV resonance in Ta and the 6.67 eV resonance in U foils at a fixed angle of scattering (θ=150°) together with the dispersion relations for atoms of mass 4, 7, 12 and 56 are shown in figure 8. In the case of hydrogeneous samples the kinematics of the scattering process dictates scattering angles of less than 90°, and hence a concentration of detectors in the forward angle direction.

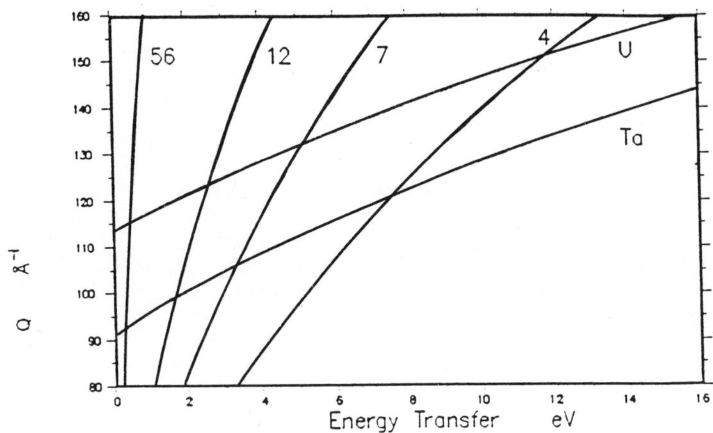

Figure 8. The (q,ω) plot for the 4.28 eV resonance in Ta and the 6.67 eV resonance in U together with the dispersion curves for recoil scattering from helium, lithium, carbon and iron.

For resonance scattering the absolute resolution is determined mainly by the resonance width of the filter and by the neutron pulse width. Improvements in the energy resolution component can be achieved by cooling the filter or alternatively by removing the Lorentzian wings of the resonance line via a difference spectra involving two foils of different thicknesses or temperature [12]. Such gains are however offset by the longer counting times involved in order to regain the required statistical accuracy. The poor resolution of these spectrometers is offset to a large degree by the broad width of the recoil line which, in the impulse approximation, is directly proportional to the momentum transfer. Unlike chopper spectrometers momentum transfer values well in excess of 100 $Å^{-1}$ can be routinly obtained using the resonance method.

DINS spectrometers based on resonance techniques are just beginning to show results. It is therefore perhaps too early to clearly indicate their true scientific potential. However, figure 9 shows some recent results obtained on the eVS spectrometer on ISIS depicting the recoil scattering from polycrystalline Be encased in an aluminium container and

using the 4.28 eV resonance in Ta; both recoil scattering peaks (from Al
and Be) are clearly visible and are well separated. Also shown are the
expected FWHM of the instrument function for both eVS and for a chopper
spectrometer. Such resolutions are typical of current photon scattering

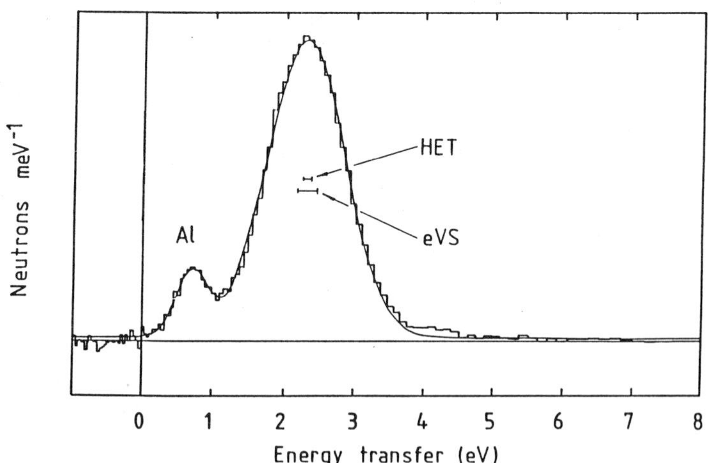

Figure 9. The DINS spectrum of Beryllium obtained using the
4.28 eV resonance in Ta [13]. Also shown are the expected instrument
resolutions appropriate to eVS and the chopper spectrometer HET.

measurements of electron momentum distributions. Studies have also been
undertaken on other light atoms including pyrolytic graphite [14,15],
liquid ^{4}He [16,17], lithium and lead as well as previous exploratory work
involving light and heavy water and hydrogen bound by various potentials
[18].

RESOLUTION EFFECTS IN MOMENTUM SPACE

In any detailed discussion on the momentum density it is important to
consider the absolute resolution in momentum space (p-space). In order to
achieve this it is necessary to consider the role of the angular
contributions to the total resolution. In a recent paper [19] calculations
have been performed to determine the resolution in momentum space for both
chopper and resonance spectrometers on pulsed neutron sources. Within the
limitations of the analytical procedure adopted the results indicate that
for large masses i.e. M>4 amu the energy resolution is the dominant factor
determining the p-space resolution. However, for recoil scattering in
hydrogenous samples the resolution of the resonance method is dominated by
the angular contribution. Moreover as the mass increases the p-space
resolution becomes progressively poorer relative to the momentum
distribution width. In the case of large q measurements i.e. resonance
techniques with q values > 50 Å^{-1}, relatively poor energy resolution can
nevertheless still provide good p-space resolution. The results have also
indicated that the advantage in energy resolution of the chopper

spectrometer over the resonance method is diminished when the comparison is made in p-space.

FURTHER EXPERIMENTAL DEVELOPMENTS

What are the future experimental developments for determining recoil spectra and extracting the momentum density using neutron techniques ? It is clear that future measurements will make extensive use of chopper spectrometers on spallation sources due to their good resolution properties and reasonable high range of q. This may equally apply to resonance spectrometers taking advantage of the more extensive range of q available. Indeed the range of q on chopper instruments is sufficient to validate the impulse approximation for a large number of masses; the shape of the recoil line can be ascertained with a high degree of precision both in terms of the counting statistics and in momentum resolution. Moreover, the large number of detector elements covering a range of scattering angles allows the recoil scattering structure to be followed as a function of q.

Resonance spectrometers on the other hand provide q values well in excess of those attainable on chopper devices i.e. >100 $\mathring{A}^{-1}$, albeit at a slightly degraded resolution, and this is illustrated clearly in figure 10.

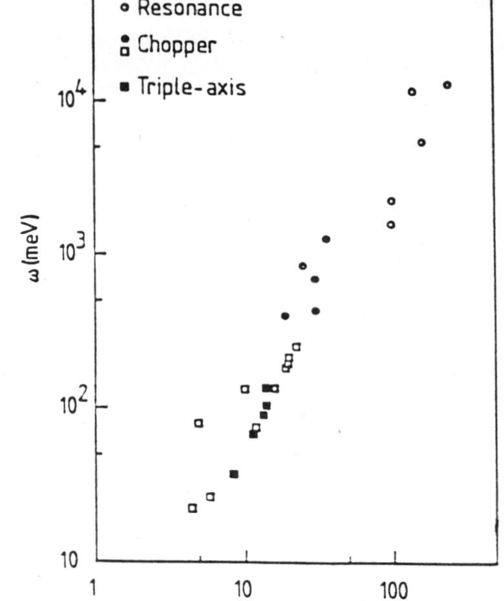

Figure 10. Energy-momentum transfer regions attained on the different types of neutron spectrometers used on reactor and pulsed neutron sources.

Moreover, the widths of the recoil scattering peaks are in general large in comparison to the total p-space resolution function. It may be also be possible to use higher energy resonance lines and thus be able to extract the momentum density at two or three E_1 values. Exploratory measurements have already been made using the 10.36 eV resonance in Ta but background problems prevents a clear analysis of the data. Nevertheless, the use of higher energy resonances could be an important development, and as previously mentioned, may lead to an improvement in the overall resolution in p-space.

Alternative ways of improving the resolution in resonance techniques have already been discussed i.e. double difference techniques, cooled resonance foils, and these will similarly be exploited. Also in the future it may be possible to reliably deconvolute the instrumental resolution function and associated multiple scattering effects from the observed recoil scattering although further studies in this area are required. The development of a good understanding of the shape of the instrumental resolution will in future studies be of paramount importance particularly when considering a detailed analysis of the recoil line shape. Finally the development of a resonance chopper device, extending the available range of a direct geometry spectrometer to beyond the 2 eV limit, is also being looked at. Such a device would be able to combine its excellent resolution characteristics of the chopper device with the high q capability found in resonance methods.

Improved instruments and new neutron sources will undoubtedly provide higher fluxes, greater momentum transfers and improved resolutions. This increase in overall performance of future instruments will allow detailed investigations of both initial [20] and final state effects, and moreover, provide a means of developing the appropriate correction procedures to give an absolute momentum density distribution. The state of DINS is similar to that encountered in photon studies some twenty years earlier. The experienced which has been gained over the last two decades will undoubtedly help the neutron scattering community in evaluating the role of the atomic momentum density distribution in condensed matter science.

ACKNOWLEDGEMENTS

The authors would like to thank the Paul Sokol and Richard Silver for the invitation to attend the Momentum Density Workshop and for the opportunity to present some of the work recently carried out on ISIS.

REFERENCES

1. M. J. Cooper (this volume)
2. S. Berko (this volume)
3. P. C. Hohenberg and P. M. Platzman, Phys. Rev. 152 198 (1966)
4. H. A. Gersch and L. J. Rodriguez, Phys. Rev. A 8 905 (1973)
5. P. Martel, E. C. Svensson, A. D. B. Woods, V. F. Sears and R. A. Cowley, J. Low Temp. Phys. 23 285 (1976)
6. O. K. Harling, Phys. Rev. Letts. 24 1046 (1970)
7. O. K. Harling, Phys. Rev. A 3 1073 (1971)
8. H. A. Mook, Phys. Rev. Letts. 32 1167 (1974)
9. R. A. Cowley and A. D. B. Woods, Phys. Rev. Letts. 21 787 (1968)
10. H. A. Mook (this volume)
11. C. G. Windsor, in Pulsed Neutron Scattering, Taylor and Francis, London (1981)
12. H. Rauh and N. Watanabe, Nucl. Inst. Meths. 228 147 (1984)
13. R. S. Holt and M. P. Paoli (to be published)
14. H. Rauh and N. Watanabe, Phys. Letts. A 100 244 (1984)
15. M. P. Paoli and R. S. Holt, J. Phys. C. 21 3633 (1988)
16. R. S. Holt, L. M. Needham and M. P. Paoli, Phys. Letts. A 126 373 (1988)
17. S. Ikeda and N. Watanabe, Phys. Letts. A 121 34 (1987)
18. A. D. Taylor, Rutherford Report RAL-84-020 (1984)
19. C. Andreani, G. Baciocco, R. S. Holt and J. Mayers, Nucl. Inst. Meths. A276 297 (1989)
20. J. Mayers, C. Andreani and G. Baciocco (accepted in Phys. Rev. B)

THE DETERMINATION OF ELECTRON MOMENTUM DENSITY DISTRIBUTIONS

Malcolm J. Cooper

Department of Physics
University of Warwick
Coventry, CV4 7AL UK

INTRODUCTION

This topic truly encompasses several experimental methods besides
Compton scattering; for example the angular correlation of positron
annihilation radiation (ACAR) and inelastic electron scattering,
sometimes referred to as electron Compton scattering. Fortunately
positron annihilation is dealt with admirably by Professor S Berko
elsewhere in this volume[1]. The theory behind Compton scattering and
potential for future studies are discussed in detail by Dr P Platzman
whose interest in and influence on this area is longstanding[2], and Dr D
Mills has concentrated on magnetic scattering[3], which is currently my
consuming passion as well as his. All this makes my task of describing
the experimental method and setting it into context that much easier. I
shall restrict myself to quoting results to illustrate the state of the
Compton scatterer's art, in the confidence that the reader will be able
to find out more about the physical interpretation elsewhere in this
volume.

Does Compton scattering deserve prominence alongside these other
methods? In the past that was questionable, principally because of the
low resolution of the technique. Now there is no doubt that the current
developments associated with instrumentation at synchrotron sources do
really promise to realise the much-vaunted potential of the method.

THE COMPTON PROFILE

The electron momentum density distribution, $n(\underline{p})$, is just as
informative a ground state function as its more familiar counterpart,
$\rho(r)$, although it requires some re-education to think about chemical
bonding or solid state cohesion in these terms. [See Williams[4] or
Cooper[5] for reviews].

Information about $n(p)$ comes from weak scattering experiments,
treated in the Born approximation, in which the target electrons are
catapulted into plane wave continuum states. This latter condition
requires a further approximation, the impulse approximation. Then, as
was pointed out by Platzman and Tzoar[6] the matrix element

$$|<f|\textstyle\sum \exp i\underline{k}.\underline{r}|i>|^2 \ , \qquad\qquad 1$$

which describes the scattering process, reduces to an integral over n(p) the final form of which depends on precisely what is measured. For example if the incident and scattered photon energies and momenta alone are known the double differential cross-section yields the so-called Compton profile, $J(p_z)$, which is the projection of the momentum distribution along the scattering vector i.e.

$$J(p_z) = \int\limits_{p_x} \int\limits_{p_y} n(\underline{p}) dp_x dp_y \qquad 2$$

This is the analogue of the long-slit ACAR positron annihilation experiment without the complications associated with the inclusion of the wavefunction of the charged positron, but also to date without the high resolution that has been the key to the exploitation of the positron method for Fermi surface reconstruction[7]. Angular correlation measurements can, and do, yield the two-dimensional projection and photon scattering studies would have to include some information about the recoiling electron in order to match this. It has not happened yet, although coincidence experiments are planned for very high energy synchrotron sources[8]. It is much more likely that the improvement in the resolution of the one dimensional Compton profile measurement, coupled with the straightforward nature of the interpretation (the photon is a "clean" probe), will make that experiment much more competitive for $n(\underline{p})$ studies. The major success of Compton studies has been in the identification of electron-electron correlation effects beyond the local approximation[9,10] despite the modest resolution and the painstaking nature of the data analysis.

Inelastic electron scattering studies are possible in a single scattering regime in gases. With complete knowledge of incident, scattered and recoil electrons the reconstruction of n(p) shell-by-shell is possible with care[11]. This is most wonderful and most difficult!

FROM X-RAYS TO GAMMA RAYS

There is no doubt that the exploitation of Compton line shape analysis began, and almost ended, with the work of Dumond and his co-workers[12,13]. Their ingenuity in designing and building not only focussing X-ray spectrometers but also X-ray tubes specially adapted for Compton work is legendary. There is only space here for one illustration, figure 1 shows the multi-crystal spectrograph of Dumond and Kirkpatrick[13] which incorporated fifty individually aligned calcite analysing crystals.

They were well aware of the principle requirements for meaningful measurements of the line profile. To begin with Compton scattering is an incoherent process, so the count rates are low unless high flux sources can be coupled to large samples. The momentum space resolution improves as the scattering angle approaches 180° and the beam divergence is reduced. This is hardly a practicable combination but the focussing multicrystal spectrograph represented an attempt to reconcile these conflicting requirements. The role of beam divergence is illustrated in figure 2 for a set of typical energies. Taking a target resolution figure of 0.1 a.u. - 10% of the Fermi momentum in aluminium it is evident that beam divergence is a minor problem at X-ray energies but a major one at higher energies were it not for the fact that semiconductor detectors produce depressingly inferior resolution parameters (never better than 0.4 a.u.).

The focussing spectrometers of today do bear comparison with Dumond's, but then they have the advantage of thirty years of optics and detector development, not to mention X-ray sources six orders of magnitude brighter than sealed tubes. The X-ray systems developed in the

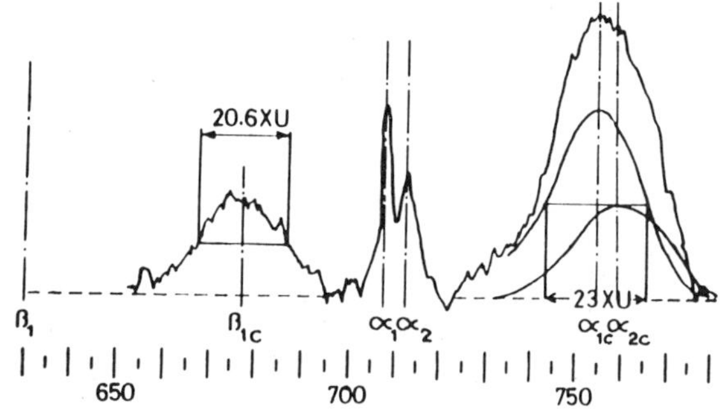

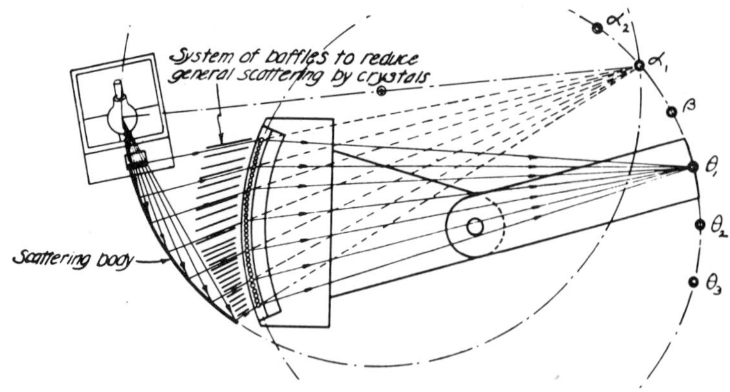

FIGURE 1.   The focussing multicrystal spectrograph of Dumond and Kirkpatrick[13] which contained 50 separately aligned calcite crystals, is shown together with a typical spectrum of molybdenum radiation scattered from graphite.   The momentum resolution is ~ 0.2 a.u.

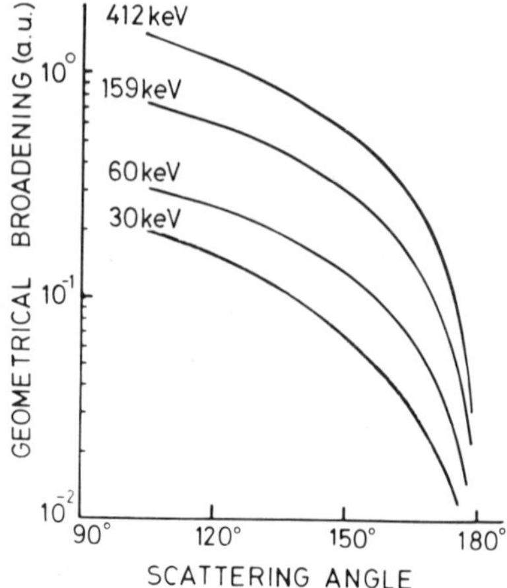

FIGURE 2.   The effect of a ± 1° beam divergence on the momentum broadening of the Compton profile at a range of typical photon energies.

mid-sixties at the renaissance of experimental work were decidedly inferior. Good resolution, achieved without the benefit of focussing X-ray optics was a prescription for miserably slow, statistically imprecise experiments. No wonder joy abounded when Eisenberger and Reed[15] led the conversion to higher energy sources and solid state detectors: there were no qualms in sacrificing resolution (worse by a factor of 2 or 3) for statistical precision (better by 1 or 2 orders of magnitude) and reliability.

In changing from 17 keV (MoKα) or 25 keV (AgKα) X-rays to 60 keV (^{241}Am), 159 keV (^{123m}Te) or 412 keV (^{198}Au) gamma rays the experiment was liberated from the restraints of photoelectric absorption as can be judged from figure 3. Furthermore the signal to noise problems associated with bremsstrahlung backgrounds and doublet sources disappeared. For elements heavier than aluminium gamma rays afford progressively increasing gains in the speed of the experiment even with source activities of tens of Curies. For the first time experiments on the 3d elements and beyond became a practicable proposition. One example should suffice: the first X-ray measurement of the Compton profile of iron[16] in 1968 took one month to accumulate $\simeq$ 50,000 counts at a signal to noise ratio of unity whereas a recent gamma ray study[10] of Ni with 200 Curie ^{198}Au sources yielded profiles with integrated intensities in excess of 10^6 counts in 50 hrs at typical signal to noise ratios of 400:1!

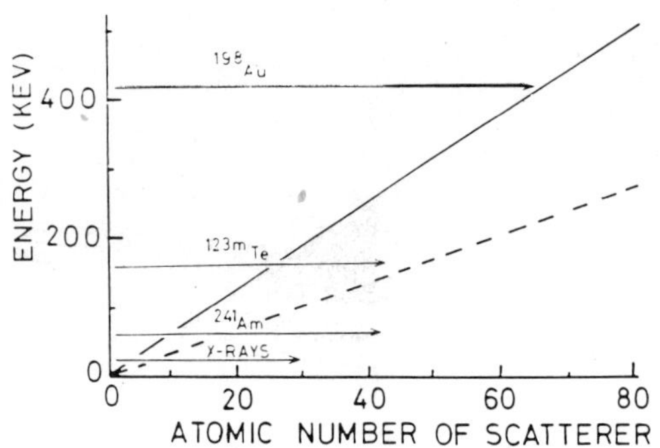

FIGURE 3. The regime of viable experiments. The diagonal solid line divides the regime to the right and above, where the Compton cross-section exceeds that of photo-electric absorption from the lower part where the situation is less favourable. The dashed line is an estimate of a similar divide between the viability (above) and failure (below) of the impulse approximation. The range of experiments performed with typical sources is shown.

The question of the validity of the impulse approximation has never appeared as a major obstacle to compare with the practical difficulties indicated above; the shift to higher energies pushed it further into the background. Theorists have worried about the interpretation of line shapes when the final electron state is not a continuum plane wave and there have been several 'exact' calculations involving more realistic final states[17,18] but experimentalists have chosen to avoid the more obvious pitfalls and calculations for typical situations predict minor defects (peak shifts and shape changes) for closed shells. The restriction of low energy X-ray and gamma ray studies to 'difference experiments' in which the principle offending core contributions are subtracted out helps to preserve the impulsive interpretation. The deviations are more important in the context of electron scattering experiments where there is a more extensive bibliography of measurements and calculations[19,20] of defects in the shift and profile asymmetries.

A bigger headache arises in the photon experiment in determining the cross-section for scattering from a moving electron. The expression that can be adopted at X-ray energies, which preserves the direct relationship between the spectral distribution and the Compton profile integral, is an inadequate approximation at the higher gamma ray energies away from backscattering[21]. Improvements involving successive iterative approximations[22] thankfully appear adequate.

The gamma ray experiments are for lazy but patient experimentalists and ardent data processors! In principle the acid test of the data analysis is the reconstruction of a symmetric profile. Unfortunately in practice the more problematical corrections are associated with asymmetric terms which may be compensatory in their effect. The situation is particularly acute for the gamma ray experiments because the profile is likely to be spread over 10-20 keV and the energy variation of those corrections listed in Table 1 becomes more critical. Multiple scattering, first acknowledged as the bete noire of Compton scattering, has gradually been tamed with the development of Monte Carlo simulations[23], implemented on more and more powerful computers. Recently there has been a realisation that the scattering within the source material produces a low energy tail that has many of the characteristics

TABLE 1. PROCESSING COMPTON DATA
GAMMA-RAY AND X-RAY CORRECTIONS COMPARED

CORRECTION	γ-RAY	X-RAY
Energy calibration	easy	easy
Background subtraction	easy-small correction	generally larger correction
Detector efficiency $\eta(E)$	difficult-wide energy range	easy
Sample absorption $\mu(E)$	easy	easy
Deconvolution	difficult-tail on response function	easy
Cross-section transformation $\dfrac{d^2\sigma(E)}{d\Omega d\omega}$	difficult	easy
Profile Normalisation	easy	more difficult
Multiple scattering in sample	difficult Monte Carlo calculation	smaller correction generally required
Source scattering contamination	very difficult	Not applicable

of a solid state detector response function, but requires a different correction procedure[24].

Some typical gamma ray data sets are collected together in figure 4. The first point to note is the small scale of the significant effects [~ 1% J(0)] and the second is the high statistical precision with which

they have been established - typically better than ¼% J(0) at low
momentum. This is especially impressive in the absolute profiles which
do not benefit from the cancellation of errors that occurs in forming
directional differences. This type of experiment is surely at its limit
when differences of this small order have to be interpreted. At this
level the validity of any correction can be called into question. It is
time to improve the sensitivity of the experiment to place less reliance
on the accuracy of the interpretative process.

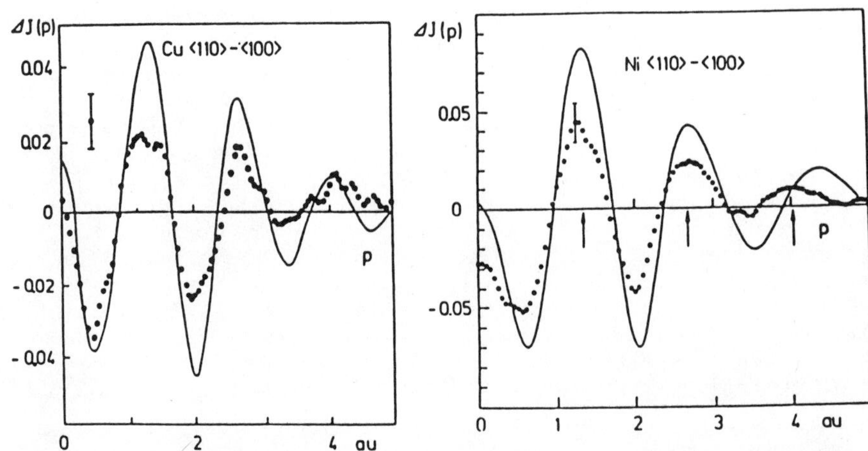

FIGURE 4.   A comparison of the Compton profile anisotropies in
copper and nickel showing the reduced scale of the
experimental data as compared with band theory
(solid lines).[10] This pervading feature is taken as
evidence for non-local correlation corrections. NB
the small scale of the anisotropies; they are less
than 1% of the profile peak heights.

INCREASING THE INFORMATION CONTENT

(a) Spin dependent measurements

     'Magnetic' Compton scattering, discussed here by Mills[3] and
originally proposed by Platzman and Tzoar[25] is an obvious example of the
process of getting closer to the physics. The poor resolution of the
solid state detector may be tolerable, at least for a while, if the
information content could be made more specific. The inelastic
scattering of circularly polarised radiation from ferromagnets affords an
opportunity to do this because the magnetic modulation of the signal
isolates a spin-dependent term that automatically eliminates the
contribution from the closed shell core.

$$\text{i.e. if } J(p_z) = \sum_i \iint [n_{i,up}(\underline{p}) + n_{i,down}(\underline{p})]\, dp_x dp_y \qquad 3$$

$$\text{then } J_{mag}(p_z) = \sum_i \iint [n_{i,up}(\underline{p}) - n_{i,down}(\underline{p})]\, dp_x dp_y \qquad 4$$

where the sum is over all the bands. Unfortunately the cross-section
yielding $J_{mag}(p_z)$ is scaled down from that yielding $J(p_z)$ by a factor of
E_1/mc^2. This implies a need for extremely large accumulated counts e.g.
a 2% magnetic effect measured to 10% statistical precision requires 10^6
photons. Therefore in view of the competition for high energy

synchrotron beam time the count rates must be high. That is not just a
pipedream, horizontal focussing has yet to be used and the inclined view
method of extracting circularly polarised photons produces fluxes which
are only a factor of 4 or 5 lower than those in the orbital plane (fig.
5). Many of the developments associated with focussing optics, discussed
later, are appropriate here. A helical Wiggler with a brightness of 10^{13}
p s^{-1} mrad^{-2} mm^{-2} in a 0.1% bandwidth, delivering a beam predominantly
circularly polarised, now being installed at the Photon Factory[26], makes
a severe dent in the practical difficulties.

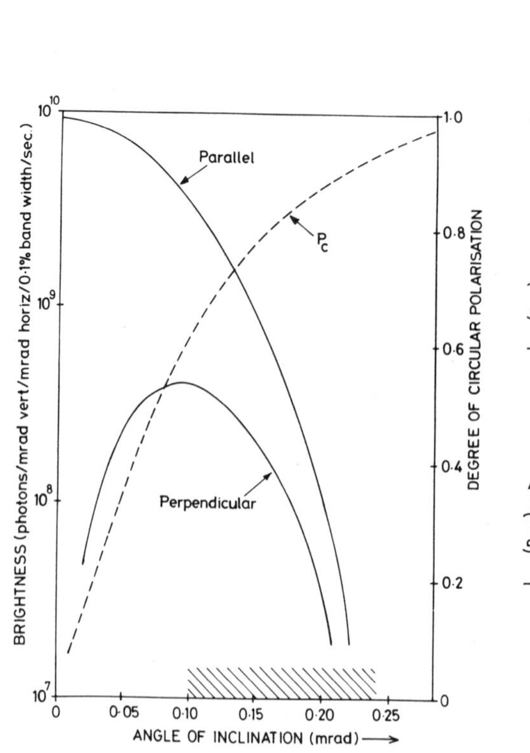

FIGURE 5. Circularly polarised flux
of 60 keV photons extracted by the
inclined view method from the Daresbury
storage ring. The angular variation of
the flux of parallel and perpendicular
components is indicated by the left hand
ordinate and the degree of circular
polarisation by the right.

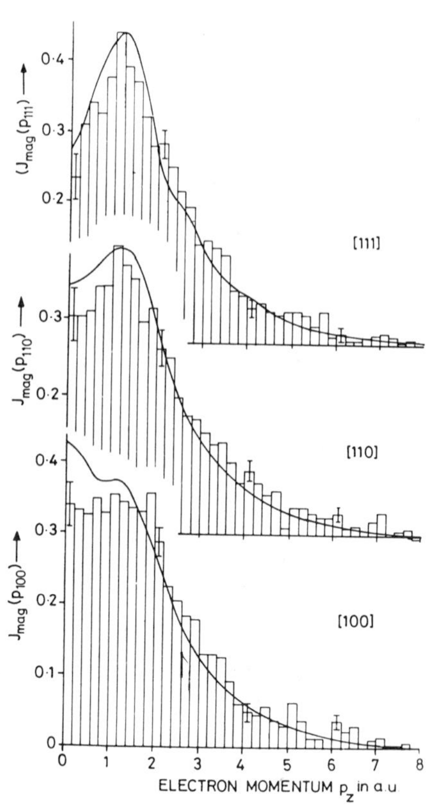

FIGURE 6. The anisotropy in
the magnetic profile of
crystalline iron measured with
circularly polarised SR. The
solid line refers to an APW
model[27].

        The enhanced information content is easily illustrated by the
directional magnetic profiles of iron, measured by the Warwick group[27]
and reproduced in figure 6: the anisotropies are on the scale of tens of
percents, not the 1% or 2% evident in total profiles such as those in
figure 4. Furthermore the separation of spin up and spin down profiles
is possible by combining $J(p_z)$ and $J_{mag}(p_z)$ — eqns. 3 and 4.

(b) Coincidence Measurements

This is something of a novelty in photon scattering experiments. To date studies have been limited to the simultaneous detection of the K-shell fluorescence X-ray and the scattered photons, thereby isolating that core contribution. Those experiments[28] have concentrated on the low energy and momentum transfer regime when the impulse approximation is invalid.

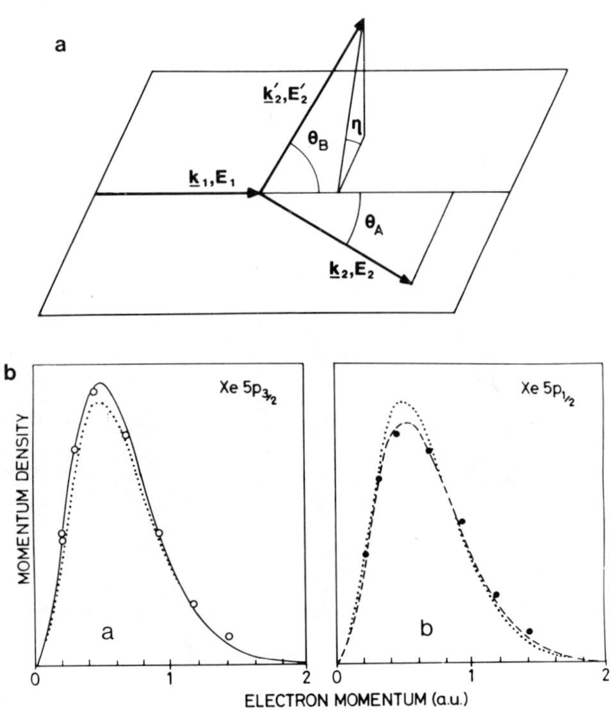

FIGURE 7(a)   Geometry of the e-2e reaction. In the actual apparatus an incident collimated monochromatic beam (k_1, E_1) is scattered by a gas target. The scattered and recoil beams (k_2, E_2, k'_2, E'_2) are detected by cylindrical mirror analyers coupled to channel electron multipliers. Their energy resolution is ~ 1% and the resolving time of the coincidence circuit is ~ 10 ns. In the symmetric coplanar geometry the experiment is performed by setting $\eta = 0$ and varying $\theta_A (=\theta_B)$.

(b)   The electron momentum density distribution for (a) the $5p^{3/2}$ (O) and (b) $5p^{1/2}$ (O) orbitals of atomic xenon as deduced from e, 2e coincidence experiments with a 1200 eV electron source. The relativistic density distributions (____) in each case are different whereas the non-relativistic prediction (...) is the same. The data are clearly closer to the relativistic calculations[29].

A more interesting coincidence experiment[8], in terms of studying n($\underline{p}$), has been proposed for the ESRF high energy beam line using 150 keV photons incident on foils sufficiently thin for multiple electron scattering to be ignored (roughly 10 nm at this energy). If information about the energy and momentum of the scattered photon and the recoil electron is recorded then n(p), not a projection of it is measured. The feasibility study for this γ-γ,e experiment suggests that true coincidence rates of 50 hr^{-1} are possible, the strays accounting for no more than 1% of that.

Although the prospect of γ-γ,e is somewhat uncertain e-2e studies of gases are well established[11,29]. They enjoy one vital advantage in that discrimination between the different atomic shells is possible by virtue of their different binding energies; electron energy analysers/guns have the resolution to tune into the particular contributions and therefore produce momentum distributions orbital by orbital. Figure 7(a) shows the kinematical arrangement and 7(b) a typical result[29]:

BACK TO X-RAYS

There is no doubt that the key to the future of Compton scattering experiments is the improvement of the resolution. Figures approaching 1/2 a.u., which is a sizeable fraction of a Brillouin zone, make the task of extracting physical sense almost impossible. Apart from unpredictable new developments associated with bolometric detectors[30] that improvement must be linked with focussing X-ray optics.

Fortunately Loupias[31] and co-workers at LURE have been following in Dumond's footsteps for some time with the instrument shown schematically in figure 8(a). It uses a double crystal monochromator, a Cauchois curved analyser and a position sensitive detector. The parameters of this and other instruments are gathered together in Table 2 but a typical set of results, on LiC_6, intercalated graphite, are shown in fig. 8(b), the count rate into the detector is a healthy 30 cps at 0.15 a.u. resolution. An imminent transfer to a wiggler beam line is expected to improve these figures at energies around 25 keV.

The resolution figure has already been bettered on a Wiggler line at the Photon Factory by Itoh et al[32] with a spectrometer of similar design (Cauchois Si422 analyser) operating at 29.5 keV. A count rate of 220 cps at Δp = 0.08 a.u. has been achieved thanks to the innovation of using an imaging plate as the position sensitive detector. This device works on the basis of laser-stimulated luminescence from a europium-loaded phosphor. Photon efficiency is almost two orders of magnitude higher than photographic film and the dynamic linear range covers five orders of magnitude! Despite the initial cost associated with reading and erasing the phosphor the adoption of imaging plates offers distinct advantages over gas-filled position sensitive detectors, especially at high energies. The combination of such a spectrometer and a helical Wiggler with a similar brightness of circularly polarised photons is a most exciting prospect.

Similar spectrometer developments are being made by Schulke[33] at Hamburg. There is interest in the whole range of the dynamical structure factor $S(k,\omega)$ from the characteristic energy loss regime where a resolution $\Delta E/E < 10^{-4}$ is possible at a count rate in excess of 100 cps with a spherically bent analysing crystal, to the high energy Compton regime where saggital focussing of the monochromater is proposed to give an inelastic scattering spectrometer which in the high energy range (40 keV) would yield Δp = 0.05 a.u. at count rates approaching 1000 cps for ESRF Wiggler parameters.

Laboratory based focussing spectrometers such as those being developed by Manninen[28] and Suortti in Helsinki[34] may remain competitive because of the advantage of full time operation at homebase. The quoted[28] figure of 10^7 p s^{-1} of $WK\alpha_1$ focussed radiation is an attractive

TABLE 2. FIGURES OF MERIT FOR COMPTON SPECTROMETERS
THE CHRONOLOGICAL DEVELOPMENT

METHOD	ENERGY (keV)	PEAK BACKGROUND RATIO	MOMENTUM RESOLUTION (a.u.)	TYPICAL INTEGRATED COUNT RATE (C.P.S.)	YEAR
MₒKα X-rays and non-focussing spectrometer	17	5:1	0.25	10	1974
^{241}Am + SSD	59.5	500:1	0.55	20	1978
^{198}Au + SSD	412	500:1	0.40	10	1979
CuKα X-rays and focussing spectrometer	8	15:1	0.07	1	1981
SR (LURE) and focussing spectrometer	10	100:1	0.15	30	1984
SR (PHOTON FACTORY) and focussing spectrometer	29.5	-	0.08	200	1988
SR (ESRF Wiggler) and focussing spectrometer	40	-	0.05	> 100	?

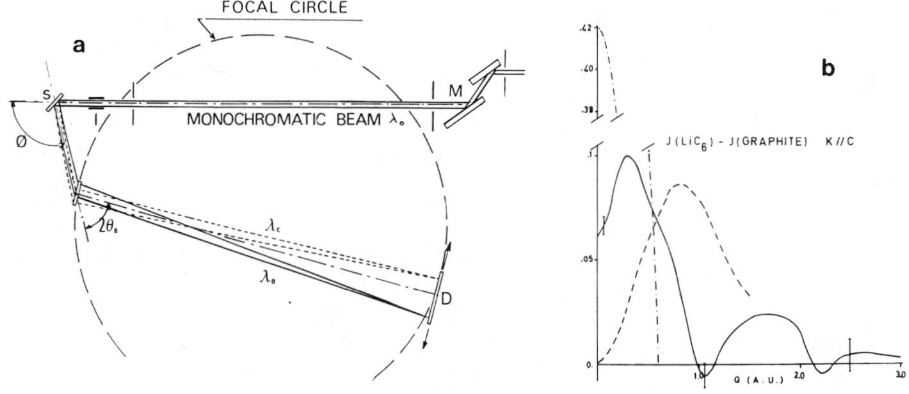

FIGURE 8(a)   Schematic diagram of the focusing Compton spectrometer installed on the LURE-DCI synchrotron ring. The white beam is monochromated by the channel-cut silicon monochromator. The curved crystal analyser and the multiwire position sensitive detector D move on the focussing circle.

(b)   Compton profile difference for k parallel to c. $J(LiC_6)$ minus $J(graphite)$. Experiment (______) is compared to a graphite Compton profile (----) and to a metal-like profile (-.-). The result shows that the additional electron is not free electron like but is delocalised as compared with the π-band.[31]

one for 60 keV Compton scattering. Storage rings are under intense
scheduling pressure and incoherent scattering studies necessarily require
longer measurement times than diffraction work, but in the long run the
advantages they offer will become overwhelming.

SUMMARY

Table 2 essentially summarises the march of progress. The wheel
has turned full circle and the instruments required to improve momentum
resolution to a physically interesting level are focussing X-ray
spectrometers that Dumond would recognise and probably improve! In
general they will only be practicable in conjunction with the bright
synchrotron sources that are now being built or designed. Gamma ray
Compton scattering has provided an opportunity to do precise experiments
which have necessitated the detailed understanding of numerous problems
associated with the interpretation of Compton spectra. It was a most
valuable exercise without which many of the important points would have
been missed. Now it is ending and there are ways forward by which this
technique can realise its full potential. I can only hope that this
brief survey conveys that positive message and provides a starting point
for further reading.

ACKNOWLEDGEMENTS

I am grateful to the organisers of this meeting for inviting my
participation and to the Science and Engineering Research Council of
Great Britain for supporting the work that has involved me in this field.
In surveying the present and proposed experimental methods I am indebted
to many colleagues for providing me with details of their current work
and future plans. In that regard I particularly wish to thank Winifried
Schulke, Seppo Manninen, Fumitake Itoh and Geneviève Loupias.

REFERENCES

1.   S. Berko, Overview of Positron Annihilation --- this volume
2.   P. Platzman, Theory and Experiment on Compton Scattering -- this
     volume
3.   D.M. Mills, Measuring the Momentum Distribution of Unpaired Spin
     Electrons in Ferromagnets Using Synchrotron Radiation --- this
     volume
4.   B.G. Williams editor Electron Momentum Determination, McGraw-Hill,
     New York 1977.
5.   M.J. Cooper, Compton Scattering and electron momentum
     determination, Rep. Prog. Phys. 48: 415 (1985).
6.   P.M. Platzman and N. Tzoar, X-ray scattering from an electron gas,
     Phys. Rev. 129: A140 (1965).
7.   D.G. Lock, V.H.C. Crisp and R.N. West, Positron annihilation and
     Fermi surface studies: a new approach.
8.   F. Bell, Electron momentum densities by inelastic gamma-electron
     coincidence measurements, in the proceedings of the Workshop on
     Applications of high energy sources, edited by A.K. Freund, ESRF
     Grenoble (1988).
9.   G.E.W. Bauer and J.R. Schneider, Nonlocal exchange correlation
     effects in the total Compton profile of copper metal, Phys. Rev.
     Letts. 52: 2061 (1984).
10. A.J. Rollason, J.R. Schneider, D.S. Laundy, R.S. Holt and M.J.
     Cooper, Electron momentum density in nickel, J. Phys. F. 17: 1105
     (1987).
11. R.A. Bonham and H.F. Wellenstein, Electron scattering in Compton
     scattering, by B.G. Williams, McGraw-Hill: New York (1987).

12. J.W.M. Dumond, The linear momenta of electrons in atoms and in solid bodies as revealed by X-ray scattering, Rev. Mod. Phys. 5: 1 (1933).

13. J.W.M. Dumond and H.A. Kirkpatrick, Multi crystal spectrograph, Rev. Sci. Inst. 1: 88 (1930).

14. P.A. Ross and P. Kirkpatrick, The constant of the Compton equation Phys. Rev. 45: 223 (1934).

15. P. Eisenberger and W.A. Reed, Gamma ray Compton scattering: Experimental Compton profiles for helium, molecular nitrogen, argon and krypton, Phys. Rev. A5: 2085 (1972).

16. M.J. Cooper and B.G. Williams, The Compton profile of iron, Phil. Mag. 17: 149 (1968).

17. P. Eisenberger and P.M. Platzman, Compton scattering of X-rays from bound electrons, Phys. Rev. A2: 415 (1970).

18. B.J. Bloch and L.B. Mendelsohn, Atomic L-shell Compton profiles and incoherent scattering factors: Theory, Phys. Rev. A9: 129 (1974).

19. F. Gasser and C. Tavard, Asymmetry of the Compton profile, the Compton defect; Chem. Phys. Letts. 79: 97 (1981).

20. F. Gasser and C. Tavard, Effet Compton et structures electroniques, J. Ch. Phys. 78: 488 (1981).

21. S. Manninen, T. Paakkari, and K. Kajantie, Gamma-ray Comtpon profile of aluminium, Phil. Mag. 29: 167 (1974).

22. R. Ribberfors, Relationship of the relativistic Compton cross section to the momentum distribution of bound electron states, Phys. Rev. B12: 2067 and 3136 (1975).

23. J. Felsteiner, P. Pattison and M.J. Cooper, Effect of multiple scattering on experimental Compton profiles: a Monte Carlo calculation, Phil. Mag. 30: 537 (1974).

24. S.O. Manninen, M.J. Cooper and D.A. Cardwell, Gamma ray source line broadening and Compton line profile analysis, Nucl. Instr. & Meths., A245: 485 (1986).

25. P.M. Platzman and N. Tzoar, Magnetic scattering of X-rays from electrons in molecules and solids, Phys. Rev. B2: 3556 (1970).

26. H. Kawata, Compton scattering experiments at the photon factory, proceedings of the workshop on Applications of high energy X-ray scattering at the ESRF, Grenoble; edited by A.K. Freund (1988).

27. M.J. Cooper, S.P. Collins, D.N. Timms, A. Brahmia, P.P. Kane, R.S. Holt and D. Laundy, Magnetic anisotropy in the electron momentum density of iron, Nature 333: 151 (1988).

28. S. Manninen, K. Hamalainen, T. Paakkari and P. Suortti, Inner shell studies using inelastic spectroscopy of low energy gamma rays, J. de Physique, coll. C9, suppl. 12: 823 (1987).

29. J.P.D. Cook, J. Mitroy and E. Weigold, Direct observation of relativistic effects in single-electron momentum distributions in Xenon outer shells, Phys. Rev. Letts. 52: 1116 (1984).

30. D. McCammon, M. Juda, J. Zhang, R.L. Kelley, S.H. Moseley and A.E. Szymkowich, Thermal detectors for high resolution spectroscopy, IEEE Trans. Nucl. Sci. 33: 236 (1986).

31. G. Loupias, J. Chomilier and D. Guerard, Lithium intercalated graphite: experimental Compton profile for state gone, J. Physique Lett. 45: L301 (1984).

32. F. Itoh, M. Sakurai, T. Sugawa, K. Suzuki, N. Sakai, M. Ito, O. Mao, N. Shiotani, Y. Tanaka, Y. Sakurai, H. Kawata, S. Nanao, Y. Amemiya and M. Ando, Proc. Synchrotron Instrumentation Conference, Tsukuba (1988), in print Nucl. Instr. & Meths.

33. W. Schulke, Inelastic X-ray Scattering with Synchrotron radiation: the scientific case, current experiments and projects, Nucl. Instr. & Meths, A246: 491 (1986).

34. P. Suortti, V. Etelaniemi, K. Hamalainen and S. Manninen, J. de Physique, Coll. C9, suppl. 12: 831 (1987).

EXPERIMENTAL DETERMINATION OF MOMENTUM DISTRIBUTIONS IN NUCLEAR PHYSICS

Donal Day

Institute of Nuclear and Particle Physics
University of Virginia
Charlottesville, VA 22901

Introduction — Quasielastic Knockout Reactions

Removal of a single nucleon from the nucleus provides information about its wavefunction and the hole state remaining. The creation of a hole state may occur though the interaction of a proton or an electron with a nucleus of A nucleons, leaving the residual system of A-1 nucleons. Experimentally, most of the effort has been in reactions with protons and electrons. The availability of intense proton and electron beams, and more recently, high duty factor electron beams have led to a wealth of data from which to extract the single particle properties of the nucleon.

Quasielastic scattering has become an important tool in the study of the single particle properties of the nucleons in the nucleus in part because of the simplicity of the reaction mechanism. A high energy particle whose wavelength is smaller than the internucleon separation, scatters off an individual (bound) nucleon. The incident projectile interacts with the nucleus as an assembly of 'free' nucleons. Consider the process in Figure 1. In this picture 0, 1, 2, and 3 refer to the incoming particle, two outgoing particles and the particle in the target nucleus that is knocked out. If we measure the $(0,1,2)$ cross section for a range of values of $\vec{k}_3$ and E_s one is able to determine the distributions of momenta and energies of nucleons in the nucleus.

Quasifree processes were first observed a long time ago [1] at the Berkeley synchrocyclotron when ^{7}Li was bombarded with 340 MeV protons. Coincidences between outgoing proton pairs were measured and found to have a strong angular correlation consistent with the assumption that the incident proton had knocked out a 'quasi free' proton moving in the nucleus. In 1957, Tyren, Hillman, and Maris [2] in Uppsala measured (using two range telescopes) both the direction and the energies of coincident protons from the reaction $^{12}C(p,2p)^{11}B$ and verified the validity of the single particle shell model. They resolved the shell structure in the summed energy spectra of the proton pairs.

In the Plane Wave Impulse Approximation (PWIA) the coincidence cross section should be of the form

$$\frac{d\sigma^{fi}}{dE_1 d\Omega_1 dE_2 d\Omega_2} = K\,S(E_s, \vec{k}_3)\frac{d\overline{\sigma}_{fr}}{d\overline{\Omega}} \tag{1}$$

where $S(E_s, \vec{k}_3)$ is the spectral function giving the probability of finding a nucleon with momentum $\vec{k}_3$ and separation energy E_s and $d\overline{\sigma}/d\overline{\Omega}$ is the (off-shell) cross section for scattering or absorption on a single nucleon. This is the general form for the cross section for knockout reactions: $(e, e'p)$, $(p, 2p)$ or (γ, p).

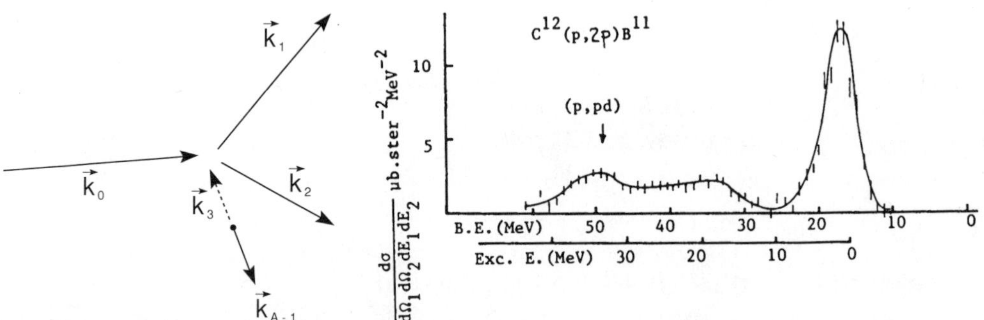

Figure 1. Kinematics of the knockout process, target at rest in lab.
$$\vec{k}_{A-1} = \vec{k}_0 - \vec{k}_1 - \vec{k}_2 = -\vec{k}_3$$
$$E_s = E_i - E_f = T_0 - T_1 - T_2 - T_{A-1}$$

Figure 2. Early (p,2p) data from Uppsala.

In the pure Independent Particle Shell Model (IPSM) the spectral function is written as

$$S(E_s, \vec{k}_3) = \sum_i |\, g_i(\vec{k}_3)\, |^2\, \delta(E_i - E_s) \tag{2}$$

where $|\, g_i(\vec{k}_3)\, |^2$ and E_i are, respectively, the momentum distribution and separation energy of the single particle state i. In the framework of the PWIA and the IPSM one can, in principle, measure the momentum distributions for the different shells and subshells in the target nucleus.

The formulation above, while intuitive, represents a considerable simplification of the process. The kinematic relations contain the implicit assumption that the asymptotic momenta are equal to the momenta right before and after the knockout process. Neglected are final state effects and, for the case (p,2p) initial state distortions as well. If these strong distortion effects are included [3,4] (under some conditions) the form of Eq. (1) remains; the spectral function is modified to become $S_{E_s}^{Dis}(\vec{k}_3)$, the *distorted spectral function*.

Much of the effort in experimental work is to minimize (or to isolate) the corrections to the PWIA such as meson exchange currents (MEC), nuclear isobar configurations (IC) or final state interactions (FSI). For example, in $(e, e'p)$, new effort [5] is being made to extract interference structure functions that are most sensitive to MEC and IC.

Quasifree (p,2p) reactions have been widely studied at medium energy laboratories and the theory has developed. The difficulties in properly handling the offshell effective t-matrix, accounting for FSI, multistep processes, and three body effects has encouraged experiments on the light nuclei ^{2}H, ^{3}He, ^{3}H, and ^{4}He [6,7,8]. It is here that $(p, 2p)$ compares favorably with $(e, e'p)$ as the NN differential cross section is only weakly dependent on energy or momentum transfer, while the elementary electron proton cross section falls quickly with q^2. Information from hadronic and electromagnetic probes complement each other and provide a deeper insight into the nuclear structure. Here I will discuss only electron scattering, leaving the reader to an excellent review of (p,2p) in Ref. [9] and the many references therein. A very recent review is Ref. [10] which gives special attention to the progress in exploiting the spin and isospin degrees of freedom.

Quasielastic Electron Scattering

Jacob and Maris [11,12] pointed out the long free mean path of an electron in the nucleus and suggested $(e, e'p)$ as a especially powerful tool to measure the single particle properties of nucleons in the nucleus.

The coincidence cross section depends on four independent functions of $\vec{q}, \omega, E_p$ (the outgoing proton energy), and α_p (the proton angle with respect to $\vec{q}$):

$$\frac{d\sigma^{e,e'p}}{dE_{e'}d\Omega_{e'}dE_{p'}d\Omega_{p'}} = \sigma_{Mott}[v_L R_L + v_T R_T + v_{LT} R_{LT} \cos\phi + v_{TT} R_{TT} \cos 2\phi] \qquad (3)$$

where the response functions, R_L, R_T, R_{LT}, and R_{TT}, contain all the nuclear structure information and are generated by the components of the nuclear electromagnetic current.

In an inclusive experiment, the proton variables are integrated over and the cross section depends only on two structure functions R_L and R_T, such that

$$\frac{d^3\sigma}{dE'd\Omega_{e'}} = \sigma_{Mott}[(\frac{q_\mu}{q})^4 R_L + (\frac{1}{2}(\frac{q_\mu}{q})^2 + \tan^2\frac{\theta_{e'}}{2})R_T]. \qquad (4)$$

This form gives the angular dependence, and from it we see that a separation of the longitudinal and transverse response functions can be made by performing measurements as fixed values of $\vec{q}$ but at different $\theta_{e'}$.

Kinematics

The reaction kinematics for $(e, e'p)$ is shown in Figure 3 for the case where the proton is detected out of the plane of the incident and scattered electrons. The incident electron emits a virtual photon of energy ω and momentum $\vec{q}$. The momentum of the scattered electron is $\vec{e'}$, detected at θ_e. The photon couples to a single proton which is knocked out of the nucleus, detected (at angle α) with momentum $\vec{p'}$.

In the coincidence experiment where $\vec{e}$, $\vec{e'}$, and $\vec{p'}$ are measured for each event,

$$\vec{p_m} = \vec{p'} - \vec{q} = \vec{p'} - (\vec{e} - \vec{e'}).$$

Similarly, the difference in the total energy before and after the scattering is given by

$$E_m = E_0 - E' - T_p - T_{A-1} = \omega - T_p - T_{A-1}$$

where E_0, E', and T_p are the initial electron, final electron, and kinetic energy of the outgoing proton, respectively. The kinetic energy of the recoiling A-1 nucleus, T_{A-1}, is immediately obtained since $\vec{p}_{A-1} = -\vec{p}_m$. The missing (separation) energy is that required to remove the nucleon from the target and includes any excitation of the residual nucleus. The measurement of the $(e, e'p)$ cross section for different values of $\vec{p}_m$ and E_m determines the momenta and energy probabilities for protons in specific shells in the nucleus. Figure 4 is an example [13] of the mapping of the spectral function possible in $(e, e'p)$. The characteristic momentum behavior of the s and p shells can be seen clearly.

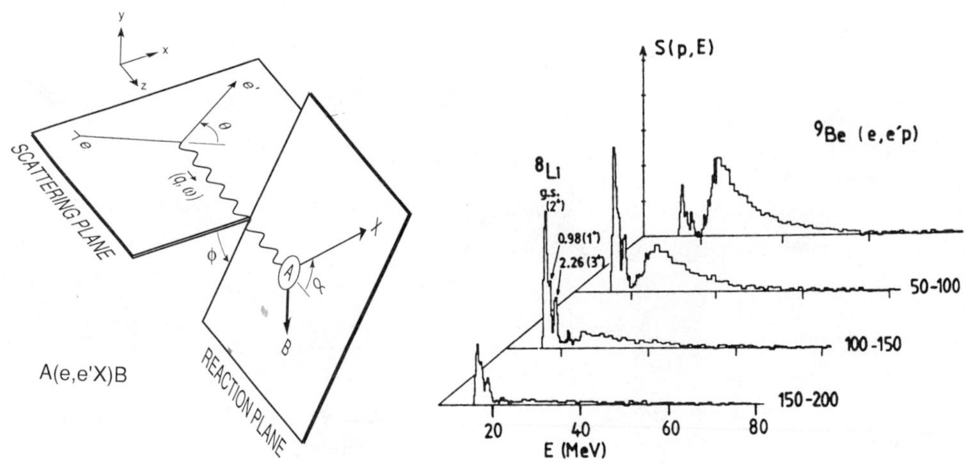

Figure 3. Momentum vectors for the $(e, e'p)$ reaction.

Figure 4. Spectral function for ^9Be from $(e, e'p)$ from Ref. [13].

Inclusive Measurements

Until just recently, inclusive quasielastic measurements have received much less attention than the exclusive experiments. The neglect is due in part because the non-coincidence measurement gives much less detailed information. However, this disadvantage must be weighed against the effort to perform the experiment – inclusive measurements are much easier to perform and understand. In addition the inclusive experiment is much cleaner — only an electron is detected with no accounting necessary for the fate of the knocked out nucleon.

From the energy momentum balance it should not be too difficult to convince oneself that inclusive measurements[14] mix up different parts of the spectral function, $S(E_s, \vec{k})$. However, the cross section in the maximum is dominated by $S(E_s, \vec{k} = 0)$, the small energy loss side of the peak by high-k, small E_s components in the spectral function.

The quasielastic peak appears as a broad bump in the spectrum of scattered electrons; the peak of the distribution appearing just at the energy loss expected for the peak of elastic scattering from a stationary, free nucleon, shifted slightly by an average separation energy. At energy losses greater than $q_\mu^2/2m_n$ the inclusive spectrum shows additional strength that can not be associated with simple single nucleon knockout.

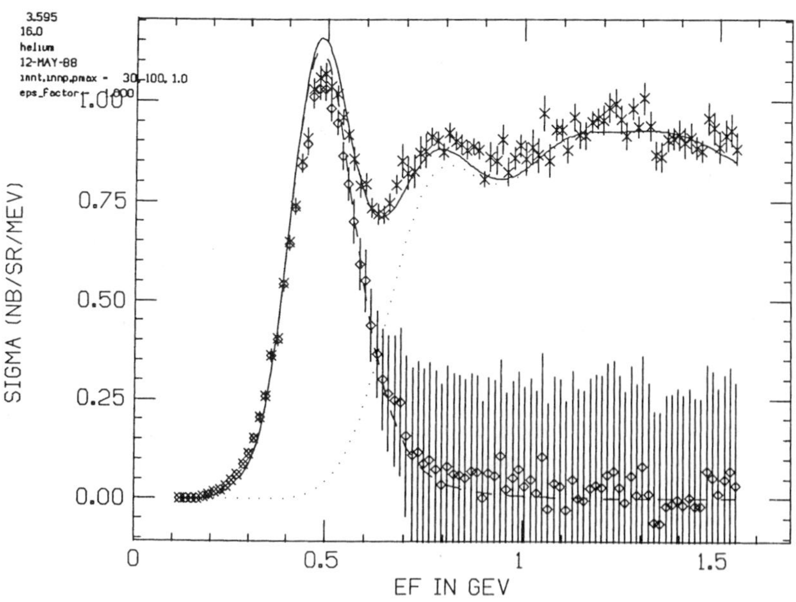

Figure 5. Inclusive cross sections on ^4He at 3.595 GeV and 16°[15].

These processes include pion production, excitation of the nucleon resonances, and deep inelastic scattering. At $q_\mu^2 \geq 1(GeV)^2$ the inelastic contribution is important under the quasielastic peak. Figure 5 shows the inclusive spectra for ^4He at 3.595 GeV and 16° from a recent experiment at SLAC-NPAS [15]. Shown is a model calculation that is the sum of the quasielastic scattering (dashes) and an inelastic contribution (dots) obtained by smearing the free nucleon inelastic response [16,17] with a realistic momentum distribution. Also shown (with large error bars) is the knockout response if the modeled inelastic background is subtracted. It is important to take this 'contamination' of the single nucleon response into account. Coincidence experiments, where the final state is determined, are the obvious way to eliminate this background.

The first systematic set of inclusive data in the quasielastic region was a product of HEPL on Stanford campus in the early 1970's. These data were well described by a Fermi gas model [18] from which the average binding energies and fermi momentum could be extracted as parameters. It was later determined that agreement of the Fermi gas model with the data was accidental. Subsequently, inclusion of final state interactions and more realistic wave functions gave an excellent description of the data.

In the early 1980's the inclusive cross sections were separated into its longitudinal and tranverse components for the first time [19,20] The longitudinal response was expected to characterize the one-body properties whereas the transverse response would be sensitive, in addition, to meson exchange currents and other two-body effects. Put another way, it was expected that the integrated sum of the longitudinal response would sum to the number of protons in the nucleus. What has been found is that while the transverse is well explained by a mean field approximation, 40% of the strength is missing in longitudinal part. The results from several nuclei are summarized by way of the Coulomb sum rule normalized to a unit point charge $\int \frac{R_L(q,\omega)d\omega}{ZG_E^2(q_\mu)}$ in Figure 6. The separated response functions can be analyzed in terms of y-scaling [21,22] in which the cross sec-

323

tion is factorized in terms of an electron-nucleon piece and the longitudinal momentum distribution,

$$\frac{d^2\sigma}{d\Omega_e d\omega} = K(Z\sigma_{ep} + N\sigma_{en})2\pi \int_{|y|}^{\infty} n(k)k\,dk \tag{5}$$
$$= K(Z\sigma_{ep} + N\sigma_{en})f(y),$$

where $n(k)$ is the nucleon momentum distribution. From Equation 4 we would expect that $f(y) = f_L(y) = f_T(y)$. However as Figure 7 shows $f(y)$ extracted [23] from the separated ^{12}C data (over a limited q-range) of [24] (contrary to the IA), $f_L(y) \neq f_T(y)$. While one is tempted to introduce a medium modified *nucleon* response to clarify this situation, results from $(e, e'p)$ [25] indicate that there is additional strength in the transverse response even in the region of the peak of the quasielastic, possibly from a two-body mechanism.

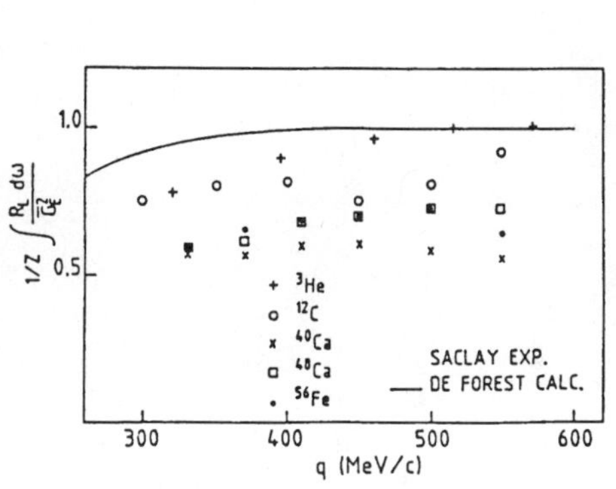

Figure 6. Longitudinal sum rule for several nuclei.

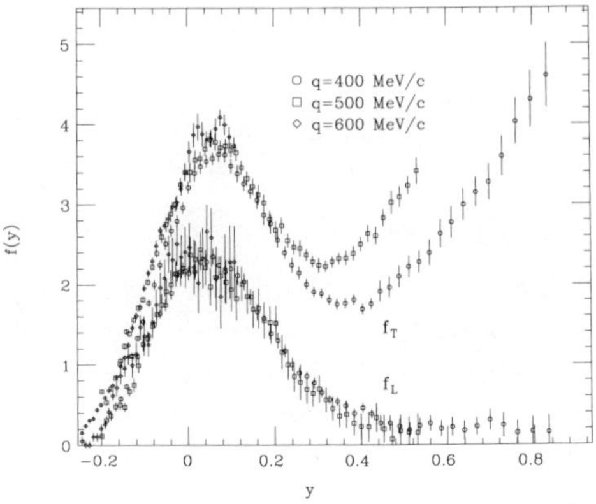

Figure 7. y-scaling for the separated response for ^{12}C from [24].

Exclusive Measurements

Three laboratories have been most active in the recent $(e, e'p)$ work – MIT-Bates, NIKHEF-K, and Saclay. An important and excellent review of $(e, e'p)$ is [3] which covers the theory, and experiment through 1983. Since that time workers at Bates and NIKHEF-K, studying reaction mechansims and high resolution spectroscopy respectively, have made important contributions. I limit my discussion here to some of the relevant progress on extracting momentum distributions, focusing on the light nuclei. It is in $(e, e'p)$ that the artificial division of the one-body knockout diagram into the nuclear part (the spectral function) and the nucleon part (the virtual photon-nucleon coupling) is exposed. The challenge of extracting single particle momentum distributions is complicated by medium modifications of the elementary interaction, MEC, spectroscopic factors, and non-nucleonic degrees of freedom. The limited application of PWIA and DWIA, for that matter, forces the experimentalist to select special kinematics in

324

an effort to either minimize those effects that render the approximations invalid, or to attempt to study those effects specifically.

Coincidence experiments on the few-body system promise to provide important information on the high momentum components in the wave function. The high momentum components are sensitive to the short range correlations between nucleons, about which very little is known. A program on $^2H(e, e'p)n$ has been completed at Saclay [26,27] out to recoil momenta of 500 MeV/c. Figure 8 is the data of Ref. [26] compared with four theoretical models for the deuteron momentum density. To distinquish among the wave functions calculated with different nucleon-nucleon potentials data has been taken in the 300 to 500 MeV/c region, where the D-state is dominant[27].

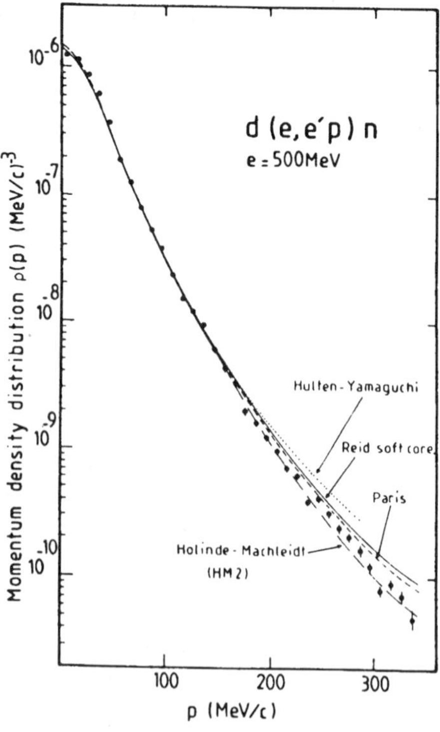

Figure 8. Data from Saclay $^2H(e, e'p)$.

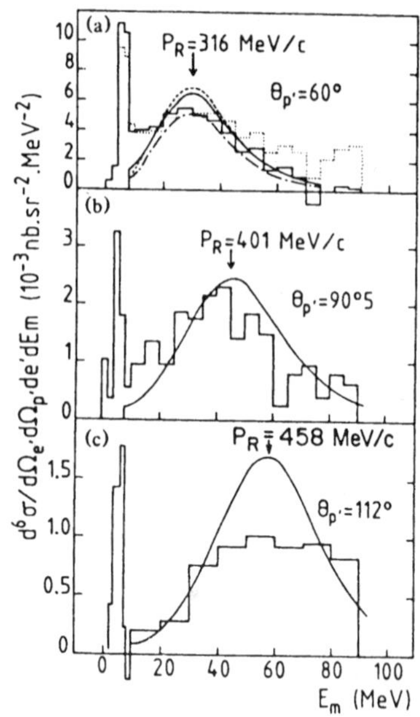

Figure 9. Distribution of missing energies showing the separation of the two- and three-body breakup channels in ^{3}He.

The proton momentum distribution in ^3He has been determined up to momenta of 310 MeV/c [28] with a missing energy resolution of $\delta E_m = 1.2$ MeV, which allowed the separation of the two- and three-body breakup channels. Very recently the Saclay group has presented [29] explicit evidence for nucleon-nucleon correlations in $^3He(e, e'p)X$, which have been measured for recoil momenta between 300 and 600 MeV/c and missing energies up to 90 MeV.

Figure 9 from Ref. [29] shows the clear separation of the pd and ppn breakup channels. The sharp peak to the left is the two-body breakup peak with a missing energy of 5.5 MeV. To the right, above 7.7 MeV is the three body breakup continuum. This broad peak

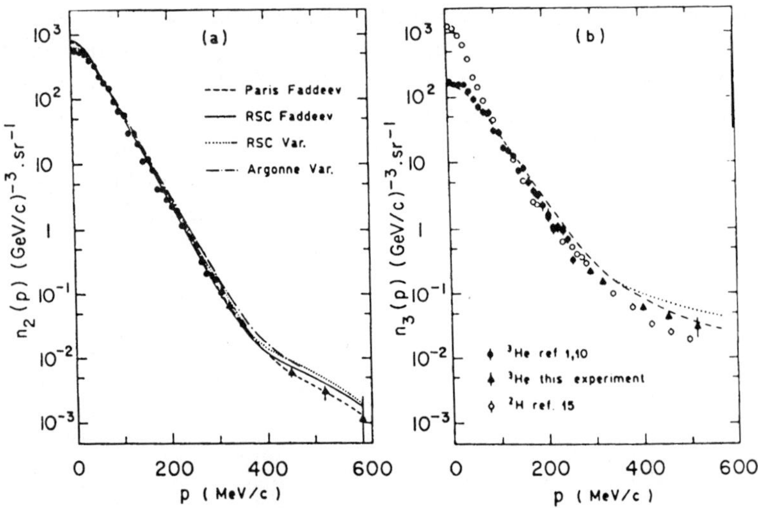

Figure 10. 'Experimental' momentum distributions for two and three body channels in ^{3}He from Ref. [29]

falls to zero at large missing energies and shifts to higher missing energies with increasing recoil momenta. The peak of this bump moves as would the peak corresponding to an $(e, e'p)$ reaction on a proton-neutron pair. The width of the bump corresponds to the motion of the center of mass of the pair.

Momentum distributions can only be extracted from measured cross section after first taking into account MEC and FSI. For the ${}^3He$ experiment under discussion, this was done by correcting the data by the difference between a calculation including MEC and FSI, and one without. The resulting 'experimental' momentum distributions are shown in Figure 10.

Experimental Setup

Electron scattering is conceptually very simple. A high energy electron beam which has been momentum selected, is directed against a target. The scattered electron's angle and momentum are determined by detectors in a magnetic spectrometer, located at a fixed angle. The flux of incident electrons is determined by nonintercepting current toroids, normalized at low beam intensites to a Faraday cup.

Spectrometers and Detectors

In the SLAC inclusive experiments, electrons are detected by the 8 GeV spectrometer [30]. It is a vertical bend spectrometer arranged as QQDDQ. The detector package, consisting of ten planes of wire chambers (for particle trajectory determination), a gas Cerenkov counter which, together with the segmented lead glass total absorption counter provides particle identification and triggering, is housed in a large concrete shielding hut. The magnets and associated shielding weighs some 750 tons. The magnets require over 2 MW of power when excited.

Together, the Cerenkov and shower counter provide a factor of 10^4 pion rejection and still have greater than 98% efficiency for detecting electrons. From measured trajectories of particles in the detector coordinate system, the original coordinates at the target are calculated. The 8 GeV spectrometer has a solid angle of 0.75 msr, and momentum resolution, $\delta P/p = 0.01$, and momentum acceptance of $\simeq 10\%$. Different electron scattering angles are selected by moving the spectrometer on its rails set in the concrete floor.

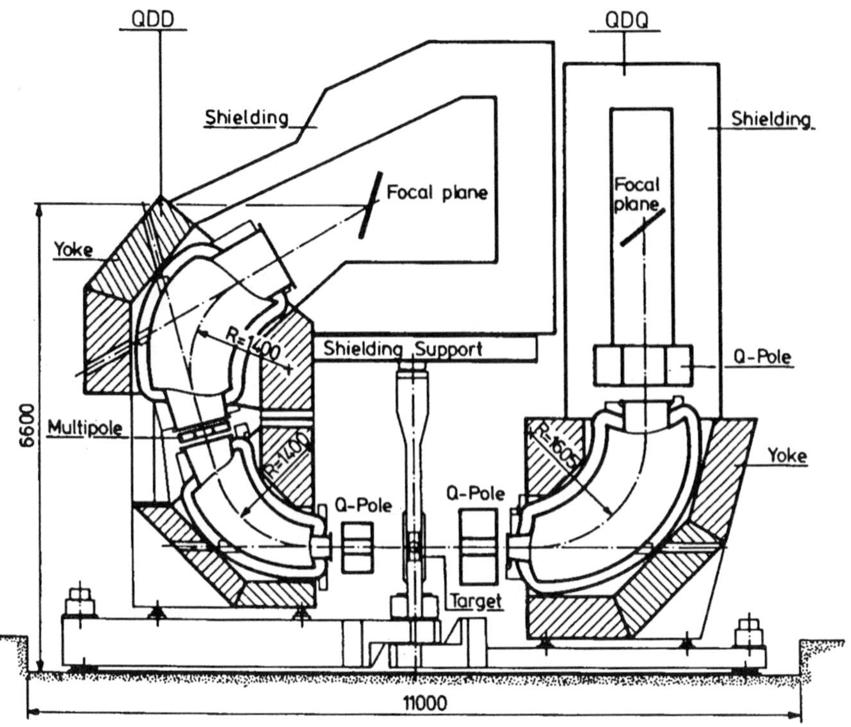

Figure 11. Layout of NIKHEF-K two-spectrometer setup. (Dimensions in mm.)

In coincidence experiments, a second spectrometer is needed to determine the energy and angle of the knocked-out proton. Shown in Figure 11 is the layout of the two spectrometer setup at NIKHEF-K [31]. Electrons are detected in the QDD spectrometer and the protons in the QDQ spectrometer. The QDD (QDQ) spectrometer has a maximum momentum of 600 (750) Mev/c, a momentum acceptance $\pm5\%$ ($\pm5\%$), a solid angle of 5.6 (17) msr, and taken together have a missing energy resolution of 0.10 MeV. This extremely good missing energy resolution is achieved by operating the beam handling system in dispersion matching mode wherein the beam energy spread reduces to the intrinsic resolution of the beam handling system ($R \simeq 3.10^{-5}$). With target thickness of about 10–20 mg/cm^2 an overall resolution of 100 kev has been achieved.

The detection system at NIKHEF-K [32] consists of 3 (4 in the QDQ) multiple wire drift chambers (MWDC), scintillators and a Cerenkov counter. The position and direction of the particle is learned from the drift chambers and the trajectory determined precisely. The Cerenkov and scintillators provide the trigger for wire chamber readout and coincidence timing. The scintillators in the QDQ are of varying thicknesses (1, 3, and 10 mm) to allow for particle identification.

Targets

The targets irradiated in nuclear physics are selected with several criteria in mind. First, of course is the specific physics goals of the effort, e.g. studies of the few body systems or to examine closed shell nuclei. All else being equal, the experimenter will opt for those targets that are of the 'drug store' character - non-toxic, non-explosive, easy to handle, and inexpensive. Unfortunately, this is not always possible - not all interesting physics can be learned from reactions on C, Al or Au.

One of the best examples of this is the recent experiment at MIT-Bates on ^3H and ^3He [33] to separate the longitudinal and transverse response functions in a long program that included elastic, quasielastic, and deep inelastic scattering. Safety was the paramount concern and dictated most of the design parameters. The need to measure extremely small cross sections, on the order of $10^{-38} cm^2/sr$, and lower limits of a few events per hour, set a minimum target thickness. Further restrictions were set by the spectrometer optics, with the resulting maximum target density of 25 mg/cm^2. In the end the target system handled 140,000 Ci of tritium in target cells operated at 225 psia. The tritium gas source was a uranium oven in which the tritium was stored as solid UH_3. Uranium is an effective pump and storage bottle for tritium as it readily combines with hydrogen at room temperature and only dissociates at elevated temperatures. A tremendous effort and expense was required to insure the safe and reliable operations of these targets in a severe environment – 25 μ A on target.

Duty Factor

Count rates are determined by the reaction cross section, the intensity of the incident beam, detector solid angles, target thickness, and detector rate handling capability. In a typical experiment, a pulsed electron beam impinges on the target, and the outgoing particles, electron and proton in the case of $(e, e'p)$, are detected in two magnetic spectrometers. The detector package (wire chambers, plastic scintillators, shower counters, total energy absorption counters) allow determination of the exact time when the event occured in the target. Precise timing is necessary in the detectors and knowledge of the path length differences in the spectrometers is crucial. The time difference between the computed times for the 'event' from the information in the two spectrometers allows the separation of real coincidences from accidentals. If the singles rates in the two spectrometers are N_1 and N_2 then the accidental rate will be

$$N_A = N_1 N_2 \frac{\tau}{DF}$$

where τ is the resolving time (the width of the Δt-distribution) and DF is the duty factor of the accelerator.

Quality factors, Q and Q', have been defined [3] for coincidence facilities. The first is a product of the two solid angles and the two momentum acceptances. To first order this is measure of the system coverage of the available phase space for the reaction in question. Q' is $Q/\tau\delta E$ where τ is the coincidence resolving time and δE the missing energy resolution. This quantity is a measure of how well the system can reject accidentals and work in a very low real count rate regime. From Ref. [3], the coincidence facilities at Saclay, NIKHEF-K, and Bates yield Q's (Q''s) of 1.37 (0.69), 0.48 (1.92), and 0.25 (0.05), respectively. The value of Q' for the planned pair of identical spectrometers at the 100% duty factor facility at Mainz will be 25.7.

Future Facilities and Experiments

Since the early 1970's, with the commisioning of high current high duty factor linacs at Saclay ($\approx 1\%$), Bates-MIT ($\approx 2\%$), and NIKHEF-K (2.0% at 500 MeV) great effort and progress has been made in coincident electron experiments. Still greater progress will be made when 100% duty factor machines under construction in the USA (CEBAF and Bates-MIT) and Europe (NIKHEF-K and Mainz) become operational in the early to mid 1990's. The additonal kinematic flexibility afforded by the intense beams ($\sim 200\mu A$) and high energies ($\geq 4GeV$) will allow new experimental programs to develop. Among these are $(e, e'n)$ (detection of a neutron in coincidence with the electron), triple coincidence experiments $A(e, e'2N)X$ and out-of-plane detection of the knocked out nucleons, allowing measurement of specific pieces of the nuclear response.

References

[1] O. Chamberlain and E. Segre, Phys. Rev. **87** (1952) 81

[2] H. Tyren, Th. A. J. Maris, P. Hillman, Il Nuovo Cimento **6** (1957) 1507

[3] S. Frullani and J. Mougey, Advances in Nuclear Physics, Vol. 14, 1984

[4] G. van der Steenhoven, Ph. D. Thesis, NIKHEF-K (unpublished)

[5] T. Tamae *et al.*, Phys. Rev. Lett. **59** (1987) 2919

[6] W.T.H. van Oers *et al.*, Phys. Rev. C **25** (1982) 25

[7] M. B. Epstein *et al.*, Phys. Rev. C **32** (1985) 967

[8] V. Punjabi *et al.*, College of William and Mary Preprint

[9] D. F. Jackson, Adv. Nucl. Phys. **4** 1971 1

[10] P. Kitching, W.J. McDonald, TH. A. J. Maris, and C. A. Z. Vasconcellos, Adv. Nucl. Phys. **15** (1985) 43

[11] G. Jacob and Th. A. J. Maris, Nucl. Phys. **31** (1962) 139

[12] G. Jacob and Th. A. J. Maris, Rev. Mod. Phys. **45** 1973 6, Rev. Mod. Phys. **38** (1966) 121

[13] J. Mougey, Nucl. Phys. **A335** (1980) 35

[14] I. Sick, Lectures at 1976 SIN Spring School, Zuoz, 1976

[15] D. Day *et al.*, Phys. Rev. Lett. **59** (1987) 427

[16] A. Bodek *et al.*, Phys. Rev. D **20** (1979) 1471

[17] A. Bodek and J. L. Ritchie, Phys. Rev. D **23** (1981) 1070

[18] E. Moniz *et al.*, Phys. Rev. Lett. **26** (1971) 445

[19] R. Altemus *et al.*, Phys. Rev. Lett. **44** (1980) 965

[20] M. Deady *et al.*, Phys. Rev. C **28** (1983) 631

[21] G. West, Phys. Rep. **18** (1975) 263

[22] I. Sick, D. Day, J. S. McCarthy, Phys. Rev. Lett. **45** (1980) 871

[23] J.M. Finn *et al.*, Phys Rev. C **29** (1984) 2230

[24] P. Barreau *et al.*, Nucl. Phys **A402** (1983) 515

[25] P. Ulmer *et al.*, Phys. Rev. Lett. **59** (1987) 2259

[26] M. Berheim *et al.*, Nucl. Phys **A365** (1981) 349

[27] S. Turck-Chieze *et al.*, Phys. Lett. **142B** (1984) 145

[28] E. Jans *et al.*, Phys. Rev. Lett. **49** (1982) 974

[29] C. Marchand *et al.*, Phys. Rev. Lett. **60** (1988) 1703

[30] D. H. Potterveld, Ph. D. Thesis, California Institute of Technology (1988) (unpublished)

[31] C. de Vries *et al.*, Nuclear Instruments and Methods, **223** (1984)1

[32] Paul Keizer, Ph. D. Thesis, NIKHEF-K (unpublished)

[33] D. Beck, Ph.D. Thesis, MIT 1986 (unpublished)

[34] L. Lapikas, Proceedings of the Fourth MiniConference, 'Nuclear Structure in the 1p Shell', Amsterdam, November 1985

CONTRIBUTED PAPERS

RESOLUTION IN DEEP INELASTIC NEUTRON SCATTERING USING PULSED NEUTRON SOURCES

Carla Andreani[@], Guilia Baciocco[+], Ron Holt[*] and Jerry Mayers[*]

[@] Dipartimento di Fisica, Universita' di Tor Vergata
Via O. Raimondo – 00173 Roma, Italy
[+] Dipartimento di Fisica, Universita' 'La Sapienza'
P. le A. Moro 2 – 00185 Roma, Italy
[*] Rutherford Appleton Laboratory, Neutron Science Division
Chilton, Didcot, Oxfordshire, U.K.

INTRODUCTION

In recoil scattering with neutrons the impulse limit [1,2] is only reached by utilising large incident neutron energies generally in excess of 1eV, which are now readily available on spallation neutron sources. Since the property of direct interest is the atomic momentum distribution, $n(\underline{p})$, (where $\underline{p}$ represents the atomic momentum), it is important to present and compare the recoil scattering data in terms of this quantity and not in terms of the measured scattering function $S(q,\omega)$ [3]. This transformation is relatively easy to undertake and, moreover, offers several distinct advantages over a comparison in time-of-flight or energy transfer.

For an isotropic system $n(p)$ is a one dimensional function. The link between $S(q,\omega)$ and $n(p)$ is provided by the property of 'y scaling' [2] where every point (q,ω) corresponds to a unique point in atomic momentum space (hereafter called p-space). Consequently any scan in (q,ω) space which crosses the recoil line gives a measurement of $n(p)$. It is generally inefficient for time-of-flight instruments to scan along a particular line in (q,ω) space but this is not necessary in impulse scattering experiments if the data are presented in terms of $n(p)$.

The advantage of analysing recoil data in this way is that for a multidetector spectrometer the detectors can be grouped together in p-space (this is only true for detectors which give a sufficiently large q that the impulsive limit is reached). There is an additional advantage in that the limit at which the impulsive regime is reached could be estimated by comparing the distribution in p-space provided by detectors at different angles. The theoretical criteria for deciding when the impulse regime is reached are still a subject of debate (see Mayers et al. [4]).

The method also allows the opportunity of comparing the performance of different instruments directly in p-space. Such a comparison has been made here on the basis of the resolution capability of two recoil spectrometers; an inverse geometry instrument, based on resonance

techniques, and a direct geometry chopper spectrometer. The various
contributions to the resolution i.e. energy, angular and time components
have been analysed independently to provide an indication as to how these
can be optimised in a particular recoil scattering experiment on a pulsed
source.

THEORY

The impulse approximation (IA) relates $S(q,\omega)$ to the atomic momentum
distribution $n(p)$ via [2]

$$S(q,\omega) = (1/\hbar) \int n(p)\, \delta\{\omega - \hbar q^2/2M - \hbar p.q\,/M\}\, dp \tag{1}$$

where $\hbar\omega$ and $\hbar q$ are respectively the energy and momentum transfers in the
scattering process, $\hbar p$ is the atomic momentum and M the atomic mass.

If we take the z axis along q equation (1) reduces to :

$$S(q,\omega) = M/q \int \delta(y-p_z)\, n(p)\, dp \tag{2}$$

where

$$y = M/q\,(\omega - q^2/2M) = M/q\,(\omega - \omega_R) \tag{3}$$

Equation (3) provides the relationship between a point in (q,ω) space and
the corresponding point, y in p-space. For an isotropic system it can be
easily shown that [2]:

$$S(q,\omega) = M/q\, J(y) \tag{4}$$

where

$$J(y) = 2\pi \int_{|y|}^{\infty} n(p)\, p\, dp \tag{5}$$

The function $J(y)$ is analogous to the well known Compton profile in
photon scattering (see the article by M J Cooper [5]; this volume). It
gives the probability that an atom has the momentum component y along the
direction of q. From equations (4) and (5) it follows that

$$n(y) = -(1/2\pi y)\, dJ(y)/dy \tag{6}$$

RESOLUTION IN P-SPACE

Consider the resolution in p-space, at the point y determined by
equation (3), for both direct and indirect geometry spectrometers. It is
important that this quantity is known so that the feasibility of the
proposed momentum measurements can be determined and the data analysed
correctly. It is assumed that the variation in resolution across the
recoil peak is small.

From equation (3)

$$y = M\omega/q - q/2 \tag{7}$$

so

$$(\partial y/\partial \omega)_q = M/q \tag{8}$$

and

$$\left(\partial y/\partial q\right)_\omega = -(M/q^2)(\omega+\omega_R) \tag{9}$$

at the recoil frequency, $\omega=\omega_R=q^2/2M$, hence

$$\left(\partial y/\partial q\right)_\omega = -1 \tag{10}$$

The total resolution is then given by

$$\Delta y^2 = \Sigma_i \left[\, M/q \;(\partial\omega/\partial x_i) - (\partial q/\partial x_i)\right]^2 \Delta x_i^{\,2} \tag{11}$$

where Δx_i denotes the errors in the experimental parameters. Several inferences can be made from equation (11). Firstly, we expect that for large atomic masses the energy resolution will be the dominant factor determining the p-resolution; since ω is independent of angle the angular resolution will be unimportant for large masses. Secondly, the FWHM of the momentum distribution for a harmonic solid is

$$\left(\Delta y\right)_{FWHM} = 2.345\;(MK_B T^*)^{1/2}$$
$$= 0.388\;(MT^*)^{1/2} \tag{12}$$

where Δy is in Å^{-1}, M in amu and T^*, the effective temperature, is in K. For liquid He, T^* is approximately determined by the zero point kinetic energy. Thus at values of M>4, where the p-resolution is dominated by the energy resolution, we expect from equation (11) that approximately

$$\Delta y/(\Delta y)_{FWHM} \propto M^{1/2} \tag{13}$$

i.e. the relative energy resolution becomes progressively poorer as the mass increases. This behaviour is illustrated in table 1 where $(\Delta y)_{FWHM}$ is listed for a series of different elements. $(\Delta y)_{FWHM}$ was calculated from the Debye model using the Debye temperatures given in table 1.

Table 1. Resolution and atomic momentum (FWHM) values for some selected elements calculated for HET (E_0=1000 meV, Θ=180°) and eVS (E_1=4280 meV, Θ=180°).

Element	Θ_D	M	T^*(77K)	Δp(HET)	Δp(eVS)	$(\Delta y)_{FWHM}$
Li	400	7	158	1.04	3.51	11.2
Be	1000	9	376	1.47	4.4	19.6
C	1000	12	376	2.11	5.65	22.7
Mg	318	24	132	4.71	10.8	19.1
V	390	51	154	10.6	22.4	30.0
Ta	225	181	107	48.7	78.1	47.0

Finally we note that the contribution to Δy arising from the energy resolution is inversely proportional to q. Thus measurements at large q values, with poor energy resolution, may still provide good resolution in p-space.

In the case of chopper spectrometers the most important contributions to the resolution arises from Δt_c, Δt_m and $\Delta\Theta$, where t_c is the time at which a detected neutron passes through the chopper, t_m is the time at which the neutron leaves the moderator and Θ is the scattering angle. Calculations show that all three contributions to the resolution i.e. $\partial y/\partial t_c$, $\partial y/\partial t_m$ and $\partial y/\partial\Theta$, are proportional to $E_0^{1/2}$ for given values of M and Θ, where E_0 is the incident neutron energy [6].

For an inverse geometry spectrometer similar calculations reveal that for scattering from samples other than hydrogen, the dominant contribution to the resolution comes from the energy term,

$$\partial y/\partial E_1 \propto E_1^{-1/2} \tag{14}$$

where E_1 denotes the final (resonance) energy. In these cases the experimental uncertainty in y is inversely proportional to q, for a given energy resolution ΔE_1.

RESULTS

Numerical computations of the components of the resolution have been made specifically for two recoil spectrometers on the pulsed neutron source ISIS; the electron Volt (resonance) spectrometer, eVS, and the High Energy Transfer (chopper) spectrometer, HET. Calculations have been performed in terms of the mass of the recoiling particle and three distinct materials have been chosen (hydrogen, helium and beryllium) as representative of the changes in resolution which can be expected.

Scattering from Hydrogen

Figure 1 shows the individual resolution contributions and the total p-resolution as a function of angle calculated for HET at an incident energy of 2000 meV and on eVS at 4.28 eV (the main tantalum resonance); the q values corresponding to the different scattering angles are also given. The resolution at 2000 meV and at lower energies is governed primarily by the energy term except for angles in excess of ~60° where the angular component begins to dominate. The uncertainties associated with the timing are comparable with the angular term for scattering angles greater than ~30°. In the case of eVS the angular term makes a significant contribution to the overall resolution, particularly for the higher energy resonance where again it is the dominant term for angles in excess of 60°. Both figures illustrate the overall improvement in p-space resolution as the scattering angle is increased due to the angular variation of $\partial y/\partial E_1$.

Table 2 provides a summary of the resolution results for both HET and eVS (at $\Theta=60°$) indicating that the resonance system is only a factor of 6-9 worse than the chopper spectrometer. This is significantly lower than might be anticipated from the energy resolution of the two systems; the energy resolution of HET is ~14 meV at 2000 meV and is therefore between 10-20 times better than that of eVS. However, it should be pointed out that much higher q values can be accessed on eVS.

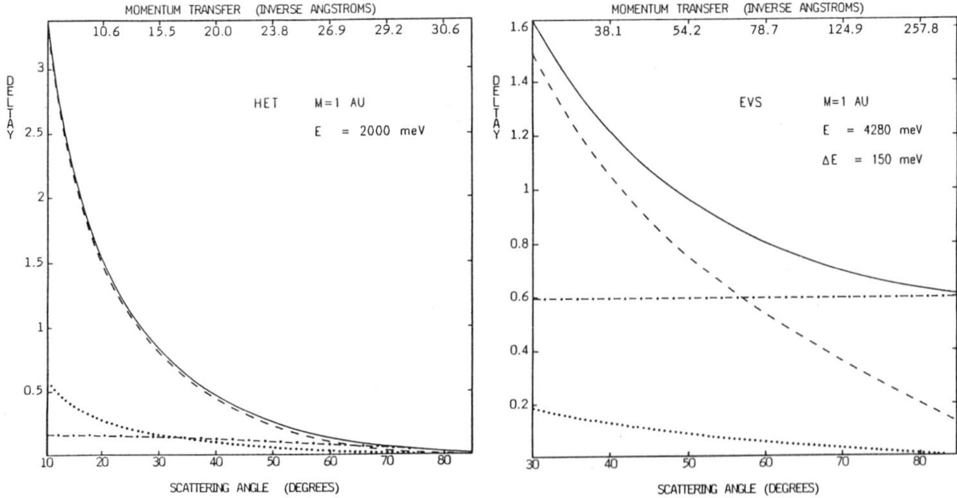

Figure 1. The total resolution (———) and the energy (– – –), angular (–·—·–) and timing (.....) contributions in hydrogen (M=1 amu) for HET and eVS recoil spectrometers.

Table 2. Selected total resolution results for recoil scattering from hydrogen, helium and beryllium using the chopper (HET) and resonance neutron absorption (eVS) spectrometers on ISIS. For eVS the results are shown for the main resonances in Ta (4.28 eV) and Sm (0.87 eV).

Mass 1	θ=60°	
Instrument	Energy (meV)	Resolution
HET	1000	0.09
HET	2000	0.14
eVS	872	0.82
eVS	4280	0.80

Mass 4	θ=160°	
Instrument	Energy (meV)	Resolution
HET	1000	0.46
HET	2000	0.65
eVS	872	3.40
eVS	4280	2.30

Mass 9	θ=160°	
Instrument	Energy (meV)	Resolution
HET	1000	1.47
HET	2000	2.07
eVS	872	6.60
eVS	4280	4.48

It is clear that the angular contribution is very important and significant improvements in the resolution can be achieved by changes in the collimation of the scattered neutrons. An alternative method of improving the resolution is by arranging the experimental geometry for 'time focussing'. This has been discussed for resonance detector spectrometers by Carpenter and Watanabe [7] and Rauh et al. [8]. They also point out the dominant effects of angular contributions to the resolution for recoil scattering from low atomic masses.

Scattering from Helium

Figure 2 shows a similar plot for both spectrometers for Helium. As a comparison the FWHM of the atomic momentum distribution in He^4 is ~ 1.9 $Å^{-1}$. At M=4 we are already approaching the regime where the p-resolution is dominated by the energy resolution of the spectrometers (see equation (11) and succeeding comments).

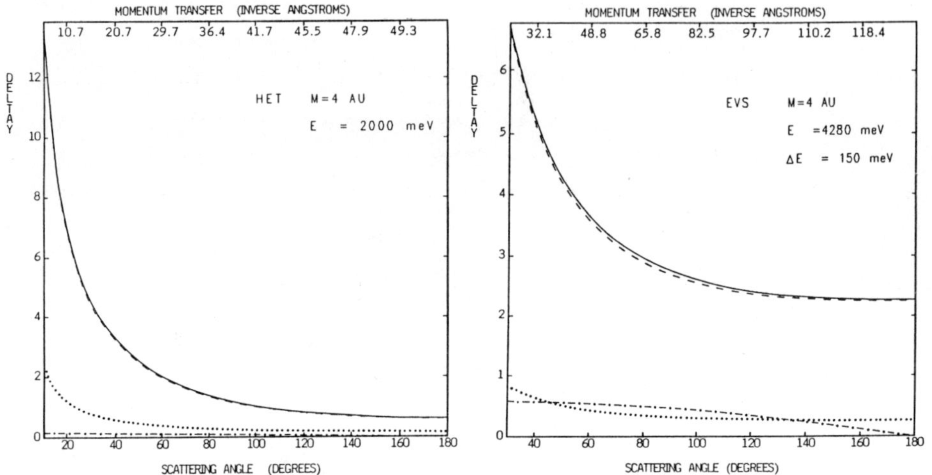

Figure 2. The total resolution (———) and the energy (− − −), angular (−·−·−) and timing (.....) contributions in helium (M=4 amu) for HET and eVS.

We find that the resolution on the chopper spectrometer is better by a factor of 4-7 (for Θ=160°, see table 2) than that for the resonance based instrument. For the latter the energy resolution dominates; the angular and timing components effectively make a similar but smaller contribution over the entire scattering angle region of interest. Improvements of the order of 20% or more could be achieved by implementing the so-called double difference method of collecting data [9] or alternatively by cooling the analysing foil to very low temperatures [10] to improve the energy resolution component.

Scattering from Beryllium

For the case of beryllium the components of the resolution are shown in figure 3. Here the p-resolution comes àlmost entirely from the energy resolution contribution and this would also be true in the case of higher

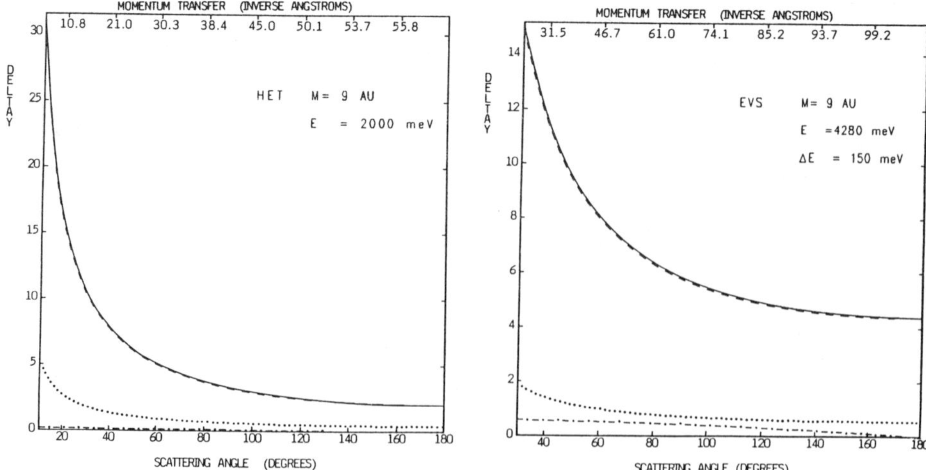

Figure 3. The total resolution (———) and the energy (– – –), angular (–·–·–) and timing (.....) contributions in beryllium (M=9 amu) for HET and eVS.

mass particles. From equation (11) we expect the p-resolution to be roughly proportional to M for M > 4. For such masses the resolution of the HET chopper spectrometer is only a factor of 2-4 better than that on eVS (for Θ=160°, see table 2) despite the large difference in energy resolution of the two instruments.

SUMMARY AND CONCLUSIONS

Calculations have been presented of the resolution in atomic momentum space (p-space), for both direct and inverse geometry instruments on the pulsed neutron source ISIS. From the general analytic forms for the resolution in momentum space (equation (11)) and from the numerical calculations it is shown that for large masses the energy resolution contribution is the dominant term and the angular resolution is relatively unimportant. Moreover, the relative resolution (i.e. resolution/width of distribution) deteriorates as the atomic mass increases.

As expected the resolution on the chopper instrument was significantly better than that for the indirect geometry spectrometer based on neutron absorption resonances. However the resolution when analysed in terms of p-space was better than expected on eVS because of the larger q values accessed. Moreover, these large q values mean that

the impulse approximation is approached more closely on eVS than HET, the
latter of which is limited to momentum transfers up to 50 Å^{-1}, rendering
a relatively simpler data analysis which is essentially free from final
state effects.

Calculations of the resolution components in recoil scattering from
hydrogen reveals that the angular term makes a significant contribution
to the resolution on both HET and eVS. Significant improvements in
resolution could be made on both instruments by implementing better
collimation of the scattered beam, or by increasing the secondary flight
path.

ACKNOWLEDGEMENTS

We would like to thank the European Economic Community Stimulation
Action scheme for funding this collaboration and to Dr M P Paoli and Dr A
D Taylor for stimulating discussions.

REFERENCES

1.   B. Tanatar, G. C. Lefever and H. R. Glyde, Physica 136B 187
      (1986) and references therein.
2.   V. F. Sears, Phys. Rev. B 30 44 (1984)
3.   S. W. Lovesey, 'Theory of Neutron Scattering from
      Condensed Matter' Vol.1 Oxford University Press (1987)
4.   J. Mayers, C. Andreani and G. Baciocco (accepted for Phys. Rev. B)
      (see this volume)
5.   M. J. Cooper (this volume)
6.   C. Andreani, G. Baciocco, R. S. Holt and J. Mayers (accepted in
      Nucl. Instr. Meths.)
7.   J. M. Carpenter and N. Watanabe, Nucl. Instr. Meths. 213, 311 (1983)
8.   H. Rauh, S. Ikeda and N. Watanabe, Nucl. Instr. Meths. 224, 469
      (1984)
9.   P. A. Seeger, A. D. Taylor and R. M. Brugger, Nucl. Inst. Meths.,
      A240, 98 (1985)
10.   H. Rauh and N. Watanabe, Nucl. Inst. Meths. A228, 147 (1984)

CONTRIBUTION OF MESON EXCHANGE CURRENTS TO THE NUCLEAR QUASIELASTIC PEAK

P. G. Blunden[†] and M. N. Butler

TRIUMF
4004 Wesbrook Mall
Vancouver, BC, Canada V6T 2A3

In quasielastic electron scattering off nuclei, there are complications above and beyond the effects of final-state interactions. As well as the constituent nucleons, virtual pi-mesons in the nucleus (from the interaction between nucleons) can also interact with the electromagnetic probe. To date, there have been only two calculations of the contributions of pions to the quasielastic cross-section. Both calculations of Meson Exchange Currents (MEC) are non-relativistic in nature. The first looks at the case where the virtual photon couples to pions, and the end effect is that one nucleon is knocked out [1]. The results are somewhat confusing, as they find that MEC's suppress the transverse response and also significanty affect the longitudinal response. Both these results are contrary to naive expectations about isovector exchange currents. The second looks at the case where two nucleons are knocked out via the pionic current [2]. We repeat calculations for both processes with one major modification.

A recent analysis of the coulomb sum rule problem in nuclear physics has shown that effects due to relativistic kinematics are important even at relatively low momentum transfers [3]. These effects are unavoidable if we believe we understand elastic electron-proton scattering. Because of this, we have chosen to use a fully relativistic formalism to study MEC's, and here we use a relativistic Fermi gas model to demonstrate the kinematic effects.

We can write the inclusive electron-nucleus cross section as

$$\frac{d\sigma}{d\Omega dE'} = \left\{ \left(\frac{Q^2}{q^2}\right)^2 R_L(q,\omega) + \frac{1}{2}\left(\frac{Q^2}{q^2} + 2\tan^2(\theta/2)\right) R_T(q,\omega) \right\}$$

where σ_M is the point electron-proton cross-section, Q^2 is the four-momentum transfer, q the three-momentum transfer and ω is the energy transfer. $R_L(q,\omega)$ is a structure function for the coupling of the nucleons to a longitudinal photon and $R_T(q,\omega)$ is the structure function for coupling to a transverse photon. The MEC contribution to $R_L(q,\omega)$ is negligible, so we will not discuss it here.

As mentioned, there are two classes of MEC effects in the quasielastic peak. The diagrams contributing to the case of one-nucleon knockout are shown in Fig. (1a-d,f). Fig.

[†] Present address: Department of Physics, University of Manitoba, Winnipeg, MB, Canada R3T 2N2

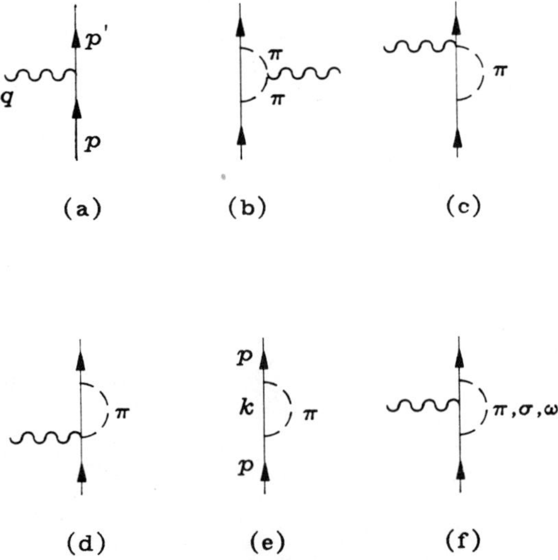

Fig. 1. (a-d) One-body knockout diagrams in this calculation, including MEC's. All nucleon propagators are over valence states. (e) is the pion self energy (for Ward Identities), and (f) is a many-body effect which is needed to satisfy the identities.

(1e) is the pion self energy, useful in that we can use Ward Identities to check this part of our calculation [4]. We actually neglect contributions from Fig. (1f) here, as this would be better treated as part of the many-body problem. Fig. (2) shows the diagrams which contribute to two-nucleon knockout (to the same level of approximation as those of Fig. (1)).

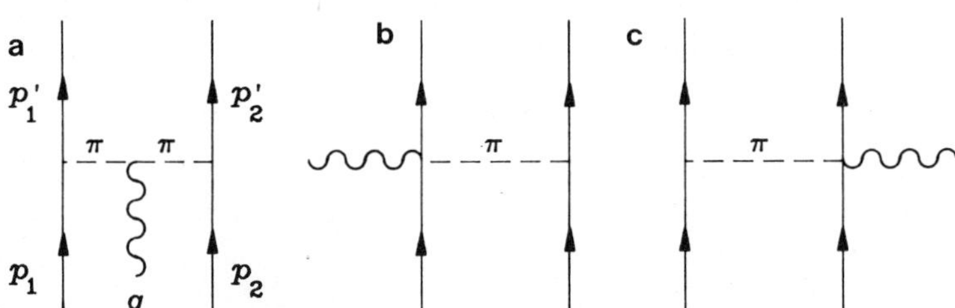

Fig. 2. Meson Exchange contributions to two-body knockout, to the same order as the diagrams of Fig. (1).

In Fig. (3) we see the results of our calculation for the transverse response of ^{40}Ca at two different momentum transfers. The data is from Ref. 5. We have used the data at the higher momentum transfer to fit our parameters. We find that $k_F = 1.0$ fm^{-1}, and an effective mass $M^* = 0.7M$ gives a reasonable fit. However, we cannot fit the lower momentum transfer data with the same parameters. This is almost certainly a sign that final state interactions are also very important. We are unable to fit the high ω side of the response. There are two reasons for this. First, the structure there is due to production of free pions from the nucleus, and the excitation of a Δ-resonance. We do not yet have this in our model. The second is that a simple Fermi gas model does not reproduce the tail of a true momentum distribution, so has insufficient strength away from the quasielastic peak.

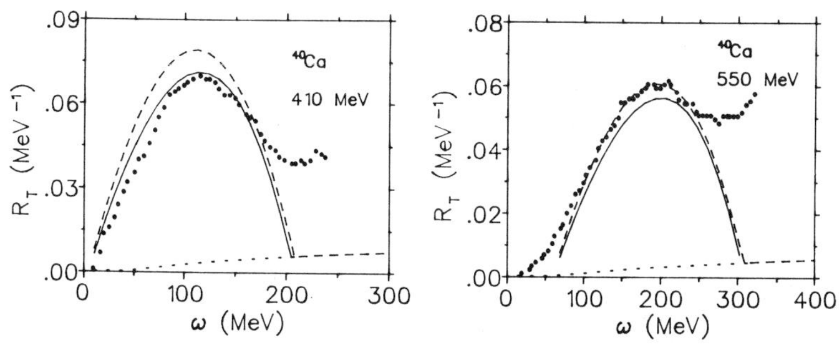

Fig. 3. A comparison of our calculation to the transverse response of ^{40}Ca at two different momentum transfers. The full line is the impulse approximation, and the dashed line includes all MEC effects in our calculation. The dotted line is the two-body knockout contribution. Parameters which give the fit at $q = 550$ MeV/c are $k_F = 1.0$ fm^{-1}, and $M^* = 0.7M$. Data is from Ref. 5.

There are two interesting observations that can be made. The first is that the location of the quasielastic peak is shifted due to the MEC's. This can have important consequences in the interpretation of quasielastic data, and how this is used in deep-inelastic quark/nucleus physics. This is quite different to what we would see in a non-relativistic calculation (shown in Fig. (4) for the one-body knockout case only), where the location of the quasielastic peak is essentially unchanged after the inclusion of MEC's. Also, the inferred Fermi momentum would be quite different to the true value due to the broadening of the peak by MEC's.

A more detailed study is currently underway using realistic wavefunctions for finite nuclei, which will look at MEC's, plus final state interactions, at the quasielastic peak. This will allow quantitative comparison to existing and expected data which in turn will help us understand the validity of the Impulse Approximation in electron-nucleus scattering.

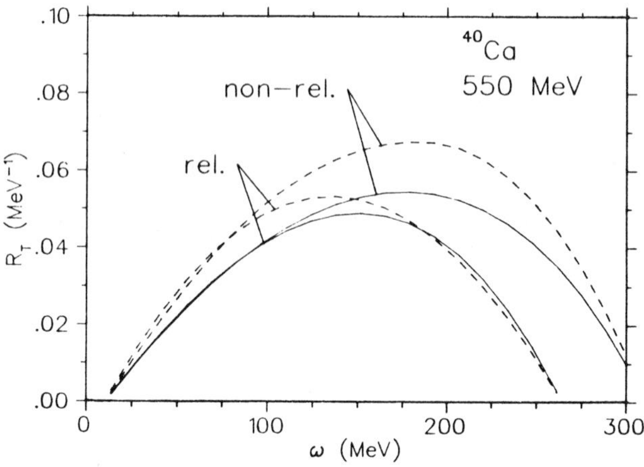

Fig. 4. A comparison of the results for one-body knockout in relativistic and non-relativistic calculations. Here $k_F = 1.3$ fm^{-1}, and $M^* = M$. The solid line is the impulse approximation, and the dashed line includes MEC's. It is clear that the behaviour of the quasielastic peak is much different in the two cases, especially in how it is modified by MEC's.

REFERENCES

[1] M. Kohno and N. Ohtsuka, Phys. Lett. **98B**, 335 (1981).
[2] J.W. Van Orden and T.W. Donnelly, Ann. Phys. **131**, 451 (1981).
[3] G. Do Dang *et al.*, Phys. Rev. **C35**, 1637 (1987).
[4] W. Bentz *et al.*, Nucl. Phys. **A436**, 593 (1985).
[5] Z.E. Meziani *et al.*, Phys. Rev. Lett. **52**, 2130 (1984).

SINGLE-PARTICLE STRENGTH AND NUCLEAR RESPONSE FUNCTIONS[†]

W. H. Dickhoff

Department of Physics, Washington University
St. Louis, Missouri 63130

M. G. E. Brand and K. Allaart

Natuurkundig Laboratorium, Vrije Universiteit
de Boelelaan 1081, 1081 HV Amsterdam, The Netherlands

A. Ramos and A. Polls

Departament d'Estructura i Constituents de la Materia
Universitat de Barcelona, E-08028 Barcelona, Spain

Abstract: The relation between the low-energy and high-energy response of nuclei is discussed. Calculations using an extended RPA method (E-RPA) show that present understanding of low-energy nuclear response functions is consistent with recent experimental evidence on the depletion of shell model states below the Fermi energy. This point is further investigated by calculating the momentum distribution and the single-particle and single-hole strength functions in nuclear matter at normal density including short-range correlations. A depletion of 10-15% is obtained for momenta below k_F. This missing strength is found as a very smooth distribution extending to very high energy. This implies that an important fraction of the particle-hole strength is not available at low energy. This disappearance of strength at low energy leads to appearance of strength at high energy and momentum transfer.

LOW ENERGY NUCLEAR RESPONSE FUNCTIONS

There is overwhelming experimental information on low-energy nuclear response functions. This information is typically analyzed with standard RPA calculations using either phenomenological or more microscopically oriented effective interactions. Clear shortcomings of this conventional approach in describing experimental data can be noted. Some of these are listed below.

(i) Magnetic strength is consistently overestimated.[1] Similar difficulties are encountered in the charge-exchange Gamow-Teller response function.[2]

(ii) The use of realistic interactions of G-matrix type leads to instabilities for the natural parity $\Delta T=0$ states ($0^+, 2^+, 3^-$), especially in heavier nuclei.[3]

[†] This research was supported in part by NSF Grant No. DMR-8519077 which also provided access to the Cray X-MP of the Pittsburgh Supercomputing Center and by NATO under Grant No. RG.85/0684 and by the Foundation for Fundamental Research of Matter (FOM) with support from the Netherlands' Organization for advancement of Pure Research.

(iii) The occurrence of many low-energy isoscalar excitations in ^{16}O and ^{40}Ca cannot be explained.[4]

(iv) At higher energies, the distribution of the strength of giant resonances is not reproduced.[5]

Significant improvement over the standard RPA approach is obtained by considering the coupling of particle-hole states (1p1h) to more complicated many-particle many-hole states. In Ref.6 this is achieved by first considering the coupling of the single-particle or single-hole to 2p1h or 1p2h states. This automatically ensures that ph propagation of such dressed propagators will involve the coupling to 2p2h states. A conserving treatment for the response is then guaranteed by the Baym-Kadanoff procedure[8] which uniquely defines the ph interaction corresponding to the considered self-energy contributions. It is then possible to formulate the ph propagator equation in terms of an eigenvalue problem corresponding to the size of the 1p1h space with an energy dependent interaction.[3] This is an efficient procedure since one only needs the 1p1h transition amplitudes for the comparison with the experimental data.

This extended RPA method (E-RPA) has been applied to ^{48}Ca using a realistic interaction with short-range correlations built in (G matrix). In ^{48}Ca one can study both the magnetic and electric response as well as the charge-exchange response of the nucleus and general conclusions about the quality of the description can de deduced. Results are shown in Fig. 1 in comparison with standard RPA results. In RPA the electric E2-response (top left) is strongly overestimated at low energy; this problem is solved in E–RPA (top right) where the strength in the lowest state is now compatible with experiment as is the general strength distribution. Nevertheless, too much energy weighted strength is found at low energy as compared to experiment.

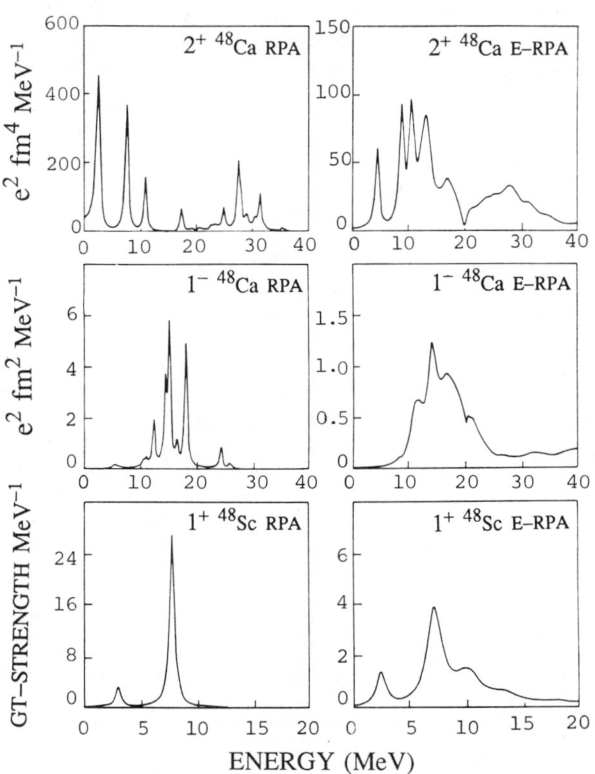

Figure 1. Low energy response functions of ^{48}Ca in RPA and E-RPA.

For the E1-response a similar improvement of the strength distribution is obtained in E-RPA (middle right) as compared to RPA (middle left). A very realistic strength distribution with an asymmetric peak is obtained together with a tail extending to the highest 2p2h energy included in the calculation (about 100 MeV). The asymmetry reflects the increasing density of 2p2h states with increasing energy and in itself is evidence for the strong coupling of 1p1h sates to more complicated states. As in the E2 case an overestimate of the energy weighted sum rule is obtained. The sum rule when integrated up to 35 MeV exceeds the classical sum rule by 16% whereas in ^{40}Ca only 80% is experimentally observed up to this energy.

Identical conclusions can be drawn for the Gamow-Teller response. Within E-RPA (bottom right) only 74% of the sum rule is found below 20 MeV which is significantly better than the RPA result of 99% (bottom left) but still higher than the experimental analysis. Even for the M1 transition to the famous 10.23 MeV state in ^{48}Ca one draws the same conclusion. In RPA 9.6 μ_N^2 is obtained whereas in E-RPA a much better 6.5 μ_N^2 is still significantly higher than the experimental 3.9 μ_N^2. The obvious conclusion from this dramatically improved description of the low-energy nuclear response is that *all* response functions still show a substantial overestimate of the strength although the distribution has become very realistic. Since the present description does not yet include the effect of ground state correlations, one concludes that these have to be included in the analysis. The size of the depletion of single-particle strength necessary to get global agreement with experiment is roughly consistent with a recent analysis of (e,e'p) measurements[8] in which a 20% depletion for the 3s½ state in ^{208}Pb is deduced.

SPECTRAL FUNCTIONS AND THE LOW AND HIGH ENERGY RESPONSE

Short-range correlations will play a dominant role in the breakdown of the conventional mean-field picture.[9] A correlated-basis functions (CBF) calculation of the momentum distribution in nuclear matter results at normal density in a depletion ranging from 13% at k = 0 to 21% at k_F.[10] Similar results are expected for finite nuclei leading to a sizable depletion of orbitals below the Fermi energy. The important question is where this single-particle strength has gone. Within a self-consistent Green's functions (SCGF) approach this question can be answered by studying the sp propagator in nuclear matter

$$g(k,\omega) = \int_{\varepsilon_F}^{\infty} d\omega' \frac{S_p(k,\omega')}{\omega - \omega' + i\eta} + \int_{-\infty}^{\varepsilon_F} d\omega' \frac{S_h(k,\omega')}{\omega - \omega' - i\eta} \tag{1}$$

where S_p and S_h represent the particle and the hole spectral function, respectively. These spectral functions provide the information on the sp strength distribution. The momentum distribution can be obtained from

$$n(k) = \int_{-\infty}^{\varepsilon_F} d\omega \, S_h(k,\omega) \quad . \tag{2}$$

The repulsive nature of the nucleon-nucleon interaction at short distances implies the necessity of including ladder diagrams in the calculation. The SCGF method then seeks a self-consistent solution between the changed properties of the particle in the medium due to such ladder summed interactions (with the other particles) and this ladder summed interaction which is generated by propagating these dressed particles.[11] A complete numerical solution is involved but an intermediate solution scheme has already provided an alternative nonrelativistic saturation mechanism due to the consistent inclusion of hole-hole propagation.[12] Also pairing can be incorporated within this scheme.[13] The occurence of pairing instabilities for realistic interactions[14] has in fact led to a treatment of only the central short-range correlations as they are present in the $^3S_1$ partial wave of the Reid soft core potential.[15] This interaction is denoted by $v_2^{l=0}$ and acts only in S waves. Investigation of the full Reid soft core with inclusion of pairing is in progress. The resulting momentum distribution for $v_2^{l=0}$ at normal density gives a depletion of 10% at k = 0 and 15% at k_F.[15] This is consistent with the neglect of tensor correlations which should lead to an additional 5% depletion.[10,16] The interesting question is where this sp strength has gone.

In Fig.2 both spectral functions are shown for k = 0.9 k_F. The hole spectral function (left) shows the typical sharply peaked behavior for momenta close to k_F. The particle spectral function (right) displays a smooth distribution of strength extending to several 1000 MeV. This is related to the strength of the short-range repulsion and depends on which interaction is used in the calculation. It is clear that strength at such high energy will not play any role in the determination of the low energy response of the nucleus. This is consistent with the discussion above on the inclusion of ground state correlations in the calculation of the low-energy response functions.

In Fig.3 the spectral functions are shown for a momentum above k_F. The peak lies now above ε_F and is broader. The strength below ε_F leads to an occupation of the order of 1%. Again a fraction of the strength is removed to very high energy. The study of

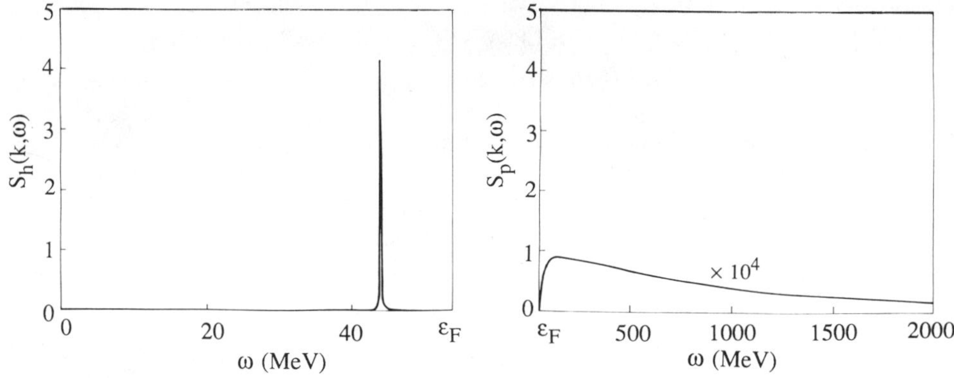

Figure 2. Hole and particle spectral function for k = 0.9 k_F (k_F = 1.4 fm^{-1}).

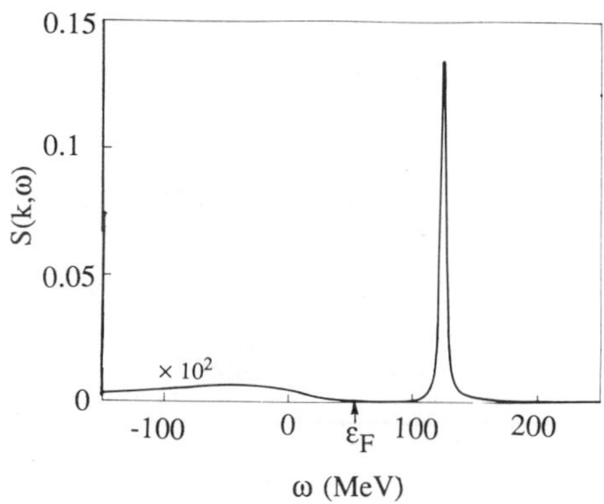

Figure 3. Same as Fig.2 for k=1.6k_F.

these spectral functions therefore reveals plainly that the ground state correlations due to high momentum (coming from high energy) components must lead to a substantial reduction of the response at low energy. With this conclusion one can draw the discussion of the calculated response for ^{48}Ca to a full circle. In the future the breakdown of the mean-field picture must be included in the theoretical calculations. Present results indicate that a dramatic improvement of our quantitative understanding of low energy nuclear phenomena is within reach.

The depletion of strength at low energy naturally has consequences for the observation of strength at high energy and momentum transfer as they are discussed in this workshop. Both Fantoni[17] and Sick[18] have stressed the importance of the inclusion of initial state correlations in the analysis of high energy and momentum transfer electron scattering. These initial correlations are represented by the hole spectral functions (see e.g., Figs. 2 and 3). One observes in general that with increasing momentum a broader distribution is obtained and in the limit of very large k the center of this distribution moves like $-k^2/2m$. The approximation of an average separation energy is therefore not very well justified.[16] The final state correlations are represented by the particle spectral function. Inclusion of a real shift due to the real part of the optical potential[18] is also not sufficient to describe these correlations. Inclusion of the imaginary part consistent with dispersion relations as it is represented here by the particle spectral functions in Figs. 2 and 3 will provide a more consistent way to analyze inclusive high energy and high momentum transfer reactions.

REFERENCES

1. A. Richter, in *Proceedings of the International Conference on Nuclear Physics*, eds. P. Blasi and R. A. Ricci (Tipografia Compositori, Bologna, 1983), Vol. 2, p.189.
2. C. D. Goodman, see Ref. 1, p.165.
3. W. Hengeveld, W. H. Dickhoff and K. Allaart, Nucl. Phys. **A451**, 269 (1986).
4. W. H. Dickhoff, in *Condensed Matter Theories*, Vol. 3, J. S. Arponen, R. F. Bishop, and M. Manninen, eds. (Plenum, New York, 1988), p. 261.
5. G. F. Bertsch, P. F. Bortignon and R. A. Broglia, Rev. Mod. Phys. **55**, 287 (1983).
6. M. G. E. Brand, K. Allaart, and W. H. Dickhoff, Phys. Lett. **B** (1988), in press.
7. G. Baym and L. P. Kadanoff, Phys. Rev. **124**, 287 (1961).
8. E. N. M. Quint *et al.*, Phys. Rev. Lett. **58**, 1088 (1987).
9. V. R. Pandharipande, C. N. Papanicolas, and J. Wambach, Phys. Rev. Lett. **53**, 1133 (1984).
10. S. Fantoni and V. R. Pandharipande, Nucl. Phys. **A427**, 473 (1984).
11. A. Ramos, A. Polls, and W. H. Dickhoff, in *Condensed Matter Theories*, Vol. 3, J. S. Arponen, R. F. Bishop, and M. Manninen, eds. (Plenum, New York, 1988), p. 319.
12. A. Ramos, W. H. Dickhoff, and A. Polls, Phys. Lett. **B** (1988), in press.
13. W. H. Dickhoff, Phys. Lett. **210B**, 15 (1988).
14. W. H. Dickhoff, in *Condensed Matter Theories*, Vol. 4, J. Keller, ed. (Plenum, New York, 1989), in press.
15. A. Ramos, A. Polls, and W. H. Dickhoff, to be published; A. Ramos, Thesis University of Barcelona (1988).
16. S. E. Koonin, *these proceedings*; M. N. Butler, Thesis California Institute of Technology (1987).
17. S. Fantoni, *these proceedings*.
18. I. Sick, *these proceedings*.

HIGH ENERGY INELASTIC NEUTRON SCATTERING FROM HYDROGEN IN CESIUM INTERCALATED GRAPHITE

G.J. Kellogg[†], P.E. Sokol[‡], and J. White[+]

[†] Lyman Laboratory of Physics, Harvard University
Cambridge, MA

[‡] Department of Physics, The Pennsylvania State University
University Park PA

[+] Research School of Chemistry, Australian National University
Canberra, Australia

INTRODUCTION

Inelastic neutron scattering at high momentum transfer, Q, can provide direct information on the the atomic momentum distribution, $n(p)$ when the Impulse Approximation is valid. These measurements are of particular interest for light molecules where the quantum zero point motion may dominate $n(p)$ and information on the local potential may be obtained.

Graphite intercalation compounds [1], which are formed when atoms or compounds are introduced between the basal planes of graphite, provide a interesting case. The intercalant materials form a layered structure along the graphite c-axis, occupying the spaces between the carbon planes. Within the intercalant layer a variety of two-dimensional structures commensurate and incommensurate with the graphite have been observed. Hydrogen molecules may be used as a probe of the local environment in the material.

INELASTIC NEUTRON SCATTERING

In the limit of large momentum transfer Q, the scattering function may be expressed in terms of the scattering of individual particles from a bound initial state, determined by the many body interactions in the sample, to a free particle final state. This is the well-known Impulse Approximation and gives

$$S(Q,E) = \int_{-\infty}^{+\infty} n(|\vec{p}|) \; \delta \left[E - E_r - \frac{\hbar \vec{Q} \cdot \vec{p}}{M} \right] d\vec{p} \qquad (2.1)$$

where $n(|\vec{p}|)$ is the distribution of atomic momenta $\vec{p}$, and $E_r = \hbar^2 Q^2 / 2M$ is the recoil energy and M is the atomic mass of the atom. The scattering shows certain characteristic features: the peak is centered at the recoil energy and the width, at constant Q, is proportional to Q. These characteristic features, while not sufficient to guarantee the validity of the IA, will be used to indicate the region where it is appropriate.

In the case of molecules with internal degrees of freedom, the IA is easily generalized if the internal excitations are decoupled from the translational degrees of freedom. The scattering function is then [2]

$$S_n(Q,E) = f_n \int_{-\infty}^{+\infty} n(|\vec{p}|) \ \delta\left[E - E_r - E_n - \frac{\hbar\vec{Q}\cdot\vec{p}}{M}\right] d\vec{p} \qquad (2.2)$$

The summation is over all internal excitations and f_n, and E_n represent the structure factor and excitation energy of the nth internal level.

The internal excitations for free H_2 are the rotational and vibrational excitations [3]. The energy of these levels are $E_{rot} = 7.35 \ meV \, J(J+1)$ and $E_{vib} = 516 \ meV (\nu + 1/2)$, where J and ν are the rotational and vibrational quantum numbers. For scattering from an initial state of $J = 0$, which is appropriate here, the major contribution arises from final states with odd J, which enter with the incoherent cross section. If vibrational excitations are not excited, the structure factor [4] for the free rotor is

$$f_J = (2J+1)C^2(0,J,J;00)\left|2i^l j_l(Qa/2)\right|^2 \qquad (2.3)$$

where J is the quantum number of the final rotational state, C is a Clebsch-Gordon coefficient, a is the inter-proton distance, and j_l is a spherical Bessel function. For trapped molecules the crystal field produces mixtures of pure rotor states and the structure factor has a form similar to (2.3).

The momentum distribution usually has a simple Gaussian form, which is specified by it standard deviation σ_p. The scattering will then also be a Gaussian, with standard deviation $\sigma_s = \hbar Q \sigma_p/M$, centered at E_r or, when internal excitations are present, a series of Gaussians centered at $(E_n + E_r)$. The kinetic energy per atom provides a convenient measure of $n(p)$ and is given by $< KE >= 3M\sigma_s/2\hbar^2 Q^2$.

RESULTS

The measurements were carried out using the Low Resolution Chopper Spectrometer [5]. The sample of $C_{24}Cs$ was contained in a cylindrical aluminum cell attached to a DISPLEX refrigerator for cooling. Normal H_2, sufficient to completely saturate the adsorbtion sites, was admitted at 105 K. Conversion of the H_2 to the $J = 0$ ground state is expected to occur rapidly at low temperatures in the paramagnetic intercalate. The scattering was measured, at 20, 100, and 150 K, with an incident energy of 1207 meV giving momentum transfers of 8 to 22 $\mathring{A}^{-1}$ for the scattering angles used.

Fig. 1 shows $S(\phi,\omega)$, the scattering function at constant angle, from the H_2 with the background from the Al cell and the $C_{24}Cs$ removed, at 20 K. The H_2 exhibits only one broad scattering peak, which may be due either to atomic hydrogen or to molecular hydrogen, with a large $< KE >$ that broadens the individual peaks so they are no longer resolved. The position of the peak center, shown in Fig 2, is proportional to Q^2 for $Q > 15\mathring{A}^{-1}$, in agreement with the IA prediction. At lower Q's, deviations from this behavior are observed which we take as characteristic of departures from the IA. Only measurements with $Q > 15\mathring{A}^{-1}$ will be used in the following analysis.

The slope of the data in Fig 2, where the IA is valid, is determined by the mass of the scattering molecule. The prediction for atomic hydrogen and the classical result for molecular hydrogen, using an effective mass tensor, are shown in the figure. The observed results are not in agreement with either prediction.

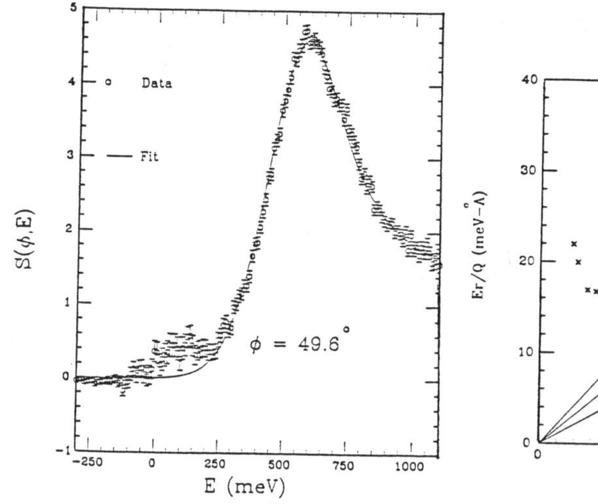

Figure 1. - Observed scattering and the fit to the data using a Gaussian $n(p)$ and the free molecule form factors.

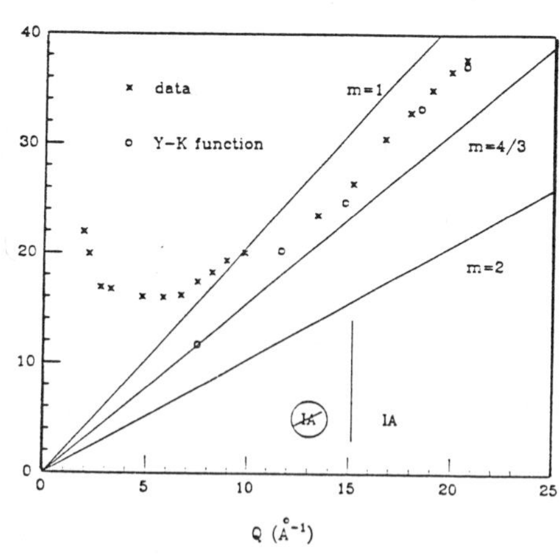

Figure 2. - Measured peak position versus Q. The solid lines are the predictions for mass 1 and 2. The line m=4/3 correponds to the classical prediction for H_2. The crosses are the quantum mechanical prediction

We may also compare with the theoretical predictions for free molecular H_2. Fig 3 shows the intensity of the various scattering peaks that contribute to the scattering in Fig 2. The peak center, which is a weighted average of the transitions, is shown in Fig 2 and is in excellent agreement with the observed peak centers. This provides

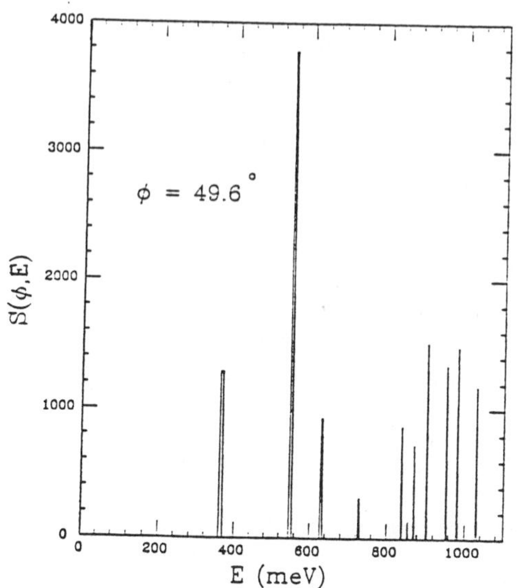

Figure 3. - Intensity and position of the rotational transitions that contribute to the fit in Figure 1.

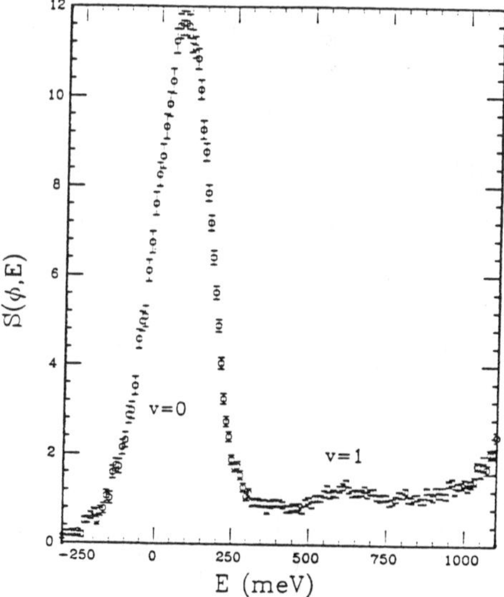

Figure 4. - Observed scattering at ϕ=13.8 showing the rotational and vibrational excitations.

strong evidence that the intercalated hydrogen retains its molecular character and, since the model is based on free H_2, that the molecular excited states approximate well to free rotors in the intercalate.

We use a Gaussian form for $n(p)$ and the free molecule form factors to extract the kinetic energy. The theoretical predictions, broadened by instrumental resolution, are compared directly to the data. The width of $n(p)$ is varied to obtain the best fit. Excellent agreement with the data is obtained, as shown in Fig 2. The fitted width is independent of Q, in good agreement with the IA prediction.

The translational kinetic energy of 450 ± 130 K, is much larger than expected based on the thermal energy alone, which would be 30 K. This is due entirely to the zero point motion of the molecule in the $C_{24}Cs$. The large magnitude of the kinetic energy indicates that the molecule is very tightly localized in the intercalate. No temperature dependence of the scattering is observed from 20 K to 150 K, which is not surprising, considering the large magnitude of the zero point kinetic energy. These results are consistent with previous measurements covering a more limited range of Q [6].

Fig 4 shows the scattering from only the H_2 at a Q of 5 $\mathring{A}^{-1}$, where transitions involving both rotational and vibrational excitations are observed. The large peak centered at ≈ 100 meV is due only to rotational transitions. The smaller peak at 600 meV is due scattering with excitation of both the rotational transitions and the first vibrational excitation at 512 meV. This peak is relatively weak so we have not attempted to extract any quantitative information. However, we note that the position of the peak is consistent with the vibrational energy levels for the free H_2 molecule. Thus, if any charge transfer to the hydrogen occurs, it does not significantly affect the vibrational frequencies.

CONCLUSIONS

We have measured the scattering from hydrogen in $C_{24}Cs$ at sufficiently large Q's that the IA may be used to interpret the scattering. The observed scattering is well described by freely rotating molecular hydrogen molecules. The translational kinetic energy of 450 ± 130 K is quite large and indicates that the molecules are tightly bound. The lack of any temperature dependence for the hydrogen scattering between 20 K and 150 K is not surprising in view of the tight binding of the molecule.

This work was supported by the NSF through Contract No. DMR-83-16979 and by OBES/DMS support of the Intense Pulsed Neutron Source at Argonne National Laboratory under DOE grant W-31-109-ENG-38. GJK and PES acknowledge the support of the Division of Educational Programs at Argonne National Laboratory during part of this work.

References

[1] Proceedings of the Fourth International Symposium on Graphite Intercalation Compounds, Tsukuba, Japan, eds. K. Nakao and S.A. Solin, Synth. Met. **12** (1985)

[2] W. Langel, D.L. Price, R.O. Simmons, and P.E. Sokol, to be published.

[3] I.F. Silvera, Rev. Mod. Phys. **52**, 393 (1980).

[4] J. U. Young and J. A. Koppel. Phys. Rev. **135A**, 603 (1964).

[5] D. L. Price, J. M. Carpenter, C. A. Pelizzari, S. K. Sinha, I. Bresof, and G. E. Ostrowski, in Proceedings of the Sixth Meeting of the International Collaboration on Advanced Neutron Sources, Argonne National Laboratory, June 1982, Argonne National Laboratory report No. ANL-82-80 (unpublished, pp 207-215).

[6] P.R. Hirst, L.M. Needham, A.D. Taylor, Z.A. Bowden, and J.W. White, Chem. Phys. Lett. **147**, 228 (1988)

INITIAL STATE EFFECTS IN DEEP INELASTIC NEUTRON SCATTERING

J. Mayers[1], C. Andreani[2], and G. Baciocco[3]

[1]Rutherford Appleton Laboratory, Oxon, OX11 0QX, UK
[2]Dipartmento di Physica, Universita di Tor Vergata, Rome, Italy
[3]Dipartmento di Physica, Universita "La Sapienza", Rome, Italy

A NEW FORMULATION OF THE IMPULSE APPROXIMATION

We develop a new approach to the impulse aproximation which was first used by Gunn et al [1]. We assume, as is consistent with the existence of a single particle momentum distribution, that the struck atom has initial and final state single particle wavefunctions $\psi_i(r)$ and $\psi_f(r)$ respectively. This is clearly valid when the environment of the atom can be approximated by a single particle potential [2] as for example in the Born-Oppenheimer approximation and can also be formally justified in the impulse limit for a many body system such as He4.[3] Then the standard expression for $S(q,\omega)$ is [4];

$$S(q,\omega) = \Sigma_f \left| \int \psi^*_i(r)\psi_f(r)\exp(iq.r)dr \right|^2 \delta(\omega - E_f + E_i) \qquad (1)$$

where q is the momentum transfer, ω is the energy transfer, E_i is the energy of the initial state, E_f that of the final state and the sum is over all possible final states.

In the impulse approximation it is assumed that the struck atom recoils as a free particle; ie. $\psi_f(r)=\exp(ik_f.r)$ and

$$S(q,\omega) = \Sigma_f \left| \tilde{\psi}_i(q+k_f) \right|^2 \delta(\omega - E_f + E_i)$$

$$= \Sigma_f \, n(q+k_f) \delta(\omega - E_f + E_i) \qquad (2)$$

where $\tilde{\psi}$ is the Fourier transform of ψ and $n(p)$ is the momentum distribution correponding to ψ_i. We define the wavenumber of the final state via

$$E_f = k_f^2/(2M) + V_f \qquad (3)$$

where M is the mass of the struck atom and V_f is a constant which approximately represents the potential energy of the final state. Then from equations (2) and (3) and converting the sum to an integral we obtain

$$S(q,\omega) = \int n(q+k_f) \delta[\omega - k_f^2/(2M) + E_i - V_f]dk_f \qquad (4)$$

With the substitution $p=q+k_f$ and using $n(p)=n(-p)$ we obtain,

$$S(q,\omega) = \int n(p) \delta[\omega - (p+q)^2/(2M) + E_i - V_f]dp \qquad (5)$$

This expression is similar to the impulse approximation (IA) [4]

$$S_{ia}(\mathbf{q},\omega)= \int n(\mathbf{p})\delta[\omega-(\mathbf{p+q})^2/(2M)+p^2/(2M)]d\mathbf{p} \tag{6}$$

Equations (6) is obtained from (5) by the approximation $E_i-V_f=p^2/(2M)$. This approximation   is valid only if the initial state can be treated as a collection of free particles.

For an isotropic system the IA reduces to [5]

$$S(q,\omega)=(2\pi M/q)\int_{|y|}^{\infty}pn(p)dp \tag{7}$$

where

$$y=(M/q)[\omega-q^2/(2M)] \tag{8}$$

The dependence of $qS(q,\omega)$ on the single variable y rather than on q and ω separately is known as known as 'y scaling' and has the consequence that at constant q, $S(q,\omega)$ is peaked at the 'recoil energy', $E_R=q^2/(2M)$ and is symmetric in ω about E_R.

Stringari [6] has suggested that the expression

$$S_s(\mathbf{q},w)= \int n(\mathbf{p})\delta[\omega-(\mathbf{p+q})^2/(2M)+\kappa_i]d\mathbf{p} \tag{9}$$

gives a better approximation than the IA at low temperatures. Here κ_i is the average kinetic energy of an atom with the momentum distribution $n(p)$. Equation (9) can be obtained from equation (5) if we assume that $\kappa_i=E_i-V_f$ ie that the average potential energy of the final state is the same as that of the initial state.

At large q values and for an isotropic system, equation (9) reduces to [6]

$$S(q,\omega)=(2\pi M/q)\int_{|y1|}^{\infty}pn(p)dp \tag{10}$$

where

$$y_1=q-[2M(\omega+\kappa_i)]^{1/2} \tag{11}$$

In this case scaling behaviour still occurs but the scaling variable is y_1 rather than y, with the consequences that the peak in $S(q,\omega)$ is at $E_R-\kappa_i$ and $S(q,\omega)$ is not symmetric about the peak.

COMPARISON WITH A DEBYE SOLID

In an isotropic harmonic system, both $S(q,\omega)$ and $n(p)$ can be calculated exactly. In figure 1 we compare an exact calculation of $S(q,\omega)$ for an isotropic harmonic Debye solid [7] with the predictions of the impulse approximation (IA) and the Stringari formulation (SIA).

We make the following observations:
1. The SIA gives a better agreement with the exact calculation than the IA, at T=0. In particular the SIA reproduces the assymmetry in $S(q,\omega)$ and the shift in the position of the recoil peak to lower energies, features which are also commonly observed in experimental data. However, the shift in the recoil peak is consistently overestimated.

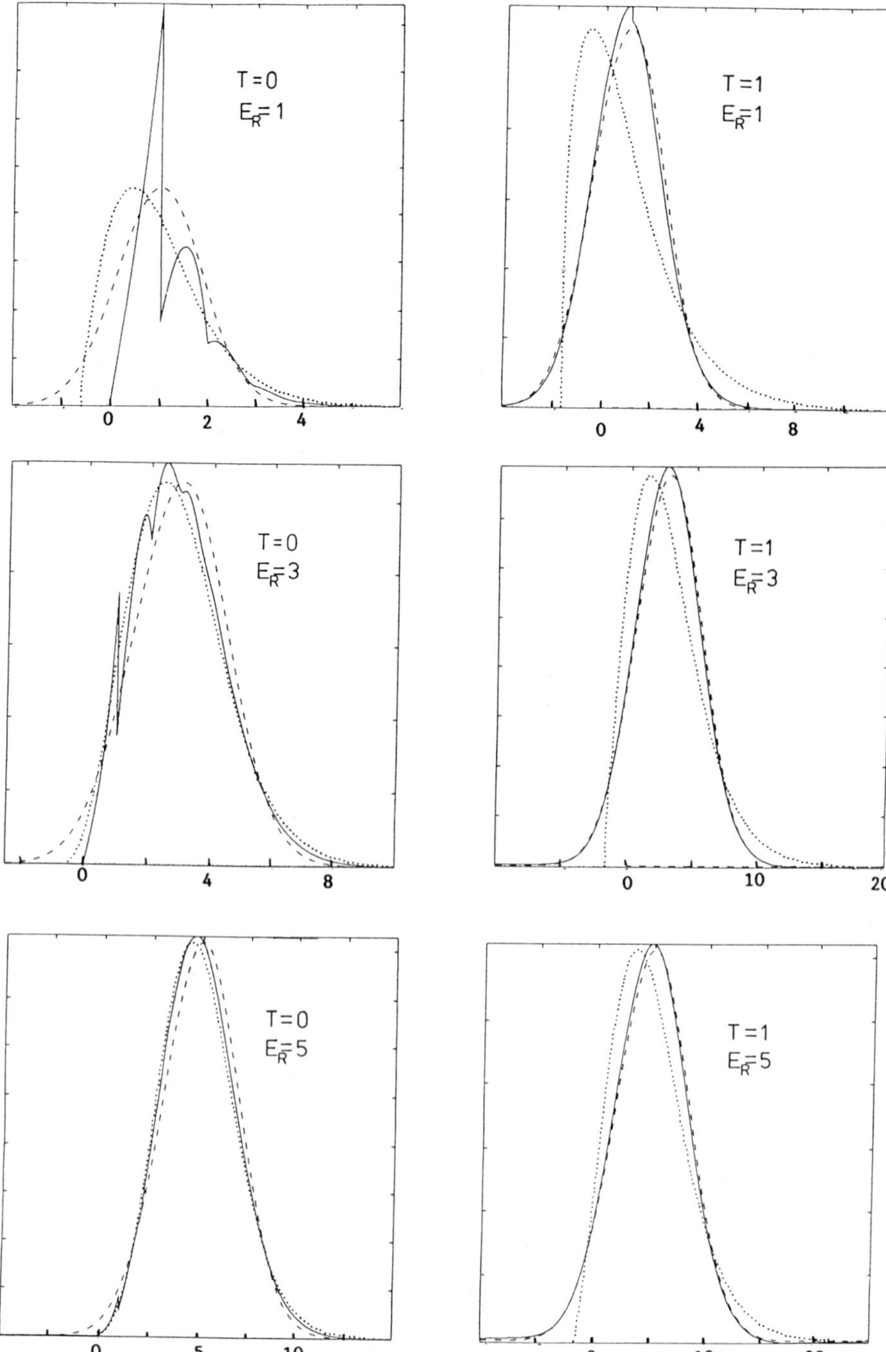

Figure 1   Comparison of numerical calculations of $S(q,\omega)$ for a Debye solid (solid line) with the impulse approximation equation (6) (dashed line) and equation (9) (dotted line). The x axis is energy transfer in units of Debye temperature. The temperature T and recoil energy E_R are in units of Debye energy.

2. The IA gives much better agreement with the exact $S(q,\omega)$ at higher temperatures.Further calculations show that the temperature at which the IA starts to be a better approximation is $\simeq\Theta_D/3$ where Θ_D is the Debye temperature.This is close to the energy of zero point motion in a harmonic system.

We interprete these obervations as follows:

1. At $T=0$,atoms have a distribution of momenta due to zero point motion although the energy of the scattering system is precisely defined. The IA predicts a finite probability that $\omega<0$, ie the system loses energy.This is forbidden in reality as the system is already in its ground state. It appears that the assymmetry and shift in position of the recoil peak are caused by the quantum nature of the initial state. The approximation of treating the initial state as a collection of free particles appears to be responsible for most of the inaccuracy in the IA. We note that Stringari has also shown that SIA gives a much more accurate description of He4 data than the IA.

The overestimate of the peak shift is caused by an underestimate of the potential energy of the final state. One would expect that the potential energy of the final state would be greater than that of the initial state, as the struck atom will tend to move from a position where the potential energy is a minimum, to regions of higher potential energy. Thus $E_i-V_f<K_i$ and the peak should occur at larger values of ω, as observed. Furthermore, the larger the potential energy in the final state, the higher the value of ω at which the peak occurs. ie final state interactions move the recoil peak to higher energies rather than lower energies as has been generally assumed.

2. The IA assumes that interactions between scattering centres affect only the momentum distribution of an apparently free gas of atoms. Clearly as the temperature is raised interatomic interactions play a less dominant role and one would expect that representing the initial state by a collection of free atoms will be a better approximation. At higher temperatures the initial state is a thermal average over allowed quantum states and processes for which $\omega<0$ are no longer forbidden.The IA then becomes a better approximation than the SIA, which replaces the thermal distribution of initial state kinetic energies by a single thermally averaged value.

CONCLUSIONS

Typical features of both experimental data and the calculation for a Debye solid presented here are assymmetry in $S(q,\omega)$ and a shift in the position of the recoil peak to lower energies.These deviations from the impulse approximation are usually attributed to final state effects.In contrast,we have shown that final state effects move the recoil peak to higher energies and that these features are caused by the quantum nature of the initial state. This conclusion is supported by figure 1 which shows that the accuracy of the impulse approximation depends strongly on temperature, ie on the properties of the target system before it interacts with the neutron.

ACKNOWLEDGEMENTS

It is a pleasure to thank Dr J M F Gunn and Professor S W Lovesey for their interest and guidance.We also thank the European Economic Community for financial support of this collaboration.

REFERENCES

1. J M F Gunn,C Andreani and J Mayers J Phys C $\underline{19}$ L835 (1986)

2. G Reiter and R N Silver Phys Rev Lett $\underline{54}$ 1047 (1985)

3. J Mayers to be published

4. S W Lovesey 'Theory of neutron Scattering from Condensed Matter',Vol 1 Oxford University Press (1987)

5. V F Sears Phys Rev B $\underline{30}$ 44 (1984)

6. S Stringari Phys Rev B $\underline{35}$ 2038 (1987)

7. J Mayers, C Andreani and G Baciocco Phys Rev B $\underline{39}$ 2022 (1989)

ON THE APPROACH TO THE SCALING LIMIT OF THE

RESPONSE OF QUANTUM GASES

A.S. Rinat

Weizmann Institute of Science, Rehovot 76100, Israel

For a non-relativistic system of particles interacting through 'regular' forces v, a series expansion has been derived for the reduced response[1]

$$\phi(q,y) \equiv (q/m)S(q,w) = \sum_{n=0}^{\infty} (m/q)^n F_n^{(i)}(y_i, [v]) \tag{1}$$

with $y_i = y_i(qw)$ some scaling variable. For a different choice y_j, satisfying $|y_i - y_j| = \mathcal{O}(q^{-1})$, one readily finds an alternative v series in y_j. $F_0(y)$ in (1) represents the asymptotic limit

$$\phi(q,y) \xrightarrow[\substack{q\to\infty \\ y\,fixed}]{} \left(4\pi^2\right)^{-1} \int_{|y|}^{\infty} n(p)pdp \tag{2}$$

where $n(p)$ is the single particle momentum distribution. $(m/q)F_1(y)$ in (1) gives the dominant correction to (2), etc. The series (1) is not a perturbative one, in that the coefficients depend on the <u>exact</u> $n(p)$, 2-particle density matrix, etc. In case v has a hard core, none of the $F_n(y), n \geq 1$, exists.

There are various recipes for handling the hard core case: One may for instance write[5]

$$v = v^{hc}, \quad r \leq a_c; \quad v = v^{reg}; \quad r > a_c \tag{3}$$

The finite t matrix corresponding to v, requires knowledge of $t^{hc} = t[v^{hc}]$[4] and $t^{reg} = t(v^{reg}) \sim v^{reg}$; the latter suffices for the $q \to \infty$ limit of ϕ. In such a treatment parts of ϕ linear in t^{hc} contribution to <u>all</u> orders in $1/q$ etc., and in particular to F_0 which is $\mathcal{O}(1)$ [2,3].

One demonstrates the following changes in the series (1):

$$F_0 \to F_0^{reg} + \delta^{(1)}F_0 + \dots \tag{4a}$$

$$F_1 \to F_1^{reg} + \delta^{(1)}F_1 + \dots \tag{4b}$$

$\delta^{(n)}F_0$ contributes to the same order in q as does F_0^{reg} and is due to nth order hard core collisions between the knocked-out and the remaining particles. For the relevant y-range only binary collisions $(\delta^{(1)}F_0)$ are retained. Likewise $\delta^{(1)}F_1$ is the $1/q$ coefficients of

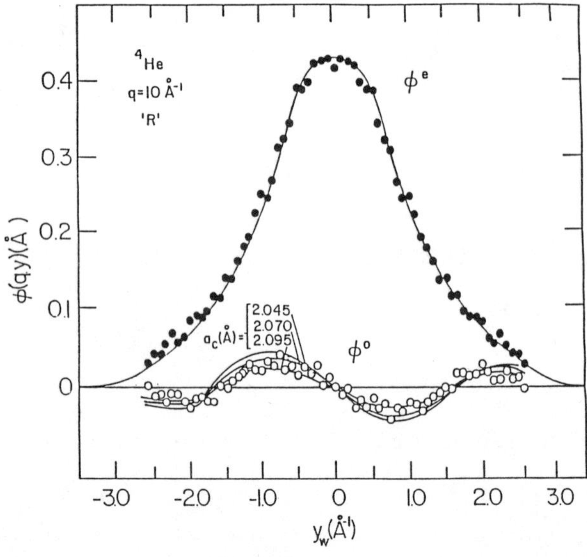

Fig. 1

$\phi(q,y)$ where states, before and after a single collision due to v^{reg}, are hard-core distorted, etc. Truncating the series as above allows (4a), (4b) to be identified as parts, even and odd in y. Only the T=1.2 K ^{4}He data for $q = 10\,\overset{\circ}{A}^{-1}$ [6] are not affected by experimental resolution and are accurate enough to permit the separation $\phi = \phi^{even} + \phi^{odd}$. Fig. 1 shows the correspondence between theory and experiment. One notices the sensitivity of ϕ^{odd} for a_c. Yet, since $|\phi^{odd}| < \phi^{even}$ for $|y| \leq 1.5\,\overset{\circ}{A}^{-1}$, the total ϕ is hardly affected by small changes in a_c. The same is the case for other data as well. Fig. 2 gives the $q = 7, 10, 12\,\overset{\circ}{A}^{-1}$ data of ref. 7) and the calculated ϕ (including condensate contributions) folded with provided resolution functions.

Finally we display in fig. 3 the pulsed neutron source data for ^{4}He, $\langle q \rangle \sim 23.8\,\overset{\circ}{A}^{-1}$ at T=3.5 K [8]. Here too ϕ^{th} is obtained after folding with a given resolution function.

In a future communication we shall include data on ^{3}He as well.

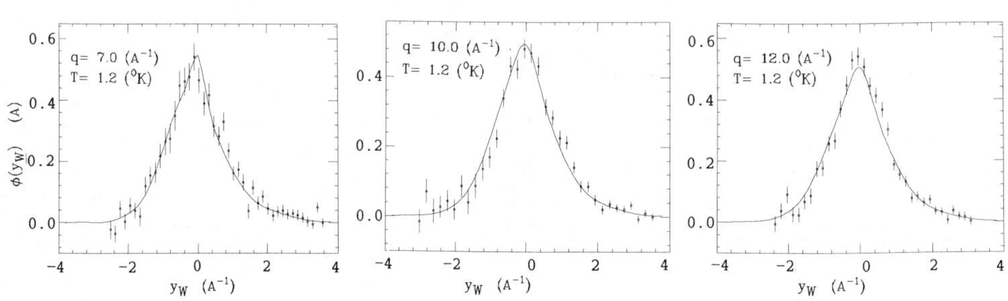

Figs. 2

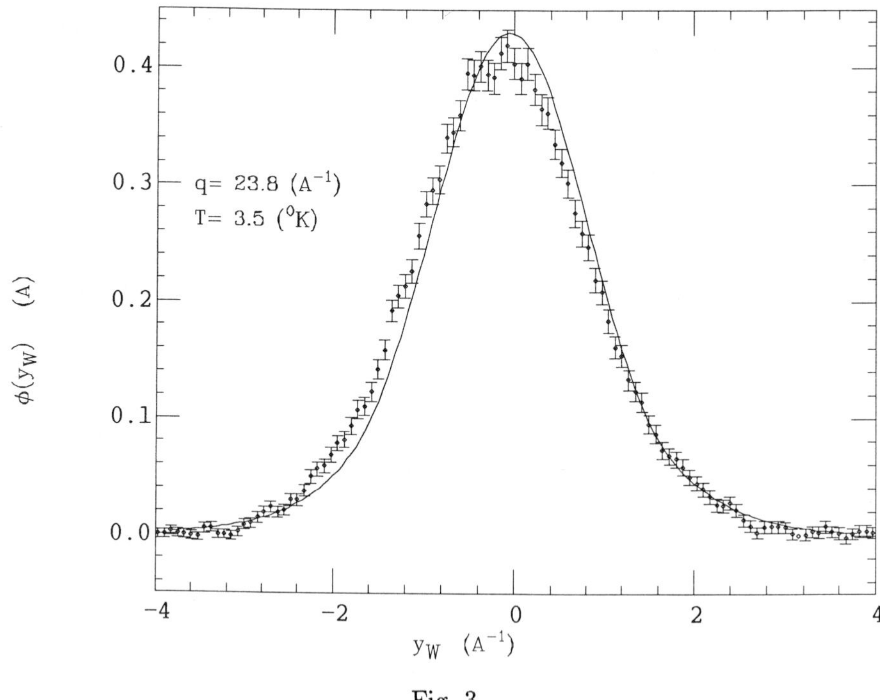

Fig. 3

References

1. H.A. Gersch, L.J. Rodriquez and P.N. Smith, Phys. Rev. A5, 1547 (1972).
2. J.J. Weinstein and J. Nagele, Phys. Rev. Letters 49, 1016 (1982).
3. S.A. Gurvitz, A.S. Rinat and R. Rosenfelder, PSI-preprint, 1988.
4. J.M.J. van Leeuwen and A.S. Reiner(Rinat), Physica, 27, 99 (1961).
5. A.S. Rinat, preprint WIS-88/31.
6. P. Martel et al., J. of Low Temperature Physics, 23, 285 (1976).
7. W.G. Stirling, E.E. Talbot, B. Tanatar and H.R. Glyde, to be publ. (J. of Low Temp. Phys. (1988)).
8. T.R. Sosnick, Harvard Ph.D. thesis (1988). T.R. Sosnick, W.M. Snow, P.F. Sokol and R.N. Silver, preprint LA-UR 88-505.

GENERALIZED MOMENTUM DISTRIBUTIONS

M. L. Ristig and J. W. Clark

Department of Physics, Washington University
St. Louis, Missouri 63130

Abstract

Generalized momentum distributions are needed in analyzing final state effects which occur in deep inelastic neutron scattering experiments on liquid ^{4}He. We elucidate their physical interpretation and extract their structure by exploring the associated two-body density matrix characterizing the ground state of a boson fluid. The formal analysis permits an accurate evaluation of these quantities by applying established many-body techniques such as the variational approach. Qualitative approximations hitherto adopted are critically reviewed.

Deep inelastic neutron scattering experiments on liquid ^{4}He are conventionally analyzed within the impulse approximation assuming that final state corrections may be ignored.[1] In this case the dynamic response of the system is entirely determined by the momentum distribution function

$$n(p) = \langle \psi | a_{\mathbf{p}}^{\dagger} a_{\mathbf{p}} | \psi \rangle \tag{1}$$

for the unit-normalized correlated ground state $|\psi\rangle$ of the quantum fluid. However, final-state effects may appreciably perturb the system.[2] For instance, the response may be influenced by physical processes where a particle with momentum $\hbar p$ does not conserve its momentum as in quantity (1) but may transfer part of it to a phonon, i.e., a density fluctuation, of the fluid (Fig. 1). In Feynman approximation the associated matrix element is $\langle \psi | \rho_{\mathbf{q}} a_{\mathbf{p}-\mathbf{q}}^{\dagger} a_{\mathbf{p}} | \psi \rangle$ with the density fluctuation operator $\rho_{\mathbf{q}} = \sum_{\mathbf{k}} a_{\mathbf{k}+\mathbf{q}}^{\dagger} a_{\mathbf{k}}$. The response function of the fluid now also depends on this more complicated function of the momenta $\hbar p$ and $\hbar q$.

A systematic tool for describing this quantity and other processes of higher order is provided by a set of generalized momentum distribution functions or, equivalently, a hierarchy of two-body, three–body, ... n–body density matrices. This set begins with the expectation value

$$n(\mathbf{p},\mathbf{q}) = \sum_{\mathbf{k}} \langle \psi | a_{\mathbf{k}+\mathbf{q}}^{\dagger} a_{\mathbf{p}-\mathbf{q}}^{\dagger} a_{\mathbf{p}} a_{\mathbf{k}} | \psi \rangle \quad . \tag{2}$$

Limiting ourselves to a boson fluid in its ground state $|\psi\rangle$, expression (2) provides information on the particle scattering process described above (Fig. 1) and on processes where a boson is created out of the Bose condensate ($\mathbf{p} = 0$) by a density fluctuation or where it is annihilated. Upon specializing to $\mathbf{q} = 0$, the function $n(\mathbf{p},0)$ describes the process (1) with one particle in orbital $\mathbf{p}$ (Fig. 2(a)).

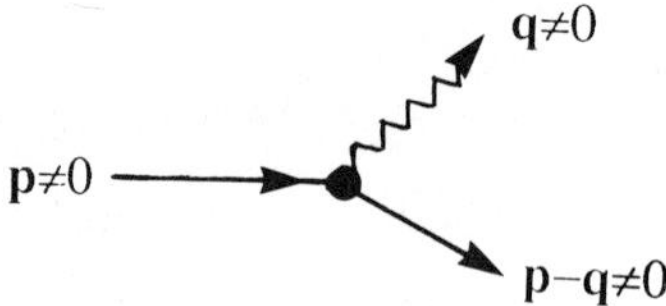

Figure 1. Scattering of a particle with momentum $\hbar\mathbf{p} \neq 0$ into the orbital $\hbar(\mathbf{p}-\mathbf{q})$ induced by a density fluctuation (a phonon).

Another example of generalized momentum distribution functions is the expectation value

$$n(\mathbf{k},\mathbf{p},\mathbf{q}) = \langle\psi\,|\,a_{\mathbf{k}+\mathbf{q}}^{\dagger}\,a_{\mathbf{p}-\mathbf{q}}^{\dagger}\,a_{\mathbf{p}}a_{\mathbf{k}}\,|\,\psi\rangle \quad . \tag{3}$$

This quantity describes effects which are caused by two-particle scattering, one-particle creation and annihilation, boson-pair creation and annihilation as well as pair scattering.

Distribution functions (2) and (3) are closely related to the two-body density matrix. This fundamental quantity is defined by the integral[3]

$$\rho_2(\mathbf{r}_1\mathbf{r}_2\mathbf{r}_1'\mathbf{r}_2') = N(N-1)\int d\mathbf{r}_3 \cdots d\mathbf{r}_N\,\psi(\mathbf{r}_1\mathbf{r}_2\mathbf{r}_3 \cdots \mathbf{r}_N)\psi(\mathbf{r}_1'\mathbf{r}_2'\mathbf{r}_3 \cdots \mathbf{r}_N) \quad . \tag{4}$$

The wave function ψ represents the unit-normalized ground state of a homogeneous system of N bosons at particle density ρ. Expression (4) is related to the generalized distribution function (3) by a Fourier transformation.[4] In similar fashion, quantity (2) may be related to the integral (4) by specializing to the case $\mathbf{r}_2 = \mathbf{r}_2'$. Introducing the function

$$\rho_2(\mathbf{r},\mathbf{r}') \equiv \rho_2(\mathbf{r}_1\mathbf{r}_2\mathbf{r}_1'\mathbf{r}_2) \tag{5}$$

of the relative distances $\mathbf{r} = \mathbf{r}_1 - \mathbf{r}_2$ and $\mathbf{r}' = \mathbf{r}_1' - \mathbf{r}_2$, we arrive at

$$n(\mathbf{p},\mathbf{q}) = \int\rho_2(\mathbf{r},\mathbf{r}')e^{-i\mathbf{p}(\mathbf{r}'-\mathbf{r})}\,e^{-i\mathbf{q}\mathbf{r}}\,d\mathbf{r}\,d\mathbf{r}' \quad . \tag{6}$$

The function $\rho_2(\mathbf{r},\mathbf{r}')$ is symmetric,

$$\rho_2(\mathbf{r},\mathbf{r}') = \rho_2(\mathbf{r}',\mathbf{r}) \quad , \tag{7}$$

and has the diagonal component

$$\rho_2(\mathbf{r},\mathbf{r}) = \rho^2\,g(\mathbf{r}) \quad , \tag{8}$$

where function $g(\mathbf{r})$ is the radial distribution function of the fluid. The sequential relation

$$\int\rho_2(\mathbf{r},\mathbf{r}')d\mathbf{r}_2 = (N-1)\rho_1(\mathbf{r}_1 - \mathbf{r}_1') \tag{9}$$

connects the matrix elements (5) to the one-particle density matrix

$$\rho_1(\mathbf{r}) = \rho_1(\mathbf{r}) = \frac{\rho}{N}\sum_{\mathbf{p}} n(\mathbf{p})\,e^{i\mathbf{p}\mathbf{r}} \quad . \tag{10}$$

The asymptotic behavior of the function $\rho_2(\mathbf{r},\mathbf{r}')$ is characterized by the properties:
$\rho_2(\mathbf{r},\mathbf{r}') \to \rho\rho_1(\mathbf{r} - \mathbf{r}')$ as $\mathbf{r}_2 \to \infty$, $\rho_2(\mathbf{r},\mathbf{r}') \to \rho^2\,n_0[1 + F(\mathbf{r})]$ as $\mathbf{r}' \to \infty$, $\rho_2(\mathbf{r},\mathbf{r}') \to \rho^2\,n_0$ as $\mathbf{r}$ and $\mathbf{r}' \to \infty$. Quantity n_0 is the condensate fraction of the boson

fluid and function F(r) is uniquely determined by the specific form of wave function ψ. It vanishes for sufficiently large distances r.

These properties permit an appropriate cluster decomposition of the matrix $\rho_2(\mathbf{r},\mathbf{r}')$,

$$\rho_2(\mathbf{r},\mathbf{r}') = \rho\rho_1(\mathbf{r} - \mathbf{r}')\{1 + F(r) + F(r') + F(\mathbf{r},\mathbf{r}')\} \ . \tag{11}$$

Functions F(r) and F(r,r') vanish for large values of r or r' and are determined by the ground state wave function. As a consequence of Eqs. (7) and (8), quantity F(r,r') is a symmetric function and has the diagonal form

$$F(\mathbf{r},\mathbf{r}) = g(r) - 1 - 2F(r) \ . \tag{12}$$

The sequential relation (9) requires

$$2\rho \int F(r)d\mathbf{r} = -1 \ , \tag{13}$$

$$\rho \int F(\mathbf{r},\mathbf{r}')d\mathbf{r}_2 = 0 \ . \tag{14}$$

The decomposition (11) establishes a clear separation of the distinct physical processes (Figs. 1 and 2) embodied in the generalized momentum distribution (2). Insertion of expression (11) into the integral (6) yields the structural result

$$\begin{aligned}
n(\mathbf{p},\mathbf{q}) = {}&\delta_{q0}(N - 1)n(p) \\
&+ (1 - \delta_{q0})(\delta_{p0} + \delta_{pq})Nn_0\, F(q) \\
&+ (1 - \delta_{q0})(1 - \delta_{p0})(1 - \delta_{pq})n'(\mathbf{p},\mathbf{q}) \ .
\end{aligned} \tag{15}$$

The first line on the right side of Eq. (15) represents the elastic process (Fig. 2a) where one particle resides in orbital $\mathbf{p}$ and $(N - 1)$ bosons act as spectators. The second line describes the creation and annihilation of bosons due to the condensate (Fig. 2(b) and 2(c)). The form factor of these processes depends on the wave vector of the phonon involved and is given by the Fourier inverse F(q) of function F(r). The particle scattering process above the condensate (Fig. 1) is represented by the last line of Eq. (15). The corresponding form factor depends on the momentum $\hbar\mathbf{p}$ of the boson as well as on the momentum $\hbar\mathbf{q}$ of the phonon and is explicitly given by

$$n'(\mathbf{p},\mathbf{q}) = \{n'(p) + n'(|\mathbf{p}-\mathbf{q}|)\}F(q) + \rho\int\rho_1(\mathbf{r}-\mathbf{r}')F(\mathbf{r},\mathbf{r}')\, e^{-i\mathbf{p}(\mathbf{r}'-\mathbf{r})}\, e^{-i\mathbf{q}\mathbf{r}}d\mathbf{r}\, d\mathbf{r}' \ . \tag{16}$$

The prime on functions n(p) and n($|\mathbf{p}-\mathbf{q}|$) indicates that the condensate portions $Nn_0\delta_{p0}$ and $Nn_0\delta_{pq}$, respectively, are to be omitted in Eq. (16).

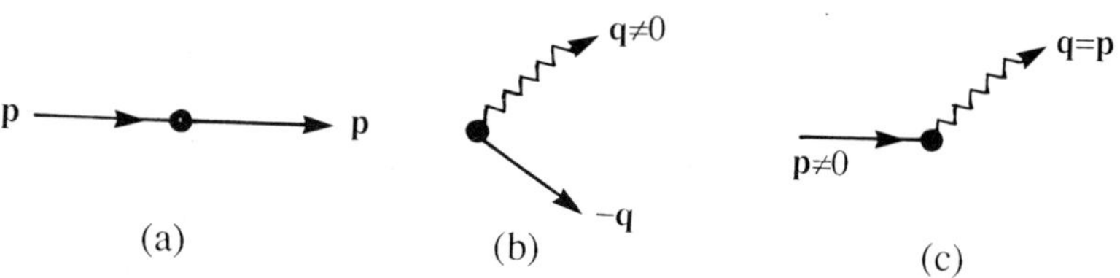

Figure 2. Contributions to the generalized momentum distribution function n($\mathbf{p},\mathbf{q}$): (a) elastic part ($\mathbf{q} = 0$) described by momentum distribution n(p), (b) creation of a particle with momentum $-\hbar\mathbf{q}$ from the Bose condensate ($\mathbf{p} = 0$) by a phonon, (c) corresponding annihilation process.

Results (15) and (16) demonstrate the symmetry property

$$n(\mathbf{p},\mathbf{q}) = n(\mathbf{p} - \mathbf{q}, -\mathbf{q}) \quad . \tag{17}$$

This property stems from the time-reversal invariance of the basic processes involved. In the Fourier picture expressed by (6), (15), and (16), the relation (8) takes the form of a sum rule for the generalized momentum distribution,

$$\frac{1}{N} \sum_{\mathbf{p}} n(\mathbf{p},\mathbf{q}) = \rho \int g(r) e^{i\mathbf{q}\mathbf{r}} \, dr \quad . \tag{18}$$

A second sum rule may be easily formulated if the ground state satisfies the condition

$$\psi(\mathbf{r}_1 \mathbf{r}_1 \mathbf{r}_3 \cdots \mathbf{r}_N) = 0 \quad . \tag{19}$$

This condition applies for fluids where the particle-particle interaction is sufficiently repulsive to prevent close encounters of two or more bosons (as is true of liquid ^{4}He). In this case,

$$\rho_2(0,\mathbf{r}') = 0 \quad , \quad F(0,\mathbf{r}') = -F(\mathbf{r}') \quad , \tag{20}$$

and

$$F(r) = -1 \quad , \quad g(r) = 0 \quad \text{as} \quad r \to 0 \quad . \tag{21}$$

Going to the Fourier picture via (6) we then obtain the sum rule

$$\sum_{\mathbf{q}} n(\mathbf{p},\mathbf{q}) = 0 \quad . \tag{22}$$

Sophisticated many-body methods are available for accurately evaluating the functions $F(r)$ and $F(\mathbf{r},\mathbf{r}')$ and the distribution function (15) for a given many-body wave function. One may follow closely the routes which have been thoroughly explored and established in numerical studies of the momentum distribution (1) and of the one-body density matrix (10). In particular, variational, stochastic, and/or correlated-basis (CBF) procedures[5,6] may be adapted to the present task. Assuming a Jastrow form for the ground state wave function, and optimizing this choice by solving an Euler-Lagrange equation, we may take reasonable first steps in a variational-CBF approach. The Jastrow ansatz allows us to exploit the formal results of Ref. 4 in evaluating the associated functions $F(r)$ and $F(\mathbf{r},\mathbf{r}')$. We have performed such a calculation for liquid ^{4}He at the experimental saturation density $\rho = 0.0218\text{Å}^{-3}$, employing the Aziz potential to describe the ^{4}He–^{4}He interaction. Our numerical results for the function $F(r)$ are displayed in Fig. 3. The sequential relation (13) is fulfilled exactly. A more detailed description of our formal and numerical analysis of the functions $F(r)$, $F(\mathbf{r},\mathbf{r}')$, and $n(\mathbf{p},\mathbf{q})$ will be presented in a forthcoming publication.[7]

At present the available literature offers and employs three rather crude approximations for the matrix $\rho_2(\mathbf{r},\mathbf{r}')$ which determines the generalized momentum distribution function (15). References 2, 8, and 9 adopt, respectively,

$$\rho_2(\mathbf{r},\mathbf{r}') = \rho\rho_1(\mathbf{r}-\mathbf{r}')g(r) \qquad \text{(Silver)} \quad , \tag{23}$$

$$\rho_2(\mathbf{r},\mathbf{r}') = \rho\rho_1(\mathbf{r}-\mathbf{r}') \sqrt{g(r)} \sqrt{g(r')} \qquad \text{(Gersch and Smith)} \quad , \tag{24}$$

$$\rho_2(\mathbf{r},\mathbf{r}') = \rho\rho_1(\mathbf{r}-\mathbf{r}') \, g(\tfrac{1}{2}|\mathbf{r} + \mathbf{r}'|) \qquad \text{(Rinat)} \quad . \tag{25}$$

These approximations all contain the one-body density matrix as a factor and involve the radial distribution function g, in somewhat different ways. By construction, all three describe correctly the elastic portion (first line) of the distribution function (15) and satisfy the sum rule (18). However, they fail in other aspects: prescription (23) breaks the symmetry requirement (7) (thus breaking time-reversal invariance). Both (23) and (24) violate the sequential relation (9). Estimate (25) conserves this condition but does not meet the second sum rule (22). The structural decomposition (11) is, in each case, poorly reproduced, in that condition (13) and/or (14) is violated. This is easily seen by comparing Eqs. (23)–(25), respectively, with decomposition (11):

$$F(r) \simeq g(r) - 1 \quad , \quad F(r') \simeq 0 \quad , \qquad F(\mathbf{r},\mathbf{r}') \simeq 0 \qquad (23')$$

$$F(r) \simeq \sqrt{g(r)} - 1 \quad , \quad F(r') \simeq \sqrt{g(r')} - 1 \quad , \qquad F(\mathbf{r},\mathbf{r}') \simeq F(r)F(r') \qquad (24')$$

$$F(r) \simeq 0 \qquad , \quad F(r') \simeq 0 \quad , \qquad F(\mathbf{r},\mathbf{r}') \simeq g(\tfrac{1}{2}|\mathbf{r} + \mathbf{r}'|) - 1 \; . \;(25')$$

To demonstrate the poor quality of the current approximations for the function F(r) or F(r'), we have plotted two of them together with the optimal Jastrow result (Fig. 3). This comparison underscores the need for a better evaluation of the functions F(r) and F(r,r') and justifies further efforts toward a truly microscopic many-body evaluation of both quantities.

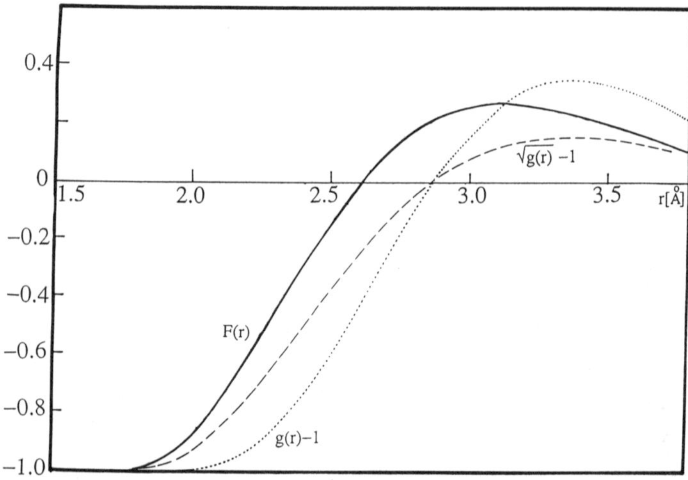

Figure 3. Optimal function F(r) describing the creation/annihilation of a ^{4}He atom with non-zero momentum, Eq. (15). The calculation is based on a wave function of Jastrow type for liquid ^{4}He at density $\rho = 0.0218\text{Å}^{-3}$, in hypernetted chain approximation. For comparison, we show also the approximate prescriptions for the quantity F(r) suggested in Refs. 2 and 8, Eqs. (23')–(25').

Acknowledgement

This research has been supported in part by the Deutsche Forschungsgemeinschaft under grant Nr. Ri 267 (MLR) and by the Condensed Matter Theory Program, Division of Materials Research, U.S. National Science Foundation, under Grant No. DMR-8519077 (JWC). MLR enjoyed the hospitality of the Department of Physics, Washington University, during a sabbatical leave from the Universitat zu Köln, West-Germany.

References

1. P. C. Hohenberg and P. M. Platzmann, Phys. Rev. **152**, 198 (1966); E. C. Svennson, in Proceedings of the 1984 Workshop on High-Energy Excitations in Condensed Matter, LA-10227; P. Sokol, in Proceedings of the 1986 Banff Conference on Quantum Fluids and Solids, Can. J. Phys. **65**, 1393 (1987).
2. R. N. Silver, Phys. Rev. B **38**, 2283 (1988).
3. E. Feenberg, *Theory of Quantum Fluids*, (Academic Press, New York, 1969).
4. M. L. Ristig, Nucl. Phys. **A317**, 163 (1979).
5. M. L. Ristig, in *From Nuclei to Particles, Proceedings of the International School of Physics, Enrico Fermi, Course LVII, Varenna, 1981* edited by A. Molinari (North-Holland, Amsterdam, 1981).
6. J. W. Clark and M. L. Ristig, "Overview of momentum distribution calculations," this volume.
7. M. L. Ristig and J. W. Clark, to be published.
8. H. A. Gersch and R. N. Smith, Phys. Rev. A **15**, 1547 (1972).
9. A. S. Rinat, Weizmann Institute, Rehovot, preprint (1988).

FINAL-STATE EFFECTS IN DEEP-INELASTIC NEUTRON SCATTERING

V.F. Sears

*Atomic Energy of Canada Limited Research Company
Chalk River, Ontario, Canada K0J 1J0*

ABSTRACT

The dynamic structure factor $S(q,\omega)$ at large momentum transfer $\hbar q$ differs from the impulse approximation as a result of interatomic interference effects and final-state interactions. In general, $S(q,\omega)$ can be expressed as the sum of a part (S_s) that is symmetric about the recoil energy $(\hbar q)^2/2m$, and a part (S_a) that is antisymmetric. To the extent that the interatomic potential can be treated as a smooth function, S_a is due entirely to final-state interactions and is of order q^{-1} while the distortion of S_s by final-state interactions is of order q^{-2}. In an earlier paper, we showed how the information about final-state interactions that is contained in S_a can be used in an approximate correction for residual final-state effects in S_s. In this paper we present a simpler derivation of this result using sum-rule methods.

1. INTRODUCTION

The determination of the momentum distribution function $n(\underset{\sim}{p})$ for a monatomic liquid from the dynamic structure factor $S(q,\omega)$ measured by neutron inelastic scattering at large momentum transfer $\hbar q$ is quite straightforward if $S(q,\omega)$ is adequately represented by the impulse approximation[1] (IA). In this case,

$$pn(\underset{\sim}{p}) = - \frac{1}{2\pi} \left[\frac{\hbar q}{m} \right]^2 \frac{\partial}{\partial \omega} S_{IA} (q,\omega) , \qquad (1)$$

where m is the atomic mass and $p = (m/\hbar q)(\omega - \omega_r)$, in which $\hbar \omega_r = (\hbar q)^2/2m$ is the recoil energy.

In general, $S(q,\omega)$ differs from $S_{IA}(q,\omega)$ as a result of the interatomic interference effects and final-state interactions that are neglected in the impulse approximation. In a system with a smooth interatomic potential these effects vanish[2] as $q \to \infty$, but for a system with a hard-core potential the final-state interactions continue[3] to give a finite contribution to $S(q,\omega)$ even in the limit $q \to \infty$. In recent years a number of authors[3-9] have stressed that liquid ^{4}He, in particular, is effectively a hard-core system for the q-values accessible to experiment, and have devised methods for including final-state interactions in the calculation of $S(q,\omega)$ at large q for this quantum liquid.

The present paper is concerned with the explicit correction for final-state interactions in neutron inelastic scattering experiments at large q. In particular, we describe a simple approximate method for determining $S_{IA}(q,\omega)$, and hence $n(\underset{\sim}{p})$, from measured $S(q,\omega)$ distributions. This is an addendum to our earlier paper[10] on this subject, and is based on the use of conventional sum-rule methods that implicitly assume a smooth interatomic potential and, hence, are not strictly valid for hard-core systems. Nevertheless, the results provide a valid approximation for the analysis of deep-inelastic neutron scattering experiments in systems where the final-state interactions produce only a small distortion of $S(q,\omega)$ as, for example, in gases or almost-classical liquids.

2. APPROXIMATE CORRECTION FOR FINAL-STATE INTERACTIONS

We begin by writing

$$S(q,\omega) = S_s(q,\omega) + S_a(q,\omega) , \qquad (2)$$

where

$$S_s(q,\omega) = \frac{1}{2} [S(q,\omega) + S(q,2\omega_r-\omega)] ,$$

$$S_a(q,\omega) = \frac{1}{2} [S(q,\omega) - S(q,2\omega_r-\omega)] , \qquad (3)$$

so that $S_s(q,\omega)$ is an even function of $\omega - \omega_r$ (symmetric) and $S_a(q,\omega)$ is an odd function of $\omega - \omega_r$ (antisymmetric). If hard-core effects are ignored,

$$S_s(q,\omega) = S_{IA}(q,\omega) + \mathcal{O}(q^{-2}) ,$$

$$S_a(q,\omega) = \mathcal{O}(q^{-1}) . \qquad (4)$$

In other words, $S_a(q,\omega)$ is due entirely to the effect of final-state interactions and is of order q^{-1}, while the correction to $S_s(q,\omega)$ for final-state interactions is of order q^{-2}. Thus, at large q, $S_s(q,\omega)$ is a better approximation to $S_{IA}(q,\omega)$ than is $S(q,\omega)$ itself. In Ref. 10 we have shown that an even better approximation is obtained by putting

$$S_{IA}(q,\omega) = S_s(q,\omega) + \alpha \frac{\partial}{\partial\omega} S_a(q,\omega) \; . \tag{5}$$

Here, the information about the final-state interactions that is contained in $S_a(q,\omega)$ is used to correct for the residual final-state effects in $S_s(q,\omega)$.

The result (5) was obtained in Ref. 10 from an approximate deconvolution of the final-state effects from $S(q,\omega)$. The method is valid provided the distortion of $S(q,\omega)$ by final-state interactons is very small, as it is in gases or almost-classical liquids. Except in the neighborhood of the condensate peak in the superfluid phase, it may be a useful approximation for liquid helium too. In this paper, we show that the validity of (5) can also be established very simply using sum-rule arguments, and the expression that we find for the coefficient α is the same as before.

We begin by defining the frequency moments

$$S_n(q) = \int_{-\infty}^{\infty} (\omega-\omega_r)^n \, S(q,\omega)d\omega \; , \tag{6}$$

so that

$$S_n(q) = \begin{cases} \int_{-\infty}^{\infty} (\omega-\omega_r)^n \, S_s(q,\omega)d\omega \; , & n = \text{even} \; , \\[2ex] \int_{-\infty}^{\infty} (\omega-\omega_r)^n \, S_a(q,\omega)d\omega \; , & n = \text{odd} \; . \end{cases} \tag{7}$$

It then follows from the expression (5) that

$$S_{IA,2n}(q) = S_{2n}(q) - 2n\alpha \, S_{2n-1}(q) \; . \tag{8}$$

In the incoherent approximation, in which one neglects interatomic interference effects, but not final-state interactions, the sum rules for $S_n(q)$ are polynomials in q^2:

$$\begin{aligned}
S_0(q) &= 1 \; , & S_4(q) &= a_{42}q^2 + a_{44}q^4 \; , \\
S_1(q) &= 0 \; , & S_5(q) &= a_{52}q^2 + a_{54}q^4 \; , \\
S_2(q) &= a_{22}q^2 \; , & S_6(q) &= a_{62}q^2 + a_{64}q^4 + a_{66}q^6 \; . \\
S_3(q) &= a_{32}q^2 \; ,
\end{aligned} \tag{9}$$

Since

$$S_{IA,2n}(q) = a_{2n,2n} q^{2n} , \tag{10}$$

it then follows from Eq. (8) that

$$a_{42} = 4\alpha a_{32}, \qquad a_{62} = 6\alpha a_{52} , \qquad a_{64} = 6\alpha a_{54} . \tag{11}$$

In other words, the sum rules are formally consistent with an expression for $S_{IA}(q,\omega)$ of the type (5) in which α is a constant, independent of q. Furthermore, (5) satisfies the $n = 0$, 1, and 2 sum rules identically for any value of α. If we choose

$$\alpha = \frac{a_{42}}{4a_{32}} , \tag{12}$$

then the expression (5) satisfies the $n = 3$ and 4 sum rules too. The remaining sum rules will, however, not be satisfied exactly.

In general,[2]

$$a_{32} = \frac{\hbar}{6m^2} \langle \Delta V \rangle , \qquad a_{42} = \frac{1}{3m^2} \langle \underset{\sim}{F}^2 \rangle , \tag{13}$$

where $\underset{\sim}{F} = - \nabla V$ is the force on the scattering atom and the brackets $\langle ... \rangle$ denote a thermodynamic average. Hence,

$$\alpha = \frac{\langle \underset{\sim}{F}^2 \rangle}{2\hbar \langle \Delta V \rangle} , \tag{14}$$

which is the same result as we found in Ref. 10.

3.   SELF-CONSISTENT PROCEDURE

The expression

$$\langle \underset{\sim}{F}^2 \rangle = \frac{2}{3} \langle KE \rangle \langle \Delta V \rangle , \tag{15}$$

in which $\langle KE \rangle$ is the average kinetic energy of an atom, is true classically[11] and also for a simple harmonic oscillator.[2] To the extent that it is a valid approximation in general, it follows from (14) that

$$\alpha = \frac{1}{3\hbar} \langle KE \rangle = \frac{\hbar}{6m} \int p^2 n(\underset{\sim}{p}) d\underset{\sim}{p} . \tag{16}$$

In this case (1), (3), (5), and (16) are a self-consistent set of equations for determining $n(\underset{\sim}{p})$ from the measured $S(q,\omega)$ distributions. We plan to apply this method to available neutron data for liquid neon and liquid helium in due course.

REFERENCES

1. P.C. Hohenberg and P.M. Platzman, Phys. Rev. $\underline{152}$, 198 (1966).
2. V.F. Sears, Phys. Rev. $\underline{185}$, 200 (1969).
3. J.J. Weinstein and J.W. Negele, Phys. Rev. Lett. $\underline{49}$, 1016 (1982).
4. T.R. Kirkpatrick, Phys. Rev. B $\underline{30}$, 1266 (1984).
5. P.M. Platzman and N. Tzoar, Phys. Rev. B $\underline{30}$, 6397 (1984).
6. G. Reiter and T. Becher, Phys. Rev. B $\underline{33}$, 4492 (1985).
7. R.N. Silver and G. Reiter, Phys. Rev. B $\underline{35}$, 3647 (1987).
8. R.N. Silver, Phys. Rev. B $\underline{37}$, 3794 (1988).
9. R.N. Silver, Phys. Rev. B $\underline{38}$, 2834 (1988).
10. V.F. Sears, Phys. Rev. B $\underline{30}$, 44 (1984).
11. P.G. de Gennes, Physica $\underline{25}$, 825 (1959).

DETERMINATION OF MOMENTUM DISTRIBUTIONS FROM DEEP INELASTIC NEUTRON SCATTERING EXPERIMENTS – A BAYESIAN STUDY

D.S. Sivia and R. N. Silver

Theoretical Division and Los Alamos Neutron Scattering Center
MS B262, Los Alamos National Laboratory
Los Alamos, New Mexico 87545

ABSTRACT

A Bayesian analysis shows that the determination of momentum distributions in quantum fluids and solids by deep inelastic neutron scattering is an extremely ill–posed problem. The argument is illustrated with the issue of the Bose condensate fraction in superfluid ^{4}He.

INTRODUCTION

There is a long history of experiments [1] aimed at measuring the momentum distributions in quantum solids and fluids by neutron scattering at high energy and momentum transfers, which is termed "deep inelastic neutron scattering" (DINS). Undeniably experiment can distinguish between various theoretical models for the neutron scattering law. However, the goal of the present paper is to show that the inverse problem of extracting $n(p)$ from the experiment is extremely ill–posed. For example, analyses of such experiments on superfluid ^{4}He [2] have claimed to confirm the existence of a Bose condensate fraction, n_o, and to determine its value. We show that available data are also consistent with $n_o=0$, and that available determinations of n_o should be regarded as model–dependent.

BAYES' THEOREM

All data analysis methods for determining $n(p)$ are, at least implicitly, based on Bayes' theorem. This is expressed in terms of the probability density function (p.d.f.), $P[n(p)|D(Y)]$, which is the conditional probability of the momentum distribution, $n(p)$, given the data, $D(Y)$. Bayes' theorem states that

$$P[n(p)|D(Y)] \propto P[D(Y)|n(p)] \times P[n(p)] \ . \qquad (1)$$

Here, $P[n(p)]$ represents our state of knowledge about $n(p)$ (or the lack of it) before we have any data – this is referred to as the *Prior* p.d.f. The data modify our prior state of knowledge through the term $P[D(Y)|n(p)]$, which is the probability of obtaining the measured data for a given momentum distribution – this is referred to as the *Likelihood* function. In the limit of independent Gaussian statistics, the Likelihood function reduces to the familiar form

$$P[D(Y)|n(p)] \Rightarrow \exp(-\frac{\chi^2}{2}) \ . \qquad (2)$$

The product of the Likelihood and the Prior p.d.f. is proportional to $P[n(p)|D(Y)]$ – the *Posterior* p.d.f., or our state of knowledge after we have measured the data. Our best estimate of $n(p)$ is given by the maximum of the Posterior p.d.f., and the reliability (error estimate) is given by its width.

Different data analysis procedures correspond to different choices for the Prior p.d.f., $P[n(p)]$. For example, the conventional procedure to determine n_o in ^{4}He [2] assumes a particular model for $n(p)$ with a few parameters based on the physics of the problem, e.g. the terms in $n(p)$ proportional to n_o involve a delta function plus associated $1/p^2$ and $1/p$ singularities. Another procedure is to arbitrarily assume a functional form for $n(p)$, such as a sum of two Gaussians. In either case, the parameters are estimated by maximizing the Posterior p.d.f., $P[n(p)|D(Y)]$. If the Prior p.d.f., $P[n(p)]$, is assumed to be a uniform function of those parameters, this is equivalent to the familiar procedure of minimizing χ^2 (i.e. maximum likelihood). Such data analysis procedures beg the question of whether the choice of Prior p.d.f. was correct. For example, how far out in p do the $1/p^2$ and $1/p$ singularities extend [3], or is there physical justification for $n(p)$ to be the sum of Gaussians? If the χ^2 for either method is acceptable, then the data shed no light on such questions. An alternative data analysis procedure is to deconvolve the data to obtain $n(p)$ which uses a different Prior p.d.f. For example, in the maximum entropy method [4] the Prior p.d.f. is taken to be the exponential of the entropy of $n(p)$ relative to a default model, which may for example be a numerical simulation of $n(p)$.

Regardless of the choice of Prior, it is the Likelihood function which incorporates the information contained in the data and it is the Likelihood function which can be evaluated exactly in terms of the experimental parameters. If the Likelihood function is a sharply peaked function of $n(p)$, then it will dominate the Posterior, i.e. no matter what our prior state of knowledge, the data force us towards the same choice of $n(p)$. If the Likelihood function is a broad function of $n(p)$, then the data have little effect on our state of knowledge and hence, the Posterior will depend crucially on the Prior. Fig. 1 shows a section through a schematic Likelihood function. The data constrain the distribution well in some directions (e.g. $n(p_2)$) but poorly in others (e.g. $n(p_1)$). The good directions are associated with large eigenvalues of the Log–Likelihood matrix ($\nabla\nabla\chi^2$), and the bad directions with small eigenvalues. In order to see how much information DINS data contain, we consider the sharpness, or otherwise, of the DINS Likelihood function.

ANALYSIS OF DINS EXPERIMENTS ON ^{4}He

In deep inelastic neutron scattering (DINS) experiments, the momentum distribution $n(p)$ is related to the experimental data, $D(Y)$, by

$$D(Y) = F_{IA}(Y) \otimes R_{expt}(Y) \otimes R_{FSE}(Y) + B(Y) + N(Y) \ . \qquad (3)$$

Here " $\otimes$ "denotes convolution,

$$F_{IA}(Y) = n_o\delta(Y) + \frac{1}{4\pi^2\varrho} \int\limits_{|Y|}^{\infty} p\, n(p)\, dp \qquad (4)$$

is the impulse approximation (IA) prediction for the neutron scattering law (the Compton profile),

$$Y = \frac{M\omega}{Q} - Q/2 \qquad (5)$$

is the scaling variable, $R_{expt}(Y)$ is the spectrometer broadening function, $R_{FSE}(Y)$ is the broadening due to corrections to the impulse approximation such as final state effects (FSE) which may also depend on Q, $B(Y)$ is background, and $N(Y)$ is noise. The problem is to infer $n(p)$ given $D(Y)$.

Using Eqs. (3–5) one may calculate the Likelihood function. Fig. 2 shows the spectrum of eigenvalues for the DINS Likelihood function for ^{4}He under the conditions of the recent experiments of Sosnick, et al. [5], which epitomize the current state–of–the–art in DINS measurements. The $R_{FSE}(Y)$ was taken from the recent theory by Silver [6]. The continuous $n(p)$ was digitized into 92 pixels for $0 \le p \le 3.4$ A^{-1} ; hence 92 eigenvalues. We see that most of the eigenvalues are very small, indicating that the Likelihood function is flat in many directions. Thus, the problem is extremely ill–posed in that many distributions will fit the data. This is particularly true at small p, which is primarily due to the form of the Compton profile, Eq. (4), rather than due to instrumental or final state broadening. Indeed, we already expect this to be true from the observation [5] that the Greens' Function Monte Carlo (GFMC) $n(p)$ [7] which does not have the correct singular behavior at small p, and the Hypernetted Chain $n(p)$ [8] which does have the correct singular behavior at small p, produce almost identical predictions for $J(Y) = F_{IA}(Y) \otimes R_{expt}(Y) \otimes R_{FSE}(Y)$.

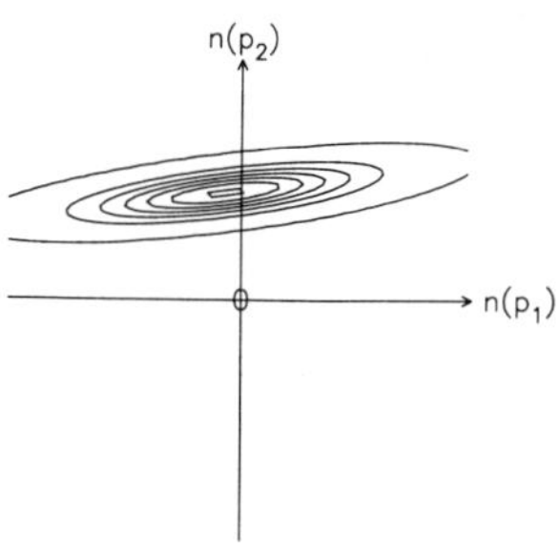

FIGURE 1 – A section through a schematic Likelihood function. The data constrain the distribution $n(p)$ well in some directions but poorly in others. The good directions are associated with large eigenvalues of the Log–Likelihood function and the bad directions with small ones.

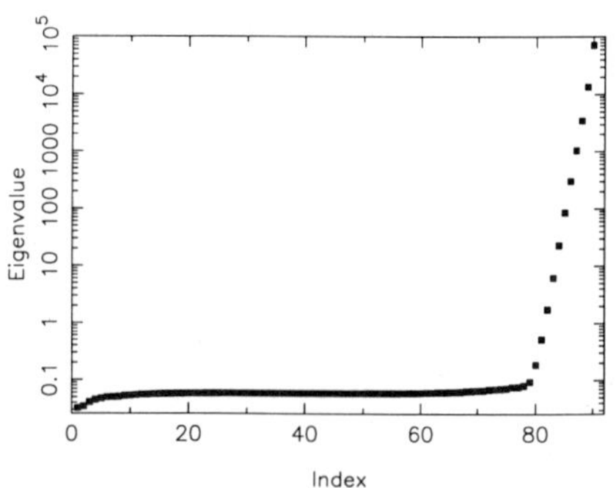

FIGURE 2 – The spectrum of eigenvalues for the Deep Inelastic Neutron Scattering (DINS) Likelihood function. Most of the eigenvalues are very small, indicating that the Likelihood function is flat in many directions.

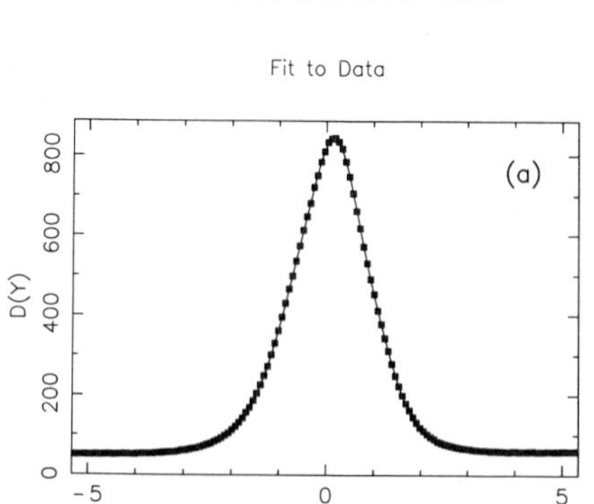

FIGURE 3(a) – Simulated DINS data using the GFMC $n(p)$ (Whitlock & Panoff, 1988) shown as the dotted line in Figs. 3(b) & (c)., having a Bose condensate of 9.2%, with 100 times better statistics than that achieved in the state–of–the–art experiments of Sosnick, et al.

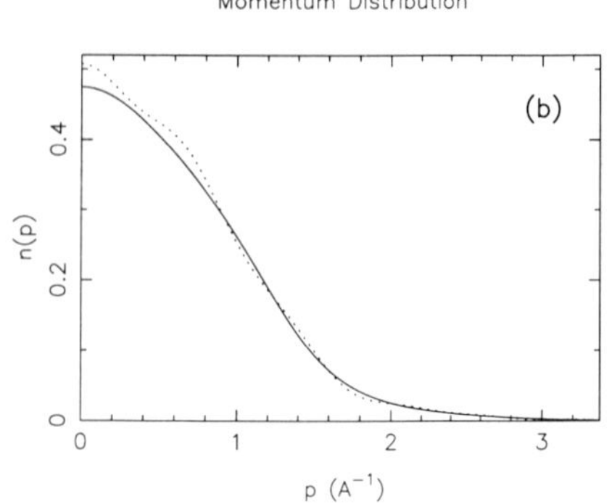

FIGURE 3(b) – Solid line is an $n(p)$ which fits the simulated data in Fig. 3(a) with a 9.8% Bose condensate. Dashed line is the GFMC prediction, which was used to create the simulated data.

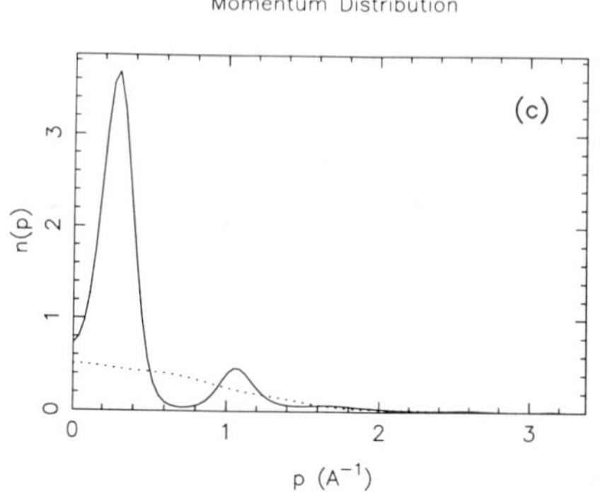

FIGURE 3(c) – Solid line is an $n(p)$ which fits the simulated data in Fig. 3(a) with a 0% Bose condensate. Dashed line is GFMC.

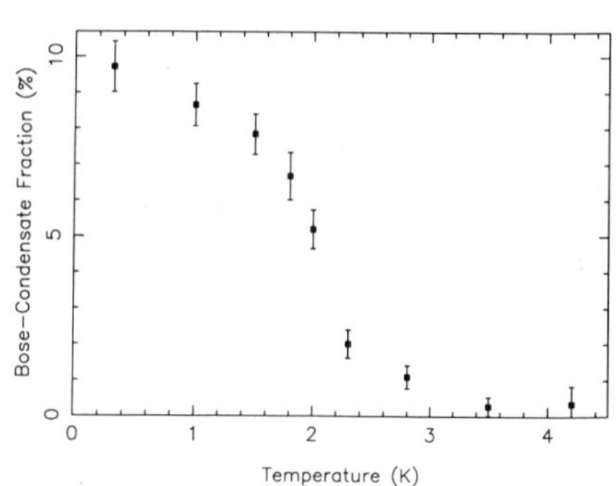

FIGURE 4 – The classical decrease of the Bose condensate fraction as ^{4}He is warmed through the λ–transition temperature. The DINS data were obtained by Sosnick, et al. (1988) and analyzed using an entropic prior with GFMC $n(p)$ at T=0 K as default model.

The ambiguity in inferring $n(p)$ from DINS data is illustrated graphically in Fig. 3 by a simulation. The mock data shown in Fig. 3(a) were created for the experimental conditions of Sosnick, et al. [5] with 100 times the number of neutrons achieved experimentally (ten times the statistical accuracy) using the GFMC $n(p)$ [7,9] as input. Figs. 3(b) and 3(c) show two very different $n(p)$ (continuous lines), both of which fit the data in Fig. 3(a). These were obtained by maximum entropy deconvolution [4] using as default models a best–fit Gaussian plus delta function for Fig. 3(b) and a best–fit Gaussian for Fig. 3(c) (which allows no Bose condensate). The GFMC had an n_o of 9.2%, to be compared with 9.8% for Fig. 3(b) and 0% for Fig. 3(c). Thus, data with statistical accuracy ten times better than currently achieved can not unambiguously establish the existence of a Bose condensate.

However, not all distributions allowed by the data make physical sense. Thus, we have extra prior knowledge, in the form of Physics, which constrains the allowed $n(p)$ much more tightly than the data alone. Fig. 4 shows such a determination of n_o, where the temperature dependent DINS data obtained by Sosnick, et al. [5] was deconvolved using an entropic Prior with the GFMC $n(p)$ (T=0 K) as the default model. The fact that n_o is found to be slightly greater than zero even above the λ–transition (although consistent with zero within errors) can be understood as a consequence of the bias of the default model toward the existence of a condensate.

CONCLUSIONS

The experimental DINS data [5] on ^{4}He show an obvious sharpening at small Y as the temperature is decreased below the λ–transition. Nevertheless, our counterexample shows that, despite prior claims to the contrary [1–2], the present generation of DINS experiments have not, and cannot be expected to, unambiguously establish the existence of a Bose condensate in ^{4}He. Prior knowledge in the form of a physical model is required to adequately constrain the inversion to determine n_o, and the available models [1–2] involve uncertainties [3] which would show up as systematic rather than statistical errors. The strongest evidence we have for the existence of a Bose condensate in ^{4}He is that *ab initio* calculations of $J(Y)$, including final state effects [6] and simulations of $n(p)$ [7–8,10], are in excellent agreement with DINS experiments [5] with no adjustable parameters, and these calculations predict the existence of a Bose condensate in ^{4}He at temperatures below the λ–transition.

More generally, the inversion of Compton profile data to extract $n(p)$ in any system is an extremely ill–posed problem, particularly at small p as can be seen from Eq. (4). Exceptional statistics, minimal final state broadening, and accurate instrumental resolution functions are required. For DINS on helium systems, it appears that significant final state broadening is unavoidable [6].

ACKNOWLEDGEMENT

Research supported by the Office of Basic Energy Sciences of the U.S. Dept. of Energy. We thank T. R. Sosnick, W. M. Snow, and P. E. Sokol for permission to use their data prior to publication.

REFERENCES

[1] See, e.g. H. R. Glyde, E. C. Svensson in *Methods of Experimental Physics*, V. 23, Part B., D. L. Price and K. Skold, eds. (Academic Press, 1987), p. 303.

[2] V. F. Sears, E. C. Svensson, P. Martel, A. D. B. Woods, Phys. Rev. Lett. **49**, 279 (1982).

[3] A. Griffin, Phys. Rev. B32, 3289 (1985).

[4] S. F. Gull, J. Skilling, IEE Proc., **131F**, 646 (1984).

[5] T. R. Sosnick, W. M. Snow, P. E. Sokol, R. N. Silver, LA–UR–88–505; P. E. Sokol, this volume.

[6] R. N. Silver, Phys. Rev. B37 (Rapid Communications), 3794 (1988); R. N. Silver, Phys. Rev. B38, 2283 (1988); R. N. Silver, this volume.

[7] P. Whitlock, R. M. Panoff, Can. J. Phys. **65**, 1409 (1987).

[8] E. Manousakis, V. R. Pandharipande, Phys. Rev. B31, 7029 (1985); E. Manousakis, V. R. Pandharipande, Q. N. Usmani, Phys. Rev. B31, 7022 (1985).

[9] The $n(p)$ shown in Fig. 3 are normalized to $\int d^3p\, n(p)/(2\pi)^3 = \varrho$, where ϱ is the density.

[10] D. M. Ceperley, E. L. Pollock, Can. J. of Physics **65**, 1416 (1987).

TIME EVOLUTION OF THE MOMENTUM DISTRIBUTION OF

BOSE GAS NEAR CONDENSATION

D.W. Snoke and J.P. Wolfe

Physics Department and Materials Research Laboratory
University of Illinois at Urbana-Champaign
Urbana, IL

This study has been motivated by the experimental searches for Bose-
Einstein condensation in systems with finite-lifetime particles, namely
excitons in semiconductors such as Cu_2O[1,2] and spin-aligned hydrogen[3,4].
In these systmes, particles with the properties of bosons can be created
which should also undergo Bose-Einstein condensation at sufficiently high
densities. Since the particles have a finite lifetime, however, one must
ask whether they will live long enough for a true Bose condensate to be
established in steady state, even if the aggregate of particles reaches
the density and average energy necessary for condensation. This question
arises because the dynamics of scattering a macroscopic number of particles
into a single quantum state are not well understood.

The basic question can be posed as follows: If a gas of bosons is
created with a nonequilibrium, noncondensed momentum distribution and is
allowed to scatter, what will n(p) look like while it moves to equilibrium,
and, in the case of a gas with greater than critical density, at what
point will a condensate appear?[5,6]

We have studied this question for two cases. In the first case, we
consider a weakly-interacting boson gas, thermally isolated from its
surroundings. We consider only two-body interaction, illustrated below:

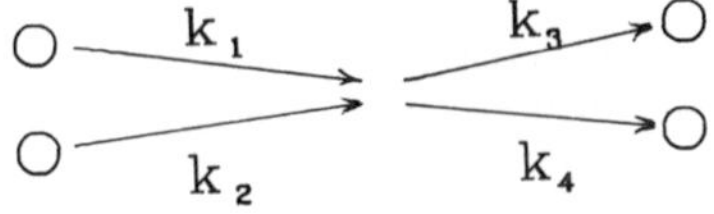

In the random phase approximation, we can write the scattering probability
of this process as

$$S(\vec{k}_1,\vec{k}_2,\vec{k}_3,\vec{k}_4) = M^2\ \delta(\vec{k}_1+\vec{k}_2-\vec{k}_3-\vec{k}_4)\ \delta(E_1+E_2-E_3-E_4) \tag{1}$$

$$\cdot f(\vec{k}_1)\ f(\vec{k}_2)\ (1+f(\vec{k}_3)\ (1+f(\vec{k}_4))$$

As shown in a derivation to be published elsewhere, this form leads to the
total scattering rate into states with momentum k of

$$\Gamma_i(E(k))dE = \int dEdE_1 dE_2 \; M^2 \; f(E_1) \; f(E_2) \; (1+f(E_3)) \; (1+f(E)) \; \Big|_{E_3=E_1+E_2-E} \quad (2)$$

$$\cdot \; \frac{(2\pi)^3}{2^4} \; \frac{1}{\frac{\partial E}{\partial k^2}\big|_k} \; \frac{1}{\frac{\partial E}{\partial k^2}\big|_{k_1}} \; \frac{1}{\frac{\partial E}{\partial k^2}\big|_{k_2}} \; \frac{1}{\frac{\partial E}{\partial k^2}\big|_{k_3}} \; [\min(k_1+k_2,k_3+k)-\max(|k_1-k_2|,|k_3-k|)].$$

In this case, we use $E(p)=p^2/2m$ without renormalization. The total scattering rate <u>out</u> of states with momentum k has a similar form. These formulas involve integration over only two energies, so that they are tractable for a computer calculation. Given any energy distribution n(E), then, we can calculate the net scattering rate dn(E)/dt for all E and add dn(E) for some small dt to get a new n(E). If we repeat this process, we can evolve n(E) in time to get "movies" of the distribution as it moves toward equilibrium. We have done all calculations on the FPS 164 and 264 array processors at the Materials Research Laboratory at the University of Illinois at Urbana-Champaign.

A low density boson gas behaves just like a classical gas, and Figure 1 shows how the distribution moves to a Maxwell-Boltzmann distribution within a few characteristic scattering times. In this case, we have used an initial distribution with constant occupation number per momentum state and a sharp cutoff at E=2.5kT. When we have used other initial distributions, however, we see similar results, that n(E) moves to a Maxwell-Boltzmann distribution within five scattering times.

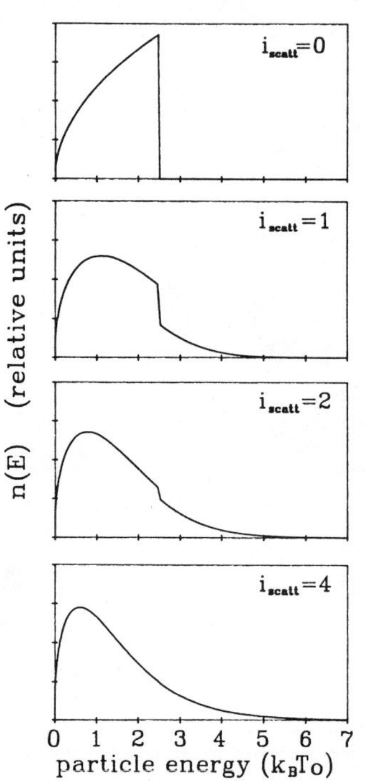

Fig. 1. Time evolution of n(E) for the case $n \ll n_c$.

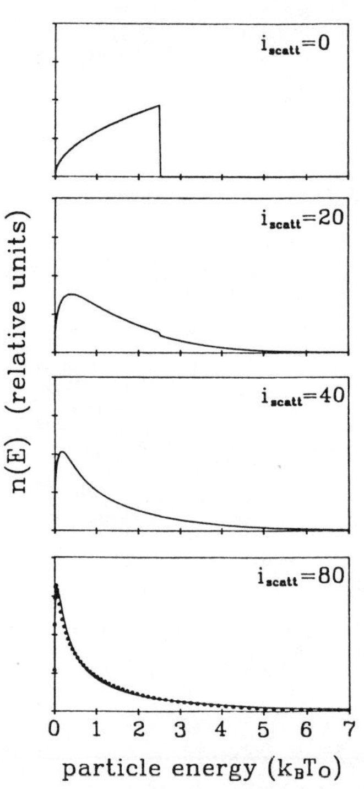

Fig. 2. Time evolution of n(E) for the case $n < n_c$.

As density is increased toward the critical density, we find that the number of scattering events required to reach equilibrium increases. When the gas has greater than the critical density, the number of scattering events per particle becomes very large. Figure 2 shows the evolution of a boson gas with greater than critical density. Although the distribution is not in equilbirium, we find that after initial transients die away, n(E) is well described by an <u>equilibrium</u> Bose-Einstein distribution of the form

$$f(\vec{k}) = f(E(\vec{k}); \mu, T) = \frac{1}{e^{(E-\mu)/k_B T} - 1} \tag{3}$$

with a value of μ which varies in time to approach zero. The dotted curve at 80 scattering times in Figure 2 is a Bose-Einstein distribution with $\mu = -0.024kT$.

A full discussion of the time variation of the shape of n(E) will be published elsewhere. We note here, though, that the condensate appears in a continuous way, as the ground state occupation number, given by $-kT/\mu$, increases to become macroscopic.

Although the number of characteristic scattering times becomes large, the time to reach condensation does not become infinite, because the overall interparticle scattering rate is enhanced by stimulated emission. Figure 3 shows the halfwidth Δ of the energy distribution as a function of time for the same case as Figure 2, where the time scale has been calibrated in terms of the classical hard-sphere scattering time τ_0. The halfwidth fits well to an exponential decay in time. From this we can conclude that the gas can reach Bose condensation within a few classical scattering times, although the actual number of scattering events per particle becomes extremely large.

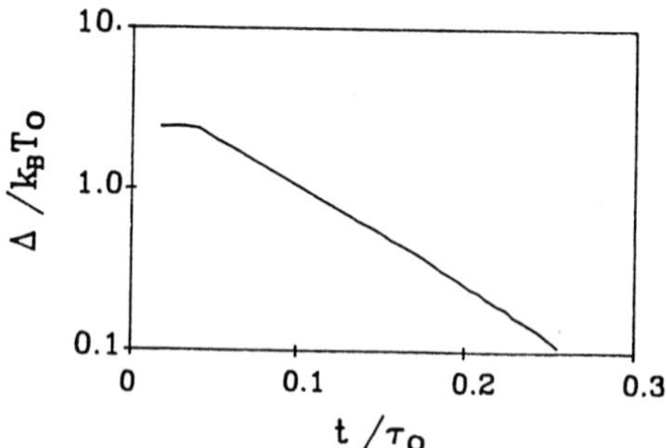

Fig. 3. The halfwidth of n(E) vs. t for the case shown in Figure 2.

The second case we have studied is an ideal Bose gas interacting with a phonon bath (planck distribution) of infinite heat capacity. We consider two kinds of interactions, illustrated below:

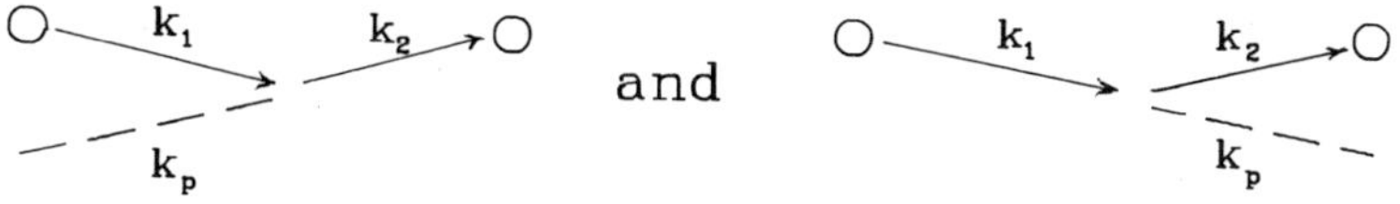

The scattering probability for these processes can be written as

$$S(\bar{k}_1, \bar{k}_2, \bar{k}_p) = M_p^2 \ [\ \delta(\bar{k}_1 + \bar{k}_p - \bar{k}_2)\ \delta(E_1 + E_p - E_3)\ f(\bar{k}_1)\ F(\bar{k}_p)\ (1 + f(\bar{k}_2))$$

$$+ \ \delta(\bar{k}_1 - \bar{k}_p - \bar{k}_2)\ \delta(E_1 - E_p - E_3)\ f(\bar{k}_1)\ (1 + F(\bar{k}_p))\ (1 + f(\bar{k}_2))\ ], \qquad (4)$$

This leads to a formula for the total scattering rates into and out of states of momentum k similar to (2), involving integration over two energies. We find dn(E)/dt for all E and evolve the distribution in this case as before. For comparison with the previous case, Figure 4 shows the halfwidth Δ of the energy distribution as a function of time, for an initial distribution the same shape as that of Figure 1 and 2, with density above the critical density. In this case, the approach to equilibrium seems to be slowing down. We conclude that for a system with both interparticle interactions and interactions with phonons (such as excitons in a semiconductor), the interparticle scattering time will always be the characteristic time for condensation.

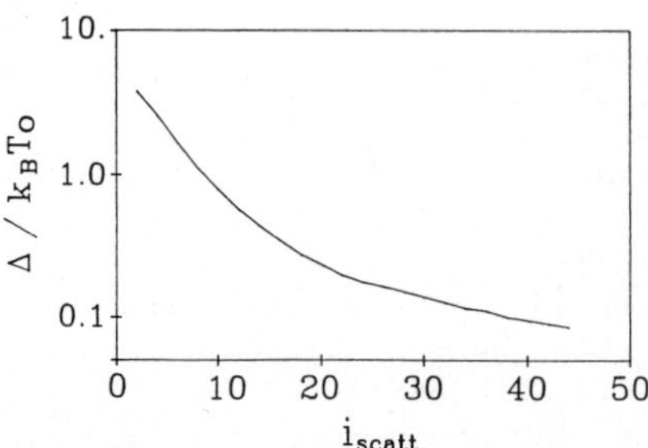

Fig. 4. The halfwidth of n(E) vs. t for the case of an ideal boson gas

For the experimental system of interest to us, excitons in Cu_2O, the estimated interparticle scattering time at T=2K and exciton density $n=10^{18}/cm^3$ is about 0.5 ns. The acoustic phonon scattering time at that temperature is about 1 ns. Since the exciton lifetimes are much greater than one nanosecond at these densities and temperatures, these calculations indicate that condensation should be possible.

ACKNOWLEDGEMENT

This work has been supported by National Science Foundation Grant NSF DMR 87-22761 with facility support by Materials Research Laboratory Grant 86-12860. D.S. was supported by a University of Illinois Fellowship.

REFERENCES

1. D. Snoke, J.P. Wolfe, and A. Mysyrowicz, PRL 59, 827 (1987).
2. For a general review of Bose condensation of excitons, see E. Hanamura and H. Haug, Phys. Rep. 33, 209 (1977).
3. For a general review of spin-aligned hydrogen, see I. Silvera and J.T.M. Walraven, Prog. Low Temp. Phys. 10, 139 (1986).
4. There also exists a possibility of Bose condensation of positronium in crystal vacancies, as mentioned by S. Berko at this workshop.
5. Previous work on the case of an exciton gas interacting with a phonon bath has been done by M. Inoue and E. Hanamura, J. Phys. Soc. Japan 41, 771 (1976).
6. Previous work on the weakly interacting boson gas has been done by E. Levich and V. Yakhot, J. Phys. A 11, 2237 (1978).

FINAL STATE EFFECTS IN LIQUID ^{4}He: AN EXPERIMENTAL TEST

[†]P.E. Sokol, [‡]R.N. Silver, [+]T.R. Sosnick, and [+]W.M. Snow

[†]Department of Physics, The Pennsylvania State University
University Park PA

[‡]Theoretical Division, Los Alamos National Laboratory
Los Alamos, NM

[+]Intense Pulsed Neutron Source, Argonne National Laboratory
Argonne, IL

INTRODUCTION

In principle, high energy inelastic neutron scattering measurements provide direct information on the atomic momentum distribution $n(p)$ when the Impulse Approximation (IA) is valid. In isotropic systems the scattering is then directly proportional to the longitudinal momentum distribution [2]

$$J_{IA}(Y) = \frac{1}{4\rho\pi^2} \int_{|Y|}^{+\infty} pn(p)dp = \int_{-\infty}^{+\infty} \int_{-\infty}^{+\infty} n(p_x, p_y, Y)dp_x dp_y \qquad (1.1)$$

which is a function of a single scaling variable $Y \equiv (M/Q)(\omega - \omega_r)$, where M is the mass of the scatterer and $\omega_r = Q^2/2M$ is the recoil energy.

In practice, however, the experimentally attainable Q's may not be large enough to reach the IA limit. Deviations from the IA due to final state scattering by neighboring atoms, known as final state effects, will distort the observed scattering. Thus, an accurate understanding of deviations from the IA is essential to accurate determinations of $n(p)$.

Liquid helium provides an excellent testing ground for studying FSE and testing theoretical predictions. Theoretical calculations of the momentum distribution are available in both the normal liquid [3][4] and superfluid [5] phases. These calculations, which are believed to be quite accurate, give good agreement with several measured properties of the liquid. In addition, $n(p)$ in the superfluid exhibits a very sharp feature, the Bose condensate peak, which should be very sensitive to FSE. Comparison of the predicted scattering obtained from the theoretical $n(p)$ using the IA to the experimentally observed scattering can be used to study deviations due to FSE.

RESULTS

The scattering from liquid helium was measured at temperatures of 0.35 K and 3.5 K at a density of 0.1469 gm/cm^3 with the PHOENIX spectrometer at the Intense

Pulsed Neutron Source at Argonne National Laboratory using a momentum transfer of 23 Å^{-1} at the recoil peak. The observed scattering is shown in Fig. 1. The empty cell scattering and a small broad component due to secondary scattering from the walls of the cryostat have been subtracted. The scattering, which has been normalized using measurements of low density helium gas, satisfies the first moment sum rule indicating that all the scattering is observed.

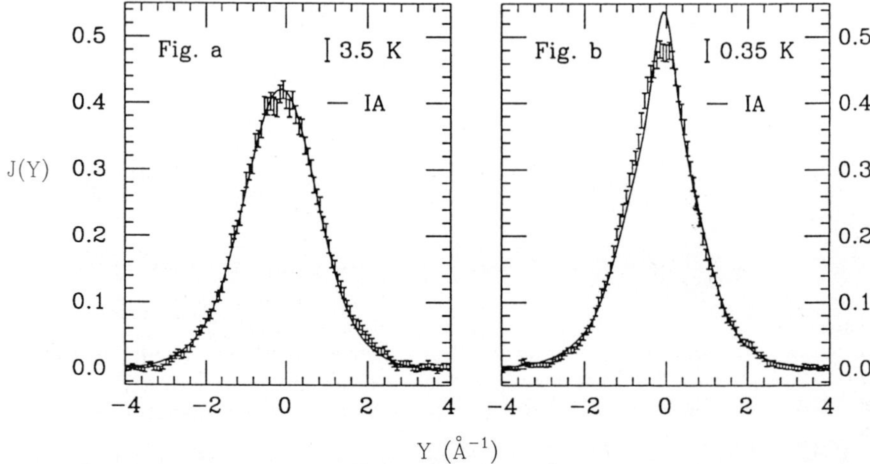

Fig. 1 Observed scattering from liquid ^{4}He. a) Normal liquid at T= 3.5 K. The line is the instrumentally broadened IA prediction using the PIMC calculation of $n(p)$. b) Superfluid at T=0.35 K The line is the instrumentally broadened IA prediction using the GFMC calculation of $n(p)$.

The observed scattering exhibits the characteristic features of scattering in the IA. First of all, it is approximately centered at $Y = 0$ and symmetric about $Y = 0$ when instrumental resolution is taken into account. In addition, the width of $J(Y)$ is independent of Q for Q's larger than 15 Å^{-1}.[5] However, while the observation of these characteristic features is consistent with the approach to the IA limit, it does not imply that the limit has been reached.[6,7]

To evaluate the applicability of the IA, the observed scattering can be directly compared to the theoretical calculations of $n(p)$. Of course, this approach implicitly assumes that the theoretical momentum distributions accurately describe the true momentum distributions in the liquid. Given the success of these calculations in reproducing many measured properties of the superfluid and normal liquid, we are confident that the momentum distribution is also quite accurate and hence that this method for studying FSE is justified.

The solid line in Fig. 1a is the predicted scattering using the IA and a PIMC calculation of $n(p)$ at T=3.33 K by Ceperley and Pollock. The calculation was carried out at a density of 0.138 gm/cm^3, which is slightly lower than the experimental density. The theoretical prediction has been convoluted with the instrumental resolution obtained from a Monte Carlo simulation of the instrument. The overall normalization of both the predicted and observed scattering are independently fixed so that no adjustable parameters have been used.[8] The agreement between the IA prediction and the observed scattering is excellent. Deviations from the IA due to FSE have little effect on the relatively broad and featureless distribution in the normal liquid.

The solid line in Fig. 1b is the predicted scattering using the IA and a GFMC calculation of $n(p)$ at T=0 by Whitlock and Panoff. Large differences between the IA prediction and the observed scattering are obvious. The deviations are largest near $Y = 0$, where the condensate peak occurs. The IA prediction has much more intensity in the peak than observed experimentally, as one would expect from final state smearing of the condensate peak.

Several theoretical calculations[9,10,11] (referred to as broadening theories in this paper) have expressed FSE as a convolution with the IA expression in Y. The scattering at finite Q then takes the form

$$J_{FS}(Y) = \int J_{IA}(Y')R(Y' - Y, Q)dY' \qquad (2.1)$$

where J_{FS} is the observed scattering including FSE and $R(Y, Q)$ is the final state broadening. The FSE broadening may then be determined by deconvoluting the instrumental broadening and $J_{IA}(Y)$ from the observed scattering.

The superfluid phase, where the deviations from the IA prediction are the largest, provides an excellent opportunity to test this procedure. Fig. 2 shows the results of deconvoluting $J_{IA}(Y)$, obtained from the GFMC calculation of $n(p)$, and the instrumental resolution from the observed scattering in the superfluid. The experimentally determined FSE broadening exhibits a sharp central peak and oscillatory tails that extend to high $|Y|$. While the general features of this curve are accurate, the detailed shape may be affected by the statistical noise in the numerical deconvolution procedure.

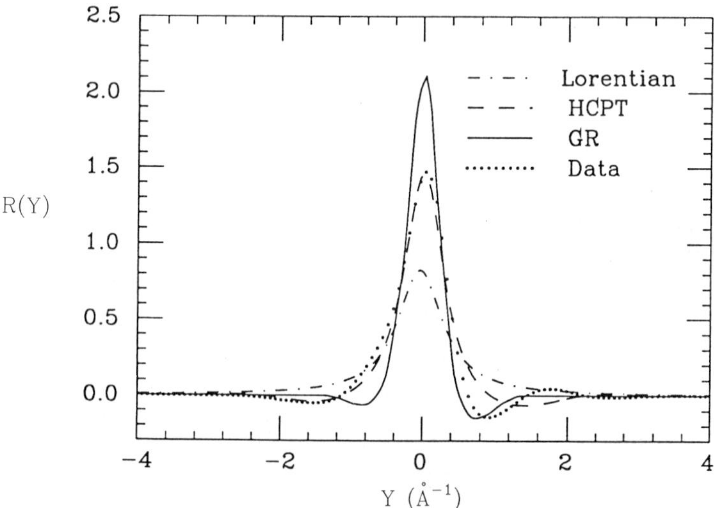

Fig. 2 Comparison of the experimentally determined final state broadening (Data) with theoretical predictions.

There are two noteworthy features in the experimentally determined $R(Y)$. First, the central peak, which is the most prominent feature, is relatively narrow. The full width at half maximum (FWHM) of this peak is 0.67 Å^{-1}, narrower than the width expected based on the measured helium-helium scattering cross section. Secondly, $R(Y)$ exhibits oscillatory tails that are both negative and positive. These tails are a natural consequence of the sum rules for incoherent scattering. In particular, the second moment sum rule requires that FSE not change the second moment of the scattering. Therefore, negative tails are required to cancel the contribution of the central peak to the second moment. The negative tails have the important consequence that FSE do not simply broaden the scattering: they move intensity around from one region of Y to another.

In principle, this procedure could be used to determine FSE broadening in the normal liquid. However, due to the statistical accuracy of the data ($\sim 3\%$) and the small effect of FSE, as seen in Fig. 1a, it is not possible to extract a final state broadening function using the same procedure as above. The experimental results require much better statistics before a broadening function can be experimentally determined in the normal liquid.

Another set of theories[12,13,14] (referred to as symmetrization theories here) expresses deviations from the IA as additive corrections to the IA result. In this approach, it is useful to split the corrections to the IA into terms that are symmetric and antisymmetric about the peak center. Then

$$J_{obs}(Y) = J_{IA}(Y) + \Delta J_{Sym}(Y, Q) + \Delta J_{Asym}(Y, Q) \qquad (2.2)$$

where ΔJ_{Sym} and ΔJ_{Asym} are the symmetric and antisymmetric corrections due to FSE.

To obtain the corrections to the IA, the theoretical prediction using the IA must be subtracted from the measured scattering after the instrumental broadening has been removed. Rather than deconvolute the instrumental resolution from the data, which is a numerically unstable procedure with noisy data, we will use a model $J_{FS}(Y)$ which is the sum of two Gaussians. The model scattering is broadened by instrumental resolution and compared with the observed scattering, and the model parameters are adjusted to obtain the best agreement.

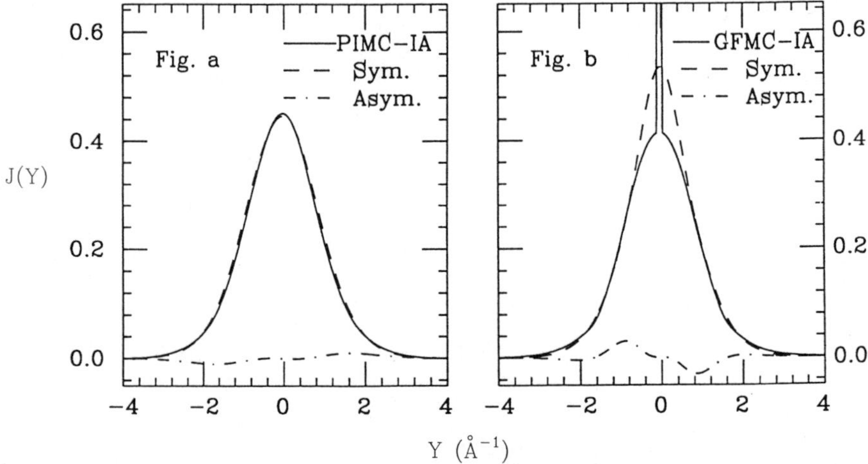

Fig. 3 Comparison of the IA predictions in the normal and superfluid phase with the model $J_{FS}(Y)$ obtained from a fit to the data. The components of $J_{FS}(Y)$ symmetric and antisymmetric about the recoil peak at $Y = 0$ are plotted separately.

The corrections to the IA in the normal liquid, obtained using $n(p)$ from the PIMC calculations of Ceperley, are shown in Fig. 3a. Both the symmetric and antisymmetric corrections are small. The maximum amplitude of the corrections is on the order of 5 % of the total peak amplitude, comparable to the statistical accuracy of the data. The small size of the corrections is not surprising, given the good agreement of the IA prediction and the observed scattering shown in Fig. 1a.

The corrections to the IA in the superfluid, obtained using the $n(p)$ from the GFMC calculations of Whitlock and Panoff, are shown in Fig. 3b. Both the symmetric and antisymmetric correction terms are now much larger than in the normal liquid. The maximum amplitude of the antisymmetric correction is now ≈ 20 % of the total peak amplitude, compared to ≈ 5 % in the normal liquid. The symmetric correction has a peak amplitude of ≈ 25 % of the total peak amplitude. In addition, it contains a negative delta function singularity with 9.2% of the total intensity, which is needed to cancel the condensate delta function in $J_{IA}(Y)$.

COMPARISONS WITH FINAL STATE EFFECT THEORIES

Theories for FSE can be tested by direct comparison to the observed scattering. The same procedure as used previously will be applied. The theoretical calculations for $n(p)$ are used to obtain the scattering in the IA. The results are then corrected for FSE, using the appropriate theory, and broadened by instrumental resolution. These results may then be compared directly to the observed scattering to evaluate the theories for FSE.

The earliest theory, by Hohenberg and Platzman, predicted a Lorentzian final state broadening. Several other theories have predicted similar behavior. The width of the FS broadening is

$$\Gamma(Q) = \rho\sigma(Q) \tag{3.1}$$

where ρ is the density and $\sigma(Q)$ is the cross section for atom-atom scattering. The Lorentzian broadening for the experimental conditions of this work is shown in Fig. 2. Fig. 4 shows a comparison of the theoretical and experimental results in both the normal and superfluid phases. In both cases the broadening is far larger than observed experimentally. The theoretical results are not in agreement with the experimental observations using a Lorentzian broadening function with the width in (3.1).

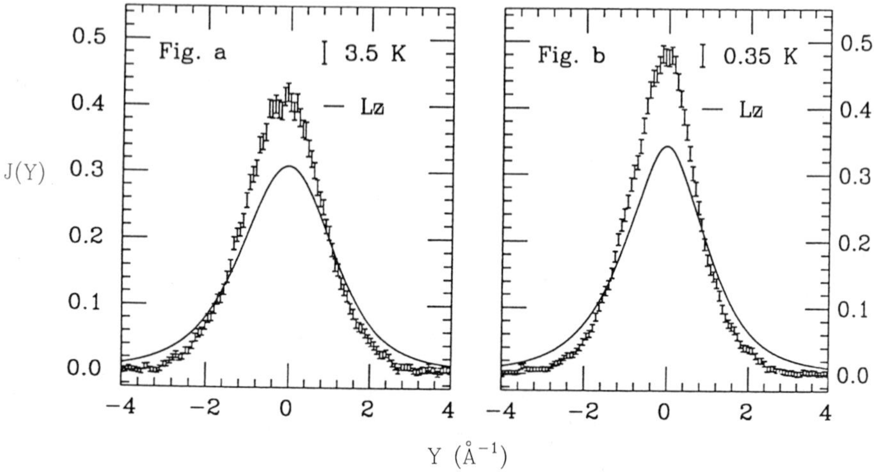

Fig. 4 Comparison of the observed scattering with the theoretical predictions, converted to J(Y) and broadened by instrumental resolution, using a Lorentzian broadening function.

Gersch and Rodriguez have obtained a final state broadenig that is not
Lorentzian. Instead it contains a narrow central peak and negative tails, as required
by the second moment sum rule. The calculated broadening is shown in Fig 2. Fig. 5
shows a comparison of the predicted and observed scattering. The agreement in the
normal liquid is excellent. However, in the superfluid phase, the predicted intensity
at the peak is too large.

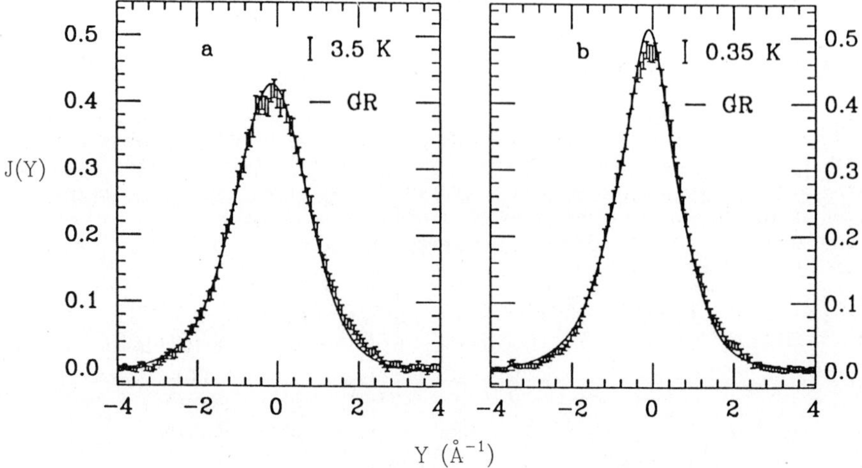

Fig. 5 Comparison of the observed scattering with the theoretical predictions, converted to J(Y) and broadened by instrumental resolution, using the broadening calculated by Gersch and Rodriguez.

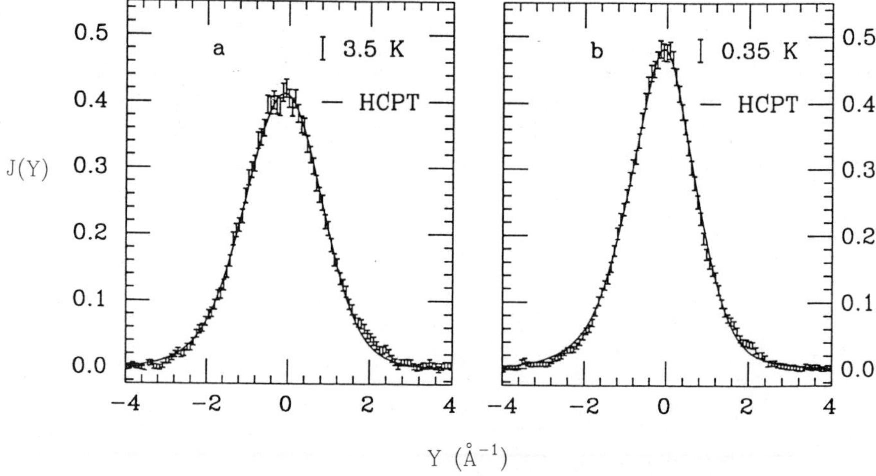

Fig. 6 Comparison of the observed scattering with the theoretical predictions, converted to J(Y) and broadened by instrumental resolution, using the broadening calculated by Silver.

Most recently, Silver has calculated a broadening similar to that of Gersch and Rodriguez using a method called Hard Core Perturbation Theory (HCPT). The broadening for the experimental conditions used in this measurement is shown in Fig 2. A comparison of the predicted and observed scattering using Silver's theory is shown in Fig. 6. The agreement is excellent in both the superfluid and normal liquid phases.

Symmetrization theories for FSE express differences from the IA as additive corrections to the IA result. These corrections are then split into terms that are either symmetric or antisymmetric about the peak center. The theories estimate the antisymmetric term to be much larger than the symmetric term. If the antisymmetric corrections are dominant, then most of FSE can be removed by simply symmetrizing the data. The removal of the antisymmetric term leaves both the first and second moments of the data unchanged, so this procedure does not violate the second moment sum rule for incoherent scattering.

In the normal liquid, as shown in Fig. 3a, both the symmetric and antisymmetric corrections are small. The symmetrization treatment of FSE corrections does not modify the second moment of the data. Thus, it will have little effect on the broad, nearly gaussian momentum distribution in the normal fluid, and its results will be consistent with the data.

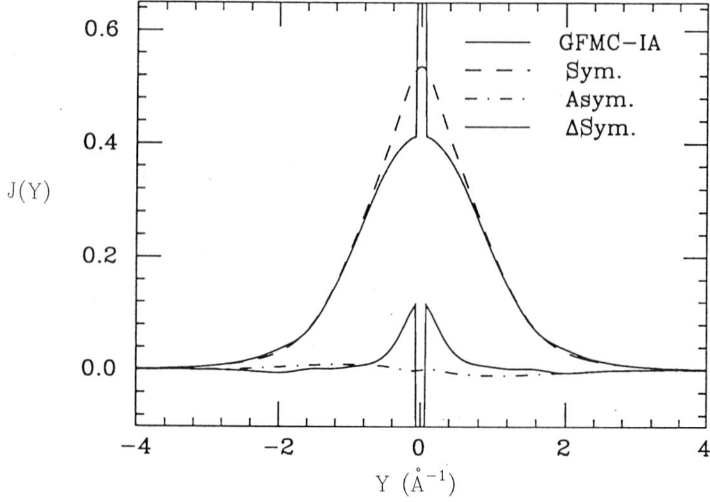

Fig. 7 Symmetric and antisymmetric components of Silver's HCPT prediction for the superfluid scattering versus the IA prediction using the T=0 K GFMC momentum distribution.

In the superfluid, as shown in Fig. 3b, the symmetric and antisymmetric corrections are much larger than in the normal fluid. The symmetric correction consists of a negative delta function singularity with 9.2% of the total intensity superimposed on a positive piece with a peak amplitude of approximately 25% of the total peak amplitude. The negative delta function is required to cancel the condensate delta function that appears in $J_{IA}(Y)$.

The symmetrization theories claim that the IA prediction is reproduced when the FSE broadened momentum distribution is symmetrized. If Silver's HCPT and the symmetrization theories were both correct, then symmetrizing the HCPT prediction should reproduce the IA. In Fig. 7 the IA prediction using the T=0 K GFMC momentum distribution is compared with HCPT and its symmetric and antisymmetric components. Symmetrizing the HCPT prediction does not reproduce the IA. The two treatments therefore yield inconsistent results in the superfluid phase. Since, as shown above, HCPT provides an accurate description of FSE broadening in the superfluid phase, it follows that the symmetrization approach fails.

This figure also shows why the symmetrization procedure fails in the superfluid. The observed scattering contains no feature with a width comparable to the instrumental resolution. Thus, the delta function singularity in $J_{IA}(Y)$ due to the condensate must be removed and replaced with a broadened peak if the IA corrections are expressed as additive terms. The dominant correction to $J_{IA}(Y)$ is then large and symmetric, and it cannot be removed by symmetrizing the data.

CONCLUSIONS

We have carried out a test of various theories for final state broadening by comparing the observed scattering in the normal and superfluid phases of liquid helium with theoretical predictions for the scattering, using current calculations for the momentum distribution as input. Implicit in our test is the assumption that the theoretical calculations of the momentum distribution are accurate. Our conclusions regarding the FSE theories are valid only to the extent that the theoretical calculations are accurate.

We find that the calculations of Gersch and Rodriguez, Silver, and the symmetrization procedures provide a good description of the FSE for the observed scattering in the normal liquid where $n(p)$ is broad and featureless. The only theories which fail in the normal liquid are those that predict Lorentzian broadening. These theories predict a scattering that is much broader than observed experimentally.

In the superfluid phase, only the calculation of Silver provides an accurate description of the final state broadening. The other models of FSE do not accurately describe the observed scattering in the superfluid phase. Inelastic neutron scattering experiments at other values of Q could test more detailed features of FSE theories.

It is a pleasure to acknowledge many enlightening discussions with Drs. J.M. Carpenter and A. Rinat. This work was supported by NSF grant DMR-8704288 and OBES/DMS support of the Intense Pulsed Neutron Source at Argonne National Laboratory under DOE grant W-31-109-ENG-38. TRS and WMS acknowledge the support of the Division of Educational Programs at Argonne National Laboratory.

References

1. G.B. West, Phys. Rep. **18C**, 263 (1975); V.F. Sears, Phys. Rev. B **30**, 44 (1984).
2. D.M. Ceperley and E.L. Pollock, Phys. Rev. Lett. **56** 351 (1986).
3. D.M. Ceperley and E.L. Pollock, Can. J. Phys. **65**, 1416 (1987).
4. P.A. Whitlock and R. M. Panoff, Can. J. Phys. **65**, 1409 (1987).
5. P.E. Sokol, Can. J. Phys. **65**, 1393(1987).
6. J.J. Weinstein and J.W. Negele, Phys. Rev. Lett. **49**, 1016 (1982).
7. R. N. Silver and G. Reiter, Phys. Rev. B **35**,3647(1987).
8. T.R. Sosnick, W.M. Snow, and P.E. Sokol, to be published.
9. P.C. Hohenberg and P.M. Platzman, Phys. Rev. **152**, 198 (1966).
10. H.A. Gersch and L.J. Rodriguez, Phys. Rev. A **8**, 905 (1973); L.J. Rodriguez, H.A. Gersch and H.A. Mook, *ibid.* **9**, 2085 (1974).
11. R. N. Silver, in Proceedings of the 11th International Workshop on Condensed Matter Theories, Oulu, Finland, 1987, Plenum Press; Rapid Communications, Phys. Rev. B **37**, 3794 (1988).
12. V.F. Sears, Phys. Rev. B **30**, 44 (1984).
13. A.S. Rinat, Phys. Rev. B **36**, 5171 (1987).
14. S. Stringari, Phys. Rev. B **35**, 2038 (1987).

MOMENTUM DISTRIBUTIONS OF LIQUID ^{4}He IN AEROGEL GLASS

W.M. Snow[†] and P.E. Sokol[+]

† Intense Pulsed Neutron Source, Argonne National Laboratory
Argonne, IL

+ Department of Physics, The Pennsylvania State University
University Park PA

INTRODUCTION

The physics of liquid and solid ^{4}He in restricted geometries has motivated a number of interesting experiments. Recent experiments include detailed measurements of the phase diagram for bulk liquid in vycor [1], showing a suppression of the superfluid transition and elevation of the melting pressure, and measurements of the superfluid fraction in vycor, aerogel, and xerogel glasses near the lambda point [2], in which critical exponents differ from the pure ^{4}He values. Many striking features in several of the experiments on helium in restricted geometries are poorly understood.

We have performed inelastic neutron scattering measurements of liquid helium in aerogel glass above and below the superfluid transition for two samples of different porosities. The kinetic energy (KE) of the confined liquid is the same as that of the bulk liquid in the normal phase, but is clearly higher than the bulk values in the superfluid phase. The observed scattering in the superfluid phase is more peaked than in the normal phase: consistent with the presence of a Bose condensate. An estimate of the condensate fraction using a modification of a method due to Sears [3] yields values consistent with those estimated for the bulk liquid.

Aerogel is a porous silicia glass formed by the supercritical drying of a colloidal silicia-ethanol mixture [4]. The pores span a wide range of sizes (unlike vycor and xerogel, where the pore distribution is more uniform) with a typical open volume fraction in excess of 90%. This high porosity favors aerogel for neutron scattering studies since the size of the helium signal is increased relative to background scattering from the glass is increased. Beam attenuation effects are also negligible for silicia glass.

The dynamic scattering function $S_{IA}(Q,\omega)$ in the Impulse Approximation (IA) is directly related to the momentum distribution $n(p)$. In isotropic systems such as a liquid, $(Q/M)S_{IA}(Q,\omega)$ is a function of a single scaling variable $Y \equiv (M/Q)(\omega - \omega_r)$, where M is the mass of the scatterer and $\omega_r = Q^2/2M$ is the recoil energy. The scattering is then directly proportional to [5]

$$ J_{IA}(Y) = \frac{1}{2\rho\pi^2} \int_{|Y|}^{+\infty} pn(p)dp = \int_{-\infty}^{+\infty} \int_{-\infty}^{+\infty} n(p_x, p_y, Y)dp_x dp_y \qquad (1.1) $$

which is the longitudinal momentum distribution. $J_{IA}(Y)$ exhibits several features which are characteristic of the IA. It is symmetric about $Y = 0$ and depends on Q only through the scaling variable Y.

RESULTS

The measurements were carried out using the PHOENIX spectrometer at the Intense Pulsed Neutron Source at Argonne National Laboratory using a mean momentum transfer of 23 Å^{-1}. Two sets of measurements were performed. The first set of measurements was at temperatures of 1.45, 2.0, 2.2, and 2.5 K; the second set was at temperatures of 0.7 and 2.3K using the less porous aerogel. All measurements were performed at the saturated vapor pressure of the bulk liquid.

After subtracting the background signal from the sample cell and the aerogel, two scattering components are visible. In addition to the helium peak centered at $Y = 0$, a much smaller broad component about five times as wide as the helium peak is present. Part of this broad component [3] is due to secondary scattering from the walls of the cryostat and the radiation shields. The remainder can be interpreted as scattering from an adsorbed layer of helium tightly bound to the pore walls. We were not able to separate these two contributions accurately. However, both components are essentially flat over the region of the helium peak and do not change the shape of the peak in the region of interest. We subtract this broad component from the data to obtain the scattering from the liquid inside the aerogel.

Figure 1 shows the liquid scattering seen at 0.7 and 2.3K. The scattering is clearly close to the IA predictions that it be symmetric and centered at $Y = 0$. The extra scattering in the superfluid near $Y = 0$ relative to the normal liquid scattering is also seen in the bulk liquid data, where it is believed to be due to the presence of a Bose condensate.

The data in Figure 1 still contain the effects of instrumental resolution and broadening due to final state scattering of the recoiling helium atom from its neighbors, called final state effects(FSE). To remove these effects we characterize the true scattering in the IA using a model $J(Y)$ which is the sum of two gaussians. The model scattering is convoluted with instrumental resolution, determined from a Monte Carlo simulation of the instrument response, and FSE, using a recent theory [6] for the bulk liquid by Silver. The amplitudes, widths, and common center of the Gaussians are varied to obtain the best fit to the observed scattering. Information on the momentum distribution is then obtained from the model distribution.

Table 1 shows the KE calculated from the model $J(Y)$ along with values for the bulk liquid obtained using the same data analysis procedure [7]. The bulk liquid data in the table were taken at a constant density of 0.147 gm/cm^3, which is close to the saturated vapor pressure density near T_λ. The absolute values for the KE are sensitive to the tails of $J(Y)$ and have an error of$+20$, -10%[7]. Nevertheless, since both of these experiments were performed on the same instrument and used the same data analysis procedure, we believe we are justified in making relative comparisons. The kinetic energies of the confined liquid are close to the bulk values in the normal fluid, but are systematically higher than the bulk values in the superfluid.

We can also attempt to estimate the size of the condensate fraction using a modification of a method due to Sears which associates the increase in scattering at small Y with the appearance of a condensate. Figure 2 shows the integrated intensity of the model $J(Y)$ as a function of a cutoff momentum for the lowest and highest temperatures for each porosity. The superfluid data clearly show more intensity at low momenta, consistent with the development of a finite condensate fraction.

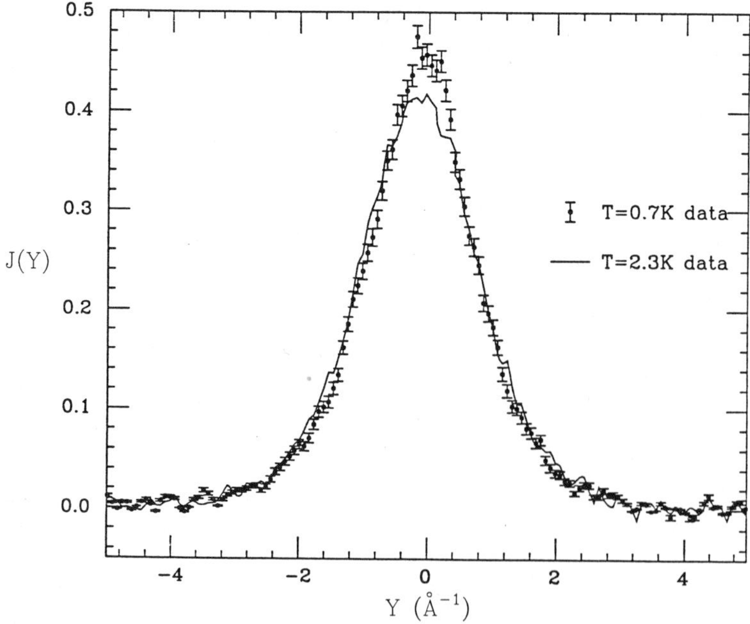

Figure 1 - Measured scattering from liquid ^{4}He in aerogel at 0.7K and 2.3K.

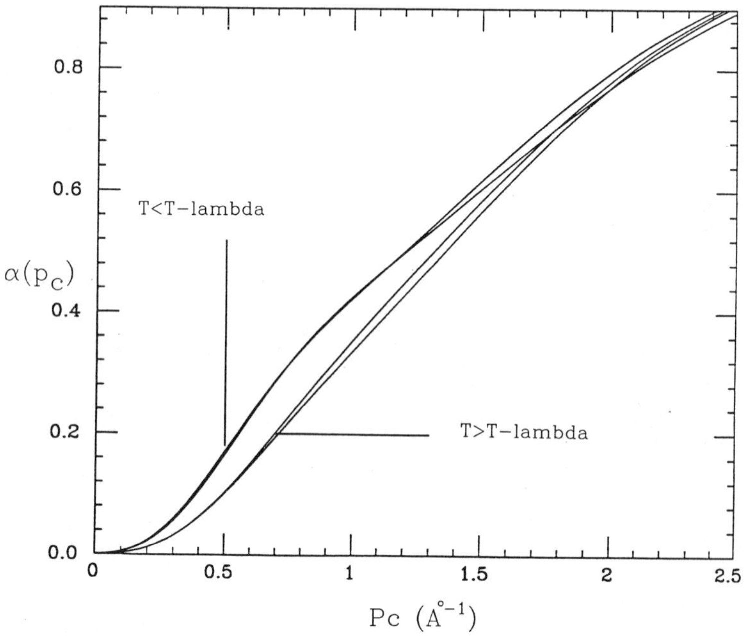

Figure 2 - Integrated area $\alpha = \int_0^{p_c} n(p)p^2\,dp$ of the two Gaussian model momentum distributions as a function of the cutoff momentum p_c for the confined liquid in the normal and superfluid phases.

The magnitude of the condensate fraction may be estimated from the integrated intensities in Figure 2 if the singular behavior induced by the condensate is known. We make use of the calculation by Griffin which has given reasonable agreement for measurements in the bulk liquid. This implicitly assumes that the singular behavior, due to coupling of long wavelength phonons to the condensate, is not affected by the confining medium. Table 2 lists the inferred values for n_0 for both the bulk and confined liquids.

Table 1. Kinetic energy and condensate fraction for bulk liquid and the confined liquid in the high porosity (HP) and low porosity (LP) samples.

Temp. (K)	KE Bulk	KE HP	KE LP	n_0 Bulk	n_0 HP/LP
0.35	13.3			8.9	
0.7			15.7		9.0
1.0	14.5			7.5	
1.5	14.5	15.3		6.8	8.3
2.0	14.8	15.6		3.7	3.4
2.2		15.7			
2.3	16.1		16.5		
2.5		16.4			
2.8	16.6				

The magnitude and temperature dependence of the condensate fraction for the confined liquid is not noticeably different than for the bulk liquid. This is consistent with the small depression of the superfluid transition observed for liquid helium in aerogel. However, it is rather surprising considering the difference between the kinetic energy of the confined and bulk liquids. The smaller change in KE at T_λ would imply a smaller value for the condensate fraction.

CONCLUSION

Measurements of the high energy inelastic neutron scattering from normal and superfluid ^{4}He in a confined geometry are consistent with the predictions of the IA, when final state effects are included. The observed scattering consists of a broad component, representing atoms localized on the pore walls, and a narrow component, representing the liquid in the pores. The scattering from the liquid component is quite similar to the scattering from the bulk liquid. The inferred condensate fraction has the same magnitude and temperature dependence as in the bulk liquid. However, the kinetic energy of the confined liquid is higher than that of the bulk liquid in the superfluid phase, suggesting that the decrease in the high Y tails of the observed scattering is not as large as in the normal liquid.

ACKNOWLEDGEMENTS

This work was supported by NSF grant DMR-8704288 and OBES/DMS support of the Intense Pulsed Neutron Source at Argonne National Laboratory under DOE grant W-31-109-ENG-38. WMS acknowledges the support of the Division of Educational Programs at Argonne National Laboratory.

References

[1] E.D. Adams,Y.H. Tang, K. Uhlig, and G.E. Haas, J. Low Temp. Phys. **66**, 85 (1987).

[2] M.H.W. Chan, K.I. Blum, S.Q. Murphy, G.K.S. Wong, J.D. Reppy, Phys. Rev. Lett. **61**, 1950 (1988).

[3] P.E. Sokol, T.R. Sosnick, and W.M. Snow, These proceedings and to be published.

[4] J.H. Mazur and C.M. Lampert, SPIE**502**,123 (1984)

[5] G.B. West, Phys. Rep. **18**C, 263 (1975); V.F. Sears, Phys. Rev. B **30**, 44 (1984).

[6] R.N. Silver These proceedings and Phys. Rev. B**37**, 3794 (1988)

[7] T.R. Sosnick, thesis, Harvard University (unpublished)

WRAP-UP

SUMMARY: WORKSHOP ON MOMENTUM DISTRIBUTIONS

Ralph O. Simmons

Department of Physics and Materials Research Laboratory
University of Illinois at Urbana-Champaign
Urbana, Illinois 61801 U.S.A.

This has been an extraordinary Workshop touching many branches of physics. The Workshop has treated momentum distributions in fluid and solid condensed matter, in nuclei, and in electronic systems. Both theoretical and experimental concepts and methods have been considered in all these branches. A variety of specific illustrations and applications in physical systems have been presented. One finds that some common unifying themes emerge. One finds, also, that some examples are available to illustrate where one branch is more mature than others and to contrast where expectations for future progress may be most encouraged.

I. INTRODUCTION

The concept for this Workshop is an interesting one. A portion, of course, is by now not too unusual. The mutually complementary and supportive characters of nuclear problems and of liquid helium problems, studied, in more generality, in the guises of nuclear matter and of quantum fluids, respectively, have long been exploited fruitfully by many-body theorists. But it is a bit of a reach to include momentum distributions in electron systems. And it is nearly unique to bring together experimentalists in the ostensibly different fields of nuclear physics and condensed matter physics, along with theorists. Success of such an experiment is not assured.

Following the conception, our organizers, Sokol and Silver, have shown good taste in organization. This has made the Workshop successful. It also simplifies the task of summarization. For nuclei, low-energy properties and electromagnetic probes have been emphasized, rather than the structure and properties of particular nuclei or exotic states or reactions in scattering. For atoms and molecules, the same idea has guided choice of participants and of their interests. In addition, in both these cases, there has been mention of the complementary nature of, on the one hand, $(e,e'p)$ studies for example, and on the other, of $(e,2e)$ scattering. These useful references can be found in the respective papers, and they provide links to broader views of the respective areas of physics. Finally, for condensed matter systems, the focus has been on those, such as liquid helium, which have provided so many rich phenomena for scientific study in the past. They clearly continue to fascinate for their ability to provide prototypes for study of both Fermi and Bose interacting systems over a wide range of parameters.

Altogether in this Workshop there have been over two dozen scheduled speakers, about 20 hours of formal presentations and questions, plus a plenary group discussion led by the session chairpersons. There has been an extensive display of posters. And there have been countless different discussions among groups of various sizes. A challenge to summarize!

Was the Workshop worthwhile? Were some of the central problems discussed in the fields touched? Have there been common themes identified? Were insights gained to apply in another area? Has there been a clear vision of future prospects? Work yet to be done?

It seems to me that the answers to these questions are generally positive. So this summary is not a requiem, not a memorial summary of an area already fully explored, described, and ready for placement in a corner of the physics Pantheon. Rather, discussions have been lively, with points both of agreement and of dispute. Some matters are enough advanced so that they may be approaching canonical form. Others are still unfolding. Altogether, it appears that prospects can be regarded with hopeful enthusiasm for future important developments.

II. SCATTERING METHODS

Scattering methods have a grand history in probing the structure and dynamics of matter. The genius of Rutherford, in confirming a nuclear model for the atom, through Geiger and Marsden's experimental results on alpha particle scattering, gives an inspiring first example.

We heard a comprehensive and provocative talk by West, whose 1975 review article[1] summarized the scaling framework within which many of the Workshop discussions have taken place. He has told us what led him to think in terms of scaling variables. For example, y-scaling naively reflects the fact that the nucleus is made of nucleons, that liquids are made of atoms, etc. But a plot of experimental data versus y for fixed q^2 removes the "trivial" physics, so that the approach to scaling, which reflects the "interesting" physics (e.g. correlations and dynamics), can be examined. And he has pointed out the ways this point of view links the present subject areas to each other and to other active areas of study, such as high-energy physics.

In Fig. 1, a very schematic diagram sketches my impressions of the status of experimental research areas touched in momentum distribution physics during this Workshop. First, it shows the area of electron scattering in nuclear physics as "mature." This maturity is not to be understood in any pejorative sense. Rather, we have heard of the remarkable refinement of electron scattering studies to the point that crosssections can be studied over many decades in magnitude and scaling tests can be made with very high sensitivity. Next, X-ray Compton scattering from electronic systems, an early area of fruitful n(p) investigation in the hands of DuMond and Kirkpatrick,[2] a field which enjoyed some quantitative attention later on with gamma-ray sources, is now in the process of rebirth and extension, as powerful new synchrotron X-ray sources become available. Another fruitful route to the study of electronic systems has been the use of positron annihilation techniques, which have been the subject of continual development and refinement; these too have recently gained power with the advent of new accelerator sources.

Condensed matter liquid n(p) studies with neutrons are probably at a 'teen stage of development (companion studies at low to moderate momentum transfers can be judged mature). Adolescence seems more appropriate than maturity for an area which is still open to questions about the possible importance of final-state interactions, about whether the impulse approximation is applicable in a given experiment, and in which the experimental tools are still in a state of development. Finally, it may be said that n(**p**) neutron studies of the solid state are in a promising childhood. Childhood, because so few have been completed and because it is often not yet clear whether there will be big surprises in the results. Promising, because the recoil regime gives a quite new analytical tool which can be used to selectively pick out single-particle or quasi-particle properties. Condensed matter solid studies were represented here by a few examples. As is often true in solid state physics, future benefits will depend in large measure upon the choices made by particular scientists of suitable problems and systems for study.

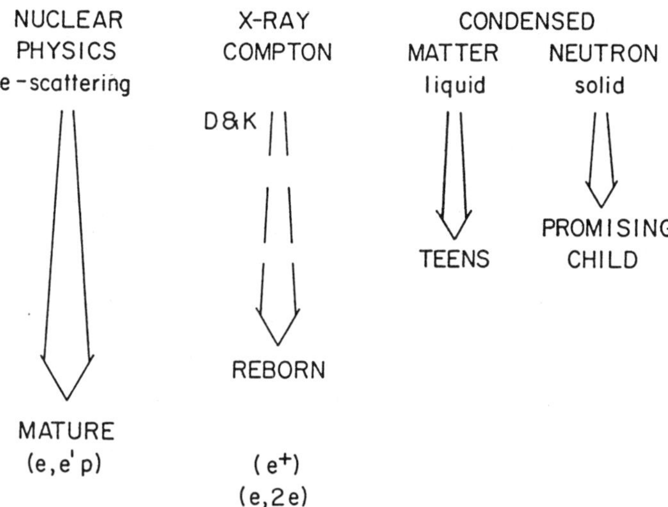

Figure 1.   Schematic diagram indicating development of the fields of experiment touched in this Workshop on Momentum Distributions. "D & K", Dumond and Kirkpatrick, are recognized for their pioneering place in n(p) studies. The parentheses at the bottom indicate complementary work.

III. ELECTRON SCATTERING NUCLEAR PHYSICS

The Workshop heard from Sick and from Day how quasi-elastic electronnucleus scattering has been highly developed. As a tool for the study of nuclei it has several useful characteristics. The electromagnetic interaction is a relatively clean probe which can be used to investigate distributions of charge density and of momentum.[3] Besides using magnetic spectrometers, nuclear scientists now make use of particle identification with the use of Cerenkov counters (for the velocity), wire chambers (for individual particle trajectories) and calorimeters (for the total energy). Pion backgrounds have been brought down to the 10^{-4} level, which permits study of the quasi-elastic peak in remarkable detail. Both inclusive and exclusive reactions can be studied. Scaling can be used to investigate the q-dependence of the nuclear form factor. Sick has told us about the considerable work on inclusive (e,e') reactions, which yield global information, and pointed out the increasing emphasis upon exclusive studies.

Experimental problems exist, in the form of the pervasive need for radiative corrections to the data. Sick has assured us that the treatment of the real part of the final-state interaction can probably be taken care of. At the higher energies there are also meson exchange corrections and proton off-shell effects to be taken into account. In exclusive reaction studies, charge exchange is a problem and it appears the observed strength is significantly lower than expected from theory.

Some interesting parallels can be drawn between inclusive scattering results and neutron scattering results from liquid helium. For nuclei (compared to the helium case, finite objects!) good information is now available for low- to moderate-momentum components of the nuclear wave function. Most information is now desired about high-momentum components, which are generated by the characteristics of the nucleon-nucleon interaction at very short distances. Unfortunately, measurement of the scattering strength on the large energy transfer side of the quasi-elastic peak is complicated in electron scattering from nuclei.

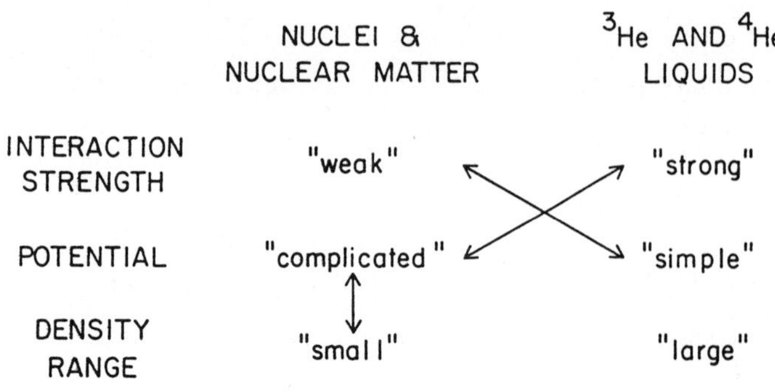

Figure 2. A matrix representing the varied complementary aspects of study of nuclei and nuclear matter versus the helium liquids as quantum fluids.

There has been fine complementarity in the theoretical study of nuclear matter and of the quantum fluids ^{4}He and ^{3}He. Figure 2 illustrates the relationships schematically. Compared to the quantum fluids, nuclear matter is a relatively weakly interacting system. The helium deBroglie wavelength is a large fraction of the interparticle spacing, and the helium potential has a relatively large hard core. On the other hand, the inter-nucleon forces are not just the straightforward pair interaction of the heliums. The way to progress in the resulting situation can be compared to night pedestrian travel on a cluttered city street. The street is a way toward a destination (the detailed understanding of the dynamics of many-body systems). But from time to time there are piles of garbage which appear on the near side, which makes it desirable to cross over and proceed along the other side for a while, as one progresses.

From Sick and from Day, we have heard of much fine work and its interpretation, including studies on ^{3}He of heroic scale at SLAC. As for prospects for the future, we have heard of the kinds of work expected to be possible at instruments at Mainz, in the Federal Republic of Germany, and eventually at CEBAF, now under construction in Virginia.

IV. ELECTRONIC SYSTEMS IN CONDENSED MATTER PHYSICS

From Platzman, we got an authoritative perspective about final state effects in the photon excitation of electronic systems, and a rapid review of past results from selected parts of the periodic table. He illustrated the prospects for future knowledge to be gained from experiments at the new X-ray sources, for example about electron correlations, about minority spin momentum distributions in polarized systems, and about various solids through the exploitation of interference effects.

For high-energy photon studies, applicability of the impulse approximation is not in question. Rather, experimental problems come in the production of photoelectrons (although the existence of photoelectric absorption edges can be turned to advantage in suitably chosen circumstances), in pervasive multiple scattering, and in the rather modest experimental resolution attainable in most studies to date. New detectors will probably help.

The future of hard X-ray studies of momentum distributions and related physical phenomena appears very bright. Already, insertion devices have been used in existing medium energy storage rings to produce intense synchrotron radiation in the range above 10 keV, tunable radiation which is exceedingly well-collimated and for which the degree of polarization can be selected.

Magnetic Compton scattering is now possible, and future helical wigglers will provide incident circular polarization for unprecedented studies. Mills and Cooper, in separate presentations, told us about both an impressive beginning in this area and about ambitious plans for future instruments and studies. Already four years ago, at the Los Alamos HEECM Workshop[4], there were predictions of future dominance of synchrotron radiation in the investigation of magnetic structures of solids, a subject long the exclusive purview of slow neutron scattering. One heard echoes of Khrushchev's famous remark "We will bury you."

The final engineering design of the Advanced Photon Source (APS) at Argonne National Laboratory is being worked out. It, and the other high energy machines in Europe and elsewhere, carry immense promise for future development of photon scattering techniques for investigating the structure and dynamics of electrons in condensed matter.

Berko cheerfully reminded us that all of these experiments are done in momentum space, before he pointed out the rich variety of techniques using the positron as a probe. Positrons are not a soft probe, and (for condensed systems) with them one has a surface to contend with, and, for intrinsic n(p) studies, competition from defect traps. The results, nevertheless, are truly impressive in scope and in detail. Many-body effects are important but do not seem to destroy applicability of the technique to subtle problems such as Fermi surface analysis in complex crystals, especially when companion calculations are available to assist in data interpretation.

V. NEUTRON n(p) STUDIES IN CONDENSED MATTER PHYSICS

The helium liquids and solids have provided an inexhaustible source of puzzles and surprises in condensed matter physics. A particular fascination has been the search for experimental verification of London's prediction that superfluidity in liquid ^{4}He is related to the appearance of a quantum Bose condensation of the fluid. An attractively direct means is the measurement of the momentum distribution in the liquid, because the condensate presumably contains particles having very small momentum components corresponding to macroscopic delocalization in the bulk fluid. At the same time, there has evolved realization that in actual neutron scattering experiments, both final state effects and the presence of several p = 0 singularities in n(p) complicate matters.

After the Workshop banquet, Svensson told us this entertaining yet sobering saga of "The Quest for n(p)" in superfluid ^{4}He, as one of the most active participants. Earlier, Sokol presented the studies of ^{4}He by Sostich, Sokol, and Silver which include both use of a dedicated neutron spectrometer at IPNS and a new theoretical estimation of final state corrections to be made to such data. And Mook reviewed the Oak Ridge work not only on superfluid ^{4}He but also on the strongly interacting Fermi fluid ^{3}He, an experimentally much more demanding specimen owing to its enormous absorption of slow neutrons. Some of the most spirited discussions and comments of the Workshop were provoked by these presentations. It is clear that Svensson's choice of "quest" is an entirely appropriate term for the process, both in the sense that the search is arduous and in the Medieval sense that it is worthy in itself to be questing.

It is fair to say that all Workshop participants regard these quests as interesting but that not all were convinced that they are finished, even for fluid ^{4}He, the subject of such ingenious efforts over the years. Direct verification of a number of expected characteristics of the momentum distribution of ground state fluid ^{4}He still appears incomplete and the dependence of the overall neutron recoil widths upon momentum transfer q is still matter of debate, particularly in ^{3}He. Because the plenary discussions of the Workshop are not recorded in these Proceedings, it is perhaps worth noting that they did not avoid basic questions. Among those to be heard during this session were: "Is n(p) an observable?" "How do you find singularities experimentally?" [For the purpose of final state corrections] "what if n(p) is not known?" "Ah, *that's* the problem!"

Not much was said about the current experimental situation in the solid heliums, but Glyde pointed out the usefulness of considering both solids and liquids in his analysis of the approach to applicability of the impulse approximation.

Hydrogen has been an important subject of neutron scattering studies from the earliest days of neutron research. At first, of course, the emphasis was upon fundamental studies of the neutron-proton interaction. But the work soon expanded because of the practical importance of hydrogen and a variety of hydrogen-containing compounds in science and technology. A large scattering cross-section has made studies of small amounts of hydrogen practicable. The Workshop heard from Herwig about recent neutron spectroscopy on solid and gaseous para-hydrogen, and about dilute hydrogen embedded in a solid argon host, and from Hempelmann and Reiter, respectively, about the successes of both experimental and theoretical investigations of the microscopic dynamics of (atomic) hydrogen in metals by neutron scattering.

Experimental investigation of momentum distributions in intermediate ranges of momentum transfer has of course been previously possible with reactor sources of neutrons, particularly by scientists in national laboratories. Traditional triple-axis machinery has been joined, we have heard from Mook, by time-of-flight cross-correlation techniques which in principle allow exact deconvolution of the data. In practice, it appears that one is not free of the familiar trade-offs between resolution and counting rate.

The new element in neutron $n(\mathbf{p})$ studies has become the operation of chopper spectrometers and resonance detector (eV) spectrometers at pulsed spallation sources. These instruments can reach large momentum transfers, to 25 A^{-1} and above. The Argonne IPNS has been operating seven years, and as the Workshop convenes the first IPNS operation with a new, enriched uranium target shows the neutron flux increased by a factor near 2 1/2. The Rutherford-Appleton ISIS has both direct- and inverse-geometry spectrometers in operation and development. And the Los Alamos LANSCE has plans for instrumentation capable of $n(\mathbf{p})$ studies. Holt has told us of considerations, in principle and in practice, which limit such instruments. It turns out that the resolution is not so bad, but that the suppression of neutron background is a difficult challenge which presently seems to limit the investigation of very high-momentum components of $n(\mathbf{p})$.

Unlike synchrotron sources, new neutron sources do not promise improvements of many orders of magnitude in flux. Nevertheless, there is much ground as yet unexplored, both in the techniques and their applications to condensed matter systems. For techniques, it is worth noting, for example, the rich variety of conceptual design studies for new neutron instruments which were produced as part of the German SNS facility proposals. And as an example of applications, it will be interesting to see further studies of the approach to the impulse approximation, in view of the various expectations of this presented to the Workshop by Silver and by Glyde.

VI. THEORY AND NUMERICAL SIMULATION

Parallel to these diverse experimental developments, we have seen increasing power and confidence in theoretical understanding of many-body systems, and impressive implementation of sophisticated numerical simulations. These studies not only give new quantitative results but also yield new insights. They also help guide experimentalists in the reduction and interpretation of their data. It is notable that during this Workshop trenchant comments and constructive practical suggestions to the experimentalists often came from the theorists. Many of the $n(\mathbf{p})$ experimentalists' goals for understanding are not confined to the particular systems chosen for study. They often have some generality of application. At the same time, during this Workshop it has not been unusual for experimentalists to pose challenging questions to the theorists for their consideration or to question some of the assumptions made in their analysis. It is to be hoped that this active interaction between theorists and experimentalists in this field will continue.

Clark's comprehensive "Overview" of theories and calculations replaces whatever summary I, as an experimentalist, might make of theoretical accomplishments and prospects.

His Overview can be read elsewhere in this volume.[5] Here I confine myself to several observations and comments.

It seems to me that as simulations and calculations with ever more powerful computers are possible, such as those described by Panoff, by Ceperley, and by Manousakis, respectively, and as expectations of their realism are increased, a question by Platzman during one session is increasingly relevant. The question is "How do you know you have thought of everything?" It is true, of course, that the simulators have the advantage that the influence of a particular interaction or effect can be turned off or on at the will of the simulator, in order to explore the nature and magnitude of such influences. But the possibility of compensating opposite influences, which thereby mask effects, should not be forgotten.

An experimental advantage that the heliums have is the possibility of studies over an extremely large range of densities. And for many purposes it appears that the precise choice of potential is not significant. But diamond-anvil studies of helium compressed to densities as much as six times the density at which it first condenses suggest that triplet forces are important under those conditions.[6] Helium provides an interesting system for investigation of the nature and processes of melting and crystallization. Ceperley showed us, in a kind of homage to Feynman, that path-integral methods, when pursued with good taste, give us new insights about both 3D and 2D helium.

I would have liked to hear more about shadow wave function applications in GFMC calculations, some discussion about properties of helium droplets, and more comments about consequences arising because nuclei are finite. But time was limited. Not too limited, however, for a "Workshop" atmosphere truly to flourish in the spirited widespread discussion provoked by Koonin's presentation about final state effects in nuclear scattering studies. Finally, it was a little surprising to me that we did not hear more about the application, to nuclear problems, of density functional methods, which have been used for ground state problems in electronic structure with some success.

VII. WHAT WOULD WE LIKE TO KNOW ABOUT n(p) IN ... ?

The speakers with this title each emphasized that it was a commissioned title, and suggested other forms for a relevant question. Nevertheless, each presentation was provocative.

Hetherington reminded us, with some wit, of the general arguments of Penrose and Onsager about the existence of a Bose condensate in ground-state liquid ^{4}He, and listed the questions: "What is n_0?" What is the overall scale and shape of n(p)?" "What is the character of the singularities near p = 0?" "Is there an exponential tail at p Æ ∞?" "What is the value of <KE>?" During the ensuing discussion, the interesting question of a possible condensate in solid ^{4}He came up, as well.

Fantoni told us why, in nuclear investigations, the spectral function S(k,E) assumes a greater importance than it does presently in condensed matter physics. In nuclear investigations the choice of scaling variable is nontrivial, but it is important for the pursuit of valid inferences from the data. He emphasized the equivalence of p(k) and p(r) in nuclear studies, and reiterated the need for more information on high-momentum components of n(k) [= n(p)].

Doniach presented a provocative vision of his hopes for future electron system research in this area, once new X-ray sources and the right high-resolution detectors become available. Strongly correlated Fermion systems provide special challenges and opportunities for new understanding of important cases where correlations dominate what is going on: heavy Fermions; Mott insulators; high T_C superconductors. His discussion focussed upon the physical origins of the effects in these systems which, in contrast to fluid helium, have a low effective density. I found interesting his homage to Feynman through the use of path-integral arguments, and which applied to narrow-band problems provoked a lot of Workshop discussion. And his analogies to Landau Fermi-liquid theory were a useful way to portray the relative state of theoretical developments here. He emphasized the need for high resolution in future experiments aiming to elucidate these physical origins; experimentalists did question possible background problems to be faced.

VIII. CONCLUSION

A common theme of the Workshop, in all branches of physics considered, was emphasis upon current knowledge of the momentum wave-function of the system studied. The knowledge is sometimes extensive, sometimes sketchy, depending upon the branch. Also common was interest in quasi-particle properties and the discussion of prospects for extension of quantitative work on correlation effects. An agreed goal is the detailed comparison of response functions, or of dynamical scattering factors, obtained from theory and from experiment, respectively, over a range of momentum transfers or other physical conditions. Approximations in the respective theories may complicate what could otherwise be straightforward transformations between descriptions of the systems in coordinate and in momentum space. Behavior in the limit of the impulse approximation is of interest, of course, but in practice this cannot necessarily be achieved, and, besides, one may also be specifically interested, in part, with aspects of few-body or of collective behavior which are not evident in that limit.

The poster sessions and working groups provided a natural extension of Workshop activities and entry to detailed discussions. For example, on display were further examples of beautiful applications of positron annihilation techniques to electron structures of solids, and on the other hand, a variety of posters on practical matters related to neutron and electron scattering experiments.

Overall, the Workshop showed that considerable progress has been accomplished since these topics were discussed as part of the 1984 Los Alamos HEECM Workshop.[4] It is not just progress made possible as exciting capabilities of new experimental facilities are realized. It is progress in theoretical understanding and simulation. And, finally, it is the increasing realization of the benefits of keeping aware of developments in the other branches of physics which, although different in subject matter, share similar intellectual problems.

IX. ACKNOWLEDGEMENTS

Thank are due to the sponsors of this Workshop: the Argonne National Laboratory Division of Educational Programs, the Argonne Intense Pulsed Neutron Source Division (Bruce Brown, Director), and the University of Chicago. Without their support, the Workshop could not have taken place. Thanks are especially due to the organizers, Paul Sokol and Richard Silver, whose energy and intiative brought together very active and busy scientists from an unusual range of fields. The program that they arranged provided extensive opportunities for both formal and informal discussion. Of course, it was these spirited contributions by the Workshop participants themselves which actually made the Workshop worthwhile.

REFERENCES

1. G. B. West, Electron scattering from atoms, nuclei and nucleons, *Physics Reports* 18:263 (1975).
2. J. W. M. Dumond, Linear momenta of electrons in atoms and in solid bodies as revealed by X-ray scattering, *Revs. Mod. Phys.* 5:1 (1933).
3. "Electron scattering in nuclear and particle science," AIP Conference Proceedings **161**, C. N. Papanicolas, L. S. Cardman, R. A. Eisenstein, eds., American Institute of Physics, New York (1987).
4. "Proceedings of the 1984 Workshop on High-Energy Excitations in Condensed Matter," R. N. Silver, ed., Report LA-10227-C, Los Alamos National Laboratory, New Mexico (1984).
5. J. W. Clark, Overview of momentum distribution calculations, (present volume).
6. P. Loubeyre, Three-body exchange interaction in dense helium, *Phys. Rev. Lett.* 58:1857 (1987).

PARTICIPANTS

Ercan Alp	Argonne National Laboratory
Giulia Baciocco	Universita La Sapienza
Arun Bansil	Northeastern University
Alexsandar Belic	University of Illinois–Urbana
Stephen Bennington	University of Birmingham
Stephan Berko	Brandeis University
Robert Blasdell	University of Illinois–Urbana
Malcolm Butler	TRIUMF
John M. Carpenter	Argonne National Laboratory
Carlo Carraro	California Institute of Technology
David M. Ceperley	University of Illinois–Urbana
John W. Clark	Washington University
Milton W. Cole	Pennsylvania State University
Roberto Colella	Purdue University
Malcolm J. Cooper	University of Warwick
Bogdan Dabrowski	Argonne National Laboratory
Donal B. Day	University of Virginia
Willem H. Dickoff	Washington University
Sebastian Doniach	Stanford University
A. Fabrocini	University of Pisa
Stefano Fantoni	University of Pisa
Kenneth D. Finkelstein	Argonne National Laboratory
Frank Y. Fradin	Argonne National Laboratory
Joseph N. Ginocchio	Los Alamos National Laboratory
Henry R. Glyde	University of Delaware
A. Griffin	University of Toronto
Edvard Heiberg	University of Chicago
Rolf Hempelmann	Los Alamos National Laboratory
Kenneth W. Herwig	Pennsylvania State University
Jack H. Hetherington	Michigan State University
Ron Holt	Rutherford Appleton Laboratory
Steven E. Koonin	California Institute of Technology
Chung Loong	Argonne National Laboratory
Efstratios Manousakis	Florida State University
Jerry Mayers	Rutherford Appleton Laboratory
Desmond McMorrow	Edinburgh University
Peter E. Mijnarends	Netherlands Energy Research Foundation
Dennis M. Mills	Argonne National Laboratory
Herbert A. Mook	Oak Ridge National Laboratory
Harold Myron	Argonne National Laboratory

Jon T. Nelson	University of Illinois – Urbana
Robert M. Panoff	Clemson University
Steven C. Pieper	Argonne National Laboratory
Phillip M. Platzman	AT&T Bell Laboratories
David L. Price	Argonne National Laboratory
George F. Reiter	University of Houston
Avraham S. Rinat	Weizmann Institute of Science
Manfred L. Ristig	University of Koln
Varley F. Sears	Atomic Energy of Canada Limited
Ryoichi Seki	California State University – Northridge
Gerhard Senger	Washington University
Peter B. Shaw	Pennsylvania State University
Ingo Sick	University of Basel
Richard N. Silver	Los Alamos National Laboratory
Ralph O. Simmons	University of Illinois – Urbana
Sunil K. Sinha	Exxon Research & Engineering Company
Devinderjit Sivia	Los Alamos National Laboratory
Kurt Skold	University of Uppsala
Lars C. Smedskjaer	Argonne National Laboratory
David W. Snoke	University of Illinois – Urbana
William M. Snow	Harvard University
Paul E. Sokol	Pennsylvania State University
Caroline K. Stahle	Stanford University
William G. Stirling	University of Keele
Warren F. Stubbins	University of Cincinatti
Eric C. Svensson	Atomic Energy of Canada Limited
Bilal Tanatar	University of Illinois – Urbana
Andrew D. Taylor	Rutherford Appleton Laboratory
David N. Timms	University of Warwick
Kalliopi Trohidou	Rutherford Appleton Laboratory
Yong Wang	Pennsylvania State University
Geoffrey B. West	Los Alamos National Laboratory
Paula A. Whitlock	New York University
Robert B. Wiringa	Argonne National Laboratory